P8-11 - full
P8-19- SHRINK or full
P8-152 - INCREASE
P8 165 - full

Brostow/Corneliussen
Failure of Plastics

Failure
of Plastics

edited by
Witold Brostow
and
Roger D. Corneliussen

With 445 Illustrations and 51 Tables

Hanser Publishers, Munich–Vienna–New York

Distributed in the United States of America by
Macmillan Publishing Company, Inc., New York
and in Canada by
Collier Macmillan Canada, Ltd., Toronto

Distributed in USA by
Scientific and Technical Books
Macmillan Publishing Company, Inc.
866 Third Avenue, New York, N.Y. 10022

Distributed in Canada by
Collier Macmillan Canada Distribution Center,
539 Collier Macmillan Drive, Cambridge, Ontario

Distributed in all other countries by
Carl Hanser Verlag
Kolbergerstr. 22
D-8000 München 80
West Germany

CIP-Kurztitelaufnahme der Deutschen Bibliothek

Failure of plastics / ed. by Witold Brostow;
Roger D. Corneliussen. — Munich; Vienna;
New York: Hanser, 1986.
  ISBN 3-446-14199-5

NE: Brostow, Witold [Hrsg.]

ISBN 0-02-947510-4 Macmillan Publishing Co., Inc., New York

Library of Congress Catalog Card Number 85-62010

Cover: Kaselow design, München

Printed in Germany by Universitätsdruckerei H. Stürtz AG, Würzburg

Index of Authors

George C. Adams
Polymer Products Department, Experimental Station,
E.I. du Pont de Nemours & Co. Inc.
Wilmington, DE 19898, U.S.A.

Jean-Claude Bauwens
Service de Physique des Matériaux de Synthèse, Université Libre de Bruxelles
Avenue Adolphe Buyl, 87, 1050 Bruxelles, Belgium

Heinz-Eberhard Boden
Barmer Maschinenfabrik AG
Leverkuser Straße 65, D-5630 Remscheid 11, West Germany

Witold Brostow
Department of Materials Engineering, Drexel University
Philadelphia, PA 19104, U.S.A.

Norman Brown
Department of Materials Science and Engineering, University of Pennsylvania
Philadelphia, PA 19104, U.S.A.

Roger D. Corneliussen
Department of Materials Engineering, Drexel University
Philadelphia, PA 19104, U.S.A.

Ralf Michael Criens
Department of Materials Engineering, University of Duisburg
Lotharstrasse 1–21, D-4100 Duisburg 1, West Germany

Ulf W. Gedde
Department of Polymer Technology, The Royal Institute of Technology
S-10044 Stockholm, Sweden

Jan-Fredrik Jansson
Department of Polymer Technology, The Royal Institute of Technology
S-10044 Stockholm, Sweden

Hans-Henning Kausch
Laboratoire de Polymères, Ecole Polytechnique Fédérale de Lausanne
32, chemin de Bellerive, 1007 Lausanne, Switzerland

Vernal H. Kenner
Department of Engineering Mechanics, The Ohio State University
Columbus, OH 43210, U.S.A.

Rainer Knausenberger
Institut für Kunststoffverarbeitung, RWTH Aachen
Pontstrasse 49, D-5100 Aachen, West Germany

Josef Kubát
Department of Polymeric Materials, Chalmers University of Technology
S-41296 Gothenburg, Sweden

Arnold Lustiger
Polymer Science and Technology Section, Battelle Columbus Laboratories
505 King Avenue, Columbus, OH 43201, U.S.A.

Shiro Matsuoka
Plastics Research and Development Department, Bell Laboratories
505 Mountain Avenue, Murray Hill, NJ 07974, U.S.A.

Georg Menges
Institut für Kunststoffverarbeitung, RWTH Aachen
Pontstrasse 49, D-5100 Aachen, West Germany

Abdelsamie Moet
Department of Macromolecular Science, Case Western Reserve University
Cleveland, OH 44106, U.S.A.

Hüter-Georg Moslé
Department of Materials Engineering, University of Duisburg
Lotharstrasse 1–21, D-4100 Duisburg 1, West Germany

Yoshihito Ohde
Department of Engineering Sciences, Nagoya Institute of Technology
Gokisomachi, Shouwaku, Nagoya, 466, Japan

Hiroshi Okamoto
Department of Engineering Sciences, Nagoya Institute of Technology
Gokisomachi, Shouwaku, Nagoya, 466, Japan

Kenneth J. Pascoe
Department of Engineering, University of Cambridge
Cambridge CB 2 IPZ, Great Britain

Sture Persson
Skega AB, S-93402 Ersmark, Sweden

Michael R. Piggott
Department of Chemical Engineering and Applied Chemistry, University of Toronto
Toronto, Ontario, Canada M5S IA4

Mikael Rigdahl
Paper Technology Department, Swedish Forest Products Research Laboratory
P.O. Box 5604, S-11486 Stockholm, Sweden

Ernst Schmachtenberg
Institut für Kunststoffverarbeitung, RWTH Aachen
Pontstrasse 49, D-5100 Aachen, West Germany

Roelof P. Stejn
Engineering Technology Laboratory
E.I. du Pont de Nemours & Co. Inc., Wilmington, DE 19898, U.S.A.

Bengt Stenberg
Department of Polymer Technology, The Royal Institute of Technology
S-10044 Stockholm, Sweden

Leendert C.E. Struik
Plastics and Rubber Research Institute of TNO
P.O. Box 71, 2600 AB Delft, The Netherlands

Henrik Sundström
Department of Polymer Technology, The Royal Institute of Technology
S-10044 Stockholm, Sweden

Björn Terselius
Department of Polymer Technology, The Royal Institute of Technology
S-10044 Stockholm, Sweden

T.K. Wu
Polymer Products Department, Experimental Station,
E.I. du Pont de Nemours & Co. Inc.,
Wilmington, DE 19898, U.S.A.

Joseph Zimmerman
P.O. Box 4042, Greenville, DE, 19807, U.S.A.

We dedicate this book
to the memory
of a great scientist, humanitarian,
our teacher and friend

Paul John Flory

Witold Brostow and Roger D. Corneliussen

Figure on the front cover:

We show on the cover one item pertaining to computer simulation of competition between chain relaxation and crack propagation (cf. Section 10.4). The horizontal coordinate is time, beginning on the left with the moment of breaking a bond between two segments of a polymer chain. The vertical axis shows the distance between adjacent segments. The continuous line shows oscillations of a segment adjacent to the broken bond, the broken line similarly oscillations of the next segment. In the present case oscillations subside with time, and a new stable position is found. In some other simulations breaking of a series of neighboring bonds and crack propagation were observed. (From W. Brostow and D.P. Turner, *J. Rheology* 1986, **30**, August issue).

Foreword

The Society of Plastics Engineers is pleased to sponsor and endorse "Failure of Plastics". The subject matter is both of significant importance within the plastics industry and of great interest to users of plastics materials. The authors are highly respected and well known in SPE for their expertise in the subject. Dr. Roger Corneliussen has been affiliated with the SPE educational seminar program for many years and his presentation on "Failure Mechanism in Plastics" has been among the most popular and well attended programs.

SPE, through its Technical Volumes Committee, has long sponsored books on various aspects of plastics and polymers. Its involvement has ranged from identification of needed volumes to recruitment of authors. An ever-present ingredient, however, is review of the final manuscript to insure accuracy of the technical content.

This technical competence pervades all SPE activities, not only in publication of books but also in other areas such as technical conferences and educational programs. In addition, the Society publishes four periodicals — *Plastics Engineering, Polymer Engineering and Science, Journal of Vinyl Technology* and *Polymer Composites* — as well as conference proceedings and other selected publications, all of which are subject to the same rigorous technical review procedure.

The resource of some 25,000 practicing plastics engineers has made SPE the largest organization of its type in plastics worldwide. Further information is available from the Society at 14 Fairfield Drive, Brookfield Center, Connecticut 06805.

Robert D. Forger
Executive Director
Society of Plastics Engineers

Preface

Failure has been a serious problem in the use of materials since the beginning of recorded history. To a large extent, the development of materials science and engineering has resulted because of serious failures. This is no less true for the newer materials such as polymeric ones. Unfortunately, such failures will become even more important as the number of critical engineering applications of polymers increases. The problem is especially difficult: polymeric materials are sensitive to processing, and affected by the environment, time, and temperature often in an "unpredictable" manner.

The present book was organized with this situation in mind. Although at first we thought of writing the whole book ourselves, the breadth of this area made the task very difficult. Instead, we decided to invite experts to contribute. We are very pleased at the caliber of those who accepted the challenge. They come from three continents. We believe this volume will give the practicing engineer as well as the researcher insight into each of the pertinent areas. To a considerable extent, the result is a forum for presenting relatively new, consistent, and exciting approaches to the entire field of failure phenomena.

An attempt has been made to make the chapters as uniform as possible from the point of view of symbols, terminology and units in particular. Each of us was frustrated many times perusing proceedings of conferences, even exciting ones, when the same quantity was discussed under different names by different authors, with no connections between contributions. Here, our authors have mutually read the contributions made before submitting the final versions of their respective chapters. Referees have also paid attention to possible repetitions, introduction of cross references, and uniformity. At the same time, we wanted to preserve the intent of individual writers. A delicate task of reviewing the entire volume, taking into account these factors, has been performed by David P. Turner of Drexel University. It is thanks to these efforts that we have a volume with much more coherence than the average collective work.

In addition to those named above, we would like to thank Mr. Joachim Spencker, the Publisher, and also Dr. Edmund H. Immergut, Consulting Editor for Hanser, for their initiative, as well as for their cheerfully given help and advice. Working with them and with our large international team of authors and experts has been a rare pleasure.

Failure, to some extent, is the dark side of engineering. We hope this book will help the polymer community deal successfully with these problems in a positive and satisfying manner. Putting together what we already know ought to lead to a coherent view of failure. Such a view would greatly aid the prevention of future unpredictable failures of polymeric materials.

Witold Brostow and Roger D. Corneliussen
Philadelphia, January 1986

Acknowledgments

The authors gratefully acknowledge permissions to reproduce copyrighted materials from a number of sources. Every effort has been made to trace copyright ownership and to give accurate and complete credit to copyright owners, but if, inadvertently, any mistake or omission has occurred, full apologies are herewith tendered.

Contents

Part B   Failure Phenomena

12   Physical Aging: Relation Between Free Volume and Plastic Deformation   235
by Jean-Claude Bauwens

13   Estimation of Long-Term Behavior From Short-Term Tests   259
by Georg Menges, Rainer Knausenberger and Ernst Schmachtenberg

Part C   Specific Systems

20   Dielectrical and Dynamic Mechanical Properties of Rubbers 393
by Sture Persson and Bengt Stenberg

21   Knit-Lines in Injection Molding and Mechanical Behavior 415
by Ralf M. Criens and Hüter-Georg Moslé

List of Important Symbols

The quantities listed here are those recommended by international learned unions and/or used by the majority of the authors.

a	area
$a_i \ (i=1, 2, \dots, N)$	constants
$a(t)$	acceleration
a_T	shift factor
e_s	surface energy
f	frequency
h	height; depth; the Planck constant
k	the Boltzmann constant
l	length
l_{max}	length at maximum deflection
l_0	original length
l_f	final length
m	mass
n	refractive index
n^*	number of defects
$\bar{r}$	stress vector; polar coordinate
r	degree of polymerization; radius
s	specific shear strength; cluster size
t	time
t^*, t_R, t_i, etc.	relaxation time
v	specific volume; velocity
v_c	craze velocity
v^*	characteristic volume
w	work; weight
W_v	wear volume
x, y, z	Cartesian coordinates
A	area
A_i	coefficients
B_i	coefficients
C_i	coefficients
$D(t)$	tensile creep compliance
D_m^*	maximum value of dynamic relaxation
E	the Young modulus = modulus of elasticity
$E(t)$	tensile stress relaxation modulus
F	force
F^+	maximum relaxation rate
G	the Gibbs function; names "free energy" or "free enthalpy" should not be used
G_{I_i}	interfacial fracture energy
G_s	shear modulus
G_s'	storage modulus
G_s''	loss modulus
H	enthalpy
H'	hardness; yield pressure
I_R	intermittant relaxation
J	Joule (unit)
J	J-integral
J_0	dielectric constant at limiting frequency

$J(t)$	shear compliance
K	degree Kelvin
K_c	critical stress intensity factor
K_{Ic}	plane strain fracture toughness
K_I	stress intensity factor
K_t	stress concentration factor
L	load
L_i	contour length of i-th polymer segment
L_e	end-to-end distance of a segment
L_σ	stress transfer length
N_A	the Avogadro number
P	pressure
R	gas constant
S	entropy
T	thermodynamic ($=$absolute) temperature
T_g	glass transition temperature
U	energy
V	volume
W	toughness; wear
α	isobaric expansivity (also called cubic or thermal expansion coefficient; the first name given is recommended)
α'	relaxation parameter
α^*	geometric factor; material constant
β	polarizability
β'	coefficient
γ	shear strain
$\tan\delta$	loss tangent
$\tan\delta_{max}$	maximum loss tangent
γ	surface tension
ε	linear strain $=$ relative elongation $=$ engineering strain
ε_a	engineering strain of the amorphous region
ε_i	permittivity
ε'	permittivity
ε''	dielectric loss factor
ε^*	complex permittivity
ε_0	permittivity of vacuum $=$ electric constant
ε_r	relative permittivity
η	viscosity
κ_T	isothermal compressibility
λ	wavelength; uniaxial deformation
μ	dipole moment; friction coefficient
$\mu(t)$	absorbed energy
ν	the Poisson ratio
ρ	mass density (not number density)
ρ_{cl}	cross-link density
σ	engineering stress
σ_0	initial stress at time $t=0$
σ_t	true stress
τ	shear stress
ω	circular frequency

1 Introduction to Failure Analysis and Prediction

Witold Brostow
Roger D. Corneliussen

Department of Materials Engineering, Drexel University
Philadelphia, PA 19104, U.S.A.

The capacity of a polymer to accommodate large deformations without sustaining rupture or other permanent changes finds immediate explanation in the diversity of configurations that its chains may adopt.

(Paul J. Flory in *The Science of Macromolecules*[1])

1.1 Introduction

The field of failure analysis and prediction is a vast one. Failure of a polymeric component can be brought about by various means: mechanical, thermal energy, different kinds of radiation, and/or effects produced by the environment. These means affect different states of polymeric materials differently; compare the effects of tension on a rubber and on a glassy polymer.

Some readers of this book might want to see a wider panorama, and some might need first of all an answer to a specific question. Both types of readers have been accommodated. This chapter contains a guide to the field—and thus also to the book. The overview is extended further in the following Chapter by Kenner. The remaining chapters in Part A are concerned with topics generally labeled as general concepts, including viscoelasticity and fracture mechanics. Part B deals with a variety of failure phenomena. Finally, in Part C we have an important Chapter on processing by Criens and Moslé, and—in view of what we have said above—contributions on specific types of behavior and structures: rubbers, fibers, and composites.

A survey of the field shows the existence of a diversity of approaches and points of view on certain topics. This is typical for any "living" discipline—before it becomes "classic" when most important problems have been solved and there is little left to investigate. Rather than take sides, we have decided to present this diversity. As an example, we have two chapters on physical aging—by Struik and by Bauwens. Their points of view coincide in a number of respects—but by no means in all. Therefore, before tackling a specific subject, the reader would do well to consult the Subject Index since several points of view may have been presented. If this is the case, the reader can choose an approach which fits best his/her own knowledge and experience, and/or is the most promising in the longer run.

Subsequent sections of this Chapter contain the succinct overview mentioned. The last Section contains information about the organization of the book; while perusing it is by no means a must, it will help the reader in accessing and absorbing the information more efficiently.

1.2   Stress and Strain: From Elasticity to Viscoelasticity

Since so many modes of failure result from mechanical means, stress-strain behavior represents a convenient reference. The simplest case of uniaxial stress-strain curve under a constant strain rate is discussed by Brown in Section 6.2.

While in general we are relating more complicated behavior to simple cases such as the uniaxial tension, this book is not a substitute for a textbook on mechanical properties of materials or of viscoelasticity. A reader who wants to know more about aspects of these fields not directly related to polymer failure can consult first a textbook of materials science and engineering, for instance,[2] and then a textbook of viscoelasticity such as the one by Aklonis and MacKnight.[3]

We shall now discuss two special stress-strain situations which are simple and convenient in analyzing more complex states; therefore, an explanation is needed at an early stage. The first of those is plane strain. An example of producing such a state by compression is shown in Fig. 1. As usual in any given direction the stress $\sigma = F/s$, that is the force F per unit area s. We have a negative stress σ_x (the sign depends on a convention) acting along the x axis. With the index $i = x, y$, or z corresponding to the three Cartesian axes, we have the engineering strain ε_i along any of the axes defined as

$$\varepsilon_i = \frac{l_i - l_{0i}}{l_{0i}} \tag{1.2.1}$$

where l_i is the current length of the specimen along the i axis, while l_{0i} is the original length before the stress-producing force F was applied. It is evident from the Figure that a negative ε_x along with a positive ε_y are produced, while $\varepsilon_z = 0$. It is this last property which defines the *plane strain*.

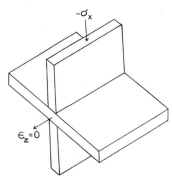

Fig. 1.2.1. An example of a plane strain produced by compression.

There is one situation important in failure analysis when a plane strain can occur: around the tip of a crack in a relatively thick specimen. A special situation arises when the specimen volume V is a constant; for an incompressible material with

$$l_{0x} l_{0y} l_{0z} = l_x l_y l_z \tag{1.2.2}$$

we have a relation between the three principal stresses:

$$\sigma_2 = \frac{\sigma_x + \sigma_y}{2} \tag{1.2.3}$$

The second special state is that of *plane stress*. It is shown schematically in Fig. 2 and occurs in deformed thin sheets. We have nonzero principal stresses σ_x and σ_y in the sheet plane. The stress σ_z along the axis normal to the sheet surface is almost $-$

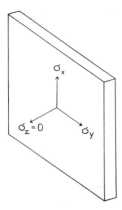

Fig. 1.2.2. An example of a plane stress in a thin sheet specimen.

although not exactly – equal to zero. For most purposes, however, the assumption of the plane stress $\sigma_z = 0$ is adequate.

Elasticity and *plasticity* are exhibited by materials of many kinds; by contrast, *visco-elasticity* is hardly a concern for instance in metals, but is very typical for polymers. These three kinds of responses to mechanical loading are explained for instance in Chapter 12 of [2]. Very briefly, in a purely elastic state the strain returns to zero after the force F has been removed. In the case of plasticity a remnant finite strain exists after the removal of the F force. Viscoelasticity can be characterized as a concomitant elastic (Hookean solid) and viscous (flowing liquid) response. Then the deformation and the corresponding strain can be decomposed into elastic and viscous components. In dynamic tests the former are called in-phase or storage, while the latter are called out-of-phase or loss components. The ratio of loss to storage components of strain (and of related quantities) is the loss tangent. In a purely elastic material under uniaxial tension the Young modulus E is simply equal to σ/ε; since the process is one-directional, we have dropped the Cartesian coordinate indices. Just like strain, in a viscoelastic material, E can be decomposed into storage and loss components. For a discussion of this separation see Section 2.2. Time dependence of modulae can be dealt with in terms of relaxation times, as discussed by Matsuoka in Chapter 3 as well as by Kubát and Rigdahl in Chapter 4.

1.3   Fracture Mechanics, Impact Behavior and Fatigue

As noted briefly in the beginning of Chapter 7, fracture mechanics (FM) has been developed for metals. Only later was an effort made to transfer the use of the same concepts to polymers. We do need FM; without it many aspects of mechanical behavior and failure of polymers would be much less clear than they are now. Using it, however, we have to be aware of the fact that *connectedness* of atoms in macromolecular chains has not been taken into account; see the motto of the present Chapter. Thus, the maximum stress that a crack-containing sheet can sustain without crack growth is given for metals by Eq. (7.4.9). An analogous stress for polymers is defined by Eq. (7.9.1); there is an *ad hoc* correction for the fact that the material is not completely elastic. FM does not provide us with methods of predicting the correction. An effort to better adapt FM to the polymer field is taking place, however; see also Section 18.5.

An import aspect of the mechanical response of materials is fatigue – leading to failure at relatively low stresses. In our book, fatigue is first dealt with in Section 2.7.

An exciting method of experimentally studying fatigue is described by Adams and Wu in Chapter 8; their instrumented impact testing procedure enables the determination of quite a few pertinent parameters. Finally, we have an entire Chapter 18 on this subject by Moet.

We have just mentioned instrumented impact testing, and in fact we have a total of three chapters on impact behavior. In Chapter 10 there is a theory connecting the stress-concentration factor, a purely geometric parameter, with behavior of macromolecular chains and with temperature dependence of this behavior. Thus, methods of connecting FM concepts with polymer chain characteristics can be developed. Clearly, we need more work to widen such bridges. There are considerable hopes in computer simulations, which are already helping us to understand the so far obscure aspects of mechanical behavior of polymers.[4, 5]

1.4 Chain Scission

Failure of a whole polymeric specimen can hardly occur unless it is preceded by scission of individual chains. Reasons for this include chain entanglements and the fact that a single chain can go through amorphous as well as crystalline regions. The scission can be brought about by mechanical as well as by other means, such as the presence of certain chemicals, thermal energy or sunlight. For this reason we have a Chapter 5 on scission and related problems by Kausch. The Chapter has been preceded by a book by Kausch on polymer fracture.[6] At the same time, we follow our policy enunciated above of presenting different points of view. Thus, in Chapter 22 by Zimmerman, there are concepts on fibers which do not coincide entirely with those of Kausch.

"Mixed" phenomena, when more than one factor producing chain degradation and scission is present, are important. Thus, strain activates thermo-oxidation of rubber during fatiguing, accelerates photo- and thermo-oxidation of polyethylene, and so on.

We therefore have an entire Chapter 14 on mechano-chemical phenomena by Terselius, Gedde, and Jansson.

1.5 States of Behavior: From Glasses to Flowing Liquids

In this brief overview we do not intend to even name all important aspects of polymer failure covered in the present volume. Many of them are evident from the list of contents. Moreover, as noted above, Chapter 2 has a character of a more detailed overview, of failure mechanisms in particular.

We shall, however, discuss now different types of polymer behavior, rubbers in particular. Chapter 20 on rubbers by Pearsson and Stenberg provides an unusually quantitative treatment on the subject. Dielectric vulcametry deals with permittivity and dielectric loss factor, quantities analogous to the respective mechanical properties named above in Section 1.2. Chapter 20 covers dielectric as well as dynamic mechanical behavior of rubbers.

It is important to note that calling a polymer a rubber refers to a state, not to a type of material. Consider a deformation such as tensile relaxation, say 30 s after a uniaxial load was imposed upon a specimen. That is, we are studying a time-dependent Young's modulus, usually called the tensile relaxation modulus, defined as

$$E(t) = \frac{\sigma(t)}{\varepsilon} \qquad\qquad (1.5.1)$$

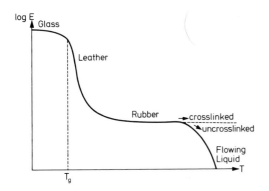

Fig. 1.5.1. Dependence of the tensile relaxation modulus E of an elastomeric material on temperature T for a fixed relaxation time t. Conversely, dependence of ln E on log of time t at a constant temperature is also represented by a curve of this shape.

Specifically for the time $t = 30$ s, taking a number of identical specimens at various temperatures T, one can construct a curve of the tensile modulus $E(30 \text{ s}, T)$. Since E can change by several orders of magnitude, it is usually convenient to work with log E. Of course, 30 s is only an example, but it is important that the same time pertains to all temperatures.

A typical curve of this kind for an elastomeric material is shown in Fig. 1. At low temperatures the material is a glass: the modulus is relatively high, fracture when it occurs is brittle. Then from the glass-transition temperature T_g up to, approximately, $T_g + 30$ K we have the leathery state with retarded high elasticity. A further temperature increase produces a further increase in free volume v^f in the material, that is more space for polymer chains and their parts to move; see again the motto of this Chapter. From the region of retarded high elasticity we go to instantaneous high elasticity in the rubbery state – characterized by a plateau of the curve. Now, two possibilities exist. If the material is cross-linked and we continue to increase the temperature, there is no significant change; the plateau persists. By contrast, if the cross-linking is absent, higher T will bring about a viscous flow.

We conclude that one material can be a glass, or a leather, or a rubber, or a flowing liquid – depending on an external parameter such as temperature. The word "leather" refers here to the state defined above, and does not imply a connection to natural leather. A further warning on terminology is also necessary. Some polymer scientists and engineers use the words "rubber" and "elastomer" interchangeably; others limit the former to materials in the rubbery state, while calling an elastomer a material which exhibits the behavior shown in Fig. 1. Thus, according to the second definition, we can have an elastomer in the glassy state.

If we apply an inverse procedure, that is for a constant tensile strain and now a constant temperature we study the changes of log E in function of time (more conveniently of log t), the resulting curve is of *the same shape* as that shown in Fig. 1. While the horizontal coordinate is different, we can see again the glassy, leathery, rubbery, and liquid states. Somebody conducting an experiment on a time scale of a second would conclude that he is dealing with a glass. If the experiment would last many hours, the same material would exhibit properties of a rubber; the initial "protest" of the material against the imposition of a strain had been short-lived, the macromolecular chains adapted themselves to the deformation (that is new length), and now only the rubbery type of response is detected. In other words, in a 100-year operation data will not be recorded more often than, say, once a week. Most of what we have just called protest of the material took place immediately after $t = 0$ when the constant strain had been imposed. The $E(t)$ curve will miss important action, since after one week this action subsided.

If we cannot record the varying parameters at short time intervals, and at the same time we don't want to miss the action in the beginning of the process, there might be a way out: lowering the temperature considerably. At a lower T there will be less total volume, and, therefore, less free volume than before. Changes in polymer conformations and the resulting changes in properties will be slowed down, and we might be able to catch these changes on a time scale convenient for us. A procedure inverse to this one is also in use: if a process at the temperature of interest is very slow, we can increase the temperature and thus observe what we want without waiting for too long. In both procedures, we change the temperature so as to change free volume. It is the latter which is the key to the situation.

The facts just outlined form the basis of an essential principle called variously time-temperature equivalence, time-temperature correspondence, time-temperature superposition, and also the method of reduced variables. The principle is discussed in particular in Sections 2.3 and 4.2; its applications appear in various places throughout the book.

Below the glass-transition temperature T_g the material is in a nonequilibrium state. The equilibrium state would be a crystal, but a crystal is of course difficult to achieve because of entanglements, interpenetrations and other complicated conformations in a set of long macromolecular chains. The volume of a crystal is smaller than the volume of a glass. Therefore, the glassy material at least *tries* to get closer to the equilibrium state by a slow volume contraction. The change of properties produced by contraction is called *physical aging*. As already noted, we have two chapters on this topic, by Struik and by Bauwens. An immediate question is: Is the free volume-time-temperature connection also applicable below T_g? A definite answer to this question has been provided by Struik; see Section 11.2.

1.6   Organization of the Book

In this volume the basic unit is a section in a chapter. Equations, inequalities, and sequences are consecutively numbered together within sections. Thus, Eq. (10.5.16) means Eq. (16) in Section 5 in Chapter 10 — this is useful for localizing equations quickly. In Section 10.5, the same equation may be simply referred to as Eq. (16).

The same system of numbering is used for tables and figures. However, references are numbered within *chapters*. Notes are numbered together with references.

As noted in the Preface, a considerable effort has been made to maintain consistency in terminology, symbols, and units. The international system of units, generally known under its French abbreviation SI, has been adopted. Since some American authors (and readers ...) are used to units which the metrologists call *contrary to SI,* in a number of instances we have nonmetric units together with their SI equivalents. Thus, the reader has a choice, and at the same time can compare the relative magnitude of the numbers. Incidentally, extensive conversion tables of units have been provided in Appendix 3 of [2]. We have encouraged our authors to use the *quantity calculus* for the headings in tables and as labels on the coordinate axes. This eliminates some repetition, permits the use of the ordinary rules of algebra when dealing with physical quantities, and eliminates situations such as "$\log (cm^{-3})$." If we have a volume $V = 12\ cm^3$, then we should not use "$\ln V$" but rather $\ln (V/cm^3)$. Of course, $\ln V$ is acceptable in an equation showing a relation between different quantities, at the stage when units have not been specified.

For terminology and symbols we have followed recommendations based on SI provided by the International Union of Pure and Applied Chemistry and by the International Union of Pure and Applied Physics. These recommendations have been discussed for instance by McGlashan.[7] We have made just one exception: the names "number-average molecular weight," or "molecular weight" (M_n, $\bar{M}$, M) are of course incorrect from the SI point of view. The correct name, relative molecular mass, however, is seldom used by polymerists. We have consciously permitted this one violation of the SI rules.

The structural and organizational rules described above have been arrived at by discussions between the publisher, editors, authors, referees, experts in specific fields, and other readers of the manuscript of the volume. To conclude this chapter, let us state one important conclusion of this fairly large international team: problems of failure of polymer materials will be eventually solved by concomitant and coordinated efforts along the lines of theory, computer simulations as well as experiments.

References and Notes

1. P.J. Flory, in *Outlook for Science and Technology,* National Research Council of the U.S., p. 370, Freeman, San Francisco 1982.
2. W. Brostow, *Science of Materials.* Wiley, New York-London 1979; W. Brostow, *Introducción a la ciencia de los materiales,* Limusa, México, D.F. 1981; W. Brostow, *Einstieg in die moderne Werkstoffwissenschaft,* Carl Hanser, München-Wien 1985.
3. J.J. Aklonis and W.J. MacKnight, *Introduction to Polymer Viscoelasticity,* 2nd edition, Wiley, New York 1983.
4. M.L. Booy, W. Brostow, H.C. Rogers and A.D. Rollett, *Proc. Symp. Mater. Res. Soc.* 1986.
5. W. Brostow and D.P. Turner, *Proc. Symp. Mater. Res. Soc.* 1986.
6. H.H. Kausch, *Polymer Fracture,* 2nd edition, Springer, Heidelberg-New York 1986.
7. M.L. McGlashan, *Pure Appl. Chem.* 1970, **21**, 2; M.L. McGlashan, *Ann. Rev. Phys. Chem.,* 1973, **24**, 51.

2 Failure Mechanisms

Vernal H. Kenner

Department of Engineering Mechanics, The Ohio State University
Columbus, OH 43210, U.S.A.

2.1 Introduction

The purpose of this chapter is to delineate the several *mechanical failure mechanisms* which are of importance when polymers are employed in applications where load-bearing capability is required. This task is undertaken from an engineer's point of view wherein "failure" may be constituted either by failure to support the imposed loads or by exceeding prescribed deformation limits. While recognizing that some of the elements of the present discussion are treated (in greater detail) in other chapters of this volume, they are nonetheless treated here in an introductory fashion in an attempt to provide a more comprehensive overview of failure mechanisms.

A logical approach, from a mechanist's point of view, is first to consider deformation as it may limit the function of a structure, or portion thereof, and then to proceed to the case of deformations leading to *rupture* (*creep rupture*). The discussion of *deformation* will also become a necessary ingredient in the description of fracture in polymers. Finally, we build using the ideas of the section dealing with fracture to discuss *fatigue failures*.

In proceeding through these topics we will first describe, in fairly general terms, the principal physical phenomenon. Then we will document the appropriate mathematical representation(s), to the extent that they are available, and thus elucidate the material-describing parameters which, at least ideally, define the material vis a vis a particular mechanical failure mode. We also describe the experimental means of measuring such parameters and point out the limits of the analytical models as they are revealed by comparison to experimentally obtained data.

2.2 Deformation

In engineering practice the stiffness of a material (or, alternatively compliance, i.e., the reciprocal of stiffness) is the mechanical property which is relevant to the question of determining the deformation of a body subjected to load. For the majority of applications involving metals this property may be considered to be independent of both time and temperature. If loads are known and the material exhibits linear behavior for the stress levels of interest then, always in principle and often in practice, the strain and, in turn, the displacements may be determined if suitable *"elastic constants"* are known. For example, in the well-known case of linearly isotropic materials, the stiffness in uniaxial tension (Young's modulus, E) and the Poisson's ratio, v, are sufficient to characterize the material. The constitutive relationship is called generalized Hooke's law and it may be presented in terms of several combinations of (always) two independent elastic constants. We record here one form of the constitutive relation, namely

$$\sigma_{ij} = \lambda^* \, \delta_{ij} \, \Delta + 2G_s \, \varepsilon_{ij} \tag{2.2.1}$$

where λ^* and G_s are elastic constants (Lamé constants; G_s is also called the shear modulus) and where σ_{ij} represent the stress and strain tensors, respectively. The indices i and j take on values 1, 2, and 3 and $\delta_{ij}=1$ if $i=j$ and zero otherwise. The symbol Δ represents the cubical dilatation and is equal to $\varepsilon_{11}+\varepsilon_{22}+\varepsilon_{33}$. Relations between λ^* and G_s and the usually measured elastic constants E and v are given in most elasticity texts.[1] The latter constants may be found through time-independent experiments – most typically a uniaxial tension test with transverse contraction of the specimen also measured.

The nature of polymeric materials is to exhibit time-dependent deformation as a response to constant loads, i.e., creep. For materials exhibiting linear response a theory of linear viscoelasticity may be constructed[2,3] which parallels the theory of linear elasticity. However, instead of elastic constants, viscoelastic material functions which are time or frequency dependent are now required to characterize material response to loading. For example, a simple tension creep test, with load held constant and time dependent strain $\varepsilon(t)$ measured, provides the creep compliance

$$D(t)=\frac{\varepsilon(t)}{\sigma} \tag{2.2.2}$$

while a uniaxial relaxation test, with strain held constant and time-dependent stress $\sigma(t)$ measured, defines the relaxation modulus

$$E(t)=\frac{\sigma(t)}{\varepsilon} \tag{2.2.3}$$

Analogous tests in pure shear (typically, torsion) generate the shear creep compliance

$$J(t)=\frac{\gamma(t)}{\tau} \tag{2.2.4}$$

and the relaxation modulus in shear

$$G_s(t)=\frac{\tau(t)}{\gamma} \tag{2.2.5}$$

Here γ is the shear strain and τ is the shear stress.

Constitutive relations now reflect the fact that the current stress (or strain) depends not only on present strain (or stress) but on all previous history as well. The viscoelastic constitutive relation corresponding to Eq. (2.2.1) is

$$\sigma_{ij}(t)=\delta_{ij} \int_{-\infty}^{t} \lambda^*(t-\tau) \frac{\partial \varepsilon_{kk}}{\partial \tau} d\tau + 2 \int_{-\infty}^{t} G_s(t-\tau) \frac{\partial \varepsilon_{ij}(\tau)}{\partial \tau} d\tau. \tag{2.2.6}$$

Several other forms are recorded elsewhere.[2]

Where the relation between compliance and stiffness in elasticity is one of simple reciprocity, the relationship between the compliance and stiffness functions of linear viscoelasticity involves convolution integrals. For example, the functions $D(t)$ and $E(t)$ are related through the formula

$$t=\int_{\delta}^{t} E_s(\tau) D(t-\tau) d\tau \tag{2.2.7}$$

In addition to the creep and relaxation functions defined above, frequency-dependent material functions may also be defined. For example, if time dependent but harmonic uniaxial stress is imposed on a rod, then the corresponding strain can be found through a stress-strain relation like Eq. (2.2.6). The relationship between stress and strain is then

$$\sigma(t) = E(i\omega)\,\varepsilon_0 e^{i(\omega t + \delta)} \tag{2.2.8}$$

Here $E(i\omega)$ is the complex modulus (sometimes called the dynamic modulus) and δ is the phase angle by which strain lags the imposed stress. $E(i\omega)$ is typically decomposed into

$$E(i\omega) = E'(\omega) + i E''(\omega) \tag{2.2.9}$$

where E' and E'' are referred to as the storage modulus and the loss modulus, respectively. Then also

$$\delta = \delta(\omega) = \tan^{-1} \frac{E''(\omega)}{E'(\omega)} \tag{2.2.10}$$

Here $\tan \delta$ is called the loss tangent. Of course, the frequency-dependent representations of constitutive properties are related to the previously defined time-dependent functions; useful details of such relationships are given in References 2 and 4–6.

The actual determination of deformations in the case of prescribed loads in a material whose viscoelastic response functions are known requires the solution, in principle, of a traction boundary value problem. Christensen[2] provides the general theoretical considerations involved in this process. Often the suitability of a given material, insofar as excessive deformation in a particular application is concerned, may be gauged approximately from a knowledge of the *viscoelastic response function(s)*. In either case, the response functions are required for a duration equal to the design life.

Ferry[5] gives a comprehensive description of various experimental configurations utilized for the collection of viscoelastic response functions along with numerous (mostly pre-1970) references. Several descriptions of more recent work are given in References 7–11.

2.3   Effects of Temperature and Other Parameters on Deformation

The viscoelastic functions which describe polymeric materials are characteristically much more sensitive to temperature than are the corresponding elastic constants for most metals. The effect of temperature is often central to determining the deformation which a structure utilizing polymeric materials will undergo and must be routinely evaluated. Thus, in viscoelastic material characterization, tests at several temperatures in the range of interest are usually carried out. Such data are typically represented in the form of a master curve (e.g., a master creep curve) generated by shifting creep curves for several different temperatures to produce a single curve at some reference temperature, T_0. This procedure, known as time-temperature superposition or "*method of reduced variables*,"[5] is represented mathematically by*

$$D(T, t) = D(T_0, t/a_T) \tag{2.3.1}$$

The parameter a_T is called the temperature shift factor and is determined empirically by the amount of the time shift required to align creep curves. Illustration of the procedure is provided here by shear creep compliance curves (Fig. 2.3.1) for a structural

---

* Here for simplicity, we ignore the effects of temperature on both material density and the rubbery modulus; for details see Ferry.[5]

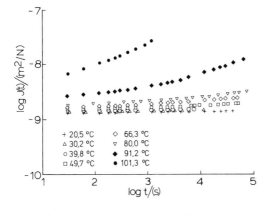

Fig. 2.3.1. Shear creep compliance for a structural adhesive, FM-73, at several temperatures (from Ref. 8).

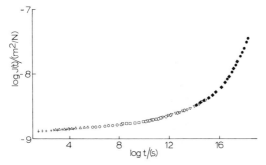

Fig. 2.3.2. Master shear creep compliance curve for FM-73 derived from Fig. 2.3.1 for a temperature of 20.5° C (from Ref. 8).

adhesive [8] consisting of a rubber-modified epoxy. These have been shifted horizontally to produce the master creep curve of Fig. 2.3.2; the corresponding shift factor is shown in Fig. 2.3.3. In this instance the effect of the vertical shift for temperature was insignificant while that for density change was yet smaller, and neither is incorporated in the master curve, Fig. 2.3.2.

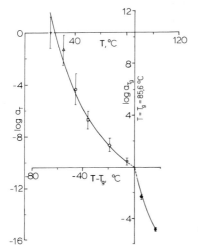

Fig. 2.3.3. Temperature shift factor for FM-73 associated with the master creep curve of Fig. 2.3.2 (from Ref. 8).

The two fit curves in Fig. 2.3.3 represent separate fittings — above and below the glass transition temperature, T_g, of FM-73 of the equation

$$\log a_T = \frac{-C_1(T - T_0)}{(C_2 + T - T_0)} \tag{2.3.2}$$

to the experimentally derived shift factor a_T. This equation, due to Williams, Landel, and Ferry[12] and known as the WLF equation, may be inferred from considerations of molecular mobility relative to free volume.[5,12] Its usefulness is primarily in the range $T_g < T < T_g + 100°$ C, and for this range, if the reference temperature T_0 is taken as T_g, the constants C_1 and C_2 are found to be approximately 17.4 and 51.6, respectively, for numerous polymers.[5] Additional discussion of the WLF-equation is given in Chapter 10.

An important practical aspect of the time-temperature superposition procedure is that it provides a basis of accelerated testing programs for polymeric materials. Thus, as an example, the creep compliance of the adhesive FM-73 may be inferred from the master curve (Fig. 2.3.2) for times which are many decades greater than the duration of any single creep test associated with the creep curves of Fig. 2.3.1. The validity of such an extrapolation must always be carefully examined. In addition to the shifting process resulting in a smooth master curve, Ferry[5] notes the following additional requirements, namely, that the same values of a_T must superpose all of the viscoelastic functions and that the temperature dependence of a_T must have a form consistent with experience. Materials which satisfy the time-temperature superposition criteria delineated above are referred to as thermorheologically simple materials.

A second practical consideration in evaluating deformation characteristics for polymeric materials is the *amount of moisture absorbed,* the consequent stresses and deformations induced directly, and the resulting changes in mechanical properties. The absorption of moisture induces swelling in unconstrained polymeric materials. When such swelling is restricted by mechanical constraints and/or when moisture concentration gradients are present, stresses analogous to thermal stresses are produced. These stresses are treated in an elastic formulation in a manner analogous to thermal stresses.[13-16] In elastic materials these stresses are time dependent only in the sense that the moisture diffusion is time dependent. The viscoelastic problem may also be treated as a (viscoelastic) thermal stress problem if the effects of moisture on material functions are ignored. Moisture-induced stresses play important roles in the failure of adhesive joints and composite materials. Elastic/viscoelastic studies of the stresses in lap joints have recently been carried out utilizing analytical techniques[17-19] and finite element methods.[20]

Of perhaps more consequence vis a vis the evaluation of deformations is the effect of moisture on the viscoelastic functions. Absorbed moisture is found to accelerate creep in a manner similar to elevated temperature.[21-24] It has been found that a master creep curve can be constructed from the results of several tests at different levels of absorbed moisture and, further, that a *moisture-concentration-dependent shift factor,* at least for one material, can be correlated with the temperature-dependent shift factor on the basis of volume change.[24] Frequently the volume change associated with particular stress states is also found to shift the time scale of the viscoelastic functions.[25] For example, Fillers and Tschoegl[9] found that uniaxial stress relaxation experiments with superposed hydrostatic pressure produced a series of pressure-dependent relaxation curves which then yielded a master relaxation curve and a corresponding pressure-dependent shift factor relation. Knauss and Emri[26] accounted for material nonlinearity on the basis of time shifted properties due to the dilatation associated with uniaxial stress.

2.4 Yielding

At sufficient values of stress some polymers, like some metals, exhibit a transition from recoverable deformation to permanent deformation which is called yielding. Often in engineering applications, this is unacceptable and thus represents a mode of failure. As a physical phenomenon, yield is also often central to the study of either fracture or fatigue, which are discussed in the sequel. In metals yield is associated with shearing motions and stresses and proves to be substantially independent of hydrostatic stress. In polymers two primary mechanisms for yield are identified; one, shear yielding, is essentially a constant volume process while the second, crazing, is fundamentally associated with dilatation.[27]

While shear yield in polymers is essentially a constant volume process (like metals), the process does depend on the deviatoric stress[28] (unlike metals). Modified yield criteria originating in applications to metals have been used successfully. Sternstein and co-workers[29] have used a modified maximum shear stress (modified Tresca, or Mohr-Coulomb) criterion and later a modified von Mises[30] criterion. These state that yield occurs when

$$\tau_{max} = \tau_0 - \mu_s \sigma_m \tag{2.4.1}$$

or when

$$\tau_{oct} = \tau_0 - \mu_s \sigma_m, \tag{2.4.2}$$

respectively. Here τ_{max} and τ_{oct} are the maximum shear stress and the octahedral shear stress, respectively, while τ_0 is the yield stress in pure shear, μ_s is a constant which characterizes the pressure sensitivity of the yield stress and σ_m is the mean normal stress. Equation (2.4.1) was first proposed for soils and, in that context, μ_s is thought of as an internal friction coefficient. Values for μ_s for a number of polymers are given by Radcliffe[28] and range from 0.03 to 0.25. Equations (2.4.1) and (2.4.2) imply that the yield stress in tension will be greater than that in compression, a difference long recognized. While these equations indicate the possibility of yield in triaxial tension, fracture will always foreclose this as a practical possibility. Shear yield may, but need not necessarily, occur in bands as thin as 1 μm or less. A further discussion of the problem is given in Chapter 6.

Crazes have been discussed in several papers.[31-34] These are highly localized plastic zones where strain levels are of the order of 100%. They require a tensile field for existence and, in uniaxial tension will constitute a planar region perpendicular to the tensile stress with a thickness typically less than 1 μm, which is many orders smaller than the other dimensions of the craze. In general stress fields, the craze will be oriented perpendicular to the maximum tensile stress. The craze material exhibits lower density and greater porosity than the surrounding material. The craze boundaries are highly reflective, while the craze itself has a lower refractive index, making these zones easily identifiable in transparent polymers. Oxborough and Bowden[35] give the craze criterion

$$\varepsilon_1 = \frac{1}{E} B(t, T) + \frac{C(t, T)}{3\sigma_m} \tag{2.4.3}$$

where ε_1 is the maximum tensile normal strain, E is Young's modulus, and $B(t, T)$ and $C(t, T)$ are parameters reflecting time and temperature dependence. Sternstein and Ongchin[30] used a slightly different relationship and replaced $E\varepsilon_1$ in Eq. (2.4.3) by the difference in principal stresses (for plane stress). For plane stress, shear yield and craze yield may each be represented in the principal stress plane as depicted in Fig. 2.4.1. While (as shown in this Figure) yielding will be of the craze type for any

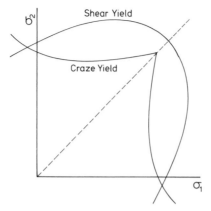

Fig. 2.4.1. Schematic of yield criteria for shear and craze (Figure after Bucknell[27]).

biaxial tension-tension stress field, mixed principal stresses may generate shear yield. Further, since the parameters of Eq. (2.4.3) depend on test conditions, changes of temperature or loading rate may alter the yield mechanism. This alteration is consistent with experience.[32] We note finally, although time dependence is not included in the representations of shear yield, Eqs. (2.4.1 and 2.4.2), this effect is observed. Brinson,[36] reporting on polycarbonate, observed shear yielding in the apparent absence of crazing which showed a mild time dependence — approximately 20% over four decades of strain rate. These data fitted the equation

$$\sigma_{yi} = B_1 + B_2 \log \frac{\dot{\varepsilon}}{\dot{\varepsilon}_0} \tag{2.4.4}$$

where B_1 and B_2, and $\dot{\varepsilon}_0$ are constants for a given material; this strain rate equation, where the dot indicates a time derivative, has also been used for metals.[37]

2.5   Rupture: Ultimate Properties

Polymers rupture under sufficiently high stresses. Such separation may occur either after substantial deformation or, as a consequence of crack growth, at insignificant levels of gross deformation. The latter case is best analyzed according to the methodology of fracture mechanics, and we defer it to the next section. We treat the former case here.

Like yield, rupture for polymeric materials exhibits a dependence on time. This dependence is associated with molecular motion, and it is found that an interrelation exists between time scale and temperature, as described in Section 2.3 for the viscoelastic functions.[5, 38] Thus, master plots of rupture stress, σ_b, versus reduced time to rupture, t_b/a_T, and plots of rupture strain, ε_r, versus reduced time may be generated. Smith[39, 40] then crossplotted from such (pairs of) plots to eliminate time and to generate a failure envelope for a particular material. Such an envelope is then independent of time and dependent on temperature only through the ratio of absolute temperatures to account for change in rubbery modulus. This failure envelope is commonly referred to as a Smith plot. Schematic representations of these plots are given in Figs. 2.5.1, 2.5.2, and 2.5.3. For the Smith plot, proceeding to higher loading rates (shorter loading times) or, equivalently, from higher to lower temperatures corresponds to moving counterclockwise around the failure curve. This plot also conveniently represents the failure

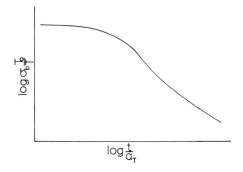

Fig. 2.5.1. Breaking stress dependence on reduced time, after Smith.[39]

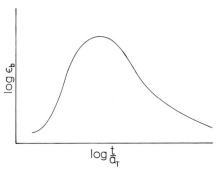

Fig. 2.5.2. Breaking strain dependence on reduced time, after Smith.[39]

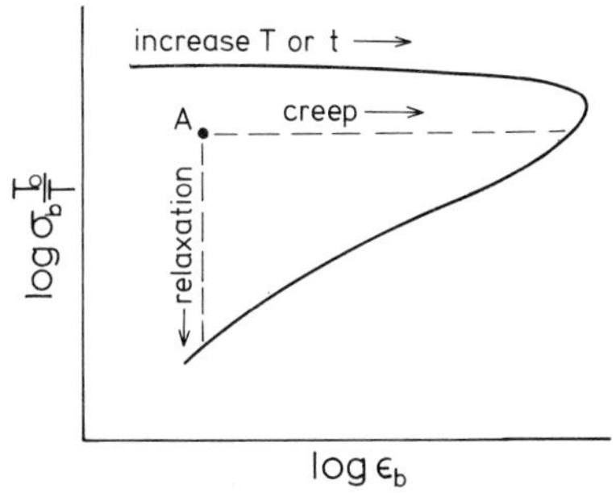

Fig. 2.5.3. Smith plot produced by crossplotting from Figs. 2.5.1 and 2.5.2, after Smith.[39]

condition ultimately attained from a particular test condition for several important experiments. For example, if the initial stress/strain condition is represented by point A in Fig. 2.5.3, then a creep test, which produces constant nominal stress, will terminate in failure corresponding to the point horizontal to the right of point A. Likewise, the failure stress and strain in a stress-relaxation test, for which strain is held constant, can be determined by intersecting the failure envelope directly below point A. We observe that failure in stress relaxation must always occur at greater time than failure in creep for corresponding initial conditions. While the Smith plot was initially developed and utilized to represent the data from experiments on elastomers at temperatures greater than T_g,[5, 38, 39] the format is currently also used to present tensile failure data for a greater variety of plastics and conditions, e.g., thermosets at temperatures below T_g.

Mathematical models of the creep rupture process exist to determine the time to rupture. These are either based on the activated rate process theory[41, 42] or are of a phenomenological nature.[43, 44] A recent exposition of creep rupture equations is given by Barton and Cherry,[45] in the context of discussing the practical problem of predicting the life expectancy of polyethylene pipe. In relation to this problem, the microscopic aspects of creep rupture are described in detail in Chapter 16.

2.6   Fracture

In this section we deal with the brittle fracture of polymers. For this case it is sometimes acceptable to ignore the time-dependent aspects of this problem and to evaluate these plastics in the context of an elastic analysis. Since, furthermore, the elastic analysis is, to a large extent embedded in the attack on the more complex time-dependent problem, we deal first with it. This treatment will provide, in capsule form, some results which are presented in Chapter 7 in greater detail. At the outset of this section we note several reviews exist which deal with the fracture of polymers. Among these are review articles by Knauss,[46] Williams,[47] and Kinloch[48]; a book by Kausch[49]; and collections of survey papers edited by Rosen,[50] Kausch e. a.,[51] and Andrews.[52] These sources contain extensive bibliographies. There is, of course, Chapter 5 by Kausch in this volume.

Griffith[53] recognized that the relatively low strength of solids, compared to values expected in view of molecular level considerations, was due to the presence of flaws, or cracks, in the materials. Using an energy approach he deduced (see Section 7.3) that the fracture stress, σ_f, in a large plate subjected to uniaxial tension and containing an elliptical flaw of length $2a$ is given by

$$\sigma_f = \frac{2E e_s}{\pi a} \tag{2.6.1}$$

where e_s is the surface energy per unit area associated with the formation of new surface.

Examination of the stress distributions around the tip of a crack (the near field solution) in a linearly elastic solid led to the definition of the stress intensity factor by Irwin[54, 55] and Williams.[56] Such factors define the near field stresses for the three possible deformation modes at a crack tip (see Fig. 7.5.1) through equations like

$$\sigma_{ij} = \frac{K_I}{(2\pi r)^{1/2}} f_{ij}(\theta) \tag{2.6.2}$$

for mode I (extension) loading with similar sets of equations for mode II (shear) loading and mode III (antiplane shear, or tearing) loading.[57] The stress intensity factor depends on both load (far field stress) and geometry and gives the severity of the stress singularity at the crack tip. Thus, for arbitrary bodies and loads, if the stress intensity factors are the same, the stresses in the vicinity of the crack tip are the same. Further, since these near field stresses may reasonably be assumed to control the growth of the crack, the values of, e.g., K_I at which fracture occurs, K_{Ic}, is a material property which measures resistance to fracture. K_{Ic} is called the mode one fracture toughness.

An energy balance along the idea of Griffith but generalized to the loading of an arbitrarily shaped elastic body leads to the definition of quantities usually called strain energy release rates, G_I, G_{II}, and G_{III} associated with the three loading modes of Fig. 7.5.1. The strain energy release rates are not unrelated to stresses in the body, and, for example, for mode I loading in plane strain, it can be shown[57] that

$$G_I = \frac{1 - \nu^2}{E} K_I^2. \tag{2.6.3}$$

For plane stress,

$$G_I = \frac{K_I^2}{E}. \tag{2.6.4}$$

The strain energy release rate at which fracture occurs, called the critical strain energy release rate, G_{Ic} (for mode I), may then likewise be taken as the material property that describes a material's resistance to fracture. For an ideally elastic material the surface energy e_s and G_{1c} are related by[47]

$$G_{1c} = 2e_s \tag{2.7.6}$$

G_{1c} then, is the energy necessary to grow the crack per unit area.

The stress intensity factor is most frequently calculated by numerical stress analysis methods,[58] although analytical solutions exist for some simple geometries.[57] For birefringent polymers the stress intensity factor may be determined photoelastically.[59] An extensive compilation of stress intensity factors is given by Tada et al.[60] While stress intensity factors are usually determined by numerical or analytical stress analysis methods, the strain energy release rate lends itself to laboratory determination. It is readily shown[47, 57] for example, that for mode I loading,

$$G_I = \frac{P^2}{2B} \frac{dD}{dh} \tag{2.6.6}$$

where P is the load, D is the compliance of the body containing the crack, B is specimen thickness and h represents the crack length. Laboratory measurements may easily be made to take advantage of Eq. (2.6.6)[61] (see Section 7.14).

In view of the above considerations, then, the fracture resistance parameter may be either the critical stress intensity factor or the critical strain energy release rate, and if elastic properties are known, these are interchangeable. Experiments to determine K_{1c} require the availability of the results of stress intensity factor calculations. Several standard geometries are used for which the stress intensity factors are available. For example, ASTM E 399,[62] written for fracture toughness testing of metals, prescribes several geometries and presents associated stress intensity factors. Williams[47] discusses several geometries which have been used for polymeric materials; see also Sections 7.15–7.17.

An important restriction on the use of the above methodology for dealing with fracture arises when it is observed that Eq. (2.6.2) implies that, for any real material, the stresses at the crack tip must exceed the elastic range. Thus, there must be a region of yielded material near the crack tip. If this region is small relative to the size of the near field in which Eq. (2.6.2) applies, then (e.g., for mode I loading) K_I still governs the stresses at the crack tip. Also, therefore, K_{1c} is an appropriate material constant to characterize fracture.

Estimates of plastic zone sizes may be made. If the material is assumed to yield according to the von Mises condition then, for plane stress, a roughly semicircular zone of radius

$$r_p = \frac{1}{2\pi} \frac{K_I^2}{\sigma_y^2} \tag{2.6.7}$$

is produced for plane stress; the plane strain zone size is smaller than this[47] (see Section 7.8). A model considered more applicable to polymers—at least when crazing is observed near the crack tip—is that of a very thin craze or process zone in front of the crack tip. In this zone the material is considered to be yielded and, if the postyield stress is assumed constant, such a model leads to a process zone length

$$r_p = \frac{\pi}{8} \frac{K_I^2}{\sigma_y^2} \tag{2.6.8}$$

This process zone model originates with Dugdale[63] and Barenblatt[64]; it has been utilized in studying time-dependent crack growth in polymers by letting σ_y depend on time.[47]

When the nonelastic zone size is not small compared to specimen thickness, or when the material is not linearly elastic, K_I (or G_I) is not an appropriate parameter for characterizing fracture resistance. The J-integral of Rice,[65] defined by Eq. (7.20.1) and here repeated,

$$J = \int_\Gamma \left(Z\,dy - T\frac{\partial u}{\partial x}\,ds \right) \tag{2.6.9.}$$

was developed to handle such cases. Here Z is the strain energy density, T is the stress vector, u is the displacement vector, ds is an element of arc length and Γ is any closed contour followed counterclockwise. If Γ is taken to include the crack tip, and the material is linearly elastic, it develops that $J = G$. Thus J may be thought of as a generalized energy release rate for crack propagation. The J-integral approach has proven useful for characterizing fracture in metals sufficiently ductile to preclude the application of ASTM E 399[62]; a test procedure in this instance is delineated in ASTM E 813.[66]

While three modes of loading have been described above, the great preponderance of testing for all materials has been conducted for mode I. This limitation of experience is even more pronounced for polymeric materials and it is apparent that study of mixed-mode loading for polymers would be desirable. Palaniswamy and Knauss[67] discuss the mixed-mode problem in detail and offer some experimental data for a rubbery polyurethane.

Since all polymers exhibit viscoelastic effects to some degree, the above elastic analysis must always be applied with discretion. Williams[68] notes that initial stress intensity factors for certain brittle polymers, e.g., polystyrene and a polyester resin are quoted free of test rate (for "short-term" loading) since only mild rate dependence exists. In the case of nonignorable time dependence, the fracture process is usually broken down to consider 1) an initiation or incubation phase in which the crack, because of time-dependent material properties, proceeds to a critical state while remaining stationary and 2) a creep, or slow, or stable, crack growth stage in which slow extension of the crack occurs. Williams[47] presents an approximate analysis of initiation for the special case of only moderate viscoelastic behavior which may be represented by a power law. For relaxation modulus of the form

$$E(t) = E_0\,t^{-n} \tag{2.6.10}$$

where E_0 and the real exponent n are material constants, the critical value of strain energy release rate G_{I_c} becomes proportional to $E(t)$.[69] Under the simplifying assumptions described above, slow crack growth may also be analyzed. The critical stress intensity for growth may also be analyzed; the critical stress intensity for growth K_c is related to crack speed $\dot{a}$ by

$$K_c \propto \dot{a}^n$$

A recent comprehensive treatment of both initiation and creep crack growth analyses is given by Kanninen and Popelar.[70] In the case of large plastic or process zone sizes, analyses along the lines of the J-integral are necessary; recent developments in this area have been made by Schapery.[71]

2.7   Fatigue

It has long been observed in engineering practice that failure may occur as the consequence of repeated loadings, any one of which would be insufficient to produce failure by itself. This type of failure is known as fatigue. The early analysis of fatigue

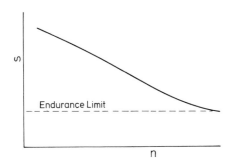

Fig. 2.7.1. The $s-n$ diagram produced by plotting alternating breaking stress, s, as a function of the number of cycles, n, producing failure.

in metals led to the $s-n$ diagram, much used by designers, wherein the alternating stress is plotted as a function of the number of cycles required to produce failure (Fig. 2.7.1). For some materials, but not all, a stress threshold exists below which fatigue failure does not occur; such a threshold is referred to as the endurance limit of the material.

While the traditional approach outlined above provides a useful tool for design, it fails to address the detailed mechanics of the fatigue process either from a continuum or a microscopic point of view. The availability of the fracture mechanics methodology described in the previous section is now often taken advantage of in the study of fatigue crack propagation. Paris[72] and Paris and Erdogen[73] found, in uniaxial stress fatigue, that the most important stress variable was the difference between the maximum and minimum values of a cyclically applied stress. In terms of the stress intensity factor K_I defined above, they wrote

$$\frac{dh}{dn} = A\,(\varDelta K_I)^m \tag{2.7.1}$$

where $\dfrac{dh}{dn}$ is the crack advance per cycle, $\varDelta K_I$ is the change in stress intensity factor and A and m are constants. The use of the stress intensity factor in Eq. (2.7.1) reflects, as before, the reasoning that, regardless of details of the fatigue process at the crack tip, such a process will be controlled by the near field elastic stress distributions. As was the case for fracture, restrictions on the size of the process zone will need to be met.

A treatise by Hertzberg and Manson[74] discusses in considerable detail the application of the ideas noted above to the fatigue of plastics. Yet more current collections of papers on the fatigue of plastics are available in special issues of the *International Journal of Fracture*[75] and *Polymer Engineering and Science*.[76] In the former volume we note, in particular, review papers on fatigue and fatigue crack growth by Sauer and Richardson[77] and by Radon,[78] respectively. A brief description of the special characteristics of the fatigue process intrinsic to polymers follows; additional coverage of this topic is to be found in Chapter 8 and in Chapter 18 which is devoted to this problem.

We consider first the fatigue of unnotched specimens as did Hertzberg and Manson.[74] This type of test, long used to evaluate the fatigue characteristics of metals, employs alternating stresses generated in either a tensile specimen or a beam of circular cross section subjected to alternating bending and results in the $s-n$ curve described above. The important factor which emerges when this test is applied to polymers is that hysteretic dissipation due to the viscoelastic character of these materials results in the generation of *heat* in a volume of material subjected to oscillating stresses.

As a consequence, test results are frequency dependent and it is possible for the temperature to rise sufficiently so that failure is a result of melting or excessive deformation if the temperature rises to the vicinity of the glass transition. This is called *thermal fatigue*.

The endurance limit, as delineated by failure in either the mechanical or thermal mode, is reduced as test frequency is increased. The mode of failure is clearly specimen shape and size dependent. For example, a bending fatigue specimen (where peak stress occurs only at the surface) will operate at a lower temperature than a uniaxial test specimen (where the peak stress occurs throughout the cross section) due to heat transfer considerations. Because of different ratios of surface area to volume, even geometrically similar specimens may give different results as the scale is changed.

In contrast to the results for fatigue of unnotched specimens, the fatigue life of notched specimens, defined by the crack propagation rate, often increases with increased test frequency $\left(\text{i.e., } \dfrac{dh}{dn} \text{ is decreased} \right)$. In this instance nominal stresses in the specimen are usually quite low and significant heating only occurs in the process zone. The large surrounding volume of cooler material mitigates for a stable, albeit elevated, temperature at the crack tip. Since this local increase in temperature lowers the yield strength and increases viscous deformation, it results in crack tip blunting and thus reduces the effective stress intensity factor which, according to the Paris equation, Eq. (2.7.1), results in lower crack growth rates.

2.8 Dynamic Problems

The various types of failure for polymers described to this point have been discussed in a context of quasi-static loading. Loading at high rates is, generally speaking, associated with greater apparent stiffness and higher yield strength. While these are positive traits from the designers point of view they are often associated with brittle failure, i.e., low fracture toughness. As we have seen in looking at other facets of polymer failure, the basic methods of evaluating the impact resistance of polymers have been carried over from established methods for metals. These include the Charpy and Izod impact tests wherein the energy loss of a mechanical striker loading, and fracturing, the material specimen of established standard geometry, is mesured. While the striker's lost energy may, nominally, be associated with the energy necessary to fracture the test coupon, these tests are still difficult to evaluate in terms of analyses and/or parameters derived from first principles defined in Section 2.6 – either the strain energy release rate or critical stress intensity factors. The reader is referred to Chapters 8, 9, and 10 of this volume or a review article by Reed[79] for as full discussion of impact testing for polymeric materials.

2.9 Closure

In the above discussion we have delineated several important ways in which polymers may cease to meet their (engineering) function. In this final section it is noted that failure modes in two of the increasingly important applications of polymeric materials, namely, as adhesives and as the matrix component of composite materials, have not been discussed. The scope of this review did not permit this. We comment only that the significant fraction of the mechanics necessary to study these problems is embedded in the above discussion. For a recent, and substantive review of the mechanics of adhesive joints the reader is referred to the work of Kinloch[80]; for an excellent starting point for a study of failure in composites see Chapters 23 and 24 of this volume.

References

1. I.S. Sokolnikoff, *Mathematical Theory of Elasticity,* 2nd ed., Chapter 3. McGraw-Hill, New York 1956.
2. R.M. Christensen, *Theory of Viscoelasticity, An Introduction,* Academic Press, New York 1971.
3. M.E. Gurtin and E. Sternberg, *Arch. Ration. Mech. Anal.* 1962, **11**, 291.
4. J.J. Aklonis and W.J. MacKnight, *Introduction to Polymer Viscoelasticity,* Chapter 1, 2nd Ed., Wiley-Interscience, New York 1983.
5. J.D. Ferry, *Viscoelastic Properties of Polymers,* 3rd ed., Wiley, New York 1980.
6. F.R. Schwarzl, in *Deformation and Fracture of High Polymers,* edited by H.H. Kausch, J.A. Hassell and R.I. Jaffee, p. 47, Plenum Press, New York 1973.
7. C.Y.-C. Lee, L.R. Denny and I.J. Goldfarb, *Am. Chem. Soc. Polymer Prepr.* 1983, **24**, 139.
8. V.H. Kenner, W.G. Knauss and H. Chai, *Exper. Mech.* 1982, **22**, 75.
9. R.W. Fillers and N.W. Tschoegl, *Trans. Soc. Rheol.* 1977, **21:1**, 51.
10. G.M. Smith, R. Bierman and S.J. Zitek, *Exper. Mech.* 1983, **23**, 158.
11. M.L. Williams, N.H. Wackenhut, R.D. Marangoni, N.R. Basavanhally, E.F.M. Winter and C.C. Yates, *Mechanical Spectroscopy for Epoxy Resins,* Technical Report AFWAL-TR-81-4070, Air Force Wright Aeronautical Laboratory, Dayton, Ohio.
12. M.L. Williams, R.F. Landel and J.D. Ferry, *J. Amer. Chem. Soc.* 1955, **77**, 3701.
13. R.A. Schapery, *J. Compos. Mater.* 1968, **2**, 380.
14. J.C. Halpin and N.J. Pagano, Technical Report AFML-TR-68-395, Air Force Materials Laboratory, Dayton, Ohio 1968.
15. G. Marom, *Polymer. Eng. Sci.* 1977, **17**, 799.
16. D. Cohn and G. Marom, *Polymer. Eng. Sci.* 1978, **18**, 1001.
17. Y. Weitsman, *Int. J. Solids Struct.* 1979, **15**, 701.
18. Y. Weitsman, Technical Report AFWAL-TR-81-4121, Air Force Wright Aeronautical Laboratory, Dayton, Ohio 1981.
19. M-H.R. Jen and Y. Weitsman, in *Advances in Aerospace Structures and Materials,* edited by S.S. Wang and W.J. Renton, ASME Publications AD-1, New York 1981.
20. J. Romanko and W.G. Knauss, *Fatigue Behavior of Adhesively Bonded Joints,* Vol. I, Technical Report AFWAL-TR-80-4037, Air Force Wright Aeronautical Laboratory, Dayton, Ohio 1980.
21. H. Fujita and A. Kishimoto, *J. Polymer. Sci.* 1958, **28**, 547.
22. K. Ninomiya, *J. Colloid. Sci.* 1959, **14**, 49.
23. R.D. Maksimov, E.A. Sokolov and V.P. Mochalor, *Polymer. Mech.* 1976, **11**, 334.
24. W.G. Knauss and V.H. Kenner, *J. Appl. Phys.* 1980, **51**, 5131.
25. J.D. Ferry and R.A. Stratton, *Kolloid-Z.* 1960, **171**, 107.
26. W.G. Knauss and I.J. Emri, *Comput. Struct.* 1981, **13**, 123.
27. C.B. Bucknell, *Adv. Polym. Sci.* 1978, **27**, 121.
28. S.V. Radcliffe, in *Deformation and Fracture of High Polymers,* edited by H.H. Kausch, J.A. Hassell and R.I. Jaffee, p. 191, Plenum Press, New York 1973.
29. S.S. Sternstein, L. Ongchin and A. Silverman, *Appl. Polym. Symp.* 1968, **7**, 175.
30. S.S. Sternstein and L. Ongchin, *Am. Chem. Soc. Polymer Prepr.* 1969, **10**, 1117.
31. S. Rabinowitz and P. Beardmore, *Crit. Rev. Macrom. Sci.* 1972, **1**, 1.
32. D. Hull, in *Deformation and Fracture of High Polymers,* edited by H.H. Kausch, J.A. Hassell and R.I. Jaffee, p. 171, Plenum Press, New York 1973.
33. C.B. Bucknall, *Toughened Plastics,* Applied Science Publ., London 1977.
34. A.M. Donald and E.J. Kramer, *J. Appl. Polymer Sci.* 1982, **27**, 3729.
35. R.J. Oxborough and P.B. Bowden, *Phil. Mag.* 1973, **28**, 547.
36. H.F. Brinson, in *Deformation and Fracture of High Polymers,* edited by H.H. Kausch, J.A. Hassell and R.I. Jaffee, p. 397, Plenum Press, New York 1973.
37. R.L. Thorkildsen, in *Engineering Design for Plastics,* edited by E. Baer, Reinholt Book Co., New York 1964.

38. R.F. Landel and R.F. Fedors in *Fracture Processes in Polymeric Solids,* edited by B. Rosen, p. 361, Interscience Publishers, New York 1964.
39. T.L. Smith, *J. Polymer Sci.* 1956, **20**, 447.
40. T.L. Smith, Technical Report ASD-TRD-62-572, Aeronautical Systems Division, Wright Patterson Air Force Base, Dayton, Ohio 1962.
41. S. Glasstone, K.J. Laidler and H. Eyring, *The Theory of Rate Processes,* McGraw-Hill, New York 1941.
42. B.D. Coleman, *J. Polymer Sci.* 1956, **20**, 447.
43. O.D. Sherby, *Acta Metall.* 1962, **10**, 135.
44. F.R. Larson and J. Miller, *Trans. Am. Soc. Mech. Eng.* 1952, **74**, 765.
45. S.J. Barton and B.W. Cherry, *Polymer Eng. Sci.* 1979, **19**, 590.
46. W.G. Knauss, *Appl. Mech. Rev.* 1973, **26**, 1.
47. J.G. Williams, *Adv. Polymer Sci.* 1978, **27**, 67.
48. A.J. Kinloch, *Metal. Sci.* 1980, **14**, 305.
49. H.H. Kausch, *Polymer Fracture,* Springer, Berlin 1978.
50. B. Rosen (ed.), *Fracture Processes in Polymeric Solids,* Interscience Publishers, New York 1964.
51. H.H. Kausch, J.A. Hassell and R.I. Jaffee (eds.), *Deformation and Fracture of High Polymers,* Plenum Press, New York 1973.
52. E.H. Andrews (ed.), *Developments in Polymer Fracture-1,* Applied Science Publishers, Ltd., London 1979.
53. A.A. Griffith, *Phil. Trans. Royal Soc. London* 1921, **A 221**, 163.
54. G.R. Irwin, *J. Appl. Mech.* 1957, **24**, 361.
55. G.R. Irwin, in *Handbuch der Physik,* Vol. VI, Springer, Berlin 1958.
56. M.L. Williams, *J. Appl. Mech.* 1957, **24**, 104.
57. P.C. Paris and G.C. Sih, in *Fracture Toughness Testing,* ASTM STP 381, p. 30, American Society for Testing and Materials, Philadelphia 1965.
58. G.C. Sih (ed.), *Methods of Analysis and Solutions of Crack Problems,* Noordhoff, Leyden 1973.
59. A.S. Kobayashi, in *Experimental Techniques in Fracture Mechanics,* edited by A.S. Kobayashi, p. 126, Iowa State University Press, Ames, Iowa 1973.
60. H. Tada, P.C. Paris and G. Irwin, *The Stress Analysis of Cracks Handbook,* Del Research Corporation, Hellertown, Pennsylvania 1973.
61. J.E. Strawley and W.F. Brown, in *Fracture Toughness Testing,* ASTM STP 381, p. 133, American Society for Testing and Materials, Philadelphia 1965.
62. "Standard of Test Method for Plane-Strain Fracture Toughness of Metallic Materials," ASTM E 399-81, *Part 10, Annual Book of ASTM Standards,* p. 592, American Society for Testing and Materials, Philadelphia 1982.
63. D.S. Dugdale, *J. Mech. Phys. Solids* 1960, **8**, 100.
64. G.I. Barenblatt, *Adv. Appl. Mech.* 1962, **7**, 56.
65. J.R. Rice, *J. Appl. Mech.* 1968, **35**, 379.
66. "Standard Test Method for J_{Ic}, A Measure of Fracture Toughness," ASTM E 813-81, *Part 10, Annual Book of ASTM Standards,* p. 822, American Society for Testing and Materials, Philadelphia 1982.
67. K. Palaniswamy and W.G. Knauss, *Mechanics Today* 1978, **4**, 87.
68. J.G. Williams, *Proc. Royal Soc. A* 1980, **347**, 39.
69. R.A. Gledhill and A.J. Kinloch, *Polymer* 1976, **17**, 727.
70. M.F. Kanninen and C.H. Popelar, *Advanced Fracture Mechanics,* Chapter 3, Oxford University Press 1985.
71. R.A. Schapery, "Correspondence Principles and a Generalized J Integral for Large Deformation and Fracture Analysis of Viscoelastic Media," Report MM 4665-83-7, Texas A and M University, College Station, Texas 1983.
72. P.C. Paris, Ph. D. Dissertation, Lehigh University, Bethlehem, PA 1962.
73. P.C. Paris and F. Erdogen, *J. Bas. Eng. Trans. ASME Ser. D,* 1963, **85**, 5428.

74. R.W. Hertzberg and J.A. Manson, *Fatigue in Engineering Plastics,* Academic Press, New York 1980.
75. H.W. Liu (ed.), *Int. J. Fract.* 1980, **16**, 479.
76. C.L. Beatty, (ed.), *Polymer Eng. Sci.* 1982, **22**, 921.
77. J.A. Sauer and G.C. Richardson, *Int. J. Fract.* 1980, **16**, 499.
78. J.C. Radon, *Int. J. Fract.* 1980, **16**, 533.
79. P.E. Reed, in *Developments in Polymer Fracture-1,* edited by E.H. Andrews, p. 121, Applied Science Publishers, London 1979.
80. A.J. Kinloch, *J. Mater. Sci.* 1982, **17**, 617.

3 Nonlinear Viscoelastic Stress-Strain Relationships in Polymeric Solids

Shiro Matsuoka

Plastics Research and Development Department, Bell Laboratories
505 Mountain Avenue, Murray Hill, NJ 07974, U.S.A.

The primary objective of this part of the work is to introduce a scheme for generating a stress-strain curve at any strain rate, temperature, pressure, in tension, shear, or compression, and the annealing history, from a set of stress-strain data stored in file, or from the data which the reader may have obtained. The computational scheme is simple, but the physical reasoning is complex. For this reason, those who are only interested in generating useful data on engineering properties may read only the short summary of the scaling procedure in each section, which describes the procedure for scaling up or down a stress-strain curve according to the change in the values of parameters such as the rate of strain, temperature, pressure, tensile to shear deformation, and physical aging.

The bulk of discussion is centered around the physical justification for assuming a particular form or forms of mathematical formulas for describing nonlinear viscoelastic behavior of polymeric solids. The discussion might help in gaining insight to the complexity of physics of viscoelsticity and the relaxation process in general.

3.1 Single Relaxation Time Function

The simplest case of viscoelasticity pertaining to a real material is found in amorphous rubbery polymers within a strain range of a few percent. In such a case neither the relaxation time nor the elastic modulus will depend on the strain magnitude. A plot of the relaxation modulus, $E(t)$, versus time obtained at one strain level is applicable to all strain levels. Since such a situation can be described by a set of linear differential equations, this behavior is called linearly viscoelastic. A stress-strain curve can be calculated from the experimentally obtained relaxation modulus data by applying the Boltzmann superposition principle. In this calculation, the stress at time t is obtained as the sum of all "little" stresses, $\delta\sigma$, that are the results of many incremental relaxing stresses each of which was started at a different time, or t' minutes ago with an incremental strain $\delta\varepsilon$, or $\delta\sigma(t) = E(t')\,\delta\varepsilon$, where $E(t')$ is the relaxation modulus after the lapse of time t'. Dividing the time interval from 0 to t into equal time increments δx, so that $t' = t - x$ and $\delta\varepsilon = (d\varepsilon/dx)\,\delta x$, the incremental stress $\delta\sigma$ at t is given by

$$\delta\sigma = E(t-x)\frac{d\varepsilon}{dx}\,\delta x \tag{3.1.1}$$

and the total stress, $\sigma(t)$, is given by the integral

$$\sigma(t) = \int_0^t E(t-x)\frac{d\varepsilon}{dx}\,dx \tag{3.1.2}$$

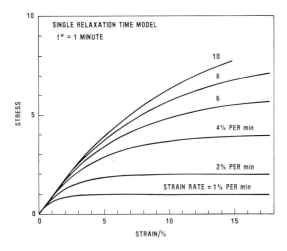

Fig. 3.1.1 Stress-strain curves for linear viscoelastic single relaxation time.

This is the well-known convolution integral. Let us consider a simplified case, though not a realistic one, with a single relaxation time, t^*, i.e.,

$$E(t) = E_0 \exp(-t/t^*) \qquad (3.1.3)$$

where $E(t)$ is the relaxation modulus, and E_0 is the initial elastic modulus. If we now consider a special case where the strain is made to increase at a constant rate, $\dot{\varepsilon}$, then Eq. (3.1.2) is solved to obtain the stress-strain relationship

$$\sigma(\varepsilon) = E_0 \dot{\varepsilon} t^* (1 - \exp(-\varepsilon/\dot{\varepsilon} t^*)) \qquad (3.1.4)$$

where $\dot{\varepsilon}$ is the strain rate. The stress-strain curves calculated with various $\dot{\varepsilon}$ from Eq. (3.1.4) are shown in Fig. 3.1.1, which exhibit the following two important features: (1) that increasing either the strain rate, $\dot{\varepsilon}$, or the relaxation time, t^*, by the same multiplication factor will result in the identical stress-strain curve, and (2) that by varying $\dot{\varepsilon}$ or t^*, the stress-strain curve will grow or shrink both in the horizontal and vertical directions by the same amount, maintaining the same congruent shape as shown.

3.2 Consideration of Distribution of Relaxation Times

Glassy polymers, which are not linearly viscoelastic and do not exhibit a single relaxation time, nevertheless exhibit stress-strain curves with one or both of the above two features, as evident from the curves for polyvinyl chloride PVC shown in Figs. 3.2.1 and 3.2.2. The congruent feature is obvious in Fig. 3.2.2 with the data obtained at different temperatures with the same rate of strain. With different strain rates, as shown in Fig. 3.2.1, the curves definitely grow taller, but a shift in the horizontal direction is slight.

There are two major differences between the experimental curves of Figs. 3.2.1 and 3.2.2 and the theoretical single relaxation curves of Fig. 3.1.1. The first of these differences is that the "size" of the stress-strain curve in Fig. 3.1.1 grows proportionately with the strain rate, whereas the size of the experimental curve grows in a manner much less sensitive to a change in the strain rate, such as to some power, n, of the strain rate, with n being less than 1. Secondly, the curves in Fig. 3.1.1 asymptotically approach the respective horizontal lines, whereas the experimental curves undergo the stress maximum.

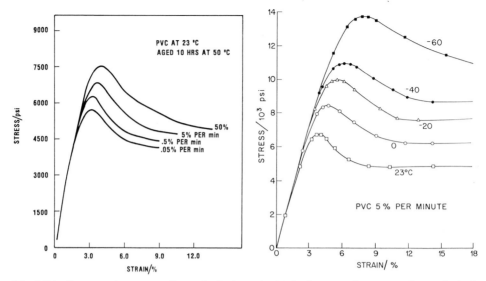

Fig. 3.2.1. Stress-strain curves for polyvinyl chloride with various strain rates ($1 \text{ psi} = 6.9 \times 10^3 \text{ N/m}^2$).

Fig. 3.2.2. Stress-strain curves for polyvinyl chloride at various temperatures ($1 \text{ psi} = 6.9 \times 10^3 \text{ N/m}^2$).

The first of these, i.e., the linear versus the less pronounced dependence of stresses and strains on the strain rate, is the consequence of the differences between a single versus broadly distributed relaxation times. With real polymers, the time dependence of stress relaxation or creep is known to be much more gradual than can possibly be represented with a single relaxation time, manifesting a broad distribution of relaxation times. To represent such a spread in the distribution of relaxation times, the technique of substituting the time by the power of time is often used. For example, the relaxation modulus in Eq. (3.1.3) is replaced by

$$E(t) = E_0 \exp(-(t/t^*)^\beta) \qquad (3.2.1)$$

where β is an empirically determined number of less than unity; t^* is also an adjustable parameter, and its value should not be a priori identified with the average relaxation time. This formula is known as the Williams-Watts equation[1] and its analogue for the complex dielectric permittivity, J^*, is frequently used,

$$J^* = J_\infty + \Delta J \, Li\left(-\frac{d}{dt}\left(\exp(-(t/t^*)^\beta)\right)\right) \qquad (3.2.2)$$

where ω is the circular frequency, ΔJ is the difference in the dielectric constant at zero and a limiting high frequency and J_∞ is the dielectric constant at the limiting high frequency, and Li denotes the one sided Fourier transform, or a pure imaginary Laplace transform. Other forms of modifying the single relaxation time expression into a distributed relaxation time function are the Cole-Cole formula

$$J^* = J_\infty + \Delta J/[1 + (j\omega t^*)^\beta] \qquad (3.2.2\,\mathrm{a})$$

and the Cole-Davidson formula

$$J^* = J_\infty + \Delta J/(1 + j\omega t^*)^\beta \qquad (3.2.2\,\mathrm{b})$$

both of which predate the Williams-Watts formula. All of the above formulas essentially modify the dimensionless quantity ωt^* for the single relaxation time formula by raising

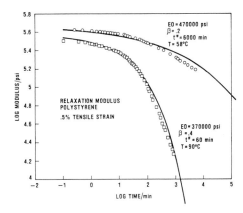

Fig. 3.2.3. Relaxation modulus of polystyrene at 58 and 90° C; both experiments were started after waiting for 15 hours at respective temperatures ($1\,\text{psi} = 6.9 \times 10^3\,\text{N/m}^2$).

to the power of β, $(\omega t^*)^\beta$. These formulas are of course applicable to the dynamic mechanical compliance in their identical forms.

The value of β is often quoted to be near $0.3 \sim 0.6$ for polymeric glasses[2] with respect to Eq. (3.2.2), and ca. 0.3 for the creep function,[3]

$$J(t) = J_0 \exp(t/t^*)^\beta \tag{3.2.3}$$

In Fig. 3.2.3, the relaxation modulus curves for polystyrene at $90°\,\text{C}$ and $58°\,\text{C}$ are fitted with the Williams-Watts formula of Eq. (3.2.1). The two curves are not superposable by shifting along the log time axis, and there is no way for describing both curves without using different values for β. The temperature dependence of β heretofore has not been recognized, primarily because most analyses have been concerned with data obtained at limited high temperature ranges where they could be collected relatively quickly. The temperature dependence of β seems to arise from the fact that any Williams-Watts type equations are two parameter equations of β and t^*. As such, almost any gradually and monotonically changing function can be fitted. Notably, the slope, $-n$, of the log E-log t plot would be

$$-n = \frac{d\log E}{d\log t} = -\beta\left(\frac{t}{t^*}\right)^\beta \tag{3.2.4}$$

hence, different pairs of β and t^* can generate the same slope $-n$ on the log-log plot at a given t, but the "width" along the abscissa that can fit with this n will become narrower with the greater β. The use of n above in place of β in Williams-Watts equations results in the better approximation. In Fig. 3.2.4, the relaxation modulus is calculated from Eq. (3.2.1) with $\beta = 0.4$ and with t^* as the parameter changing from 60 to 6×10^5 minutes. It is evident that none of these curves can be fitted with the $58°\,\text{C}$ curve in Fig. 3.2.3, even if an additional shift is allowed in the vertical direction. The shift of the relaxation time from 60 to 6000 minutes from $90°\,\text{C}$ to $58°\,\text{C}$ in Fig. 3.2.3 corresponds to the activation enthalpy of 35 kcalories (1 calorie $= 4.184$ J). This comparison is meaningful *only* because the two samples have undergone equivalent thermal histories, as will be discussed in the section dealing with the termodynamic aging process and as discussed in Chapter 12 by Bauwens. At this point, it will suffice to mention that the value of 35 kcalories is consistent with the results of a number of other viscoelastic and thermodynamic measurements made on glassy polymers and will be discussed in more detail.

The relaxation modulus of glassy polymers at room temperature is typically fitted with $\beta = 0.03$, as shown in Fig. 3.2.5. This is an order of magnitude smaller than the value of $\beta = 0.4$ near T_g. The value 4×10^7 minutes for t^* was obtained by shifting

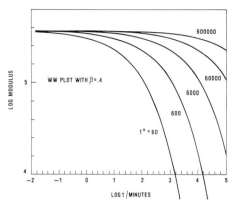

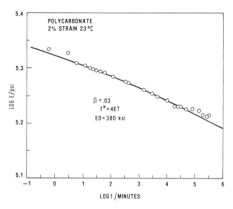

Fig. 3.2.4. Relaxation modulus with Williams-Watts formula, Eq. 3.2.1, with $\beta = 0.4$.

Fig. 3.2.5. Experimental relaxation modulus for polycarbonate at room temperature fitted with Williams Watts formula, Eq. 3.2.1 with $\beta = 0.03$ (1 psi $= 6.9 \times 10^3$ N/m^2).

from 1 minute at T_g of 147° C. Figure 3.2.6 is the illustration of the relaxation modulus calculated from Eq. (3.2.1) with $\beta = 0.03$. The negative of the slope n of the $\log E$ versus the $\log t$ plot is equal to β only when $t/t^* = 1$. From this graph, it can be seen that the Williams-Watts formula with $\beta = 0.03$ will never be able to fit higher temperature data, as $\beta = 0.4$ will never be able to fit the room temperature data.

The slope of *log* relaxation modulus vs *log* time for polycarbonate is plotted against temperature in Fig. 3.2.6. The functional form for the data follows Eq. (3.2.4) closely when the temperature dependence of t^* is considered. The temperature dependence of β as discussed above also seems to follow the same trend as for this n. Thus β and n must be considered to depend on temperature in glassy polymers, indicating that the distribution of relaxation times changes with the temperature. This behavior is distinctly different from the temperature dependence of the viscoelastic behavior of amorphous polymers above T_g, where the distribution of relaxation times remains approximately the same, while the whole spectrum is shifted with the temperature

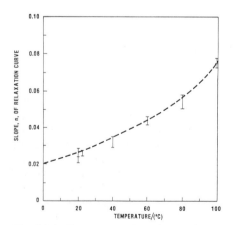

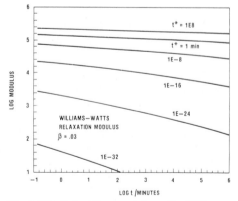

Fig. 3.2.6. Temperature dependence of parameter n which is related to the distribution of relaxation times (polycarbonate).

Fig. 3.2.7. Relaxation modulus with Williams-Watts formula, Eq. 3.2.1 with $\beta = 0.03$.

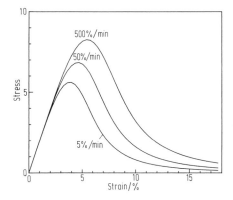

Fig. 3.3.1. Stress-strain curves with Eq. 3.3.4.

change along the logarithmic time axis. The value of β for polystyrene at $90°$ C near T_g is 0.4, and it is about the same value as the values found from creep and other experiments, *all conducted near T_g*. The temperature dependence of β and n turns out to be an important aspect of our model for describing the mechanical behavior of both glassy and crystalline polymers. The method for evaluating β and n at different temperatures will be explained subsequently. One generalized approach with the Williams-Watts formulas would be to substitute the dimensionless quantity (t/t^*) in the single relaxation time formulas with the quantity $(t/t^*)^{\beta}$. We have pointed out that this β should be considered to depend on the temperature via the temperature dependence of t^* so that the use of the parameter n instead of a fixed value β is better applicable over the wide range of temperatures. We now modify the stress-strain formula for single relaxation time as described by Eq. (3.1.4) in the above manner to obtain a Williams-Watts equivalent of stress-strain formula by substituting $\dot{\varepsilon}t^*$ by $(\dot{\varepsilon}t^*)^n$, i.e.,

$$\sigma = E_0(\dot{\varepsilon}t^*)^n\,(1 - \exp(\varepsilon/\dot{\varepsilon}t^*)^n)) \tag{3.3.1}$$

For small strains this equation fits the curves plotted in Fig. 3.3.1. Both of the features about the stress-strain curves mentioned in 3.2 are manifest, i.e., the near-congruency of the curves which grow with $\dot{\varepsilon}^n$. This equation is to be the basic model for scaling the stress-strain curves throughout this chapter.

3.3 Strain-Rate Dependence of Viscoelastic Behavior of Glassy Polymers

The Williams-Watts formula of Eq. (3.2.1) for the relaxation modulus is a linear viscoelastic function if t^* is assumed to be independent of strain. The stress-strain relationship can be obtained by substituting this relaxation modulus into the convolution integral of Eq. (3.1.2). When $\beta = 0.03$ is used, the convolution integral becomes practically a straight line over a wide range of $\dot{\varepsilon}$ or t^*, as shown in Fig. 3.3.2. This result means that the curvature in a stress-strain plot of the glassy polymers with β of ca. 0.03 must necessarily be due to the *nonlinear* viscoelastic effect, even if some ad hoc linear superposition procedure along the log time axis *seems* to work.

The strain-dependent relaxation modulus data for polycarbonate is shown in Fig. 3.3.3. The log-log plot can be superimposed by incorporating the vertical and horizontal shifts which depend on the strain. The horizontal shift is to account for the shortened relaxation time due to the strain, and the vertical shift is to account

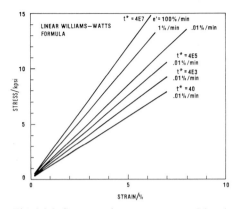

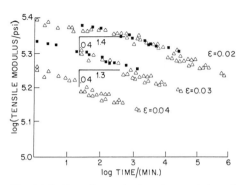

Fig. 3.3.2. Stress-strain curve generated by the convolution integral of the linear Williams-Watts modulus, Eq. 3.2.1.

Fig. 3.3.3. Experimental relaxation modulus of polycarbonate at room temperature exhibiting nonlinear viscoelasticity $(1 \text{ psi} = 6.9 \times 10^3 \text{ N/m}^2)$.

for the decreased rigidity due to the strain. This will be explained in more detail subsequently. The nonlinear viscoelastic relaxation modulus for polycarbonate is given by the equation

$$E(\varepsilon, t) = E_0 \exp(-C_1 \varepsilon) \exp[-(t/t^*)^\beta \exp(\beta C_2 \varepsilon)] \tag{3.3.2}$$

where the values of $C_1 \approx 10$ and $C_2 \approx 250$ are derived from the strain-induced increase in entropy and free volume.[4] The stress-strain curves are obtained by integrating Eq. (3.3.2) and are shown in Fig. 3.3.4 with the strain rate as the parameter. The stress-strain curves grow almost totally in the vertical direction. This is a consequence of the relatively flat relaxation curve, i.e., the small value of $\beta = 0.03$. The maximum stresses are found to increase by $(\dot{\varepsilon}_1/\dot{\varepsilon}_2)^n$, with $n = 0.023$. This value of n is obtained from Eq. (3.2.4) by letting $\beta = 0.03$ and $t/t^* \approx 10^{-3}$. In Fig. 3.3.5, experimental curves are shown for comparison. Phenomenologically, the relaxation modulus is relatively flat in this temperature range (see Fig. 3.3.3) and the total shift due to the strain

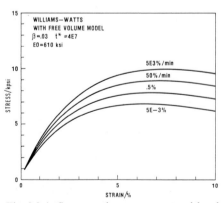

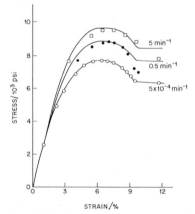

Fig. 3.3.4. Stress-strain curve generated by the convolution integral of the modified nonlinear relaxation modulus, Eq. 3.3.2.

Fig. 3.3.5. Stress-strain curves for polycarbonate at various strain rates $(1 \text{ psi} = 6.9 \times 10^3 / \text{m}^2)$.

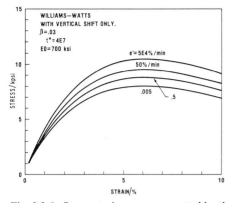
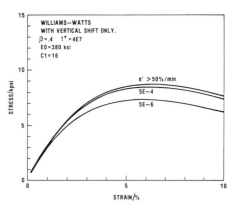
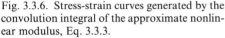

Fig. 3.3.6. Stress-strain curves generated by the convolution integral of the approximate nonlinear modulus, Eq. 3.3.3.

Fig. 3.3.7. Stress-strain curves generated from Eq. 3.3.3 except assuming $\beta = 0.4$ instead of 0.03.

can be approximated by taking up the strain dependence with the shift entirely in the vertical direction only. By so doing, Eq. (3.3.2) can be simplified considerably and we obtain

$$E(\varepsilon, t) = E_0 \exp(-C_3 \varepsilon) \exp[-(t/t^*)^\beta] \qquad (3.3.3)$$

This modulus with $\beta = 0.03$ is integrated and the stress-strain curve is shown in Fig. 3.3.6. The maximum point does not shift along the horizontal axis, but the general feature is well preserved. In comparison, if $\beta = 0.4$ is used (with an enormous shift in t^*, of course), as shown in Fig. 3.3.7, then the curves will not shift at all when $\dot{\varepsilon} \geq 50\%$ min, while below $\dot{\varepsilon} < 5 \times 10^{-4}\%$ per min the curves begin to shift rapidly. This is another assurance that the value of $\beta = 0.4$ cannot be fitted to data at room temperature with the change in t^* alone, and instead, β changes by an order of magnitude from T_g to ca. $T_g - 100°$ C.

The stress-strain curve shown in Fig. 3.3.5 can be approximately represented by

$$\sigma(\varepsilon)|_{\dot{\varepsilon}} = E_0 \exp(-C_3 \varepsilon) \exp[-(\varepsilon/\dot{\varepsilon} t^*)^\beta] \, \varepsilon \qquad (3.3.4)$$

which can be reduced to the form of Eq. (3.3.1) for small ε by expanding the exponential term. Eq. (3.3.4) is plotted in Fig. 3.3.1.

The logarithm of the secant modulus, σ/ε, is nearly a straight line when plotted against the strain, as verified by the experimental data shown in Fig. 3.3.8. From Eq. (3.3.4) it can be shown that

$$\left. \frac{d\sigma}{d\varepsilon} \right]_{\dot{\varepsilon}} = \varepsilon \left[\frac{\partial \sigma}{\partial \varepsilon} \right]_{\dot{\varepsilon}} + E\left(\varepsilon, \frac{\varepsilon}{\dot{\varepsilon}} \right) \qquad (3.3.5)$$

The first term on the right is the vertical shift factor for Eq. (3.3.4) and it is negative. The second term is the relaxation modulus in which t is substituted by $\varepsilon/\dot{\varepsilon}$. Because of the first term, this stress-strain curve *can* have a negative slope and therefore the natural stress maximum can exist.

The existence of the maximum in the stress-strain curves for glassy polymers is a real material-dependent phenomenon and not a result of the reduction in the cross sectional area or the necking. The existence of the maximum is primarily the consequence of reduced elastic rigidity due to increased strain, rather than the consequence of reduced relaxation time, though both of these phenomena occur. With the small values of β and n, the shift of relaxation times alone will not produce the stress maxi-

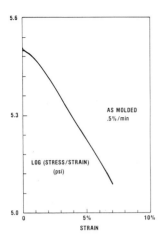

Fig. 3.3.8. Log secant modulus (σ/ε) versus strain data for polycarbonate being straight line as predicted by Eq. 3.3.4.

mum, as explained in relation to Fig. 3.3.2. A question arises then whether β or n would remain small as the strain is increased and the relaxation time is reduced. If in fact β or n did depend on strain, and if in fact the distribution of relaxation times did revert back to that of near T_g, i.e., $\beta \approx 0.4$, then the stress maximum will emerge purely as a result of the shortened relaxation time alone.

The experimental data show that the value of n remains constant nearly up to the yield strain, but unfortunately no data exist at the levels immediately beyond the yield. However, the data well beyond the yield in the steady state elongation regime, where the necking is taking place, show that n is the same low value as it was before the yield, and not a large value such as 0.4. Thus the flat shape of the relaxation spectrum is most likely maintained independently of stress or strain levels at a given temperature. This steady-flow stress, such as observed in PVC in Figs. 3.2.1 and 3.2.2, can also be scaled by the same scaling factor, which will be shown below. Thus, one parameter scaling scheme can be applied to the entire stress-strain curve including the regime of elongation-by-necking. We used the engineering stress, i.e., the force divided by the initial zero-strain cross-sectional area for all tensile data. The scaling factor, being empricial in origin, can include the change in the cross-sectional area with the change in strain or temperature. The effects from the nonuniformity in strain throughout a tensile test specimen are also built into the scaling factor and, as long as the pattern of the distribution of the strain remains unchanged, this scaling scheme will automatically include such an effect and will predict a realistic stress-strain relationship under widely varying conditions.

Scaling Rule No. 1 Strain Rate Effect for Glassy Polymer

To obtain a stress-strain curve at a strain rate $\dot{\varepsilon}$ from the stress-strain data obtained at $\dot{\varepsilon}_0$, multiply each stress by $R_1 = (\dot{\varepsilon}/\dot{\varepsilon}_0)^n$, where n is calculated from n_0, tabulated for many polymers in Table 3.4.1.

3.4   Temperature Dependence of Stress-Strain Curves

The congruency of the stress-strain curves for PVC in Fig. 3.2.2 and polycarbonate in Fig. 3.4.1 with the temperature as the parameter turns out to be a typical feature for many glassy polymers. The congruency means that stress and strain will increase

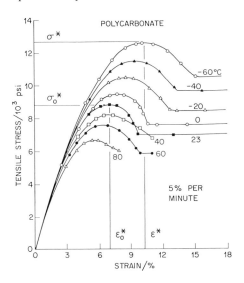

Fig. 3.4.1. Experimental points and the calculated (by scaling rule) stress-strain curves for polycarbonate $(1 \text{ psi} = 6.9 \times 10^3 \text{ N/m}^2)$.

or decrease by the same scaling factor. The temperature dependence of the relaxation process in glassy polymers is complicated because of the temperature dependence of the distribution of relaxation times, and also because of the effects of the thermal history. There is no doubt, however, that the relaxation processes in glassy polymers are temperature dependent even if the free volume (or the fictive temperature) is deliberately kept constant through the controlled manipulation of their thermal history. We have found[4] that the temperature dependence of the aging rate for the isofree volume nonequilibrium polymeric glasses is of the Arrhenius type with a constant activation enthalpy of ca. 35 kcalories (1 calorie = 4.184 J).

Superimposition of this Arrhenius process over the Williams-Landel-Ferry equation for free volume[5] has been successfully applied to the analysis of the aging process and the shift in mechanical and dielectric relaxation spectrum for several polymeric glasses. It is a curious fact that the Arrhenius term disappears above T_g in the equilibrium state, whereas below T_g the free volume form of the Williams-Landel-Ferry (WLF) equation *without altering* the coefficients must be applied together with the Arrhenius term to fit the data. Thus it must be concluded that the Arrhenius term arises only when the rate of rearranging the conformations is severely impaired. Empirically, the emergence of the Arrhenius term and the temperature dependence of the distribution of relaxation times are coincidental at T_g, and continue to persist at lower temperatures. This is probably related to the nature of the β process in glass. Taking all of these bits of information about the glassy polymer, we can assume that the dependence of the relaxation process in a polymeric glass can be separated into the history dependence and the temperature dependence. The history dependence is expressed by free volume, excess entropy, or fictive temperature. This aspect will be discussed in detail in the later section of this chapter. At present, we will be concerned with the glassy polymer with the equivalent thermomechanical history. The temperature dependence of such a polymer is described by the Arrhenius equation

$$\ln \frac{t^*}{t_0^*} = \frac{H}{R} \left(\frac{1}{T} - \frac{1}{T_0} \right) \qquad (3.4.1)$$

where R is the universal gas constant, t^* and t_0^* are the relaxation times at temperatures T and T_0° K, and H is the enthalpy of activation.

The contribution from the temperature variation toward the stress-strain curve, then, is described by the equation for the scaling factor R_0

$$R_0 = \left(\frac{\dot{\varepsilon}}{\dot{\varepsilon}_0}\right)^n = e^{n\frac{H}{R}\left(\frac{1}{T}-\frac{1}{T_0}\right)}$$

$$(3.4.2)$$

where $\dot{\varepsilon}$ in this case is the *equivalent* strain rate at T_0, when in fact the strain rate of $\dot{\varepsilon}_0$ was employed at T. Thus the logarithm of the scaling factor is proportional to n and $1/T$, or

$$\ln R_0 = \ln\left(\frac{\sigma}{\sigma_0}\right) = n\frac{H}{R}\left(\frac{1}{T}-\frac{1}{T_0}\right)$$

$$(3.4.3)$$

It is evident from Eq. (3.4.3) that the only way that n can remain independent of temperature is for the equivalent stress between the two curves, such as the corresponding yield stresses, to depend on the temperature in the Arrhenius manner. Experimental data point out that such is not the case, and far from it. From the experimental curves of Figs. 3.2.2 and 3.4.1, and many more to follow, the maximum or the yield stress is found to decrease in proportion to the negative of the temperature, following the equation

$$R_0 = \frac{\sigma(T)}{\sigma(T_0)}\bigg]_{\varepsilon_0} = \frac{T_c - T}{T_c - T_0}$$

$$(3.4.4)$$

where T_c is an empirically derived temperature at which the maximum stress is extrapolated to reduce to zero. This is shown in Fig. 3.4.2 for four glassy polymers.

Equation (3.4.4) above can be derived if the stress is to reduce the free energy of activation for the relaxation process, following the concept of the rate process, or

$$H - T_c S = H - C_4 \sigma - TS$$

$$(3.4.5)$$

where C_4 is an empirical constant. The above formula implies that at temperature T, the addition of stress σ will cause the material to flow in the manner possible with an infinitesimally small stress at temperature T_c. The straight line relationship as dictated by Eq. (3.4.4) is supported by the plot for several glassy polymers shown in Fig. 3.4.2. T_c is higher than T_g by about $70°$ C. When T_c and n at T_0 are empirically determined, the activation enthalpy can be evaluated from the formula

$$H = \lim_{T \to T_0}\left[\ln\frac{T_c - T}{T_c - T_0}\bigg/\frac{n}{R}\left(\frac{1}{T}-\frac{1}{T_0}\right)\right]$$

and

$$(3.4.6)$$

$$H = \frac{R T_0^2}{T_c - T_0}\frac{1}{n_0}$$

as obtained by combining Eqs. (3.4.3) and (3.4.4) and utilizing de l'Hospital's rule, where n_0 is n at T_0, and the temperature must be in Kelvin degrees. The temperature dependence of n can now be determined from the formula

$$n = \frac{R T^2}{H(T_c - T)}$$

$$(3.4.7)$$

by the similar method. The temperature dependence of n has been shown in Fig. 3.2.6 for polycarbonate. These parameters are shown in Table 3.4.1. Crystalline polymers

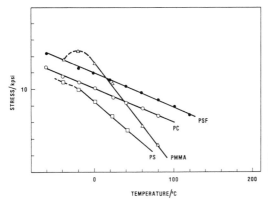

Fig. 3.4.2. Yield stress at 5% per minute versus T for glassy polymers (1 ksi = 6.9 MN/m^2).

have been included in the table. It turns out that the stress-strain curves of crystalline polymers are sharply different from those of glassy polymers, as neither the congruency of the stress-strain curves nor the temperature dependence of Eq. (3.4.4) is observed. However, many basic features of the scaling technique described heretofore on glassy polymers are applicable to crystalline polymers and, as it will be shown subsequently, the scaling of the viscoelastic stress-strain curves for crystalline polymers is also possible.

Scaling Rule No. 2 Temperature Effects on Glassy Polymer

To obtain a stress-strain curve at temperature T from the curve obtained at T_0, multiply each stress *and* strain by $R_0 = (T_c - T)/(T_c - T_0)$ where T_c is tabulated for several polymers in Table 3.4.1. The parameter n is calculated from Eq. (3.4.7) $n = RT^2/H(T_c - T)$.

Table 3.4.1

Polymer	$T_c/°C$	n_0	$H/(kcal)$	Crystalline C or Glassy G
ABS	170	0.0287	41.5	G
polystyrene	176	0.0250	45.8	G
polycarbonate	217	0.027	33.4	G
polyethylene (high density)	145	0.080	25.4	C
polyethylene (low density)	120	0.050	48.0	$C(G, T < -35°\,C)$
polypropylene	170	0.060	29.7	$C(G, T < 0°\,C)$
NORYL	160	0.032	40.0	G
PVC	100	0.037	45.5	G
Nylon	190	0.076	21.6	$C(G, T < 20°\,C)$
polysulfane	295	0.025	23.8	G
PBT	110	0.048	42.0	$G(C, T > 85°\,C)$

The values of H, the activation enthalpy, for glassy polymers are similar, ca. 30 to 45 kcalories. This is approximately the value involved in the physical aging process as measured by the decrease of volume or enthalpy as well as in the dielectric relaxation process in glassy polymers with equivalent thermal histories.

As the reader probably has recognized, the scaling technique is recommended to be centered around room temperature. The reason for this is a practical one; the region where the most accurate prediction is needed is around room temperature. However, because the emphasis has been placed on room temperature behavior, the accuracy in the extrapolation to a temperature near T_g is sacrificed. If such a requirement should arise, the stress-strain data must be obtained at different strain rates at or near the desired temperature, and the values of Table 8.4.1 will be altered, in some cases substantially.

If T_g is taken as the reference temperature, then

$$\ln\left(\frac{t^*}{t_g^*}\right) = \frac{H}{T}\left(\frac{1}{T} - \frac{1}{T_g}\right) \qquad T < T_g \tag{3.4.8}$$

and by combining with Eq. 3.4.2, sufficiently accurate stress-strain curves may be obtained in this temperature range. However, it must be cautioned that in this temperature range, the relaxation process is markedly influenced by thermal aging, and a proper account, following the procedure which will be explained in a subsequent section must be taken.

The four stress-strain curves in Fig. 3.4.3 were calculated by applying the convolution integral to the nonlinear Williams-Watts type formula of Eq. (3.2.1) in which the dependence of t^* with strain has been accounted for. The temperature dependence of β has been included, following Eq. (3.4.8). The initial t^* has been calculated at each temperature, starting with 1 minute at $T = T_g = 147°\,C$ with the activation enthalpy of 33 kcalories. The vertical shift factor, $(1-2\nu)/f$ is 10 at 23 and 60° C, but 9 at $-20°\,C$ and 8 at $-60°\,C$. Thus, by knowing only the T_g, it is possible to construct a reasonable stress-strain curve for any glassy polymer for any temperature. The objective of this study, however, is not to produce theroetically valuable but numerically less accurate stress-strain curves by utilizing a few universal parameters, but to generate useful engineering data. For this reason, it is recommended that the reference temperature be placed at 23° C.

The stress-strain curves for polysulfone are shown in Fig. 3.4.4. This glassy polymer exhibits the viscoelastic behavior qualitatively very similar to polycarbonate, but it is less temperature dependent and the yield strain at room temperature is about the same as that of polycarbonate in spite of the fact that its T_g and T_c are both substantially higher. The stress-strain curves for the solid solution of polyphenylene oxide and polystyrene are shown in Fig. 3.4.5. These curves do not undergo a maximum in the stress, but they appear more like Eq. (3.4.1), and go into the steady plastic flow process at yield. The essential features for the scaling rule are all present, and the curves

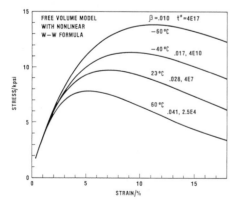

Fig. 3.4.3. Stress-strain curves generated from convolution integral of Eq. 3.3.2 in which the strain dependence of relaxation time, t^*, was assumed to follow the free volume formulas of Eq. 3.4.1 and 3.4.2.

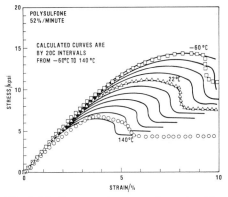

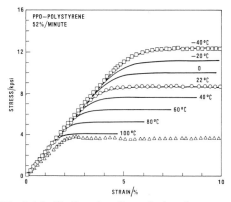

Fig. 3.4.4. Stress-strain curves for polysulfone calculated by the scaling rule. Experimental points at only two extreme temperatures are shown, but all data in the intermediate temperatures agree with the curves (1 psi $= 6.9 \times 10^3$ N/m^2).

Fig. 3.4.5. Similar plots for polyphenylene oxide–polystyrene mixture as Fig. 3.4.4.

at various temperatures can be quite accurately reproduced from the 22° C curve by scaling with the parameters included in Table 3.4.1.

The stress-strain curves for polystyrene are shown in Fig. 3.4.6. This and the next glassy polymers are exceptional in that their strain dependence of the relaxation modulus is slight, and hence they appear as if nearly linearly viscoelastic, as exemplified by the calculated curves without the vertical shift in Fig. 3.3.2. In this case the range of strain is very small, and the material is "brittle."

The stress-strain curves for polymethyl methacrylate are shown in Fig. 3.4.7. Below room temperature there is almost no vertical shift for the stress relaxation modulus of this glassy polymer, and a linear viscoelastic function such as Eq. (3.2.1) can be utilized. The scaling is done with the center of congruency at 1.5% strain rather than at the origin for the preceding polymers. Below 20° C, a brittle failure occurs and

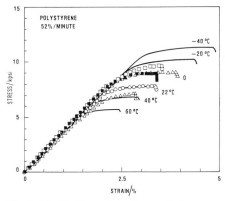

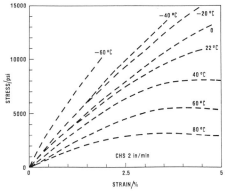

Fig. 3.4.6. Similar plot for polystyrene as Fig. 3.4.4.

Fig. 3.4.7. Similar plot for polymethyl methacrylate, which is an exception to the scaling rule in the text. This behavior can be fitted better with Eq. 3.2.1 with no vertical shift as illustrated in Fig. 3.3.2.

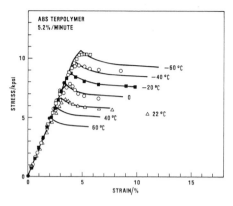

Fig. 3.4.8. Acrylonitrile-butadiene-styrene (ABS) terpolymer, with polybutadiene as the rubbery phase spheres dispersed in the glassy matrix, instead of; behaves more like a glassy polymer than the
two components typified by the crystalline
polymer (1 psi $= 6.9 \times 10^3$ N/m^2).

the complete curves cannot be observed. The brittle failure is less favored when the
molecular weight of the polymer is high, and whether brittle or ductile failure will
ensue at a given strain rate at a given temperature can be established by another
scaling rule incorporating the molecular weight and the critical stress. In case of this
polymer sample (with this particular molecular weight) the brittle-to-ductile transition
is observed at a temperature above 22° C but below 40° C, with the strain rate of
5% per minute. At a higher strain rate and/or for a lower molecular weight, the brittle
failure will occur at a higher temperature. This will be discussed in the appropriate
chapter of this volume dealing with the time- and temperature-dependent failure processes.

The stress-strain curves for an acrylonitrile-butadiene-styrene system known as ABS
are shown in Fig. 3.4.8. The structure of this polymer composite is well known, that
is, the rubbery polybutadiene spheres with acrylonitrile grafted skin are dispersed in
the matrix of glassy acrylonitrile-styrene copolymer. The features of the stress-strain
curves are basically those of a homogeneous glassy polymer, and as it will be shown,
very different from a composite of hard and soft components exemplified by the semicrystalline polymers with a rubbery amorphous phase. However, the threshold of the
brittle failure behavior is substantially extended toward the lower temperature as compared to unmodified styrene-acrylonitrile copolymer.

The Computer Program

The computer program for the scaling of the stress-strain curves for glassy polymers
is summarized below. To determine the value of n_0 at T_0, the minimum of two stress-
strain curves are obtained at strain rates of $\dot{\varepsilon}_1$ and $\dot{\varepsilon}_2$ percent (or inches if same geometry)
per minute. Take the yield or the maximum stresses σ_1 and σ_2 from the two curves.
The value of n_0 is obtained from the formula

$$n_0 = \log\left(\frac{\sigma_2}{\sigma_1}\right)\bigg/\log\left(\frac{\dot{\varepsilon}_2}{\dot{\varepsilon}_1}\right) \tag{3.4.9}$$

Next, to obtain T_c, at least another stress-strain curve must be obtained at another
temperature, T_3, but preferably at one of the strain rates with which the data has
been obtained at T_0, say $\dot{\varepsilon}_1$. Then obtain the yield or maximum stress σ_3 from this
curve, and using a variation of Eq. (3.4.4) T_c is obtained, i.e.,

$$\frac{\sigma_3}{\sigma_1} = \frac{T_c - T_3}{T_c - T_0} \tag{3.4.10}$$

and the enthalpy of activation H can be obtained from Eq. (3.4.6). Now we have the basic parameters, n_0, T_0, T_c, and H. The computation of the stress-strain curve at a new strain rate $\dot{\varepsilon}$ at a new temperature T can be accomplished by the following procedure:

1. Input T and $\dot{\varepsilon}$
2. Input n_0, T_0, T_c, H
3. Calculate n at T from Eq. (3.4.7)
4. Read data file $X(I)$, $Y(I)$ for strain, stress obtained at $\dot{\varepsilon}_1$, T_0
5. Calculate R_0 and R_1 at T, $\dot{\varepsilon}$ from Eq. (3.4.4) and

$$R_1 = \left(\frac{\dot{\varepsilon}}{\dot{\varepsilon}_1} \right)^n \qquad (3.4.11)$$

6. IF for T, $R_3 = R_0$
   IF for $\dot{\varepsilon}$, $R_3 = 1$
7. Do loop for I
   $X = X(I) * R_3$
   $Y = Y(I) * R_1$
   PLOT X, Y

3.5 Crystalline Polymers

We have not yet considered the complete scaling rules for glassy polymers, such as the effect of aging, prediction of shear stress-strain curves from tensile stress-strain curves, the uniaxial compression, and the effects of hydrostatic pressure. Before going on to these aspects, however, it is probably expedient to discuss the scaling technique for the tensile stress-strain curves in crystalline polymers, because the actual mechanics for the scaling procedure is quite similar except for one very important difference on the scaling of the strain.

Experimentally it is found that the stress-strain curves for crystalline polymers obtained at different strain rates and temperatures are not congruent. This fact alone may seem to dash off any hope of generating a new and simple scaling rule comparable to that for glassy polymers. As it turns out, a scaling rule utilizing a different procedure works well with crystalline polymers. Although the congruency rule does not apply to the stress-strain curves for crystalline polymers at various strain rates and temperatures, the following features are applicable: (1) the stress levels increase with the power of strain rates, $\dot{\varepsilon}^n$, as was true of glassy polymers, (2) at the same time, the strains decrease with the negative power of strain rates, $\dot{\varepsilon}^{-n}$, *unlike* glassy polymers, (3) the stress levels decrease linearly with $T(T_c - T)$ instead of $T_c - T$, as for glassy polymers, and (4) the values of n and H are about the same orders of magnitude as for the glassy polymers and are determined from experimental data in a manner similar to that used for glassy polymers. Thus, the scaling of stress-strain data in crystalline polymers can be accommodated by multiplying the stress by R_1 and dividing the strain by R_1, where R_1 is obtained from Eq. (3.4.11). Figure 3.5.1 is an example of calculated curves, starting with a stress-strain relationship described by Eq. (3.3.4). In Figs. 3.5.2 and 3.5.3 examples of high and low density polyethylenes are shown, one at one strain rate but at different temperatures, and the other at different strain rates but at the same temperature. In both cases, the lines drawn were actually calculated by scaling the stresses up by multiplying by R_1 but at the same time scaling the strains down by *dividing* by R_1. The stress-strain equation for glassy polymers where the congruency rule was observed to hold is *not* applicable to the crystalline polymers.

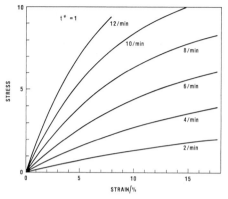

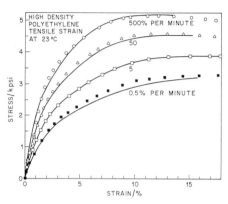

Fig. 3.5.1. Stress-strain curves calculated by the scaling rule for crystalline polymers, using a single relaxation time model.

Fig. 3.5.2. Stress-strain curves for high density polyethylene at various strain rates, fitted by the curves generated by the sealing rule for crystalline polymers (1 psi = 6.9 × 10³ N/m²).

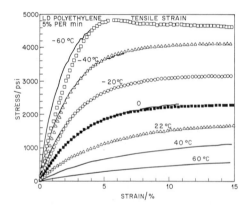

Fig. 3.5.3. Stress-strain curves for low density polyethylene at various temperatures, fit curves generated by the scaling rule for crystaline polymers (1 psi = 6.9 × 10³ N/m²).

3.5.1   Effect of Strain Rate

To construct a stress-strain model of a multiphase body based on morphological structure would be very complex and difficult, and it is not intended here. For some linear properties, there are models proposed by Takayanagi,[6] McCullough,[7] and Boyd,[8] but they are not applicable to our case involving nonlinear viscoelasticity. However, some reasonable qualitative argument can be generated in support of our empirical scaling procedure. It is well known that both crystalline and amorphous (rubbery) regions are significantly viscoelastic, with the crystalline region exhibiting a substantially greater relaxation time which provides the rigid skeletal structure, while the amorphous regions provide the time dependent aspects. In such a case, going back to the primitive single relaxation model of Eq. (3.1.4),

$$\sigma = E_0 \,\dot{\varepsilon}\, t^* (1 - \exp(-\varepsilon/\dot{\varepsilon}\, t^*)) \qquad (3.1.4)$$

the $\dot{\varepsilon}\, t^*$ is related to the amorphous regions and E_0 is related to the crystalline region. Now E_0 is also viscoelastic, with much greater t^*, and will be affected by the strain rate, as shown in Fig. 3.3.2. In such a case, the effect of increasing the strain rate is to increrase E_0, in case of the single relaxation model proportionally to $\dot\varepsilon$, but with distributed relaxation times proportional to $\dot\varepsilon^n$. Thus the scaling for the stress can be attributed to the response by the crystalline regions to the change in the strain rates. Given this situation, however, with amorphous regions which can react relatively fast to the strain energy built up due to the mechanical work done on the body, a criterion for determining the course of stress and strain may be established on the basis of the equivalence of recoverable energy. Such a criterion will require that the energy, U, be as nearly independent of change in $\dot\varepsilon$ or t^*, where

$$U = \int_0^{\varepsilon^*} \sigma \, d\varepsilon \qquad (3.5.1)$$

where ε^* is the corresponding strain for the same U among different stress-strain curves at different rates of strain. This energy criterion will be discussed in detail in Subsection 3.6.1. Thus, this scaling rule establishes the corresponding states on the energy basis, in contrast to the glassy state where the criterion was based on the relaxation times as they change with the increasing strain.

Scaling Rule No. 3 Strain Rate Effects on Crystalline Polymers

Multiply the stress by the scaling factor $R_1 = (\dot\varepsilon/\dot\varepsilon_0)^n$, and divide the strain by the same scaling factor R_1, when the rate of strain is changed from $\dot\varepsilon_0$ to $\dot\varepsilon$.

3.5.2 Effect of Temperature

Although the effect of temperature on the yield stress seems to be almost the same for crystalline polymers as for glassy polymers, a close examination reveals that such is not the case. The yield stress for three kinds of crystalline polymers is plotted against the temperature in Fig. 3.5.4. While such a plot was a straight line for the glassy polymer, these curves are definitely curved. When the data are plotted in a slightly different form, i.e., $\sigma T/T_0$ versus T, where T, T_0 are in degrees kelvin they become straight lines approaching the respective melting temperature, T_m, as shown in Fig. 3.5.5. Thus for crystalline polymers, the scaling factor R_0 for the temperature

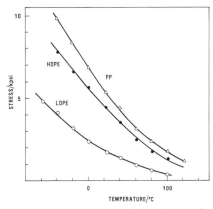

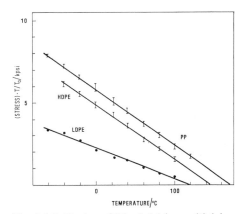

Fig. 3.5.4. Temperature dependence of yield stress for crystalline polymers taken at 5%/minute (1 psi $= 6.9 \times 10^3$ N/m²).

Fig. 3.5.5. Replot of Fig. 3.5.4 by multiplying the stress by T/T_0.

is slightly different from Eq. (3.4.4), i.e.,,

$$R_0 = \frac{\sigma(T)}{\sigma(T_0)}\bigg]_{\varepsilon_0} = \frac{T_0(T_c - T)}{T(T_c - T_0)} \tag{3.5.2}$$

This formula can be derived if the stress will increase the activation entropy, i.e.,

$$H - T_c S = H - T(C_5 \sigma - S) \tag{3.5.3}$$

where C_5 is an empirical constant.

Similarly, from the empirically determined value of n_0 at T_0, the activation enthalpy H is determined by the formula

$$H = \frac{R T_0 T_c}{n_0(T_c - T_0)} \tag{3.5.4}$$

and the temperature dependence of n is determined from the formula

$$n = \frac{R}{H}\frac{T T_c}{T_c - T} \tag{3.5.5}$$

T_c can be taken to be T_m, but our empirically determined values are shown in Table 3.4.1.

Scaling Rule No. 4 Temperature Effects on Crystalline Polymers

The scaling factor $R_0 = T_0(T_c - T)/T(T_c - T_0)$ is used to multiply the stress and to divide the strain to scale from the data obtained at T_0 to the stress-strain curve at T. The temperature dependence of n is obtained from Eq. (3.5.5). Among all crystalline polymers we have studied in which the amorphous regions are in the rubbery state, $T > T_g$, this scaling rule has been found to apply. In the temperature and the strain rate range where amorphous regions become glassy, behavior reverts back to the model of congruency for glassy polymers, i.e., the congruent stress-strain curves. T_g of polypropylene is observed at about $0°$ C by the specific heat measurement made at $10°$ C/minute. This transition is clearly observable in Fig. 3.5.6 where the scaling formula for crystalline polymers was used above $0°$ C from data taken at $23°$ C, and

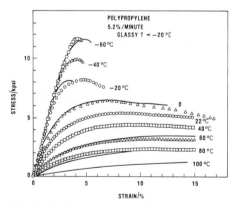 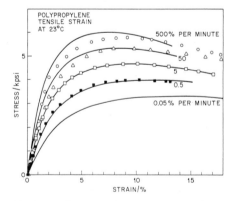

Fig. 3.5.6. Stress-strain data for polypropylene at various temperatures are fitted (1) with the scaling rule for crystalline polymers above $0°$ C but (2) with the scaling rule for glassy polymers below $-20°$ C.

Fig. 3.5.7. Stress-strain data for polypropylene at various strain rates can fit the scaling rule for crystalline polymers, but a use of R_1 for the stress only, also works.

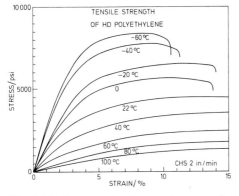

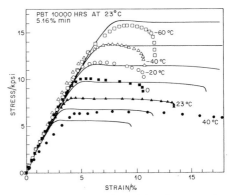

Fig. 3.5.8. Stress-strain data and curves for high density polyethylene at various temperatures, using the scaling rule for crystalline polymers.

Fig. 3.5.9. Even though polybutylene terephthalate is a semicrystalline polymer, its T_g for the amorphous phase is near 80° C, and, therefore, the scaling rule for glassy polymer fits better.

the formula for glassy polymers was used for -20, -40, and $-60°$ C curves based on the $-40°$ C curve. When comparing various data obtained with different strain rates at $23°$ C, however, polypropylene behaves in a manner similar to high and low density polyethylenes, indicating the behavior typical of the crystalline rubber composites, as shown in Fig. 3.5.7. The stress-strain curves of high density polyethylene at various temperatures are shown in Fig. 3.5.8.

Polybutylene terephthalate (PBT) is a semicrystalline polymer whose T_g is about 80° C. Therefore, according to the rule found applicable to polypropylene, it should behave like a glassy polymer below the T_g. Curves in Fig. 3.5.9 show the congruency rule to hold for strains up to the yield range, as the slope remained nearly constant as the temperature was varied. This feature is reminiscent of the styrene-PPO mixture shown in Fig. 3.4.5. The overall elongation beyond the yield, as briefly discussed before, is a function of the molecular weight, and this part did not follow the congruency rule.

Computer Program for Crystalline Polymers

The computer program for scaling the stress-strain curves for crystalline polymers above the T_g of the amorphous phase is summarized below. To determine the basic parameters n_0, T_0, T_c, and H, the procedures described by Eqs. (3.5.2) through (3.5.4) can be used, since they are determined only from stress variations not involving the strains.

1. Input T and $\dot{\varepsilon}$
2. Input n_0, T_0, T_c, H
3. Calculate n at T from Eq. (3.5.5)
4. Read data file $X(I)$, $Y(I)$ for strain, stress obtained at $\dot{\varepsilon}_1$, T_0
5. Calculate R_1 at $\dot{\varepsilon}$, T from Eq. (3.5.2) and (3.4.11)
6. Do loop for I
   $X = X(I)/R_1$
   $Y = Y(I) * R_1$
   PLOT X, Y

3.6   Stress-Strain Behavior in Shear

3.6.1   Strain Energy During Constant Rate Deformation

The shear deformation can be divided into tensile and compressive components. The tensile component tends to shorten the relaxation time rapidly and accelerates the process further. The compressive component only responds reversibly and elastically, and this process would not result in an increase in the relaxation time. Thus the overall relaxation time is predominantly determined by the rapidly shifting tensile component. The shift of the relaxation time in shear, as it will be shown, can be calculated if the shift due to the tensile deformation is known. Actually, Robertson's model[9] for the strain-induced shift in the relaxation time can be identified with this model on the molecular level.

In order to consider a model for scaling shear stress-strain curves from tensile stress-strain curves, we will use Eq. (3.3.4). The slope of the plot in Fig. 3.3.8 is -6.60 at $\dot{\varepsilon}$ of 0.5% per minute. Thus when $n \ll 1$, the secant modulus of the stress-strain curve is nearly equal to the relaxation modulus $E(\varepsilon, t)$ in which the time t has been replaced by $\varepsilon/\dot{\varepsilon}$.

Equation (3.3.4) can be considered to describe a nonlinear *elastic* behavior dealing with strain as the sole variable for the conservative part of the viscoelastic deformation energy. For a conservative system, there is the elastic constitutive relationship

$$\underset{\sim}{\sigma} = 2\mu \, \underset{\sim}{\varepsilon} + \lambda^* \, \varepsilon_{\kappa\kappa} \, \underset{\sim}{1} \tag{3.6.1}$$

where " $\sim$ " denotes a tensor notation, μ is the generalized secant modulus which depends on the strain and λ^* is the Lammé constant.* Rosenberg and Matsuoka[10] have shown that the introduction of the second invariant of the deviatoric strain

$$\Pi_\varepsilon = \tfrac{1}{2} \, \underset{\sim}{\varepsilon}_{ij} \, \underset{\sim}{\varepsilon}_{ij} \tag{3.6.2}$$

into the strain terms in Eq. (3.6.1) satisfies the energy requirements for the conservative component, when the generalized modulus, μ, is patterned after the tensile secant modulus in Eq. (3.3.4), i.e.,

$$\mu = \mu_0 \exp(-C\Pi_\varepsilon^{1/2}) \tag{3.6.3}$$

The two special cases are derived from the above general equation, i.e., the uniaxial tension and the simple shear

$$\sigma_{11} = \varepsilon_{11} \, \mu_0 \exp(-C(1+v) \, \varepsilon_{11}/\sqrt{3}) \quad \text{tension} \tag{3.6.4}$$

and

$$\sigma_{12} = \gamma\mu_0/2(1+v) \exp(-C\gamma/2) \quad \text{shear} \tag{3.6.5}$$

where σ_{11}, σ_{12} are tensile and shear stresses, ε_{11} is the tensile strain, and $\gamma = 2\,\varepsilon_{12}$ is the shear strain, because the following is true

$$\Pi_\varepsilon^{1/2} = \frac{1+v}{\sqrt{3}} \, |\varepsilon_{11}| \quad \text{tension} \tag{3.6.6}$$

$$\Pi_\varepsilon^{1/2} = |\varepsilon_{12}| = \gamma/2 \quad \text{shear} \tag{3.6.7}$$

---

* Note: λ is not a constant since we are dealing with a nonlinear case. However, the hydrostatic term is less important mechanically than the pressure effect entering as the thermodynamic intensive quantity and can be neglected in this case.

3.6.2 Scaling Tensile Curve to Shear

Using the above pair of formulas, we can scale a tensile stress-strain curve to obtain a corresponding shear stress-strain curve by the following procedure:

(1) Multiply the tensile strain by the scaling factor, R_{sx}
$$R_{sx} = 2(1+v)/\sqrt{3}$$

(2) Multiply the tensile stress by the scaling factor, R_{sy}
$$R_{sy} = 1/\sqrt{3}$$

(3) Optionally, if the engineering stress-strain is used for tensile data, i.e., the force divided by the *initial* area, then the tensile stress in (2) may be further *divided* by $(1-2v)$* (tensile strain).

The equivalence in the rates of strain in tension and shear are established by the factor $2(1+v)/\sqrt{3}$ to correct the constant C, but this results in an error of order of 1% for the stress, hence it can be generally ignored. Figure 3.6.1 demonstrates how well the scaling scheme works for polycarbonate. The same scaling method was applied on high density and low density polyethylene samples, and excellent agreement was obtained as shown in Figs. 3.6.2 and 3.6.3. It is important to emphasize at this point that this scaling scheme also works for crystalline as well as for glassy polymers, even where the shape of the curve does not conform to Eq. (3.3.4). This point suggests that the conservative energy criterion is a generally applicable criterion in a wide variety of viscoelastic materials being deformed at a constant strain rate.

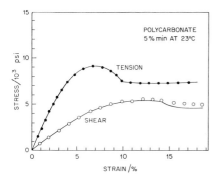

Fig. 3.6.1. Shear stress-strain curve is generated from the tensile data following the scaling rule for shear. Though slightly, the shear data also undergoes a maximum.

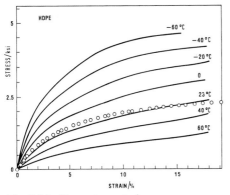

Fig. 3.6.2. Shear stress-strain curves are generated from tensile data and compared with shear data for high density polyethylene.

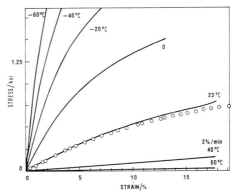

Fig. 3.6.3. Similar curves for low density polyethylene as Fig. 3.5.7.

Scaling Rule No. 5 From Tension to Shear

Multiply the stress by R_{sy} and the strain by R_{sx} for all polymers.

Computer Program for Scaling From Tension to Shear

The computer program for scaling the stress-strain curves in tension to obtain the stress-strain curves in shear is summarized below.
1. Input T and $\dot\gamma$
2. Input n_0, T_0, T_c, H
3. Calculate n at T from Eq. (3.4.4)
4. Read data file $X(I)$, $Y(I)$ for tensile strain, stress obtained at $\dot\varepsilon_1$, T_0
5. Calculate R_1 at $\dot\gamma$, T from Eq. (3.4.4) using $\dot\varepsilon = 2\,\dot\gamma$
6. $R_{sx} = 2*(1+v)/\mathrm{SQR}(3)$
   $R_{sy} = 1/\mathrm{SQR}(3)$
   Do loop for I
   If crystalline, $X = X(I)/R_1^* \, R_{sx}$
   If glassy, $X = X(I)* \, R_1^* \, R_{sx}$
   $Y = Y(I)* \, R_1^* \, R_{sy}$
   PLOT X, Y

3.6.3   Effects of Hydrostatic Pressure on Deformation, Uniaxial Compression

It was already mentioned that a mechanical strain in tension or shear reduces the relaxation time quickly, in the exponential order of strain, whereas it would be a very time consuming process to increase the relaxation time by cooling or by compression. The effect of pressure on the relaxation time of glassy polymers should be considered on the basis of the equivalent thermal history, as we have done on the effect of temperature by separating out the Arrhenius term in Eq. (3.4.3). Thus the best way of analyzing the effect of pressure on the viscoelastic behavior is to superimpose the hydrostatic pressure on the specimen which is deformed in shear or tension. The most confusing and the worst method for analysis would be to attempt to analyze data on uniaxial compression without sorting out these factors; the lateral walls of the sample are exposed to the atmosphere while the interior is subjected to the pressure distribution which depends on the stress distribution.

Tensile stress-strain experiments under hydrostatic pressure have been conducted for a number of commercially available polymers by Staats-Westover and Vroom.[11] An example for polycarbonate is shown in Fig. 3.6.4. Those curves maintain the congruency rule, and with the increase in the pressure, the stress-strain curves become large both in horizontal and vertical directions, similar to the case with the decrease in temperature. Thus a scaling ratio similar to Eq. (3.4.3) can be generated

$$\ln R_1 = \ln\left(\frac{\sigma}{\sigma_0}\right) = n\,\frac{V^*}{R\,T_0}\,(p - p_0)\tag{3.6.8}$$

where V^* is the volume of activation, p is the pressure, and the subscript 0 refers to the reference condition. Also, in place of Eqs. (3.4.4) and (3.4.7), the following formulas are used

$$R_1 = \frac{\sigma(p)}{\sigma(p_0)} = \frac{p - p_c}{p_0 - p_c}\tag{3.6.9}$$

and

$$n = \frac{R\,T}{V^*(p - p_c)}\tag{3.6.10}$$

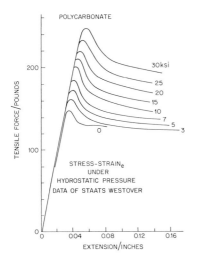

Fig. 3.6.4. Stress-strain data under hydrostatic pressure by Staats-Westover.

where the critical hydrostatic pressure, p_c, is the extrapolated value for the yield stress to vanish; it is a large negative value, -45 kpsi; 1 kpsi ≈ 6.9 MPa $= 6.9$ J·cm^{-3}. The quantity $V^*/R\,T_0$ is found to be 6.9×10^{-4} psi^{-1} for polycarbonate. V^* is about 250 cc, equivalent of 1 mole of the repeating unit, which is a reasonable quantity considering that the activation enthalpy of the process is 31 k calories. The parameters have been extracted from Staats-Westover's data and are shown in Table 3.6.1.

Table 3.6.1

Polymer	$p_c/(ksi)$[a]	$n\,V^*/R\,T_0/(10^{-5}\ psi^{-1})$
Polycarbonate	-45	1.73
ABS	-32	1.66
PTFE	-17	3.50
PCFE	-27	2.60
Phenoxy	-77	1.13
Polypropylene	-26	2.38

[a] 1 ksi $= 6.9$ MPa.

The values of n_0 were not obtained, but those values obtained at atmospheric pressures at different rates can be used.

Uniaxial compression loading is different from the above experiment, because the distribution of hydrostatic pressure is varied with stress. A crude but simple model has been considered here, in which the relaxation time is controlled by the shear deformation upon which hydrostatic pressure is superimposed. Since the shear and tensile stress-strain curves are uniquely related to each other as demonstrated, the uniaxial tensile stress-strain curve, under the increasing hydrostatic pressure should approximate a symmetrical image about the origin of a uniaxial compressive stress-strain curve. This is demonstrated for polycarbonate in Fig. 3.6.5. From values of the parameters shown for polycarbonate in Table 3.6.1, the compressive curve is equivalent to the tensile stress-strain curve conducted at 10% per minute under the hydrostatic pressure of 11 kpsi. Thus, the uniaxial stress is approximately equal to the tensile stress under

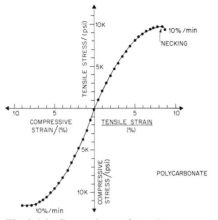

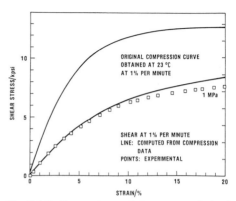

Fig. 3.6.5. Comparison of tensile stress-strain data for polycarbonate and uniaxial compressive data. The difference is accounted for by the scaling rule for pressure.

Fig. 3.6.6. Comparison of shear curve derived from the compression data *without* correcting for the hydrostatic pressure data, using only the scaling rule from tension to shear, for polyoxymethylene.

the increasing hydrostatic pressure whose absolute value is equal to the stress itself. Acquisition of the pressure parameters need not be solely dependent on the elegant but highly specialized equipment such as Staats-Westover's, at least for the purpose of scaling only; these parameters can be obtained by comparing the tensile and uniaxial compressive data, provided a proper adjustment is made for the strain-dependent changes in the cross-sectional area by the factor $(1-2v\varepsilon_{11})$. In Fig. 3.6.6, the shear stress-strain curve was obtained for polyoxymethylene by scaling the uniaxial compression curve, by using the scaling formula from *tension* to shear without correcting for the pressure effect, therefore without modifying for the addition of hydrostatic pressure. Thus the calculated curve has resulted in overestimation as compared with the actual shear data.

Scaling Rule No. 6 From Tension to Uniaxial Compression

Multiply the tensile stress, after correcting for the area change by multiplying with $(1-2v\varepsilon)$, with the factor $(\sigma-p_c)$ according to Table 3.6.1.

3.7   Effect of Physical Aging on Viscoelasticity

3.7.1   Increase of Relaxation Time During Aging

There is a whole Chapter 12 by Bauwens in this volume devoted to the subject of thermodynamics of aging. It is well known that polymer glasses typically undergo the physical aging process accompanied by the decrease of volume, entropy, and enthalpy during which time the dielectric relaxation spectrum can be observed to shift to the longer relaxation time. Although the thermodynamic treatment of the aging process is complex and the quantitative prediction for all types of history is difficult, a simple physical model can be followed to enable a sufficiently accurate prediction of the shift in viscoelastic functions due to the thermal history. Kovacs[12] has shown that a free volume model such as is used in the WLF-equation can be utilized in

the formulation of the isothermal aging process following a sudden change in temperature from above to below T_g, i.e.,

$$-\frac{df}{dt} = kf \tag{3.7.1}$$

where f is the free volume fraction and k is the rate constant for the volume contraction.
This rate constant is a rapidly diminishing function of the free volume function, and,
in fact, it is within the magnitude of the reciprocal of the characteristic dielectric relaxation time, t^*, i.e.,

$$t^* = t_0^* \exp(1/f - 1/f_0) \tag{3.7.2}$$

where the subscript 0 denotes some reference time during the steady state contraction
process. As the volume continues to contract during aging, t^* continues to grow greater.
Now, the rate of increase of t^* is given, from Eqs. (3.7.1) and (3.7.2), by

$$\frac{d}{dt}\left(\frac{1}{k}\right) = \frac{dt^*}{dt} = \frac{1}{f} \tag{3.7.3}$$

Since f is in the order of 10^{-2}, Eqs. (3.7.1), (3.7.2), and (3.7.3) are approximately
satisfied if

$$\frac{t^*}{t} = \frac{1}{f} - 1 \tag{3.7.4}$$

from which follows

$$\frac{d\log t^*}{d\log t} = \frac{1}{1-f} \approx 1 \tag{3.7.5}$$

The slope of $\log t^*$ versus $\log t$ plot will eventually become nearly 1, provided that
(1) the time t must become an order of magnitude greater than the initial relaxation
time, t_0^*, to satisfy the steady state condition and (2) the temperature must *not* be
near T_g where the glassy system is near equilibrium. If either of the above two conditions
is not met, the slope of the $\log t^*$ versus $\log t$ plot will be *less* than 1. This approximation
is concerned only with the case where the glassy state is sufficiently removed from
the equilibrium (liquidus) state. Near T_g, the equilibrium state, f_∞, can be reached
well before the free volume fraction reaches near zero. When f_∞ is not zero, Eq. (3.7.1)
must be revised as

$$-\frac{df}{dt} = k(f - f_\infty) \tag{3.7.6}$$

and

$$\frac{dt^*}{dt} = \frac{1}{f} + \frac{t}{f^2}\left(-\frac{df}{dt}\right) - 1 = \frac{1}{f} - \frac{f_\infty}{f} \tag{3.7.7}$$

and

$$\frac{d\log t^*}{d\log t} = \frac{1}{1-f}\left(1 - \frac{f_\infty}{f}\right) < 1 \tag{3.7.8}$$

Thus, if a glass is near the equilibrium liquidus line, the rate of increase of the relaxation
time is slowed down, primarily because the relaxation time depends on the free volume.
Thus, experiments conducted in the high temperature range near T_g, where results
are obtainable in a relatively short time, will not show the results that are representative
of a typical aging process taking a longer time. The typical result is obtained following
Eq. (3.7.5), expressing the rule that a decade increase in the logarithmic time for aging

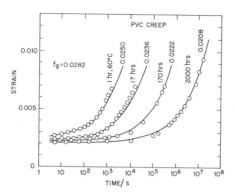

Fig. 3.7.1. Creep data for PVC after aging for different periods of time. The parametric numbers are the fractional free volume at start of creep experiments, respectively.

will increase the characteristic relaxation time by a decade. This is clearly demonstrated by the creep curves shown in Fig. 3.7.1. A series of creep experiments were conducted at 60° C for polyvinyl chloride, each following a different time of aging, varied by about a decade, at the same temperature. It can be observed that an increase of logarithmic aging time by one decade will shift the creep curve by one decade, i.e., the shift factor, or the logarithmic ratio of relaxation times to the time spent for aging, is unity as described by Eq. (3.7.5).

3.7.2   Scaling for Physical Aging

The effect of physical aging on viscoelastic stress-strain curves is illustrated in Fig. 3.7.2 where the stress-strain curves of samples having undergone various thermal histories are compared. Curves were picked among the PVC stress-strain curves run at 23° C after having undergone exposure to 60° C for 1, 10, and 100 hours, with the strain rates of 5, 0.5, and 0.05%/minute, respectively. They all superimpose on top of each other. The curve for the "as-molded" sample, run at 50%/minute, also superimposed on top of the others. This graph clearly supports the rule described by Eq. (3.7.5) for the relaxation time and the aging time to be proportional to each other. As already pointed out, however, the aging will slow down and then cease to continue as equilibrium (liquidus) state is approached when annealing is performed

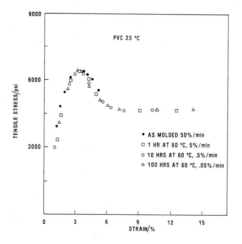

Fig. 3.7.2. Stress-strain curves for PVC samples each aged one decade longer are superimposed when each test was conducted at one decade slower strain rate (1 psi = 6.9×10^3 N/m^2).

at a very high temperature near T_g. Also, aging will not take place if the original sample has undergone a substantial aging already, i.e., if the relaxation time at the aging temperature is substantially greater than the time to be spent on aging. According to Eq. (3.7.4), time t must be *at least* within one or two orders of magnitude of t^* (since $f \sim 10^{-2}$) in order to observe an aging effect. The "as-molded" sample in Fig. 3.7.2 was injection molded, and can be considered as having been rapidly cooled from the molten state, so its equivalent relaxation time at 60° C happened to be comparable to 0.1 hour. However, had this "as-molded" sample been cooled much more slowly, its relaxation time at 60° C would have been much greater. For example, suppose that its history were comparable to 10 hours at 60° C. In such a case, annealing the sample for 1 hour or 10 hours at 60° C would have made no difference on the stress-strain curves; they would have superposed on top of each other *without* changing the rate of strain, and the difference would have become observable only when the aging period exceeded 10 hours at 60° C. Thus it is of utmost importance to know the equivalent thermal history of a sample that is being tested before the prediction can be made on the effect of physical aging. To do this, first subject a sample to a prescribed thermal history, typically 100 hours at a temperature, T_a, about 25° C below the value of T_g determined by the differential scanning calorimetry at ca. 1° C per minute, and obtain the stress-strain curve at 5% per minute. Now all other curves can be calibrated against this value, by using the scaling factor, R_2, for the thermal history

$$R_2 = \left(\frac{t_a}{t_{a0}}\right)^n_{T_a \dot{\varepsilon}_0} = \frac{\sigma}{\sigma_0} \tag{3.7.9}$$

where t_a is the time for aging, as recalculated to the reference condition with the temperature T_a and the strain rate of $\dot{\varepsilon}_0$. The power n should be the value at the temperature of mechanical testing, and not at the temperature of aging; it should be the values in Table 3.4.1 if the test was run at 23° C. The equivalent annealing time can be converted to any temperature of annealing by the use of the formula

$$\ln\left(\frac{t_{a,T}}{t_{a,T_{a0}}}\right) = \frac{H}{R}\left(\frac{1}{T} - \frac{1}{T_{a0}}\right) \tag{3.7.10}$$

where the same value of H as the one in Table 3.4.1 may be used. The equivalent annealing time is the minimum time spent on that specimen at that temperature. Any subsequent annealing time must be added to this time after being converted to the equivalent time at the temperature of aging being considered. Any *additional* aging, such as leaving on the shelf for an extended period of time or subjecting it to annealing

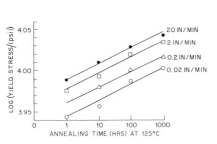

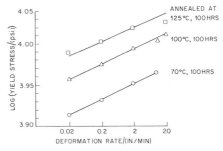

Fig. 3.7.3. The dependence of the maximum stress on the annealing time and the strain rate for polycarbonate.

at elevated temperatures must be accounted for by adding the equivalent additional time t at T_a, i.e.,

$$R_2 = \left(\frac{t+t_a}{t_{a0}}\right)^n \tag{3.7.11}$$

where t can be calculated from Eq. (3.7.10). The interchangeability of the annealing time and the strain rate is also shown for polycarbonate in Fig. 3.7.3.

The value of n depends slightly on the aging also, because it depends on the relaxation time t^* which becomes longer with annealing. Recalling Eq. (3.2.4)

$$n = \beta \left(\frac{t}{t^*}\right)^\beta \tag{3.7.12}$$

and if t^* is to be increased by 4 decades from 10^3 to 10^7 minutes, then for $\beta = 0.03$, n will decrease from 0.024 to 0.018. In some of the creep studies on glassy polymers, we have found that such would be the case.

3.7.3 Crystalline Polymers

As already mentioned, for those polymers which include the rubbery amorphous regions, such as low density polyethylene, high density polyethylene, and polypropylene, aging primarily affects the crystalline regions, and the relaxation time of the amorphous region can be assumed to remain unaffected. Thus the stiffness factor is affected without changing the strain-dependent or the time-dependent components of the relaxation modulus. The change in the stiffness can be considered as uniquely related to the change in density, within the same class of polymers, and the change in density is proportional to the logarithmic time during secondary crystallization. The stiffness factor rises with the same nth power of the annealing time, so that only the stress need be multiplied by $R_2 = (t+t_0)^n/t_0^n$ following Eq. (3.7.12).

In fact, for branched polyethylenes with different densities, whether their density variations are due to the concentration of branches or due to the different thermal histories, the relaxation behavior coincides for samples with the same density, as can be observed from Figs. 3.7.4, 3.7.5, and 3.7.6. Thus density is a good parameter for characterizing the thermal history of crystalline polymers. On the other hand, the an-

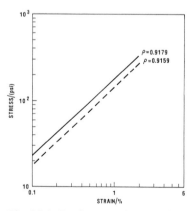

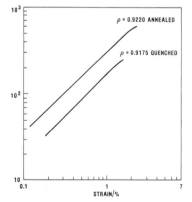

Fig. 3.7.4. Isochronous stress-strain curves analyzed from the creep curves for low density polyethylene annealed under different thermal history.

Fig. 3.7.5. Similar curves as for Fig. 3.6.6, molded under different conditions.

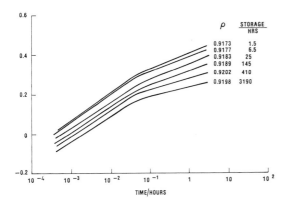

Fig. 3.7.6. Creep curves in tension according to Ogorkiewicz[14] for polyethylene samples having undergone different aging histories.

nealing time itself is a less reliable parameter. We have conducted extensive and systematic studies on aging effects on the stress-strain curves covering the temperature range from 25 to 100° C and the time range up to 10000 hours. The results show little variation in viscoelastic behavior of the samples, primarily because the quenched samples had already crystallized far enough; hence t_0 in Eq. (3.7.12) was already very large even at a high temperature.

Computer Program

The computer program to account for the effect of aging in a glassy polymer is shown below:

A. Determination of the equivalent thermal history

Compare the yield or maximum stress, σ, of your sample against the value, σ_0, of the "standard" sample which has been established as having undergone annealing for t_{a0} hours at T_{a0} °K, where T_{a0} is preferably $20 \sim 30°$ C below T_g. Then the equivalent time of annealing of your sample is t_a hours at T_{a0} °K, where $t_a = t_{a0} (\sigma/\sigma_0)^{1/n}$.

B. The effect of a subsequent aging for t hours at T °K will be calculated by Eq. (3.7.11) and then by Eq. (3.7.12), and R_2 is obtained

$$t = t_a \, e^{\frac{H}{R}\left(\frac{1}{T} - \frac{1}{T_{a0}}\right)}$$

and

$$R_2 = \left(\frac{t + t_a}{t_{a0}}\right)^n$$

and the scaling factor R_1 from the stress-strain curve of the "standard" sample is now multiplied by R_2, so that new $R_1 = (\text{old } R_1)* R_2$ and this new scaling factor R_1 is used for calculation of your stress-strain curve from the "standard" curve.

Scaling Rule No. 7 For Physical Aging

The scaling factor for aging, R_2, is obtained according to Eq. (3.7.12). Multiply the rate-temperature scaling factor R_1 by R_2, and use this product for scaling for glassy polymers. For crystalline polymers, examine the density, and if they are different, only the stress should be multiplied by R_2.

3.8   Filled Polymers

We have conducted a limited number of experiments on glass fiber-reinforced molded plastic compounds. The important factors are the concentration of the filler and the integrity of the interface between the filler and the polymer matrix. When the interface is not maintained properly through the use of a good interfacial agent, the mechanical properties are not altered substantially by the filler. The concentration of the filler with good surface characteristics will increase the modulus, as expected. Up to 40% glass by weight in polycarbonate, the zero strain modulus seems to follow McCullough's formula,[7] which is an improved version of Hashin-Shtrikmann criterion[13] for mixing two different phases.

For simple approximation, we found that scaling factors among the stress at 2% strain to be proportional to the weight percent of the filler,

$$R_f = \frac{\sigma_f}{\sigma_0} = 1 + C_6 \, w \tag{3.8.1}$$

where w is the weight fraction of the filler, and C_6 is a constant and is equal to 0.30 ksi/% of filler. σ_0 must be the *extrapolated* 0 concentration stress, and in this case it turns out to be 5 ksi rather than 4.2 ksi observed for actual unfilled polymer, as can be seen from Figs. 3.8.1 and 3.8.2. For a given composite, the stress-strain curves follow the same temperature dependence as the neat (unfilled) polymer. For glass-filled polycarbonate, the congruent stress-strain curves grow larger as the temperature is lowered, as shown in Fig. 3.8.3.

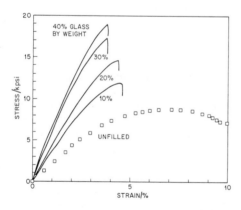

Fig. 3.8.1. Stress-strain curves of glass-reinforced polycarbonate, at 51.6% per minute at 22°C (1 psi = 6.9×10^3 N/m²).

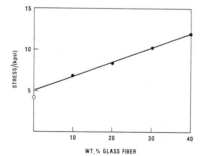

Fig. 3.8.2. 2% strain modulus versus weight percent glass content for samples in Fig. 3.7.3.

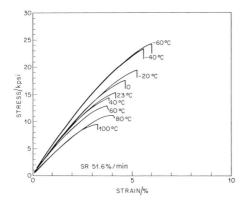

Fig. 3.8.3. Temperature dependence of stress-strain curves for glass-reinforced polycarbonate with 20% glass content shows the congruency feature.

Scaling Rule No. 8 For Rigid Fillers

Plot 2% modulus versus weight % filler, and obtain the scaling factor R_f according to Eq. (3.8.1). Use this R_f for multiplying the stress. For temperature dependence, use the same congruency rule as for unfilled glassy polymers, except the reference stress-strain curve is already brittle, as shown in Fig. 3.8.3.

3.9 Beyond Yield Point Rheology

The discussions heretofore have been confined to the stress-strain behavior within a few percent strain, because that *is* the strain range of interest for design engineers. However, what happens with stress-strain curves beyond this limited strain range is of interest when fracture properties are being considered. The aspects of failure in plastics will be dealt with elsewhere in this book. For scaling for fracture strength a new parameter involving the molecular weight will have to be introduced. The stress-strain specimens frequently and typically neck beyond this strain range, and a steady level for the engineering stress is maintained, which is proportional to the scaling factor $R_1 \sim (\dot{\varepsilon}\,t^*)^n$, Eq. (3.4.11). Thus the logarithmic stress versus logarithmic strain rate at steady state will exhibit a slope of n in tension as well as in shear, which is related to the Williams-Watts β according to Eq. (3.2.4).

How long this pseudosteady state can be maintained in tension seems to depend on the molecular weight, and polymers with inadequately low molecular weight will not be able to maintain the stress level. The mode of failure seems analogous to the cavitational failure of a rubber whose strength depends on the cross-link density.

The value of n is very small at room temperature for most solid polymers, ca. $0.02 \sim 0.05$. This value will increase as the temperature is raised, as evident from Eq. (3.4.7), first gradually but later more rapidly as T_g or T_m is approached. The value of n will eventually become equal to β, which is about 0.4 when the polymer is essentially in the rubbery state, and it will not change upon further raising of the temperature. Thus the slope of the logarithmic stress versus logarithmic rate of strain will be about 0.4 at steady state, regardless of the molecular weight. The magnitude of the stress involved in this range of viscoelastic behavior should be very high while the rate of deformation is in the range of minutes^{-1} to seconds^{-1}. As the temperature is raised further, this part of the relaxation spectrum is shifted to the very short time scale, and the part of the spectrum that arises from intermolecular interaction, commonly known as entanglement, comes to the time range relevant to the viscoelastic measurement. The distribution of relaxation times at the far end of the spectrum may be gradual if the molecular weight distribution is broad. It has about the same β as at the transition region of the spectrum which we have been discussing all along, except that the elastic constant is of the order of the modulus for the rubbery state. An example is shown in Fig. 3.9.1. Here the zero time modulus of 5×10^6 dynes/cm^2, or 5×10^5 Pa, t^* of 10^{-4} seconds, β for the Williams-Watts formula of 0.42 was assumed, and is in good agreement with the data obtained with polyethylene melt at $190°$ C. Since this is in the entanglement region of the relaxation spectrum, the zero time modulus, G_0, should be the (entanglement) plateau modulus and should correspond to the rubber-like elasticity; the value used here is ca. 735 psi (5×10^5 Pa) as compared to 300,000 psi (2×10^9 Pa) for the glassy state near the transition zone. The value of β in this case is quite close to the value for the relaxation modulus of an amorphous material near the transition zone, signifying that the distribution of the relaxation times in the terminal zone of this polyethylene with a broad molecular weight distribution is quite close to the spectrum in the transition zone. The experimental data at four different temperatures are shown in Fig. 3.9.2. The value of β seems to depend on the temperature, slightly increasing at a higher temperature.

The slope β will eventually have to reach 1 at either a high enough temperature and/or slow enough shear rate. Some polymers, probably because of the narrower molecular weight distribution, exhibit much greater values for β at a wide range of temperature. In Fig. 3.9.3, $\beta = 0.68$, $t^* = 3 \times 10^{-4}$ seconds, and $\varepsilon_0 = 5 \times 10^6$ dynes/cm^2 were assumed to fit data for polybutylene terephthalate (PBT) at $250°$ C. More complete data are shown in Fig. 3.9.4. Similar data for polycarbonate are shown in Fig. 3.9.5, also with a relatively high value, $\beta = 0.86$. For Nylon polyamide, Fig. 3.9.6, the value

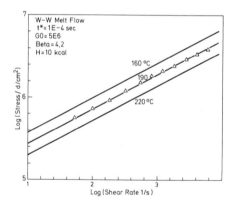

Fig. 3.9.1. Stress-shear rate curves for polymer melt are generated from the scaling rule for the stress-strain curves for polymer solids.

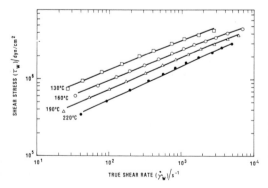

Fig. 3.9.2. Stress-shear rate data for polyethylene to be compared with Fig. 3.7.6 (1 dyne/cm^2 = 0.1 N/m^2).

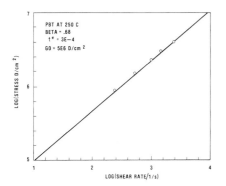

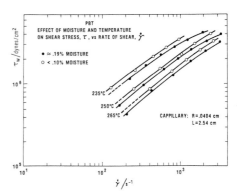

Fig. 3.9.3. Stress-shear rate curve generated to fit the flow data for polybutylene terephthalate (1 dyne/cm^2 = 0.1 N/m^2).

Fig. 3.9.4. Stress-shear rate data for polybutylene terephthalate used in Fig. 3.8.2 (1 dyne/cm^2 = 0.1 N/m^2).

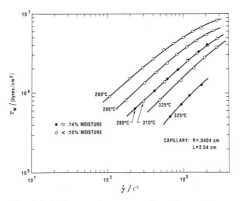

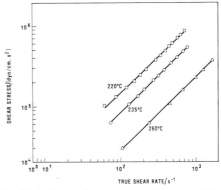

Fig. 3.9.5. Stress-shear rate data for polycarbonate (1 dyne/cm^2 = 0.1 N/m^2).

Fig. 3.9.6. Stress-shear rate data for polyamide (1 dyne/cm^2 = 0.1 N/m^2).

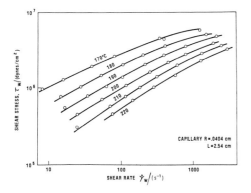

Fig. 3.9.7. Stress-shear rate data for polyvinyl chloride (1 dyne/cm^2 = 0.1 N/m^2).

of β is even greater at 0.91. Finally, in Fig. 3.9.7 are data for polyvinyl chloride copolymerized with a small amount of co-mers exhibiting $\beta = 5$.

The topic of melt rheology was not meant to be included as part of the scaling rules. It was included to shed an interesting light on the problems of melt rheology, particularly in view of the fact that the assumed values for the plateau modulus, the relaxation time and the activation energy are all reasonable values that fit data well.

Acknowledgment

The author gratefully acknowledges helpful discussions with D.S. Pearson, H. Ghoneim, T.T. Wang, and L.L. Blyler, Jr. The bulk of the data were generated by J.T. Ryan, (late) R.P. Wentz, E.H. Gilbert, C. Giniewski, S.S. Bearder, H.E. Kern, and R.F. Staats-Westover, and have been published elsewhere.

Nomenclature for Chapter 3

C, C_1, C_2, etc.	empirical constants
$D(t), D(\varepsilon, t)$	linear and nonlinear creep compliance at strain ε and time t
$E(t), E(\varepsilon, t)$	linear and nonlinear stress relaxation modulus at strain ε and time t
E_0	zero strain tensile modulus
E_f and E_p	zero strain moduli for filler and polymer
$f, \Delta f_0, f_\infty$	fractional free volume
G_s	shear modulus
H	activation enthalpy for aging and for relaxation process for solid state having undergone the same or equivalent thermomechanical history
$I^*, I_\varepsilon, \Pi_\varepsilon$	critical, the first, and the second strain invariants
$J, J^*, J_\infty, \Delta J, J_0$	shear compliance, J^* is the dynamic compliance
k	rate constant for the aging process
n, n_0	the power for scaling time, relaxation time, or strain rate for viscoelastic function; n is the slope of log relaxation time against log time at constant strain or stress
p, p_0, p_c	pressure in psi, p_c is a parameter for scaling the pressure effect on the stress
R_0	scaling factor for the temperature
R_1	scaling factor for the strain rate
R_2	scaling factor for the physical aging
R_f	scaling factor for the filler concentration
R_{sx}, R_{sy}	scaling factor to convert tensile to shear stress-strain
R	the gas constant
t	time
t_a, t_{a0}	annealing (physical aging) time
T, T_0, T_c	temperature in °K, T_c is a parameter for scaling the temperature effect on the stress
U	strain energy
$V, \Delta V, V_f, V_p$	specific volume; V_f and V_p are volumes of filler and polymer

Greek Letters

β	power for time in Williams-Watts formula
γ	shear strain
$\varepsilon, \varepsilon_0$	tensile strain
ε^*	strain corresponding to the equal rate of energy increase for constant strain rate process
$\varepsilon. \varepsilon_{kk}, \varepsilon_{11}, \varepsilon_{12}$	strans in tensor notation
η	viscosity
λ^*	Lammé constant

ν	Poissons ratio
$\sigma, \sigma_0, \sigma_1, \sigma_2, \sigma_3$	stress (usually tensile)
$\sigma, \sigma_{11}, \sigma_{12}$	stresses in tensor notation
t^*, t_0^*, t_g^*	relaxation time, t_g^* is the extrapolated value for T_g
ω	circular frequency

References

1. G. Williams and D.C. Watts, *Trans. Faraday Soc.* 1970, **66**, 80.
2. G.D. Patterson and C.P. Lindsey, *Macromolecules* 1981, **14**, 83.
3. L.C.E. Struik, *Physical Aging in Amorphous Polymers and Other Materials,* TNO Central Lab Com. No. 565, Delft 1977; also the Chapter 11 by Struik in this book.
4. S. Matsuoka, *Polymer Eng. Sci.* 1981, **21**, 901.
5. J.D. Ferry, *Viscoelastic Properties of Polymers,* 3rd edition, Chapter 11, Wiley, New York 1980.
6. M. Takayanagi, *J. Polymer Sci.* C, 113, 1964.
7. R.L. McCullough, *Polym. Composites* 1981, **2**, 149.
8. R.H. Boyd, *Macromolecules* 1984, **17**, 903.
9. R.E. Robertson, *J. Chem. Phys.* 1966, **44**, 10, 3950.
10. J. Rosenberg and S. Matsuoka, *Polymer J.* 1985, **17**, 321.
11. R.F. Staats-Westover and W.J. Vroom, *Soc. Plant Engrs. J.* 1968, **25**/8.
12. A.J. Kovacs, *Fortschr. Hochpolym. Forsch.* 1963, **3**, 394.
13. Z. Hashin and S. Shtrikmann, *J. Mech. Phys. Solids* 1962, **10**, 335.
14. R.M. Ogorkiewicz, *Engineering Properties of Thermoplastics,* Wiley, New York 1970.
15. S. Matsuoka, *Polymer Eng. Sci.* 1978, **18**, 1073.

4 Stress-Relaxation Behavior of Solid Polymers

Josef Kubát

Department of Polymeric Materials, Chalmers University of Technology
S-41296 Gothenburg, Sweden

Mikael Rigdahl

Paper Technology Department, Swedish Forest Products Research Laboratory
P.O. Box 5604, S-11486 Stockholm, Sweden

4.1 Introduction

Stress relaxation of solid materials is, together with creep deformation, the most fundamental transient experiment used for characterizing the viscoelastic properties of such materials. It relates to the stress decay that occurs with time when the solid is subjected to a constant strain. Ideally the initial elongation of the solid (straining) is considered to be performed instantaneously.

Since the first experimental account of stress relaxation of lead and a number of other materials was published by Trouton and Rankine[1] the number of works on this phenomenon has increased dramatically. This, of course, reflects the basic importance of the stress-relaxation process (and the corresponding creep deformation) in assessing the fundamental physical properties of matter and also its impact on technological applications. Within the polymer field the stress-relaxation process has been described in several reviews and excellent textbooks.[2-5]

Compared to the creep process, stress relaxation appears to provide a somewhat simpler tool for analyzing the basic flow mechanisms, since the problem of dividing the flow process in primary, secondary, and tertiary stages is avoided. Despite this, it seems that these mechanisms are still not very well understood. This will be further discussed in this Chapter, where an attempt to give a unified flow theory, based on cooperative interaction between the flow units, will also be presented.

From an engineering point of view stress relaxation may perhaps be of less general significance than creep, although it is of central importance in certain load bearing elements, e.g., screws, fasteners, seals. Furthermore, it may significantly affect physical properties of materials such as piezoelectricity,[6] birefringence,[7] and infrared absorption.[8]

The characteristic feature of a uniaxial stress-relaxation experiment is that the strain remains constant while the corresponding stress and stress rate decrease with time. Formally the time dependence of the stress can be expressed as

$$\dot{\sigma} = \frac{d\sigma}{dt} = -\phi(t, \sigma, T) \tag{4.1.1}$$

where σ is the (time dependent) stress, t the time, T the absolute temperature, and $\phi(x)$ a relaxation function. The effect of the mechanical and thermal history, e.g., physical aging, of the solid as well as any influence of the environment is here assumed to be described by Eq. (4.1.1).

For polymers, the interpretation of stress-relaxation processes appears to be a rather complicated issue. Several theories have been proposed to describe the flow process giving different forms of the relaxation function $\phi(x)$. This Chapter presents the main trends in this area and points out some far-reaching similarities in the flow behavior of various polymers and other materials. The different relaxation theories for polymers have been collected under three headings:

I Relaxation time spectra, RTS
II Stress dependent thermal activation, SDTA, and related flow theories, and
III Cooperative theories.

Before discussing these flow models it is appropriate to show a few experimental examples of typical stress-relaxation behavior of some solid polymers and other materials. Figure 4.1.1 shows the behavior of high-density polyethylene and polyisobutylene at room temperature in a $\sigma(\log t)$-diagram. The stress is given as the ratio $\sigma(t)/\sigma_0$, where σ_0 is the initial stress of the experiment, i.e., the stress at zero time ($t=0$). Several features of Fig. 4.1.1 are noteworthy. First, apart from the initial region of the curves, the $\sigma(\log t)$-relation is linear at higher values of σ/σ_0. The slope of this linear region constitutes the maximum slope of the $\sigma(\log t)$-curve. At lower values of σ/σ_0 (longer times) a marked deviation from linearity is observed, possibly indicating a change in the flow mechanism. The relative extent of the linear $\sigma(\log t)$-region is dependent on the material. For instance, this region is shorter for high-density polyethylene than for polyisobutylene.

From Fig. 4.1.1 it is also evident that for polyisobutylene the applied stress approaches zero for sufficiently long times (complete relaxation) while for polyethylene a nonzero stress level is gradually attained (incomplete relaxation). This stress level which is approached after sufficiently long measuring times is often referred to as the internal stress level, σ_i. It is to be expected that, assuming that σ_i is constant during the flow process, an expanded view of the kinetics of the relaxation is obtained if the curve is plotted as $(\sigma-\sigma_i)/(\sigma_0-\sigma_i)$ versus $\log(\text{time})$. This is done in Fig. 4.1.2, which shows the stress-relaxation behavior at room temperature of isoprene rubber, polyisobutylene, oriented and isotropic low-density polyethylene, cadmium, indium, and lead. Despite the different structures of these materials the $\sigma(\log t)$-curves have

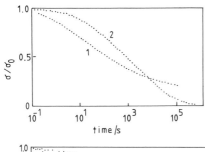

Fig. 4.1.1. Stress-relaxation curves, given as σ/σ_0 versus $\log t$ for 1 high-density polyethylene and 2 polyisobutylene at room temperature. The curve for polyethylene is an example of incomplete relaxation while complete relaxation is observed for polyisobutylene.

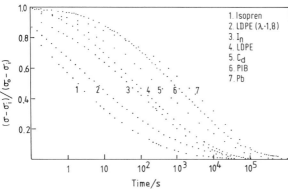

1. Isopren
2. LDPE (λ·1,8)
3. I_n
4. LDPE
5. C_d
6. PIB
7. Pb

Fig. 4.1.2. Stress-relaxation curves, given as $(\sigma-\sigma_i)/(\sigma_0-\sigma_i)$ versus $\log t$ for materials of different structures (room temperature). From J. Kubát and M. Rigdahl.[136]

a very similar shape, which might appear somewhat surprising. In particular the maximum relaxation rate of these materials, given as $F^\dagger = (-d\sigma/d\ln t)_{\mathrm{max}}$, satisfies the following relation

$$F^\dagger = (0.1 \pm 0.01)(\sigma_0 - \sigma_i) \tag{4.1.2}$$

This relation, which has been found to be valid for a large number of materials, irrespective of their structure,[9] will be discussed in detail since it points out some inherent weaknesses of the flow theories currently used.

4.2 Relaxation-Time Spectra (Linear Viscoelasticity)

4.2.1 Basic Concepts

The Maxwell model, consisting of a Hookean spring in series with a Newtonian dashpot, is the simplest model for describing uniaxial stress relaxation. The stress decay during the relaxation process is given by

$$\sigma(t) = E\,\varepsilon_0\,e^{-t/t^*} \tag{4.2.1}$$

where ε_0 is the constant, initially applied strain, E the elastic modulus of the Hookean spring, and t^* the relaxation time given by $t^* = \eta/E$ where η is the viscosity of the dashpot.

In most cases, the stress-relaxation behavior of solid polymers cannot be described by the Maxwell model. The flow process normally extends over a much wider time interval than predicted by Eq. (4.2.1). However, when the flow is governed mainly by chemical processes, as oxidative bond scission in rubber, Eq. (4.2.1) may be applicable.[10-12] This process of chemical stress relaxation is not considered in the following discussion.

Thus, for solid polymers a single relaxation time is not sufficient in describing the relaxation behavior. By coupling a number of Maxwell models in parallel the time region of flow can be broadened. The corresponding stress-time relation of this generalized model is given as

$$\sigma(t) = E_\infty \varepsilon_0 + \varepsilon_0 \sum_{i=1}^{n} E_i\, e^{-t/t_i^*}. \tag{4.2.2}$$

Here E_i is the modulus of the i:th Maxwell model and t_i^* the corresponding relaxation time. The term $E_\infty \varepsilon_0$ denotes the stress level approached after very long times, i.e., it corresponds to the internal stress discussed earlier. For practical purposes, a limited number of exponential terms in Eq. (4.2.2) are often sufficient to describe a normal relaxation curve with sufficient accuracy.[13, 14] A method for the analysis of such curves in terms of a limited number of exponentials has been proposed by Tobolsky.[5]

A more practical description of the relaxation process is achieved by introducing a continuous distribution of relaxation times. Equation (4.2.2) then takes on the following form

$$\sigma(t) = E_\infty \varepsilon_0 + \varepsilon_0 \int_{-\infty}^{\infty} H_d(t^*)\,e^{-t/t^*}\,d\ln t^* \tag{4.2.3}$$

where $H_d(t^*) = t^* E(t^*)$ is the distribution of relaxation times t^* (the relaxation spectrum). The evaluation of $H_d(t^*)$ and its possible interpretation in terms of molecular structure has been the subject of numerous investigations in the area of polymer viscoelasticity.[2-5] In passing it could be mentioned that $H_d(t^*)$ and also its counterpart for creep deformation provides a valuable link between transient properties and dynamic behavior of linear viscoelastic solids, since

$$G'_s(\omega) = E_\infty + \int_{-\infty}^{\infty} \frac{\omega^2 t^{*2} H_d(t^*)}{1 + \omega^2 t^{*2}} \, d \ln t^*$$

$$G''_s(\omega) = \int_{-\infty}^{\infty} \frac{\omega t^* H_d(t^*)}{1 + \omega^2 t^{*2}} \, d \ln t^*$$

(4.2.4)

where $G'_s(\omega)$ and $G''_s(\omega)$ are the storage and loss moduli, respectively, and ω the angular frequency.

For solid polymers and polymers in the rubbery state a "box-distribution" may often be found to fit the observed behavior with sufficient accuracy provided that the process occurs at a temperature which is not in the vicinity of any transition temperature. The box-distribution of $H_d(t^*)$ is given by

$$H_d(t^*) = \begin{cases} q & \text{for } t_1^* \le t^* \le t_2^* \\ 0 & \text{for } t^* < t_1^* \text{ and } t^* > t_2^* \end{cases}$$

(4.2.5)

where q, t_1^*, and t_2^* determine the dimensions of the "box."

The corresponding stress-time dependence is obtained from Eq. (4.2.3) as

$$\sigma(t) = \varepsilon_0 \, q [E_i(-t/t_1^*) - E_i(-t/t_2^*)]$$

(4.2.6)

Here $E_i(x)$ denotes the exponential integral function; for convenience we have put $E_\infty \varepsilon_0 = 0$.

Equation (4.2.6) is depicted in Fig. 4.2.1; the dimensions of the box have been chosen to comply with the experimentally found Eq. (4.1.2). As evident, Eq. (4.2.6) produces a $\sigma(\log t)$-curve which is linear in the main region of the relaxation process in agreement with the experimental curves in Fig. 4.1.2. On the other hand, the box distribution itself does not provide any deeper insight into the basic mechanisms underlying Eq. (4.1.2).

Recognizing the similarity in relaxation behavior of different materials, Feltham[15] suggested that a log-normal distribution of relaxation times $H_d(t^*)$, which has been introduced by Wiechert,[16] might be useful in describing the flow process in solids. In this case the spectrum has the form

$$H_d(t^*) = \frac{b}{\sqrt{\pi}} \exp[-b^2 (\ln t^*/t_m^*)^2]$$

(4.2.7)

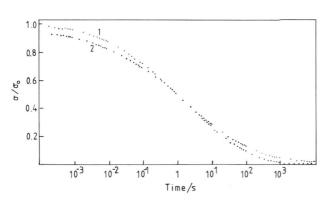

Fig. 4.2.1. Stress-relaxation curves corresponding to 1 Eq. (4.2.6) and 2 Eq. (4.2.8). The parameters of these equations have been chosen to comply with Eq. (4.1.2); $F^\dagger = 0.1 \sigma_0^*$. From J. Kubát and M. Rigdahl.[136]

where t_m^* is the relaxation time corresponding to the maximum $b/\sqrt{\pi}$ in the distribution. The corresponding stress is approximately given by [15]

$$\sigma(t)=\frac{\sigma_0}{2}\left[1+\mathrm{erf}\left(b\ln\frac{t_m^*}{t}\right)\right]\tag{4.2.8}$$

where $\mathrm{erf}(x)$ is the error function. The corresponding $\sigma(\log t)$-curve is included in Fig. 4.2.1, where the parameters of Eq. (4.2.8) again have been chosen to comply with Eq. (4.1.2). It is evident from this Figure and it has been pointed out earlier[17] that it is not possible to decide from the shape of experimental relaxation curves whether the corresponding relaxation spectrum $H_d(t^*)$ is of the box-type or the log-normal type, since the theoretical curves Eq. (4.2.6) and Eq. (4.2.8) are very similar. This may illustrate the rather formal character of the relaxation spectrum concept as applied to polymeric solids.

Another spectrum which has been suggested for natural composites, like compact bone, is the logarithmic distribution,[18] i.e.,

$$H_d(t^*)=\begin{cases}\ln t^*, & t_1^*\le t^*\le t_2^*\\ 0, & t^*<t_1^* \text{ and } t^*>t_2^*\end{cases}\tag{4.2.9}$$

The corresponding decrease in stress with elapsed time is given by [18]

$$\sigma(t)/\varepsilon_0=\ln(1/t_2^*)\,E_i(t/t_2^*)-\ln(1/t_1^*)\,E_i(t/t_1^*)+E_1^2(t/t_1^*)-E_1^2(t/t_2^*)+E_\infty\tag{4.2.10}$$

where

$$E_1^2(y)=\int\limits_{y}^{\infty}\frac{E_1(x)}{x}\,dx\quad\text{and}\quad E_1(x)=-E_i(-x).\tag{4.2.11}$$

A number of different empirical distributions $H_d(t^*)$ have also been analyzed and discussed by Gross.[19]

Before concluding this Section it should be stressed that an application of the spectral theory requires that the polymer can be assumed to be linearly viscoelastic. For amorphous polymers in the glassy state the upper strain limit corresponding to linear viscoelasticity is of the order 1%[20] while it may be considerably lower for semicrystalline polymers.[21]

4.2.2   Time-Temperature Superposition Principle for Amorphous Polymers

So far we have discussed only stress-relaxation behavior of solid polymers at a specific temperature that is not too close to any transition temperature. In general, a change in temperature has a profound influence on the physical properties of polymeric materials. This applies in particular to their viscoelastic behavior, where the effect of temperature on stress relaxation and creep of polymers has been studied in great detail.[2-5] In the following we will discuss amorphous and semicrystalline polymers separately with regard to their temperature-dependent mechanical properties.

Consider a relaxing amorphous polymer at a specific temperature, T_1. If we assume that the conditions of linear viscoelasticity are fulfilled, the relaxation behavior of the polymer is specified by representing the logarithm of the relaxation modulus $E(t)=\sigma(t)/\varepsilon_0$ as a function of $\log(\text{time})$. If the temperature is now raised to the level T_2, this will not, to a first approximation, change the shape of the relaxation curve but just shift the curve to shorter times. In other words, a change in temperature leaves the shape of the relaxation spectrum $H_d(t^*)$ unaltered, merely shifting $H_d(t^*)$ along the time axis. This is an example of the time-temperature superposition principle[22] and it can be used to shift relaxation curves (or creep curves), obtained at different temperatures, horizontally along the logarithmic time axis to yield a stress relaxation master curve at a specific reference temperature covering several decades of time.

Figures 4.2.2 and 4.2.3 provide an example of this procedure for poly(vinyl acetate) PVAC.[23] The relaxation behavior of PVAC up to 1000 minutes has been measured at different temperatures in the range 15 to 102° C (Fig. 4.2.2). Figure 4.2.3 shows the corresponding master curve, obtained by shifting the individual relaxation curves in Fig. 4.2.2. The reference temperature was 75° C. Several features of the master curve are noteworthy. First, in the glass transition region (around 30° C for this polymer) the relaxation modulus $E(t)$ decreases sharply 3–4 decades down to a rather low value in the rubbery state (ca. 1 MPa). Secondly, the relaxation rate in the rubbery state is rather slow. This rubbery plateau is due to physical chain entanglement; the length along the time axis of the plateau will increase with increasing molecular weight, as shown in Fig. 4.2.3, simply because it will take a longer time for the chains to disentangle. Beyond the rubbery plateau, the viscous flow region will be approached and the modulus will decrease very sharply again. Actually, the relaxation behavior close to the viscous flow region can in some cases be described by a Maxwellian type relation,

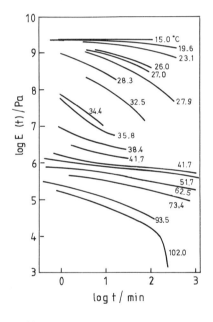

Fig. 4.2.2. Stress-relaxation curves, given as log $E(t)$ versus log t, for poly(vinyl acetate) at different temperatures.[23]

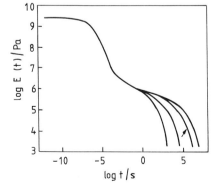

Fig. 4.2.3. Stress-relaxation master curves for poly(vinyl acetate). The curve corresponding to the data in Fig. 4.2.2 is indicated by the arrow. Note that the length of the rubbery plateau increases with molecular weight.[23]

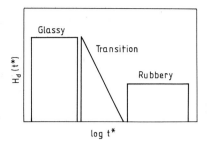

Fig. 4.2.4. Schematic drawing of the relaxation spectrum $H_d(t^*)$ versus log t^* for an amorphous polymer covering the glassy state, the T_g-region, and the rubbery plateau.

Eq. (4.2.1).[5] It should be noted that there is no discernable effect of the molecular weight on the master curve in the glassy region and in the glass transition region (the primary transition region).

In Fig. 4.2.4 a schematic drawing of the relaxation spectrum $H_d(t^*)$ corresponding to the stress-relaxation master curve of an amorphous polymer is shown. It consists of a box-wedge-box distribution. The wedge distribution, which applies to the relaxation behavior in the glass transition (T_g) region can be written as

$$H_d(t^*) = K(t^*)^{-m} \tag{4.2.12}$$

where K and m are constants. The wedge distribution will be discussed in a separate section.

A well-known method of relating the horizontal shift factors along the time axis (log a_T) to the temperature when constructing master curves for amorphous materials has been developed by Williams e.a.[24]. If the glass transition temperature, T_g, is taken as the reference temperature, the shift along the time axis can be represented as

$$\log a_T = \frac{c_1\,(T - T_g)}{c_2 + T - T_g} \quad \text{(WLF-equation)}. \tag{4.2.13}$$

The WLF-equation has been assumed to be valid for amorphous polymers at temperatures between T_g and $T_g + 100°$ C.[3] It should, however, be mentioned that some recent work indicates that the validity range of the WLF-equation actually is less than $100°$ C; see Chapter 10. The constants c_1 and c_2 can be expressed in terms of the isobaric expansivity α and the fractional free volume f through the relation between viscosity η and f (Doolittle equation)[25]

$$\log a_T = \log(\eta/\eta_g) = B\left(\frac{1}{f} - \frac{1}{f_g}\right). \tag{4.2.14}$$

Here $f = v_f/v$ is the fractional free volume; η_g and f_g denote the viscosity and fractional free volume at T_g, respectively, and B is a constant. By denoting the isobaric expansivity in the glassy and rubbery state by α_g and α_r, respectively, the temperature dependence of the fractional free volume can be expressed as

$$f = f_g + (\alpha_r - \alpha_g)\,(T - T_g) \tag{4.2.15}$$

Thus, f is assumed to be a linear function of temperature. Further, B, f_g, and $(\alpha_r - \alpha_g)$ are assumed to be constants for amorphous solids; then the WLF-constants c_1 and c_2 can be evaluated using Eqs. (4.2.14) and (4.2.15) giving $c_1 = 17.44$ and $c_2 = 51.6°$ C.[3] If a reference temperature other than T_g is used, the c_1- and c_2-values will change accordingly.

To obtain a satisfactory superposition of the individual relaxation curves when constructing the master curve in the temperature region between T_g and $T_g + 100°$ C a smaller vertical shift, a_v, is also often required. This shift is given by

$$a_v = \frac{T_g \varrho_g}{T \varrho} \tag{4.2.16}$$

if T_g is the reference temperature. Here ϱ denotes the density at the temperature T of the amorphous polymer. The vertical shift factor stems from the kinetic rubber theory.

Models, other than the WLF-equation, are available for predicting the temperature dependence of the shift factors $\log a_T$. For example, using the free volume theory and the hole theory of liquids, Utracki[26, 27] suggested a relation of the form

$$\log a_T = B_1 [\{\exp(bx) - 1\}^{-1} - \{\exp(b) - 1\}^{-1}] \tag{4.2.17}$$

where b and B_1 are constants and $x = (T/T_g)^{3/2}$. Equation (4.2.17) was used to describe the temperature dependence of the viscoelastic shift factors for styrene ionomers[27] and it provided information regarding the size of the flow elements.

For temperatures well below T_g the influence of temperature on the position of the relaxation curves along the time scale cannot be described by the WLF-equation; one explanation is that in this region the rearrangement of the molecular segments within the available free volume does not contribute to the flow to any larger extent. Instead, some authors believe that the relaxation is due to the passage of "flow units" over energy barriers. The relaxation time is thus shifted through a Boltzmann factor:

$$t^* = t_0^* \exp(\Delta G/kT) \tag{4.2.18}$$

where ΔG is the Gibbs function and k the Boltzmann constant. The reasons for the inapplicability of the WLF-equation below T_g is discussed in Chapter 10.

Numerous examples of the use of the time-temperature superposition principle exist in the polymer literature.[2-5] It would not be meaningful to give a detailed account on this subject here. As examples of specific polymers analyzed using these concepts one could mention the work of Tobolsky e.a.[28-35] dealing with polymers such as polyisobutylene, poly(methyl methacrylate), styrene-butadiene rubbers, polystyrene, natural rubber, and poly(vinyl chloride). Additional examples can be found in the Literature.[2-5, 36-39] The problem is also discussed in Section 2.3, in Chapter 10, and in other points throughout this volume.

4.2.3 Time-Temperature Superposition Principle for Semicrystalline Polymers

The time-temperature superposition principle and its applicability to semicrystalline polymers has also been tested and discussed in several reports.[3] In general, this principle appears to be less valuable in this case. This can be attributed mainly to the fact that the structure of semicrystalline polymers is rather sensitive to changes in temperature and they are linearly viscoelastic only at very low strain levels. In most situations the WLF-equation does not apply for the horizontal shifting of the curves when constructing master curves. In some cases it is not possible to construct a master curve for crystalline polymers by vertical and horizontal shifts.[40-42] In other situations time-temperature superposition is possible, at least in certain temperature regions,[42-45] but the corresponding shift factors should be regarded as more or less empirical. The strain level should also be kept sufficiently low in order to fulfill the requirements of linear viscoelasticity.[21, 42]

When the polymer undergoes a secondary transition in the temperature region covered by the stress-relaxation experiments, this will be evident in the shape of $H_d(t^*)$ or in the temperature dependence of the shift factors. In many cases the superposition principle fails in this region. Consequently, information about the transition can be

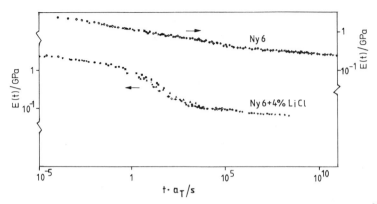

Fig. 4.2.5. Stress-relaxation master curves for PA 6 and for PA 6+4% LiCl.[50]

extracted from relaxation data. Such experiments have been made on polytetrafluoro-ethylene (PTFE),[43] which is known to exhibit transitions at room temperature, and on polybutene-1[46] and polyethylene at ca. $-130°$ C.[47] Among more recent works, the investigation of the so-called γ-band of polyethylene between $-100°$ C and $-190°$ C can be mentioned.[48] With the aid of stress relaxation the authors were able to resolve three different transitions in the γ-band characterized by different values of the activation energy.

In general, the master relaxation curves of a semicrystalline polymer do not show the significant changes in slope that amorphous polymers do even when the T_g-region is passed. Instead the relaxation spectrum is very broad and flat.[42, 43, 49] This may be due to the cross-linking effects and filler action of the crystallites which cause a reduction in segmental mobility. An example of a master curve for a crystalline polymer (polyamide 6) is shown in Fig. 4.2.5.[50] Although the stress-relaxation curves were obtained from measurements at temperatures including the T_g-region, this is not apparent in the flat master curve. If, however, 4% LiCl is added to polyamide 6 the rate of crystallization is depressed[50] due to an interaction between lithium ions and the carbonyl-oxygen groups of the polyamide. As a result of this, a well-developed glass transition region will appear in master curves of such a mixture, see Fig. 4.2.5. The significant influence of crystallinity on the shape of $H_d(t^*)$ is apparent even at very low degress of crystallinity. Note the behavior of plasticized poly(vinyl chloride)[35] where the crystalline fraction amounts to only a few percent. Another example relating to a polycarbonate is given in Fig. 4.2.6. Here the gradual disappearance of the primary

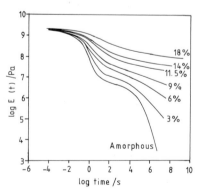

Fig. 4.2.6. Stress-relaxation master curves for polycarbonate of different degrees of crystallinity. The degree of crystallinity is indicated for each curve.[51]

transition with increasing degree of crystallinity is clearly seen.[51] Note that the effect of the crystallinity is largest in the rubbery region.

4.2.4 Linear Viscoelasticity of Copolymers, Blends (Heterogeneous Polymers)

It may be expected that heterogeneous polymeric systems like graft copolymers and blends do not obey the time-temperature superposition principle, i.e., they are not thermorheologically simple materials. In many cases this is also supported by experimental findings. For obvious reasons, the viscoelastic properties of such systems are rather complicated, and more experimental and theoretical work is required in this area, especially since the importance of such materials tends to increase.

In multiphase material, several relaxation mechanisms may be operating in temperature regions which are rather close, and a straightforward application of superposition principles may be difficult. In temperature regions where only one mechanism is active, time-temperature superposition may be applicable or it may be possible, in some cases, to separate different mechanisms by suitable techniques. This may illustrate the complexity of the situation encountered with multiphase materials. For instance, for a plasticized polyester-polystyrene system, a WLF-type equation was found to apply to master stress-relaxation curves.[52] Similar behavior was found for the rather simple system of carbon-black filled styrene-butadiene copolymers, at least in the primary transition region.[53] Even in a complicated material like poly(ethylene-g-styrene), a graft polymer which is both semicrystalline and heterogeneous, the time-temperature superposition principle could be used for temperatures between ca. -20 and ca. $130°$ C.[54] The temperature dependence of the shift factors was in this case found to be described by a Boltzmann expression, but a change in activation energy was noted at higher temperatures. This was suggested to be due to changes in crystallinity. For graft copolymers of PVAC with styrene and mechanical mixtures of PVAC and polystyrene, superposition of individual curves was possible although the WLF-equation was not always valid.[55]

For other complex materials, like tri-block copolymers, deviations from the time-temperature superposition principle were observed.[56] This was also the case with a 50/50 blend of PVAC and lightly cross-linked poly(methyl methacrylate).[57, 58] In addition, a method for constructing master curves for thermorheologically complex materials has been described[58] and stress relaxation behavior of complex materials investigated.[59, 60]

Styrene-based ionomers, which have been studied in detail,[61-63] are a rather interesting group of complex materials. For low contents of ions these materials behave as thermorheologically simple materials, i.e., the time-temperature superposition principle holds, while at higher ion contents relaxation master curves cannot be constructed. Fig. 4.2.7 shows master curves for copolymers of styrene and styrene-sodium sulpho-

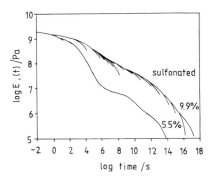

Fig. 4.2.7. Stress-relaxation curves for a copolymer of styrene and styrene-sodium sulphonate. For the copolymer containing 5.5 mole% sulphonated polystyrene the time-temperature superposition principle holds, while it breaks down when the amount of sulphonated polystyrene increases to 9.9 mole%.[63]

nate.[63] At lower ion contents a master curve can be obtained while with 9.9 mole% of the ionic component in the polymer this is evidently not possible. This effect can be interpreted in terms of a "composite" structure consisting of ion-rich clusters immersed in an ion-poor matrix.

4.2.5   Stress-Relaxation in the Primary Transition Region of an Amorphous Polymer (Damped Debye Lattice)

It has already been mentioned that in the T_g-region of an amorphous polymer the relaxation modulus decreases very sharply from a high value in the glassy state to a low value corresponding to the rubbery plateau, see Fig. 4.2.3. The viscoelastic properties in this region are not well-understood but in general the relaxation behavior can be described by a wedge distribution of relaxation times

$$H_d(t^*) = K(t^*)^{-m} \tag{4.2.12}$$

From the well-known results of Rouse[64] and Bueche[65] the $H_d(t^*)$-spectrum for a linear polymer is given by

$$H_d(t^*) = K_1(t^*)^{-1/2} \tag{4.2.19}$$

In this case a plot of log $H_d(t^*)$ versus log t^* would give a straight line with a slope of $-1/2$. This is approximately the case for e.g. polyisobutylene[66] and poly(vinyl chloride).[67]

Equation (4.2.19) is derived from a system of linearly coupled, damped oscillators, i.e., only intermolecular interactions are considered to be important. When intramolecular forces also become significant, the relaxation spectrum in the primary transition region is modified to [67-69]

$$H_d(t^*) = K_2(t^*)^{-1} \quad \text{(two-dimensional damped Debye lattice)} \tag{4.2.20}$$

$$H_d(t^*) = K_3(t^*)^{-3/2} \quad \text{(three-dimensional damped Debye lattice)} \tag{4.2.21}$$

where K_2 and K_3 are constants.

In cases where three-dimensional (inter- as well as intramolecular) interactions are at hand, a slope of $-3/2$ for the log $H_d(t^*)$-line would be expected. This has been found, e.g., in polystyrene (PS).[67, 70, 71] If a plasticizer with a solubility parameter not too close to that of PS is added the slope decreases; this may be due to a reduction of intramolecular interactions.[70, 71] Nilsson[72] points out that PS appears to be rather easy to destabilize (i.e., decrease the slope from $-3/2$) by copolymerization.

The concept of a damped Debye lattice cannot fully explain the viscoelastic properties in the T_g-region of amorphous polymers, but it may prove to be useful in providing guidelines for understanding the differences in relaxation behavior in this region of otherwise similar polymers like polystyrene and poly(methyl methacrylate).[72]

It is relatively difficult to perform stress-relaxation tests in the T_g-region due to the high relaxation rate. Different methods for analyzing data obtained in this region are discussed by Lin and Aklonis.[73]

4.3   Stress-Relaxation Behavior at Higher Strain Levels

The influence of deformation beyond the linear limit on the stress-relaxation curves has not been studied to the same extent as the linear viscoelastic behavior at small strains. It is, however, known that higher strain levels can influence the relaxation behavior significantly.[74, 75] For cellulose, an increase in elongation up to 5% shifted the curves to shorter times.[76] A shift with increasing strain, although sometimes more complicated, has also been reported for poly(ethylene terephthalate)[77] and polycarbonate[78, 79] by Titomanlio and Rizzo. By considering only the relaxable part of the stress, master curves could be constructed using time shift factors which were related to the strain level and the strain rate prior to the commencement of the test.[78]

For polyethylene, deformed to large deformations, Kubát e.a.[80] found similar shifts of the relaxation curves along the time axis. The general Eq. (4.1.2) was not, however, significantly affected by the strain level.

Matsuoka e.a.[81] have analyzed the influence of strain on nonlinear stress relaxation in polymeric glasses in detail. They found that the decrease in relaxation modulus with increasing deformation levels could be attributed to two phenomena. The curves were shifted to shorter times due to an increase in free volume (strain-induced dilatation) and downward due to a reduction in Helmholtz energy. A good agreement between predictions and experimental results was obtained.

4.4 Physical Aging

Physical aging of glassy polymers is due to the nonequilibrium state of the material when cooled through the T_g-region at such a rate that the molecules cannot acquire their equilibrium configuration corresponding to the temperature T at which the cooling is ceased $(T < T_g)$. Though the molecular mobility is very low in the glassy state, it is not negligible; molecules therefore also approach their equilibrium state in glassy materials. This slow approach to equilibrium is called physical aging. It manifests itself in several ways, e.g., as volume relaxation and changes in viscoelastic properties. Physical aging thus reflects the thermal history of polymeric material. The aging process is thermoreversible in contrast to changes where chemical reactions are involved. The phenomenon of physical aging has been studied in detail by Struik,[82] who mainly investigated its effect on creep deformation and describes it in detail in Chapter 11. His results are also applicable to stress relaxation. Struik suggested that physical aging is associated with a gradual decrease in free volume which in turn shifts the relaxation spectrum towards longer times.

The following picture may illustrate the effect of aging on the relaxation process.[81] If the relaxation times at two different temperatures T and T_0 are denoted by t^* and t_0^*, respectively, it can be assumed that they are related to the corresponding fractions of free volume as follows (cf. Eq. (4.2.14)),

$$\ln \frac{t^*}{t_0^*} = \frac{1}{f} - \frac{1}{f_0} \tag{4.4.1}$$

When the polymer undergoes the glass transition, the relaxation times will increase (decreasing temperature). Kovacs[83] studied the volume contraction of polymers near T_g and found that the rate of this contraction is slowed with time according to

$$-\frac{df}{dt} = \frac{f}{t^*} = \frac{f}{t_0^*} \exp\left(\frac{1}{f_0} - \frac{1}{f}\right) \tag{4.4.2}$$

$$\int_f^{f_0} \frac{\exp\left(\frac{1}{f} - \frac{1}{f_0}\right)}{f} df = \frac{t}{t_0^*} \tag{4.4.3}$$

The important feature of this equation is that a plot of log $t*$ versus log t is a straight line with a slope of unity.[81] This means that the relaxation times will increase in exact proportion to the time spent at the specific temperature (below T_g). This has been tested by Matsuoka e.a.[81] using stress relaxation of polystyrene. Four samples were placed in an oven at 58° C and their relaxation measured after different annealing times. The result is shown in Fig. 4.4.1. With increasing annealing time (aging time) the relaxation curves are shifted horizontally to longer times. From the corresponding master curve the shift factors were found to be the same as the logarithm of the time spent at 58° C, $d \log a_t/d \log t_a \approx 1$ where $\log a_t$ is the shift factor and t_a the annealing time.

Physical aging may thus be viewed as the result of a gradual decrease in free volume with time below T_g. Another example of the effect on stress relaxation at 25° C of aging is given in Fig. 4.4.2 for poly(methyl methacrylate) (PMMA).[84] Here the stress relaxation curves have been measured at different storage times at 25° C (after being quenched from 165° C). A master curve was then constructed, and the $d \log a_t/d \log t_a$-factor was again of the order of one, 0.65 in this case. A small vertical shift was also necessary to obtain the master curve, and it was suggested that the vertical shift factor was related to densification of the glassy polymer during aging.[84]

Physical aging appears to be a rather general phenomenon.[82] Its effects are significant in the temperature range between T_g and the highest secondary transition below T_g. Physical aging has also been shown to take place in semicrystalline polymers at temperatures above T_g. In this case the aging process is more complicated, as it is associated with both the amorphous regions of sufficient segmental mobility and with recrystallization of the imperfectly ordered phase. In general, experimental results showing the shifts of the relaxation spectrum with increasing aging time are largely similar to those found with amorphous materials.

We have restricted discussion to the situation where the elapsed time of the relaxation experiment is shorter than the aging time. The problem encountered when the polymer

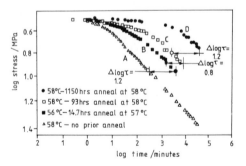

Fig. 4.4.1. Stress relaxation of polystyrene at 58° C with different annealing times (at 58° C) but with same applied strain. The horizontal shift along the time axis is the same as the logarithm of the annealing time.[81]

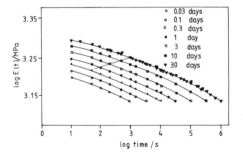

Fig. 4.4.2. Stress-relaxation curves and the resulting master curve for poly-(methyl methacrylate) at 25° C for samples aged different times at 25° C.[84]

ages during creep or relaxation (long-term behavior) has been analyzed by Struik in his book[82] and more succinctly in Chapter 11 of this volume as well as by Bauwens in Chapter 12. Finally, there are several other recent works that deal with stress relaxation and its dependence on physical aging.[85-88]

4.5 Stress-Dependent Thermal Activation and Related Flow Theories

The idea of thermally-activated motion of localized "disturbances" was suggested by Becker[89]; in principle, it can be applied to crystalline as well as to amorphous materials.[90] Its applicability to stress relaxation or creep can best be explained with reference to Fig. 4.5.1. It is assumed that a given structural element contributes to the macroscopic flow by passage over a potential barrier, shown in Fig. 4.5.1, from, e.g., site 1 to site 2. The energy necessary for this transition is supplied by thermal vibrations and also by externally applied stress. In the unstressed state the barrier height is given by the Gibbs function change ΔG, and the probability of a successful jump is proportional to exponent $(-\Delta G/kT)$. When a stress is applied, the energy barrier, in the forward direction, is lowered, i.e.,

$$\Delta G' = \Delta G - V\sigma \qquad (4.5.1)$$

where V is a constant, which has the dimensions of a volume, and is often termed the (stress) activation volume. In some cases an activation area A is introduced as $V = Ab$, where b is a displacement vector (cf. the Burgers vector for crystalline solids) and A may be associated with the area swept out by the flow element as it reaches the top of the barrier. If V is stress dependent[90] it would be more appropriate to write

$$\Delta G' = \Delta G - \int_0^\sigma V d\sigma \qquad (4.5.2)$$

For the moment we will, however, assume that V is constant.

The application of stress would thus increase the probability of successful jumps in the forward direction and the corresponding macroscopic rate equation (stress relaxation case) will be

$$\dot\sigma = \frac{d\sigma}{dt} = -A' \exp[-(\Delta G - V\sigma)/kT] = -A'' \exp(V\sigma/kT) \qquad (4.5.3)$$

where A' and A'' are constants.

Integrating Eq. (4.5.3) yields

$$\sigma(t) = -\frac{kT}{V} \ln\left(e^{-V\sigma_0/kT} + \frac{A'' V}{kT} t\right) \qquad (4.5.4)$$

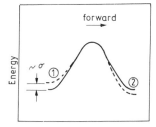

Fig. 4.5.1. When a stress σ is applied, the potential barrier for the advancement of flow is decreased in the forward direction (from site 1 to site 2).

where σ_0 is the initially applied stress at $t=0$. In a $\sigma(\log t)$-diagram this is, with exception of the very first part of the curve, an approximately straight line with the slope kT/V. Thus the exponential law, Eq. (4.5.3), produces curves which are linear with the logarithm of time, which is in accordance with experimental results, see Fig. 4.1.2. Actually, the stress in Eqs. (4.5.1 through 4.5.4) should be replaced by the effective stress $\sigma^*=\sigma-\sigma_i$, where σ_i is the internal stress. The above is of minor importance at this point, and is discussed in the next Section.

The activation volume can thus be determined from the slope of the $\sigma(\log t)$-curve or by using Eq. (4.5.3),

$$\frac{d\ln(-\dot\sigma)}{d\sigma}=\frac{V}{kT} \tag{4.5.5}$$

An exponential relation between stress rate $(\dot\sigma)$ and stress (σ) has been used extensively to describe the stress relaxation behavior of solids; as a rule this applies to metals,[13, 91, 92] but in some cases also to polymers.[93–96] The stress-dependent thermal activation theory has also been analyzed in detail in a textbook.[97] Despite the appeal of Eq. (4.5.3) there are some basic difficulties encountered in its application. As already stated the slope of the linear region, which is equivalent to the maximum slope, of the relaxation curves follows the general relation[9, 95]

$$F^\dagger=\left(-\frac{d\sigma}{d\ln t}\right)_{\max}\approx0.1\,\sigma_0^* \tag{4.1.2}$$

where $\sigma_0^*=\sigma_0-\sigma_i$. This equation is an empirical result based on experiments with different solids including polymers, metals, and molecular compounds. Since $F^\dagger=kT/V$ it follows

$$V\sigma_0^*\approx10\,kT \tag{4.5.6}$$

Equation (4.5.6) would imply that there is a numerical relation between the activation volume and the initial stress which is independent of the structure of the solid. Thus, Eq. (4.5.3) in the present form cannot give any information regarding the basic mechanism responsible for the macroscopic flow, and activation volumes determined in this way may not be "true" activation volumes.

This difficulty was recognized by White e.a.[98–101] who analyzed stress relaxation in detail in terms of the two-site model of stress-dependent thermal activation.[2] White[100] pointed out that two factors are not accounted for in Eq. (4.5.3), 1) the possibility of backward flow over the potential barrier, and 2) the population distribution of the sites available, which might not be the same for different sites. Using the two-site model, he arrived at the following stress-time relation for stress relaxation

$$\sigma(t)=\sigma_0\,e^{-ct}+\frac{D}{c}\,\sigma_0[1-e^{-ct}] \tag{4.5.7}$$

where c and D are functions of temperature but independent of stress. Equation (4.5.7) cannot describe stress relaxation of polymers since it contains only one relaxation time, but it still provides valuable information concerning the concepts used. It should be noted that Eq. (4.5.7) predicts a nonzero level of stress at very long times, i.e., an internal stress value σ_i

$$\sigma_i=\frac{D}{c}\,\sigma_0 \tag{4.5.8}$$

The internal stress is thus proportional to the initial stress. This is further discussed in the next section. Using the two-site model it was also shown that (cf. Eq. (4.5.5))

$$\sigma^* \frac{d \ln \dot{\sigma}}{d\sigma} = \text{constant} \quad \text{(regardless of structure)} \tag{4.5.9}$$

as required by experiments, but the activation volume does not enter the equation. This implies that the activation volumes in Eq. (4.5.6) are not significant here. Instead the mean activation volume could be estimated as[100, 101]

$$V = \tfrac{1}{2} c k T \tag{4.5.10}$$

where c is the slope of the straight line obtained when plotting $1/\sigma_i$ versus $1/(\sigma_0 - \sigma_i)$. The values of V obtained for acrylic polymers using this method were reasonable.[101]

Although the site model by White is very valuable for obtaining a better understanding of and deeper insight into the concepts of stress-dependent thermal activation, it cannot explain the basic character of the experimentally well-documented Eq. (4.1.2). It must be emphasized that equations similar to Eq. (4.1.2) are not exclusively obtained for stress relaxation. Similar relations can also be found for creep[82, 102, 103] and for grain boundary sliding.[104]

4.5.1   Power Law

In some cases a power law of the type

$$\dot{\sigma} = B(\sigma - \sigma_i)^n \tag{4.5.11}$$

can be used to describe the stress relaxation with sufficient accuracy. Here B and n are constants. Equation (4.5.11) is considered to be empirical but it is interesting to note that the stress dependence of the average dislocation velocity also obeys a similar power law.[105]

Integration of Eq. (4.5.11) yields

$$\sigma(t) = (\sigma_0 - \sigma_i) \left[1 + (n-1) B(\sigma_0^*)^{n-1} t \right]^{-1/(n-1)} + \sigma_i \tag{4.5.12}$$

The power law is usually applied to metals, but it has also been employed for polymers in a few cases.[99, 106–108] Very often a stress relaxation curve is linear in a $\sigma(\log t)$-diagram at shorter times, i.e., at higher values of the ratio $\sigma(t)/\sigma_0$, which is required by the exponential law, Eq. (4.5.3), while at lower relative stresses σ/σ_0 a marked deviation from linearity is observed. This latter portion of the curves can be described by a power law[107], and the transition stress (σ_{tr}) where the change from exponential to power law behavior occurs is approximately

$$\sigma_{tr} - \sigma_i \approx \frac{n}{10} (\sigma_0 - \sigma_i) \tag{4.5.13}$$

The change in flow behavior is rather clear when the rate $\dot{\sigma}$ is plotted against time in a double logarithmic plot. In the exponential flow region this is a straight line with the slope -1, while in the power law region the slope is $-n/(n-1)$. This is shown for high-density polyethylene in Fig. 4.5.2.[80]

In many situations, the entire stress-relaxation curve can be approximated by Eq. (4.5.12) with sufficient accuracy. The maximum slope $F^\dagger$ in a $\sigma(\ln t)$-diagram can then be evaluated as

$$F^\dagger = n^{-n/(n-1)} (\sigma_0 - \sigma_i) \tag{4.5.14}$$

A value of $n = 6 - 7$ then corresponds to the constant 0.1 in Eq. (4.1.2).

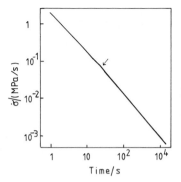

Fig. 4.5.2. The variation of $\dot{\sigma}$ with time during stress relaxation of high-density polyethylene at room temperature. The arrow indicates the transition from the exponential to the power law region of flow. The initial stress σ_0 was 20.3 MPa.[80]

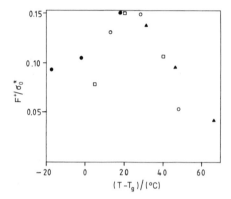

Fig. 4.5.3. The ratio $F^{\dagger}/\sigma_0$ versus $T-T_g$ for poly(vinyl chloride) containing (●) 15% dioctyl phthalate (DOP), (□) 20% DOP, (o) 25% DOP, and (▲) 30% DOP.[110]

The primary transition region of an amorphous polymer is especially interesting in this context. In this region the relaxation modulus $E(t)$ decreases very sharply and its time dependence can usually be described by a power law. The maximum slope of the log $E(t)-$log t curves, denoted m, can be correlated with the dimensionality of the corresponding damped Debye lattice. From Eq. (4.5.12) it follows that

$$n = 1 + \frac{1}{m} \tag{4.5.15}$$

In the T_g-region, the power law exponent is low, $n \leq 3$, but at higher and lower temperatures it will increase again.[109] From Eq. (4.5.14) it then follows that the $F^{\dagger}/\sigma_0$-ratio should exhibit a maximum when plotted against temperature in the T_g-region. This has been verified experimentally for plasticized poly(vinyl chloride)[110] and poly(vinyl acetate).[111] An example is given in Fig. 4.5.3.[110]

4.5.2 Internal Stress

The concept of an "internal stress" has been mentioned several times before and is referred to here as the stress level approached after very long times. In general, the internal stress will have a significant influence on the shape of the relaxation curves. In the metals field there have been several works dealing with the importance of internal stresses when studying solid state flow.[112, 113]

The internal stress, σ_i, can be determined by stress relaxation. One common method is that proposed by Li.[112] It consists of plotting $(-d\sigma/d\log t)$ versus σ for a single relaxation curve and extrapolating the curve obtained to zero stress rate. The intercept with the stress axis gives a measure of σ_i. Figure 4.5.4 is an example of such a Li-plot constructed from a relaxation curve of high-density polyethylene.[108] The internal stress value is indicated in the figure. Actually Li's method for determining σ_i is based on the power law, Eq. (4.5.11). Using the site model for stress relaxation Haworth and White[101] suggested that σ_i could be determined by plotting $[(-1/t)(d\sigma/d\ln t)]$ against σ and extrapolating down to the stress axis. The intercept is then the internal stress value. When applied to the stress-relaxation behavior of an acrylic polymer, reasonable agreement between the two methods for determining σ_i was noted.[101]

The type of internal stresses we are discussing here are "deformation induced", i.e., they are related to the initial deformation or initial stress (if the material behaves approximately linearly). In Fig. 4.5.5 the internal stress is shown versus the initial stress σ_0 for well-annealed samples of high-density polyethylene. Evidently a linear relation exists between σ_i and σ_0, but the proportionality constant is dependent on the temperature. A similar linearity, found, e.g., in acrylic polymers,[101] is also predicted by the two-site model, Eq. (4.5.8). Actually this is also in accordance with the spectral theory, cf. Eq. (4.2.3). For plasticized poly(vinyl chloride) a linear relation between σ_i and σ_0 was also observed. In this case it was suggested that the internal stress was associated with the modulus in the rubbery region.[110]

In addition to the deformation-induced internal stresses, residual stresses, i.e., stresses prevailing in the sample in the absence of external loads, can also influence the stress relaxation behavior of polymers.[101, 106] Such stresses can, for instance, be the result of the rapid cooling of the melt in the mold cavity of an injection molding machine (thermal stresses[106]). Normally, such stresses will vary over the cross section of the part, and with the geometry and cooling conditions. In principle a measure of such stresses could be obtained by plotting the maximum stress rate $F^{\dagger}$ versus σ_0 and extrapolating the resulting straight line to zero stress rate.[106, 108] The intercept with the σ_0-axis would then give a measure of the residual stresses. However, due

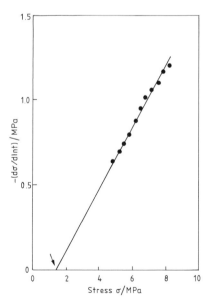

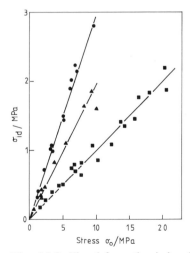

Fig. 4.5.4. Determination of the internal stress (indicated by the arrow) for high-density polyethylene at 24° C according to the method by Li.[108]

Fig. 4.5.5. The deformation-induced internal stress, σ_{id}, versus the initial stress σ_0 for high-density polyethylene at (■) 24° C, (▲) 50° C, and (●) 69° C.[108]

to the complicated structure of injection molded parts and because of the variation
of the residual stresses over the cross section, the value obtained would probably be
of little significance.[114] This is perhaps to be expected since it is not possible to characterize an unknown stress distribution with a single parameter. The extensive work
on stress relaxation of injection molded polymers by White e. a.[101, 115, 116] also indicates
that this stress-relaxation method for determining residual stresses might be less useful
than other methods and should be used with care.[114]

4.6   Cooperative Theories

As evident from the preceding sections, the flow theories described so far have
not provided any explanation of the empirical similarity in relaxation behavior expressed
through Eq. (4.1.2). A common feature of all approaches so far is that movements
of the elementary flow units are assumed to take place independently of each other.
Holzmüller[117, 118] pointed out that thermally activated processes must be cooperative.
Still a model based on cooperation is not easily formulated since the physical characteristics of the flow units taking part in the process are not known in detail and thus
cannot be uniquely related to the parameters of the theoretical formulae. It is thus
not surprising that a cooperative model is likely to be based on simplified assumptions.

Many cooperative models suggested for the description of the time dependence
of mechanical properties of solids aim at reproducing the linear relation between, e. g.,
the stress and the logarithm of elapsed time. It is then only natural that some of
the approaches are based on the concepts of stress-dependent thermal activation
(SDTA). This was, for example, the case for Borodin and Zelenev[119] who also used
nonequilibrium thermodynamics. When introducing a parameter describing the degree
of cooperation between flow units they arrived at a model which essentially reproduced
the linear $\sigma(\log t)$-process. Another flow model, also related to the SDTA concept,
is proposed for recovery after plastic deformation by Kuhlmann e. a.[120, 121] They assumed that the activation energy decreased linearly with the number of defects in
the material at a given time. Denoting the number of defects by n^*, they write

$$\frac{dn^*}{dt} \sim \exp[-(U - aVn^*)/kT] \tag{4.6.1}$$

where a is a factor converting n^* into stress and V is an activation volume.

A similar approach was also used by Overhauser[122, 123] for describing recovery
in irradiated copper. The cooperative models by several Soviet authors[124–127] are
also based on activation energy concepts; the activation energy is assumed to be linearly
related to the occupancy of the states of the flow units.

It is apparent that these approaches are similar to the SDTA theory and that they
will describe the linear relation with regard to log (time). The basic difference is that
the cooperative character of the flow is clearly evident; the probability of a "successful"
elementary event is dependent on the number of units which have not yet contributed
to the macroscopic flow. The flow rate is thus related to the degree of advancement
of the process. This is also apparent in the cooperative model for volume relaxation
proposed by Adam.[128, 129]

A somewhat different approach was taken by Bohlin and Kubát,[130] who used an
Ising-model for describing stress relaxation in polymers. They assumed that the flow
units could occupy two states, relaxed and unrelaxed. When a flow unit relaxed, the
probability that a neighboring element would relax decreased, which resulted in a broad
relaxation spectrum. Only nearest neighbor interactions were accounted for. The rate

equation could be expressed as

$$B'\frac{d\sigma}{dt}=(1-\sigma)\exp(-a-b\sigma)-\sigma\exp(a+b\sigma) \tag{4.6.2}$$

where B' is a constant and a and b are linear functions of the interaction energy. Equation (4.6.2) reproduced the linear $\sigma(\log t)$-curve but the constant 0.1 in Eq. (4.1.2) could not be explained by this model.

In the theories treated above, it is assumed that transitions of flow units between two states take place as single elementary events independent of each other. Introducing the possibility of the occurrence of multiple elementary processes creates a new class of cooperative models. We may assume that the flow units, as earlier, can occupy one of two energy levels. When the material is strained, the flow units are raised to the upper energy level and during the stress relaxation process they fall to the lower. In doing so they emit energy in the form of phonons which may induce transitions of other unrelaxed flow units. Now the phonons will form clusters (Bose-Einstein statistics) and the transition probability of the unrelaxed flow units will be related to the size of these phonon clusters. Using the theory of partitions the average number of clusters of size s in an unrestricted partition of n flow units undergoing transition can be evaluated as

$$E_n(s)=\frac{1}{e^{s/b'}-1} \tag{4.6.3}$$

where b' denotes the distribution parameter $\sqrt{\dfrac{6n}{\pi^2}}$.

The clustering of the phonons will produce a spectrum of relaxation times of the type

$$t^*,\ t^*/2,\ t^*/3,\ \dots \tag{4.6.4}$$

where t^* is the longest relaxation time. Obviously, these relaxation times are associated with clusters having the sizes 1, 2, 3, … .

Omitting the details of the derivation, the $\sigma(t)$-relation corresponding to this cooperative model[131] is approximately

$$\frac{\mu}{t^*}\sigma^*(t)=\beta'\left\{\psi\left(\beta'+\frac{\beta't}{t^*}\right)-\psi\left(1+\frac{\beta't}{t^*}\right)\right\} \tag{4.6.5}$$

where μ is a constant and β' a parameter determining the shape of the relaxation curve. $\psi(x)$ is the psi or digamma function. The maximum slope $F^\dagger$ of Eq. (4.1.2) is related to β' as[131]

$$\frac{F^\dagger}{\sigma_0^*}=\frac{1}{\ln\beta'+\gamma} \tag{4.6.6}$$

where γ is Euler's constant (0.5778 …). Equation (4.6.5) gives the linear $\sigma(\log t)$-behavior as $\psi(t)\approx\ln t$. In Fig. 4.6.1 Eq. (4.6.5) is shown to reproduce the experimental relaxation behavior of polyethylene and cellulose with good accuracy, at least in the linear region.[132]

The behavior of this model is further analyzed in Ref.[133]. It is shown there that the maximum cluster size s_m is approximately equal to the parameter β' of Eq. (4.6.5). Using this result and the spectral character of this cooperative model one arrives at

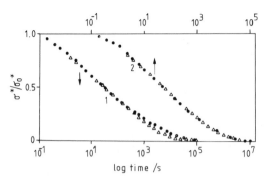

Fig. 4.6.1. Comparison between ($\triangle$) Eq. (4.6.5) and ($\bullet$) experimental stress relaxation behavior of 1 high-density polyethylene and 2 cellulose at room temperature.[132]

the following result

$$\ln \beta' \, e^{\gamma} = \frac{\pi^4}{9} - 1 \approx 9.82 \dots \tag{4.6.7}$$

Thus, the empirical relation Eq. (4.1.2) was found to have a counterpart in the cooperative model based on the occurrence of multiple elementary processes.

4.7   Final Remarks

It is obvious that the problem of describing the stress-relaxation process has many aspects, but we believe that at least some of the essential ones have been covered here. However, several important contributions, both experimental and theoretical, have not been discussed. These include the multiple integral theory[2, 134] and the BKZ-theory.[135] Both represent a mathematically somewhat more refined approach to stress relaxation. The selection of basic facts presented above intends to show, among other things, that solid state flow is not a well-understood process. It is true that a number of models exist which give a formally correct description of the relaxation process. This, however, is not a guarantee that the underlying physical assumptions are correct. This is clearly shown by the continued use of physically incompatible models when describing the same process. The reliance on formal agreement between theory and experiment, especially when the theoretical assumptions appear plausible, has resulted in considerable confusion. For instance, the well-established linearity of the relaxation process with regard to initial stress − once the basis of the theory of linear viscoelasticity − is consistently neglected when the theory of stress-aided thermal activation, which is basically nonlinear, is used indiscriminately. This situation shows that there is need for a new approach that has the capacity of accommodating the known experimental facts.

In this respect, the cooperative model based on the occurrence of multiple elementary transitions appears promising because it gives a correct value of the numerical constant in Eq. (4.1.2). As explained above, the value of this constant is independent of the nature of the relaxing solid. It is reasonable to speculate that such an insensitivity toward the details of chemical structure is the result of a cooperative phenomenon.

Acknowledgment

Financial support from the Swedish Board for Technical Development is gratefully acknowledged. Thanks are also due to Dr. C. Högfors for valuable suggestions and advice.

References

1. F.T. Trouton and A.O. Rankine, *Phil. Mag.* 1904, **8**, 538.
2. I.M. Ward, *Mechanical Properties of Solid Polymers,* Wiley, London 1971.
3. L.E. Nielsen, *Mechanical Properties of Polymers and Composites,* Vol. 1 and 2, Marcel Dekker, New York 1974.
4. J.D. Ferry, *Viscoelastic Properties of Polymers,* 3rd ed., Wiley, New York 1980.
5. A.V. Tobolsky, *Properties and Structure of Polymers,* Wiley, New York 1960.
6. T.T. Wang, *J. Appl. Phys.* 1982, **53**, 1828.
7. M.M. Qayyum and J.R. White, *Polymer* 1982, **23**, 129.
8. R.P. Wool and W.O. Statton, *J. Polymer Sci. Phys.* 1974, **12**, 1575.
9. J. Kubát, Nature 1965, **204**, 378.
10. A.V. Tobolsky, B.A. Dunell and R.D. Andrews, *Textile Res. J.* 1951, **21**, 404.
11. G.H. Hsiue and G.W. Wu, *J. Appl. Polymer Sci.* 1980, **25**, 2119.
12. M. Ito, *Polymer* 1982, **23**, 1515.
13. R. de Batist, *Rev. Deform. Behav. Mater.* 1975, **1**, 71.
14. R.B. Beevers, *Colloid Polymer Sci.* 1974, **252**, 367.
15. P. Feltham, *Brit. J. Appl. Phys.* 1955, **6**, 26.
16. E. Wiechert, *Ann. Phys. Chem.* 1893, **50**, 335.
17. J. Kubát, *D. Sc. Thesis,* Stockholm University, Stockholm 1965.
18. R.S. Lakes, *J. Rheol.* 1981, **25**, 663.
19. B. Gross, *Mathematical Structure of the Theories of Viscoelasticity,* Herrman, Paris 1953.
20. I.V. Yannas, *J. Polymer Sci. Macromol. Rev.* 1974, **9**, 163.
21. W. Retting, *J. Polymer Sci. Symp.* 1973, **42**, 605.
22. H. Leaderman, *Elastic and Creep Properties of Filamentous Materials and Other High Polymers,* Textile Foundation, Washington D.C. 1943.
23. K. Ninomiya and H. Fujita, *J. Colloid Sci.* 1957, **12**, 204.
24. M.L. Williams, R.F. Landel and J.D. Ferry, *J. Amer. Chem. Soc.* 1955, **77**, 3701.
25. A.K. Doolittle, *J. Appl. Phys.* 1951, **22**, 1471.
26. L.A. Utracki and R. Simha, *J. Rheol.* 1981, **25**, 329.
27. L.A. Utracki, *Polymer Eng. Sci.* 1982, **22**, 81.
28. J. Bischoff, E. Catsiff and A.V. Tobolsky, *J. Amer. Chem. Soc.* 1952, **74**, 3378.
29. J.R. Loughin and A.V. Tobolsky, *J. Colloid Sci.* 1952, **7**, 555.
30. E. Catsiff and A.V. Tobolsky, *J. Appl. Phys.* 1954, **25**, 1092.
31. Idem, *J. Colloid Sci.* 1955, **10**, 375.
32. A.V. Tobolsky and E. Catsiff, *J. Polymer Sci.* 1956, **19**, 111.
33. A.V. Tobolsky, *Rubber Chem. Technol.* 1957, **30**, 427.
34. J.J. Aklonis and A.V. Tobolsky, *J. Appl. Phys.* 1965, **36**, 3483.
35. N. Hata and A.V. Tobolsky, *J. Appl. Polymer Sci.* 1968, **12**, 2597.
36. K. Fujino, K. Seusku and H. Kawai, *J. Colloid Sci.* 1961, **16**, 262.
37. K. Fujino, T. Horino, K. Miyamoto and H. Kawai, *ibid.* 1961, **16**, 411.
38. K. Fujino, K. Seusku, T. Horino and H. Kawai, *ibid.* 1962, **17**, 726.
39. K. Murakami, K. Ono, K. Shiina, T. Ueno and M. Matsuo, *Polymer J.* 1971, **2**, 698.
40. A.V. Tobolsky and R.D. Andrews, *J. Phys. Chem.* 1955, **59**, 989.
41. E. Catsiff, J. Offenbach and A.V. Tobolsky, *J. Colloid Sci.* 1956, **11**, 48.
42. Tie Hwee Ng and H.L. Williams, *Makromol. Chem.* 1981, **182**, 3331.
43. K. Nagamatsu, T. Yoshitomi and T. Takemoto, *J. Colloid Sci.* 1958, **13**, 257.
44. B.A. Dunell, A.A. Jones and R.T.B. Rye, *J. Colloid Sci.* 1960, **15**, 193.
45. J.A. Faucher, *Trans. Soc. Rheol.* 1959, **3**, 81.
46. A. Tanaka, N. Sugimoto, T. Asada and S. Onogi, *Polymer J.* 1975, **7**, 529.
47. S. Fukui and T. Hideshima, *Jap. J. Appl. Phys.* 1977, **16**, 159.
48. Y. Yamada, M. Kakizaki and T. Hideshinza, *ibid.* 1982, **21**, 352.
49. T. Yoshitomi, K. Nagamatsu and K. Kosiyama, *J. Polymer Sci.* 1958, **27**, 335.

50. F.P. La Mantia, G. Titomanlio and D. Acierno, *Rheol. Acta* 1980, **19**, 88.
51. J.P. Mercier and G. Groeninckx, *ibid.* 1969, **8**, 510.
52. D. Katz and I. Steg, *Polymer* 1972, **13**, 541.
53. R. Oono, *J. Polymer Sci. Phys.* 1974, **12**, 1383.
54. J. Diamant, D.R. Hansen and M. Shen, *Polymer Sci. Technol.* 1977, **10**, 429.
55. T. Soen, T. Horino, Y. Ogawa, K. Kyuana and H. Kawai, *J. Appl. Polymer Sci.* 1966, **10**, 1499.
56. R.E. Cohen and N.W. Tschoegl, *Intern. J. Polym. Mater.* 1973, **2**, 49; 1974, **3**, 3.
57. T. Horino, Y. Ogawa, T. Soen and H. Kawai, *J. Appl. Polymer Sci.* 1965, **9**, 2261.
58. D. Kaplan and N.W. Tschoegl, *Polymer Eng. Sci.* 1974, **14**, 43.
59. G.W. Nelb, S. Pedersson, C.R. Taylor and J.D. Ferry, *J. Polymer Sci. Phys.* 1980, **18**, 645.
60. G.W. Kamykowski, J.D. Fery and L.J. Fetters, *ibid.* 1982, **20**, 2125.
61. A. Eisenberg and M. Navratil, *Macromolecules* 1973, **6**, 604.
62. A. Eisenberg and M. King, *Ion-Containing Polymers,* Academic Press, New York 1977.
63. M. Rigdahl and A. Eisenberg, *J. Polymer. Sci. Phys.* 1981, **19**, 1641.
64. P.E. Rouse, *J. Chem. Phys.* 1953, **21**, 1272.
65. F. Bueche, *ibid.* 1954, **22**, 603.
66. A.V. Tobolsky and J.J. Aklonis, *ibid.* 1964, **68**, 1970.
67. L.L. Chapoy and A.V. Tobolsky, *Chem. Scripta* 1972, **2**, 44.
68. A.V. Tobolsky, *J. Chem. Phys.* 1962, **37**, 1575.
69. A.V. Tobolsky and D.B. DuPré, *Adv. Polymer. Sci.* 1969, **6**, 103.
70. J.J. Aklonis and V.B. Rele, *J. Polymer. Sci. Symp.* 1974, **46**, 127.
71. L.L. Chapoy and S. Pedersen, *J. Macromol. Sci. Phys.* 1975, **B 11**, 239.
72. L.-Å. Nilsson, *Intern. J. Polym. Mater.* 1982, **9**, 217.
73. K.S.C. Lin and J.J. Aklonis, *J. Appl. Phys.* 1980, **51**, 5125.
74. E. Passaglia and H.P. Koppehele, *J. Polymer Sci.* 1958, **33**, 281.
75. R. Meredith and B. Hsu, *ibid.* 1962, **61**, 253.
76. D. Krieger, *Kolloid Z. Z. Polymere* 1972, **250**, 1131.
77. G. Titomanlio and G. Rizzo, *J. Appl. Polymer Sci.* 1977, **21**, 2933.
78. Idem, *Polymer* 1980, **21**, 461.
79. Idem, *Polymer Bull.* 1981, **4**, 351.
80. J. Kubát, R. Seldén and M. Rigdahl, *J. Appl. Polymer Sci.* 1978, **22**, 1715.
81. S. Matsuoka, H.E. Bair, S.S. Bearder, H.E. Kern and J.T. Ryan, *Polymer Eng. Sci.* 1978, **18**, 1073.
82. L.C.E. Struik, *Physical Aging in Amorphous Polymers and Other Materials.* Elsevier, Amsterdam 1978.
83. A.J. Kovacs, *J. Polymer Sci.* 1958, **30**, 131.
84. M. Cizmecioglu, R.F. Fedors, S.D. Hong and J. Moacanin, *Polymer Eng. Sci.* 1981, **21**, 940.
85. S.S. Sternstein, in *Treatise on Materials Science and Technology,* edited by J.M. Schultz, Vol. 10, Academic Press, New York 1977.
86. M.R. Tant and G.L. Wilkes, *Polymer Eng. Sci.* 1981, **21**, 325.
87. E. Siu-Wai Kong, G.L. Wilkes, J.E. McGroth, A.K. Banthia, Y. Mohajer and M.R. Tant, *ibid.* 1981, **21**, 943.
88. M.M. Qayyum and J.R. White, *J. Polymer Sci. Letters* 1983, **21**, 31.
89. R. Becker, *Phys. Zeitschr.* 1925, **26**, 919.
90. E. Pink, *Rev. Deform. Behav. Mater.* 1977, **2**, 37.
91. E.C. Aifantis and W.W. Gerberich, *Mater. Sci. Eng.* 1975, **21**, 107.
92. M. Grosbros, E. Dedieu and M. Cahorcau, *Phys. Status Solidi A* 1977, **42**, 449.
93. G.M. Bartenev, *Plaste Kautschuk* 1974, **21**, 481.
94. D.M. Shinozaki, G.W. Groves and R.G.C. Arridge, *Mater. Sci. Eng.* 1977, **28**, 119.
95. J. Kubát and M. Rigdahl, *ibid.* 1976, **24**, 223.
96. E. Pink, V. Bouda and H. Beck, *ibid.* 1979, **38**, 89.
97. A.S. Krausz and H. Eyring, *Deformation Kinetics,* Wiley-Interscience, New York 1975.

98. J.R. White, *Rheol. Acta* 1981, **20**, 23.
99. Idem, *Mater. Sci. Eng.* 1980, **45**, 35.
100. Idem, *J. Mater. Sci.* 1981, **16**, 3249.
101. B. Haworth and J.R. White, *ibid.* 1981, **16**, 3263.
102. J.C.M. Li, C.A. Pampillo and L.A. Davis, in *Deformation and Fracture of High Polymers,* edited by H.H. Kausch, J.A. Hassel and R.T. Jaffee, p. 239, Plenum Press, New York-London 1973.
103. N. Balasubramanian and J.C.M. Li, *J. Mater Sci.* 1970, **5**, 434.
104. Idem, *ibid.* 1970, **5**, 839.
105. W.G. Johnston and J.J. Gilman, *J. Appl. Phys.* 1959, **30**, 129.
106. J. Kubát and M. Rigdahl, *Intern. J. Polym. Mater.* 1975, **3**, 287.
107. Idem, *Phys. Status Solidi A* 1976, **35**, 173.
108. J. Kubát, M. Rigdahl and R. Seldén, *J. Appl. Polymer Sci.* 1976, **20**, 2799.
109. B. Hagström, J. Kubát and M. Rigdahl, *Intern. J. Polym. Mater.* 1981, **9**, 37.
110. J. Kubát, L.-Å. Nilsson and M. Rigdahl, *Mater. Sci. Eng.* 1982, **53**, 199.
111. J. Kubát and L.-Å. Nilsson, *ibid.* 1982, **52**, 223.
112. J.C.M. Li, *Can. J. Phys.* 1967, **45**, 493.
113. R.W. Rohde and T.V. Nordstrom, *Mater. Sci. Eng.* 1973, **12**, 179.
114. J. Kubát and M. Rigdahl, *ibid.* 1975, **21**, 63.
115. L.D. Coxon and J.R. White, *J. Mater. Sci.* 1979, **14**, 1114.
116. G.J. Sandilands and J.R. White, *Polymer* 1980, **21**, 338.
117. W. Holzmüller, *Z. Phys. Chem.* 1954, **202**, 440.
118. Idem, *Rheol. Acta* 1961, **1**, 495.
119. I.P. Borodin and Yu. V. Zelenev, *Vysokomol. Soed. A* 1968, **10**, 2256.
120. D. Kuhlmann, *Z. Phys. Chem.* 1948, **124**, 468.
121. D. Kuhlmann, G. Masing and J. Raffelsieper, *Z. Metallkd.* 1949, **40**, 241.
122. A.W. Overhauser, *Phys. Rev.* 1953, **90**, 393.
123. Idem, *ibid.* 1954, **94**, 1551.
124. O.B. Ptitsyn, *Dokl. Akad. Nauk SSSR* 1955, **103**, 1045.
125. Yu. A. Sharanov and M.V. Volkenstein, *Vysokomol. Soedin.* 1962, **4**, 917.
126. Yu.Ya. Gotlib, *Sov. Phys. — Solid State* 1962, **3**, 1574.
127. Yu.Ya. Gotlib and O.B. Ptitsyn, *ibid.* 1962, **3**, 2456.
128. G. Adam, *Kolloid Z. Z. Hochpolymere* 1961, **180**, 11.
129. Idem, *ibid.* 1964, **195**, 1.
130. L. Bohlin and J. Kubát, *Sol. State Commun.* 1976, **20**, 211.
131. Ch. Högfors, J. Kubát and M. Rigdahl, *Phys. Status Solidi B* 1981, **107**, 147.
132. J. Kubát, L.-Å. Nilsson and W. Rychwalski, *Res Mechanica* 1982, **5**, 309.
133. J. Kubát, *Phys. Status Solidi B* 1982, **111**, 599.
134. A.E. Green and R.S. Rivlin, *Arch. Ration. Mech.* 1957, **1**, 1.
135. B. Bernstein, E.A. Kearsley and L.J. Zapas, *Trans. Soc. Rheol.* 1963, **7**, 391.
136. J. Kubát and M. Rigdahl, *Rev. Deform. Behav. Mater.* 1983, **4**, 335.

5 Intersegmental Interactions and Chain Scission

Hans-Henning Kausch

Laboratoire de Polymères, Ecole Polytechnique Fédérale de Lausanne
32, chemin de Bellerive, 1007 Lausanne, Switzerland

5.1 Introduction

The principal aim of this book is to elucidate the ultimate behavior of plastic materials under stress. Failure of a solid invariably is preceded by a (local) mechanical and/or thermal disintegration. In order to understand and describe the effective mechanisms, different approaches are possible which may conveniently be identified as statistical interpretation, continuum mechanical analysis, and microstructural molecular approach. Most fracture theories combine elements of all three aspects.[1]

In this Chapter, stress transfer between molecular segments is studied. The important parameters in molecular stress transfer are easily identified. Strong intermolecular attraction potentials and a high degree of primary bond loading lead to efficient stress transfer; a large segment mobility is detrimental to it. Two aspects of intersegmental interactions have already been discussed in previous chapters: stress relaxation mechanisms (Chapter 4) and (nonlinear) viscoelastic stress-strain relationships at modest strains (Chapter 3). In this Chapter the role of chain stretching and scission in fracture of highly oriented polymers and of entanglements on stress transfer and strength of (isotropic thermoplastic) polymers are investigated.

5.2 The Role of Chain Stretching and Scission in Deformation and Fracture of Highly Oriented Polymers

Some Historical Remarks

Uncross-linked polymers cohere principally through the van der Waals interaction between their chain segments. These secondary bonds are the first to be affected in loading and deformation of an isotropic thermoplastic material. If, therefore, one wishes primarily to study thermoplastic material, or make use of the strength of the covalent or primary bonds, one has to work either with highly cross-linked resins or with highly oriented thermoplastic networks. High degrees of chain orientation are found in commercial fibers, even higher ones in ultradrawn polymers. The structure of fibers has been studied from the very beginning of polymer science, but it was only in the middle sixties that the role of chain stretching and scission in deformation and failure of fibers was clearly recognized and studied. For about 10 years there was intensive activity in this area, especially in Leningrad,[2] Salt Lake City,[3] Darmstadt,[4] London[5] and Sapporo.[6]

The results of electron spin resonance (ESR) and infrared (IR) investigations of these and other laboratories have been critically reviewed[6-9]; they also formed the bais of the author's monograph[1] on polymer fracture.

5.2.1 Phenomenology of Free Radical Formation

A partially oriented semicrystalline fiber contains mechanically quite different structural elements (Fig. 5.2.1). The uniaxial loading of such a fibrillar structure results in shearing and axial stretching of microfibrils. Within microfibril amorphous regions, crystallites, and the so-called tie segments are strained to different degrees. In several materials, especially in polyamide 6 (PA) fibers and polypropylene (PP) films, rupture of tie segments is observed by ESR and IR spectroscopy.[1, 2] There is agreement on the phenomenology of chain scission, fiber deformation and stress relaxation. Thus, it has been recognized that local molecular stresses are unevenly distributed. They evidently depend on the conformation of a strained chain segment. With respect to their role in stress transfer during the loading of a fiber, four different segmental situations may be identified (Fig. 5.2.2):

a) crystalline segments (assumed to share homogeneously the stress σ_m, acting within a microfibril in axial direction);
b) fully extended tie molecules (loaded elastically from the beginning of a load cycle according to chain modulus E_k and axial strain ε_a of the amorphous regions);
c) partially extended tie molecules which will only be stretched elastically at the end of a load cycle;

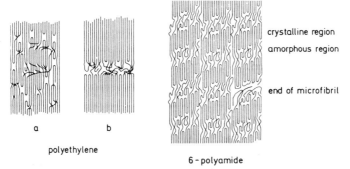

crystalline region

amorphous region

end of microfibril

a   b

polyethylene

6 - polyamide

Fig. 5.2.1. Models of semicrystalline fibers (note the intermicrofibrillar regions visible in the model of 6-polyamide).

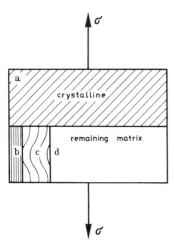

Fig. 5.2.2. Schematic arrangement of chain sections of different conformation in highly oriented fibers (of ~ 50% crystallinity). (a) crystalline regions, (b, c, and d) fully, partially, and nonextended segments in an amorphous matrix (see text for further explanation).

d) insufficiently extended molecules which will not assume a straight conformation during the complete fiber loading history.

The local axial stress ψ_i acting on a chain segment depends on its elastic elongation. If, due to bond rotation, the conformation and length of a stressed segment change, then the stress ψ_i changes (generally diminishes) as well.[1] The important insight to be gained from this consideration is that within one and the same amorphous region of a microfibril some segments may relax while others, and even adjacent ones, will not. The local stresses, therefore, are not simply a constant multiple of the average fibril stress σ_m. This is well-confirmed by the experimental data of Becht, reproduced in Fig. 5.2.3. The chain segments breaking toward the end of a strain interval obviously have not participated in the macroscopic stress relaxation (a decrease of the axial chain stress to 90% of the chain strength ψ_b increases the chain lifetime by two orders of magnitude and virtually stops the scission process). It is concluded that microfibrils do not unload through slippage and that the lateral rigidity of the crystal blocks within the sandwich-structured microfibrils must be large enough to permit the buildup of large stress concentrations. If the lateral rigidity of the crystal blocks were insufficient, the fibril would break up into submicrofibrils. If the crystal blocks break up while being strained, the tensile stresses in every part of the structure will equilibrate and be equal to the external stress. Since the external stresses σ are always much smaller than the chain strength ψ_b chain scission should be impossible. If, however, the layered structure is also preserved at large strains, stresses can be exerted which are proportional to local amorphous strain ε_a and tensile modulus E_i of the amorphous subregions of thickness L_a.

Under these conditions molecular stress concentration occurs in those highly oriented and fairly extended tie segments which interconnect two crystal blocks of the same or different microfibrils. A segment of contour length L_i and end-to-end distance, L_0 will experience ultimate stresses ψ_i only after it has assumed the most extended conformation accessible to it. The elastic stress then is

$$\psi_i = E_k \frac{L_a}{L_i}(1+\varepsilon_a) - 1 + \frac{L_0 - L_a}{L_i}(1+\varepsilon_c) \tag{5.2.1}$$

where ε_c is the crystalline strain. Chains break if $\psi_i > \psi_b$. A detailed discussion of the stressing of chains in a strained semicrystalline polymer is given in the author's monograph *Polymer Fracture*.[1]

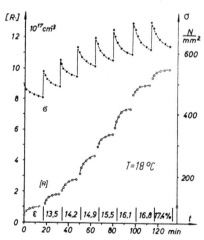

Fig. 5.2.3. Classical experiments by Becht and Johnsen on free radical production in step-straining of polyamide 6 fibers (from 1).

5.2.2 Location of Tie Molecules

A word seems to be in order with respect to the possible location of the breaking tie molecules. It was shown by Becht[4] that strain-induced free radicals were formed within the amorphous regions of the fiber. There are two types of amorphous regions: the *intra*microfibrillar ones, shown schematically in Fig. 5.2.2, and the *inter*microfibrillar ones, visible in Fig. 5.2.1. Becht and Kausch[4] deduced from the size of the long spacing (10 nm), lamellar thickness (5 nm), and degree of crystallinity (50%) that the *intra*microfibrillar regions themselves would already account for the measured degree of crystallinity. The contribution of the *inter*microfibrillar regions to the total amorphous material should be of the order of the experimental error, that is a few percent. It was observed that at the moment of rupture of a PA 6 fiber 0.3% of all amorphous segments were broken. Even if these broken chains were all *intra*fibrillar ties they would still constitute *only a fraction* of the total number of those ties. In that case, it is easily explained that the subsections of a once broken fiber show a similar (or even larger) strength if they are loaded up to fracture for a second time. The maximum number of extended ties can be approximated from the elastic modulus. Kausch[13] obtained for the above PA 6 fibers a value of $5.5 \cdot 10^{12}$ chains/cm^2 which represents 1.1% of the amorphous segments.

However, if the breaking chains were all *inter*fibrillar ties, they would account for practically the entire population of such segments within the loaded fiber. This means that in this case a once broken fiber would have been decisively weakened throughout, which is contrary to experience.[1] Convincing evidence that predominantly intrafibrillar ties break, comes from a consideration of the loading mechanism of the microfibrils. The axial stress σ_m acting in the central part of a microfibril is the resultant of the shearing forces acting on the microfibril end sections along the so-called *stress transfer length* L_σ:

$$L_\sigma = \sigma_m \cdot q/\tau \cdot u \tag{5.2.2}$$

where q is the fibril cross section, τ the shearing stress, and u the fibril circumference. Within the stress transfer length the axial stress increases from 0 to σ_m. It follows from such a distribution of stresses that the fraction $2L_\sigma/L$ of all intermicrofibrillar ties sitting in the sheared zones is subjected to *shear strains*, the remaining portion to *tensile strains*. The behavior of the latter portion is, to a first approximation, indistinguishable from that of intramicrofibrillar ties.

In addition, it seems to be highly unlikely that within the total fiber volume all microfibrils are shear loaded and displaced with respect to each other to such an extent that ties will be broken but *without* provoking premature rupture. However, the morphology of fiber fracture surfaces seems to suggest that the final fiber fracture is caused by the shear failure of intermicrofibrillar regions. A quantitative estimate of the critical fiber loads is derived from Eq. (5.2.2) assuming that shear failure is initiated if L_σ becomes larger than $L/2$

$$(\sigma_m)_{\text{failure}} \sim L\tau u/2q \tag{5.2.3}$$

The *average* tensile stress within a bundle of microfibrils is given by

$$\sigma_{av} = (L - L_\sigma)\,\sigma_m/L \tag{5.2.4}$$

Since σ_{av} should correspond to the macroscopic stress σ, one has at the moment of failure initiation

$$\sigma = \sigma_m/2 = L\tau u/4q \tag{5.2.5}$$

In the case of PA 6 fibers with a microfibril length of 200 nm, a diameter of about 30 nm and a strength of 700 MPa, one obtains failure initiation in those intermicrofibrillar regions where τ is smaller than 210 MPa. This seems to be an entirely reasonable value.

5.2.3   Infrared Observations

Apart from its conventional applications, infrared spectroscopy has been used[2, 12] to determine the axial stresses acting on a chain (through the ensuing frequency shift of relevant absorption bands) and the number of broken chains through the concentration of newly formed (oxygen-containing) end-groups. For a highly oriented isotactic polypropylene film at 30°C Wool[12] has identified the following molecular differences between stress relaxation and creep:

	Stress Relaxation	*Creep*
Number of highly stressed chains	Decreases	Increases with time
Number of intermediately stressed bonds	Increases	Decreases
Helix bands	Decrease	Increase
Orientation	Decreases	Increases
Number of broken chains	Not detectable	Large

In general, the application of IR spectroscopy to highly stressed polymers has led to the following conclusions[1, 2, 12, 14]:

— overstressed bonds are observed, the stresses extend up to the "natural upper limit", the chain strength (21 GN m^{-2} for PA 6)
— the most highly stressed segments (in PETP and PBTP) are fully oriented and extended (no gauche or "crumpled" conformations)
— the observed number of newly formed chain ends (apparently?) exceeds the number of free radicals (PE, PP).

5.2.4   Effect on Deformation

The undisputed fact of the rupture of chains in large numbers during mechanical action[1-12] is in itself neither proof nor even an indication that macroscopic stress relaxation, deformation, and fracture are a consequence of the breakage of chains. Kausch and Becht[4] noted that the observed total number of broken chains is much too small to account, by virtue of their load carrying capability, for the measured reductions in macroscopic stress. The stress relaxation within a strain step of 0.65% amounts to 60 to 100 MN/m^2. The load having been carried by the $0.7 \cdot 10^{17}$ chain segments broken during such strain step, however, is calculated to be 2.4 MN/m^2 if the tie segments had traversed just one amorphous region; it is n 2.4 MN/m^2 if they had spanned n such regions. In the case of $n=1$ the macroscopic stress relaxation obviously has an intensity 25–40 times larger than the decrease of the accumulated molecular stresses derived from the number of observed chain scission events.

This is, in fact, one of the questions which had not been unambiguously answered in 1978. Following the earlier work of Müller[15] several authors[16-22] studied consecutive stress-strain cycles using different techniques. They reported that in order to account for the irreversible energy dissipation, the number of free radicals should have been by a factor f_c between 7 and 40 times larger than actually observed. To explain such a discrepancy while basically maintaining the model, one is forced to make additional assumptions:

- Either the number of broken chains is systematically larger by f_c than the number of observed radicals, e.g., due to a Zakrewskii-mechanism.[1, 2] Careful molecular weight measurements of strained PA 6, PA 66, PETP, PP, and PE fibers[12, 20, 22] gave no convincing evidence that the hypothesis is true. However, in view of recent measurements reported by Frank[23], Veltegren[23], and Zakrevskii[23], this question still seems to be open to further discussion.
- Or the breakage of N_1 chains in a particular volume element V_1 leads to the unloading of $f_c N_1$ extended chains outside of V_1; this amounts to the consideration of a dominant submicrofibrillar structure.
- Provision must be made to account for the anelastic deformation of the *nonextended chains*. The stress-induced change of conformation of a tie segment profoundly reduces axial chain stresses (Fig. 5.2.4). In subjecting a fiber to several stress-strain cycles a large fraction of the more extended conformations ($n_k = 0$–3) formed during the first cycle are present at the beginning of the second cycle. Stress-induced "irreversible" conformational changes are considered to be mostly responsible for the differences between the first and the following stretching cycles.[10, 14, 24]

It is the common observation of the above authors[11, 16, 18–20] that the slopes of stress-strain and radical concentration-strain curves correspond qualitatively. The interpretation that 20 to 40 times as many chains break as free radicals are observed appears to be rather strong in view of the fact that no correspondingly strong decrease of the molecular weight has been reported and the load bearing capability of the fiber material outside of the immediate fracture zone does not necessarily suffer. The qualitative agreement between the slopes of stress-strain and free radical-strain curves can be sought in an amplification effect accompanying chain scission. The stress transfer to neighboring segments, the rapid release of elastic energy stored within the breaking chain, and the ensuing local temperature rise will facilitate the irreversible slippage and the extension of kinked chains under annihilation of kinks. This is practically equivalent to unloading those chains without breaking them (Fig. 5.2.4). Within the framework of this model, one would have to conclude that the breakage of one amorphous segment at maximum load, $q \psi_b$, leads to conformational changes in the sur-

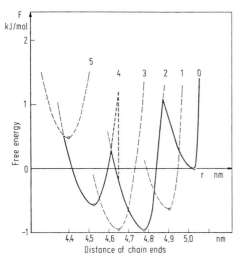

Fig. 5.2.4. Free energy of strained polyethylene kink-isomers of 5 nm extended length, parameter: number of kinks; solid line: minimum free energy of *isolated* valence-angle chain in thermal contact with surrounding, heavy broken line: in presence of external hindrances to segment rotation axial chain stresses and average chain free energy are drastically increased; entropy-elastic driving forces are not sufficient to overcome the external hindrances (from 1).

rounding segments of the same microfibrillar region resulting in local load decreases of 40 $q\,\psi_b$. In view of the fact that on the average there are more than 2000 surrounding segments (of 5 nm length in any amorphous region 400 nm^2 in cross section) the value of 40 $q\,\psi_b$ appears very reasonable.

Independently, a similar model of the microstructure of the amorphous regions of microfibrils in highly oriented semicrystalline polymers (PE, PP, PAN, PA 6) had been obtained by Marichin[24] on the basis of density, IR, NMR, and X-ray diffraction measurements.

The above hypothesis that high loads (even if they do not lead to chain scission) are correlated well with irreversible deformations[10] has been confirmed.[14, 20, 21] With highly oriented PA 6 yarn (from Enka Glanzstoff GmbH) strained to between 10 and 20% of *total* strain, Gaur[21] found about 3 to 4·10^{17} and Frank and Wendorff[20] found about 4·10^{17} spins cm^{-3} per percent of *irreversible* strain. Holland-Moritz and Siesler[14] identified an irreversible conformational transition in PBTP by rapid-scanning Fourier transform IR.

Frank and Wendorff[20] and also Stoeckel and Crist[19] indicate that the formation of free radicals in as received, highly oriented PA 6 yarn is *not* accompanied by submicrocracks (their concentration under the above conditions would be smaller than 10^{13} cm^{-3}).

In a later section some remarks will be made on the role of chain scission in the breakdown of fibrils in crazes.

5.2.5 Chain Scission and Fracture

In straining highly oriented polyamide fibers, the first free radicals are observed at a macroscopic strain of about 8 to 10%; at higher strains free radicals are produced in increasing numbers (see the radical-strain histograms in Fig. 5.2.5). Such histograms reflect the conformation distribution of the tie molecules discussed above. The primary information to be obtained from them is twofold. The number, N_f, of free radicals at fracture and the width of the distribution can be determined. The *volume* concentrations of free radicals formed in uniaxially stressed fibers range from <10^{20} m^{-3} (PMMA, PS), 0.01–2·10^{22} m^{-3} (PE, PET) to about 0.5·10^{24} m^{-3} (PA 6, natural silk). The latter value is, by a factor of 16, smaller than the theoretical limit calculated for a regular sandwich structure. The *surface* concentrations of free radicals in ground or milled polymers are of the order of 1–20·10^{16} m^{-2}, roughly two orders of magnitude below the upper limit.[1] As has been outlined above, the number of free radicals is

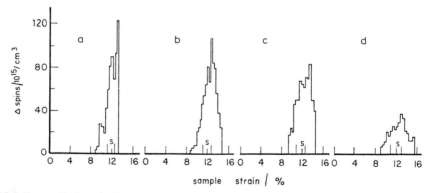

Fig. 5.2.5. Free radical-strain histograms of polyamide 6 fibers (from 25). (a) $-25°$ C; (b) $+22°$ C; (c) $+50°$ C; (d) $+100°$ C.

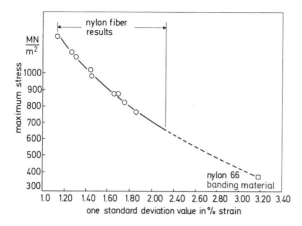

Fig. 5.2.6. Maximum stress of uniaxially loaded polyamide 6 fibers and width of free radical-strain histograms are correlated to each other because they both depend on the fibril microstructure.

taken to correspond to the number of newly formed chain ends; it is much too small to control the deformation process of the fiber and even less its fracture. However, the occurrence of chain scissions is intimately related to the fibril microstructure and it is through the microstructure that fiber strength and radical production are correlated (Fig. 5.2.6). This should be taken into consideration whenever a failure criterion is to be based on ESR data.[26] Since the formation of free radicals is as *consequence* and not an *independent cause* of fiber deformation, it does not seem advisable to base any fracture criteria on the free radical distribution curves.

The more highly oriented polyamide 6 fibers are the stronger ones. They also have a higher modulus of elasticity of their amorphous regions and a higher number of *extended* tie molecules.[27] Because of the higher fiber rigidity, chain scission begins at higher stresses.

5.3   Chain Scission and/or Disentanglement in Macroscopically Isotropic Samples

5.3.1   Chain Energy Contributions to Crack Resistance

The preceding section has made it clear that the probability of finding a highly loaded, extended chain segment in an isotropic thermoplastic polymer is rather small. Stress relaxation through segment reorientation and/or bond rotation will invariably occur at stresses close to but below chain strength, ψ_b. Chain scission in macroscopically isotropic or modestly oriented thermoplastic will only be expected, therefore, as a consequence of large *local orientation*; for examples, in the proximity of an advancing crack or within the fibrils of craze matter. Attention will be given to these two situations.

In a polymer where crack propagation is preceded by craze formation, one has the following sequence of energy consuming processes in the material ahead of a slowly advancing planar crack (mode I, Fig. 5.3.1):

— homogeneous energy elastic straining of the matrix and partial relaxation of the ensuing stresses (region A)
— initiation of nonhomogeneous deformation, i.e., plastic deformation, void formation, chain scission, or craze initiation (small band B)
— heterogeneous deformation until material disintegration through chain separation and/or scission is completed (region C)

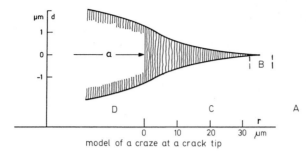

Fig. 5.3.1. Homogeneous straining (A), craze initiation (B), nonhomogeneous deformation of crazed material (C), ahead of a slowly propagating crack (D) in a glassy polymer, e.g., polymethyl methacrylate (from 1).

model of a craze at a crack tip

— retraction of strained chain segments, coils, fibrils, and matrix material adjacent to the now well defined crack (region D).

The density of stored elastic energy, $w = \sigma_y^2/2E$, in the homogeneously strained region (A) increases with decreasing distance r from the crack tip as $1/r$; the energy density is limited by the onset of nonelastic deformation. In PMMA the maximum strain energy density will be of the order of 0.1 to 0.5 MJ/m^3 which is an insignificant fraction of the cohesive energy density of 500 MJ/m^3. In this respect it is worthwhile to point out that at these stress levels, the stored elastic energy is not sufficient to achieve chain segment separation directly, prior to chain reorientation.

If a crack were to grow solely by the separation of two layers of chains to a distance beyond mutual attraction, say, 0.2 nm and solely at the expense of the stored elastic energy, it would require the cooperative transfer of the energy stored within a 200 nm layer to the boundary of that same layer. Such a process is absolutely unlikely to occur, irrespective of chain length, strength, or conformation.

The purely elastic separation of an individual chain segment becomes more likely if the thermal energy fluctuation is taken into consideration. At room temperature the thermal energy of a PMMA monomer unit reaches 25 kJ/mole; the energy to create a new surface by separation of two monomer units is $2\gamma\, l_m \sqrt{q}$, here 8 kJ/mole (surface tension γ, monomer length l_m, chain cross section q). There is, therefore, a finite probability for the thermal activation of a void which will grow by further activation to the extent repulsive forces and chain entanglements permit.

With the stress level in region A being what it is, it can be stated that the loss of stored energy through stress relaxation during crack propagation does by no means account for the observed energy dissipation in fracture surface formation. Without giving detailed proof for all environmental conditions, one may say in general that major energy dissipation begins with the onset of nonhomogeneous deformation (B). The main effect of a long chain backbone cannot be seen as raising the local stress at which plastic deformation initiates, but rather in minimizing the effect of material inhomogeneities or cracks as stress concentrators.

Thus, in nonoriented glassy polymers the onset of nonhomogeneous deformation is determined more by intermolecular attraction and chain mobility than by chain length, strength, or conformation. The latter properties come to bear, however, in regions C and D through the amount of load carried by the reorganizing "network" of chains and through the extent to which the network can be deformed before material separation occurs.

5.3.2   Disentanglement in Glassy Thermoplastics

The classic definition of entanglements as (widely) separated temporary physical junction points of chains, where effective stress stransfer is possible, was developed more than 50 years ago.[28] In the following decade, the important role of entanglements in the deformation behavior of polymers was further analyzed by Treloar[29] and Müller.[30] In a very comprehensive paper Müller[30] stressed, as early as in 1941, the importance of the stability of statistically formed entanglements (in German Verschlaufungen, Verknotungen, Haftpunkte) of molecular chains in deformed networks. The above definition includes, first, the *topological entanglements* which generally can only be resolved by force, i.e., by chain scission or at elevated temperatures after prolonged loading times. Second, it includes interaction points of different "strengths P_i" which may immobilize effectively two unloaded contacting chain segments up to a temperature $T_i > T_g$ but which are quite capable of dissociating at a slightly higher temperature (say $T_i + 10$ K). Under load, such interaction points break at much lower temperatures than the corresponding T_i and give rise to stress relaxation and flow. The mathematical description of those physical situations have been given in Chapters 3 and 4 of this volume.

Evidently, it is not always possible to give a clear-cut distinction between the two types of entanglements. The presence and stability of topological entanglements are an essential condition for the workability and macroscopic strength of uncross-linked amorphous networks. Vice versa, local disentanglement within such networks must be considered as a weakening phenomenon. In view of the competitive nature of *chain scission* and *disentanglement,* especially in the case of fibril craze disintegration, a few remarks may be made here.

The role of topological entanglements in glassy polymers evidently must be somewhat different from that in polymer melts and concentrated solutions. In particular, the following aspects have to be taken into consideration:

— the lifetime and strength of an entanglement at a temperature below T_g must be considered to be very large;
— at small deformations few molecular weight effects and consequently no entanglement effects become apparent;
— at large deformations, for examples in fibril formation during crazing, the presence of entanglements can be the limiting factor in the orientational extension; and
— the ultimate strain and energy of fracture strongly depend on the presence of topological entanglements.

There are numerous observations to support these statements.[31 – 39]

A particularly interesting phenomenon has recently been observed in Lausanne[40] — the *intrinsic crazing* (of PC, PMMA, and PS). Intrinsic crazing occurs at a high stress level and after considerable (uniaxial) deformation λ^{II} of the sample. The new crazes, called crazes II, appear throughout the total sample volume and in a large concentration. For details the reader is referred to the extensive review article of Dettenmaier.[40] Here we retain only the observation that the draw ratio λ^{II} at which crazes II were initiated, agreed rather well with the maximum extensibility of chains between entanglement points, λ^{nat} (Table 5.3.1). Thus, it must be assumed that in deforming a (PC) sample at 130° C numerous interaction points are broken, whereas the topological entanglement points are preserved. The breakdown of the entanglement network (by chain slippage or chain rupture) plays an important role for craze II formation and growth. At present, which of the two degradative mechanisms is the more important is open to discussion. There is some support[41] for the slipping hypothesis. At the same times Henkee and Kramer[60] and Donald[61] find, in agreement with the above statements, that disentanglement is important for low molecular weight samples or at higher temperatures, whereas chain scission (or complete suppression of crazing in favor of shear deformation zones) occurs for higher molecular weights or crosslinked (PS) samples.

The craze II concept has been tested by measuring craze initiation as a function

Table 5.3.1 Instability Phenomena in Amorphous Polymers: Comparison between the Extension Ratio, λ^{II}, at the Initiation of Stress-Whitening and the Maximum Extension Ratio, λ^e, of Chains between Entanglement Points (from Dettenmaier[40])

Polymer	l_0	M_e	β	λ^{nat}	λ^{II}
	nm	$\dfrac{g}{mole}$	$nm\left(\dfrac{mole}{g}\right)^{\frac{1}{2}}$	$=l_0\sqrt{M_e}/M_0\beta$	
Polycarbonate	1.1	2490	0.11	2.0	2.1
Poly(methyl methacrylate)	0.21	9150	0.076	2.7	2.7
Polystyrene	0.21	19100	0.067	4.1	*

* An isotropic sample breaks before λ^{II} is attained.
  In preoriented PS Dettenmaier[40] was able, however, to produce intrinsic crazes (crazes II).

of preorientation.[40] In fact, if PC is stretched above T_g the number of entanglements decreases with increasing extension ratio λ_1, and as a consequence λ^{nat} increases. It has been demonstrated, in agreement with the craze model developed above, that the total extension ratio λ^{II} of the samples at craze initiation also increases. A quantitative analysis of the data has been given on the basis of the entanglement model.[40]

While the mechanical and environmental aspects of crazing will be treated in Chapters 15 and 16 reference should be made at this point to the extensive work on the formation and breakdown of conventional crazes (crazes I) done in Ithaca and reviewed by Kramer.[39] As Kramer demonstrated for more than a dozen different polymers, the observed fibril draw ratios λ_{fib} correlate very well with λ_{nat}. Again, the interaction points are constantly broken during the fibrillation process, whereas the entanglements survive (initially). Also in this case an ambiguity exists as to the mechanism responsible for fibril breakdown (principally in PS). For high molecular weight samples Kramer found evidence of chain scission; for lower molecular weights and also for fatigue loading disentanglement by chain slippage seemed to prevail. This interpretation is also strongly supported by the observations of Döll[42] and Schirrer e. a.[43] namely that the time-to-breakdown of PMMA craze fibrils and the β-peak have equal activation energies (of ~ 96 kJ/mol). On the other hand, Popli and Roylance[44] detected very small molecular weight changes in a thoroughly conventionally crazed polystyrene which could indicate chain scission.

5.3.3 Crack Healing

A new way to learn something about topological entanglements opened up by the study of crack healing. The first experiments of this type were done by Kausch e. a.[45-50] and Wool e. a.[51; 52]. Based on the work of de Gennes[53] and of Doi and Edwards,[54] they proposed that crack healing of amorphous polymers slightly above T_g involves the reptational diffusion of macromolecules across the interface. Their fracture mechanics experiments have clearly shown that there is a linear proportionality between the square of the fracture toughness, K_{Ii}, of the rehealed crack zone and the square root of the rehealing time, t_p. More recent experimental work by these groups and theoretical studies by Prager e.a.[55] permit similar conclusions.

An analysis of the microscopic fracture process at the crack tip showed[47-49] that fracture proceeds through the formation and rupture of crazes. The length, s, of the crazed zones observed at the crack tip increased linearly with the chain interpenetration previously achieved during healing. The energy associated with the formation of the crazed zones corresponded very well to the fracture energy, G_{Ii}, of the rehealed samples which in PMMA amounted to 13 to 380 J m^{-2}, depending on the degree of healing. Based on the experience[56] that the simple pull out of *nonentangled* chains from a matrix requires fracture energies of 0.1 to 1 J m^{-2}, one can safely conclude that the interdiffusion of chains during rehealing must have involved the formation of entanglements.

In crack healing experiments one knows that at the moment of contacting two fracture surfaces, there are no entanglements existing *across* the interface, the interfacial fracture energy G_{Ii} is that of the adhesive bonds and must be very smnall. This has been well observed.[45-50] With progressive healing time the concentration $n(t_p, T_p)$ of entanglements per unit surface area coupling chains from opposite surface sides will increase in proportion to the average length $(\langle l^2 \rangle)^{1/2}$ of chain interpenetration

$$n(t_p, T_p) \sim (\langle l^2 \rangle)^{1/2} = (2 D t_p)^{1/2} \tag{5.3.1}$$

where D is the longitudinal chain diffusion coefficient. The nature of this diffusion is discussed in the literature.[48, 51-59] The crack healing technique used in Lausanne[46] employs compact tension specimens (see Chapter 7 for key-aspects of fracture mechanics). These specimens permitted the sending of a crack right through the rehealed interfacial zone and the measuring of the fracture toughness K_{Ii} of this zone. It was then assumed that

$$K_{Ii}^2 = \beta n(t_p, T_p) \tag{5.3.2}$$

where β is a constant. The constant β can be derived from the "completely healed" state with n_0 entanglements per unit surface area

$$\beta = K_{I0}^2 / n_0 \tag{5.3.3}$$

It follows that

$$\frac{n(t_p)}{n_0} = \frac{(2 D t_p)^{1/2}}{(\langle l_0^2 \rangle)^{1/2}} \tag{5.3.4}$$

where $(\langle l_0^2 \rangle)^{1/2}$ is the average length of interpenetration necessary to obtain full healing. Equations (5.3.1) through (5.3.4) do not permit the determination of D and $\langle l_0^2 \rangle$ independently, but they explain the experimental observation that K_{Ii} depends linearly on $t_p^{1/4}$ very well.

Two facts seem to be important with regard to the topology and number of entanglements in the interfacial region. First, it should be noted that according to the mode of entanglement formation through chain reptation *any new entanglement will necessarily be formed at the chain end* and then eventually be transferred along the chain. Thus, the stress transfer in the interfacial region of incompletely healed specimens must be assured by the newly formed entanglements of which a large proportion is sitting close to the end of the chain. Second, the local distribution of new entanglements within the former fracture surface are initially more or less irregular. This variation diminishes with healing time.

These two facts should be borne in mind when one analyzes the mechanism of craze and fibril formation in incompletely healed specimens. In the interfacial region of such specimens, there are fewer entanglements than elsewhere and they are found close to the chain ends. It must then be assumed that in the process of fibrillation especially the newly entangled chains are highly stressed and *that they are gradually losing those entanglements which are close to the chain ends*. Evidently the complete dissolution of the new entanglements in a particular cross section of a fibril will lead to the rupture of that fibril. The above assumption would conveniently explain the observed correlated between the extent of fibril deformation and rehealing.[48, 49] The entanglements formed in crack healing seem to represent a borderline case as defined in the beginning of Section 5.3.2. The entanglements are topological in nature as indicated by the fracture energy necessary for their dissolution. But the proportionality between length, *s*, of craze zone in an incompletely healed sample and depth of reptational chain interpenetration *l* suggests that slip of the entangled chains at roughly constant speed might be the parameter which controls fibril breakdown.

The above studies open the way to an understanding of a number of important technical problems: the influence of the molecular weight of a sample on its strength (see also Chapter 10), the development of defects in polymers (Chapter 2), especially under long-time loading conditions (Chapters 13 and 14), and the effect of weld lines on the mechanical behavior of injection molded samples (Chapter 21).

References

1. H.H. Kausch, *Polymer Fracture*, Springer, Heidelberg-New York 1978; 2nd ed. 1986.
2. S.H. Zhurkov, A.Ya. Savostin and E.E. Tomashevskii, *Soviet Phys.-Doklady* (Engl. Transl.), 1964, **9**, 986; S.N. Zhurkov, and V.E. Korsukov, *J. Polymer Sci. Phys.* 1974, **12**, 385.
3. L.L. DeVries, D.K. Roylance and M.L. Williams, *J. Polymer Sci.* A-1 1970, **8**, 237.
4. H.H. Kausch and J. Becht, *Rheol. Acta* 1970, **9**, 137; J. Becht, Dissertation, Technische Hochschule, Darmstadt 1971.
5. E.H. Andrews and P.E. Reed, in *Deformation and Fracture of High Polymers*, edited by H.H. Kausch, J.A. Hassell and R.I. Jaffe, p. 259, Plenum Press, New York 1973.
6. J. Sohma and M. Sakaguchi, *Adv. Polymer Sci.* 1976, **20**, 109.
7. P.Yu. Butyagin, A.M. Dubinskaya and V.A. Radtsig, *Russian Chem. Rev.* 1969, **38**/4, 290.
8. H.H. Kausch, *Revs. Macromol. Chem.* 1970, **5**/1, 97.
9. K.L. DeVries and D.K. Roylance, *Progr. Solid State Chem.* 1973, **8**, 283.
10. H.H. Kausch, Polymer Eng. Sci. 1979, **19**/2, 140.
11. D. Klinkenberg, *Progr. Colloid Polymer Sci.* 1979, **66**, 341; D. Klinkenberg, *Colloid Polymer Sci.* 1979, **257**, 351.
12. R.P. Wool, *Polymer Eng. Sci.* 1980, **20**/12, 805; R.P. Wool and R.H. Boyd, *J. Appl. Phys.* 1980, **51**, 5116.
13. H.H. Kausch, *Proceedings of the International Conference on Mechanical Behavior of Materials*, Vol. III, The Society of Materials Science, Kyoto, Japan, 1972; p. 518.
14. K. Holland-Moritz and J.W. Siesler, *Polymer Bull.* 1981, **4**, 165.
15. F.H. Müller, *Kunststoffe* 1959, **49**/2, 67; *J. Polymer Sci.* 1967, C **20**, 61.
16. K.L. DeVries, B.A. Lloyd and M.L. Williams, *J. Appl. Phys.* 1971, **42**, 4644.
17. Yu.K. Godovskii, V.S. Papkov, A.I. Slutsker, E.E. Tomashevskii and G.L. Slonimski, *Soviet Phys.–Solid State* 1972, **13**/8, 1918.
18. T. Nagamura, K. Fukitani and M. Takayanagi, *J. Polym. Sci. Phys.* 1975, **13**, 1515.
19. T.M. Stoeckel, J. Blasius and B. Christ, *J. Polymer Sci. Phys.* 1978, **16**, 485.
20. O. Frank and J.H. Wendorff, *Colloid Polymer Sci.* 1981, **259**, 70.
21. H.A. Gaur, *Colloid Polymer Sci.* 1978, **256**/10, 64.
22. R.K. Popli and D.K. Roylance, *Polymer Eng. Sci.* 1982, **22**, 1046.
23. O. Frank, Darmstadt 1983; V.A. Zakrevskii and V.I. Vettegren, Leningrad 1985, personal communications.
24. V.A. Marichin, *Acta Polym.* 1979, **30**, 507.
25. H.H. Kausch and K.L. DeVries, *Int. J. Fracture* 1975, **11**, 727.
26. D. Roylance, *Int. J. Fracture* 1983, **21**, 107.
27. S.P. Mishra and B.L. Deopura, *J. Appl. Polymer Sci.* 1982, **27**, 3211.
28. W.F. Busse, *J. Phys. Chem.* 1932, **36**, 2862.
29. L.R.G. Treloar, *Trans. Faraday Soc.* 1940, **36**, 538.
30. F.H. Müller, *Kolloid-Z.* 1941, **95**, 138, 306.

31. J.H. Golden, B.L. Hammant and E.A. Hazell, *J. Polymer Sci.* 1964, A **2**, 4787.
32. R.N. Haward and G. Thackray, *Proc. Roy. Soc.* A 1968, **302**, 453.
33. P.C. Moon and R.C. Barker, *J. Polymer Sci. Phys.* 1973, **11**, 909.
34. J.F. Fellers and B.F. Kee, *J. Appl. Polymer Sci.* 1974, **15**, 394.
35. R.P. Kusy and D.T. Turner, *Polymer* 1974, **15**, 394.
36. W. Döll, *J. Mater. Sci.* 1975, **10**, 935.
37. J.F. Fellers and T.F. Chapman, *J. Appl. Polymer Sci.* 1978, **22**, 1029.
38. B.H. Bersted, *J. Appl. Polymer Sci.* 1979, **24**, 37.
39. E.J. Kramer, in *Crazing in Polymers,* edited by H.H. Kausch, Advances in Polymer Science, Vol. 52/534, p. 1, Springer-Verlag, Heidelberg-New York 1983.
40. M. Dettenmaier and H.H. Kausch, *Polymer,* 1980, **21**, 1232; M. Dettenmaier, in *Crazing in Polymers,* edited by H.H. Kausch, Advances in Polymer Science, Vol. 52/53, p. 57, Springer-Verlag, Heidelberg-New York 1983.
41. M. Dettenmaier and H.H. Kausch, Lausanne, 1983, unpublished results.
42. W. Döll, in *Crazing in Polymers,* edited by H.H. Kausch, Advances in Polymer Science, Vol. 52/53, p. 105. Springer-Verlag, Heidelberg-New York 1983.
43. R. Schirrer, R. Lang, J. Le Masson and B. Tomatis, 4th Cleveland Symposium on Macromolecules *Irreversible Deformation of Polymers,* Case Western Reserve University, Cleveland, June 13–15, 1983.
44. R. Popli and D. Roylance, *Polymer Eng. Sci.* 1982, **22**, 1046.
45. K. Jud and H.H. Kausch, Polymer Bull. 1979, **1**, 697.
46. K. Jud, H.H. Kausch and J.G. Williams, *J. Mater. Sci.* 1981, **16**, 204.
47. H.H. Kausch, *Colloid Polymer Sci.* 1981, **259**, 917.
48. K. Jud, *PhD Thesis No. 413,* Swiss Federal Institute of Technology, Department of Materials Science, Lausanne 1981.
49. L. Könczöl, W. Döll, H.H. Kausch and K. Jud, *Kunststoffe* 1982, **72**, 46.
50. H.H. Kausch and M. Dettenmaier, *Colloid Polymer Sci.* 1982, **260**, 120.
51. P.G. de Gennes, *J. Chem. Phys.* 1951, **55**, 572.
52. M. Doi and F. Edwards, *Faraday Trans.* II 1978, **74**, 1789.
53. R.P. Wool and K.M. O'Connor, *J. Polymers Sci. Letters* 1982, **20**, 7.
54. Y.H. Kim and R.P. Wool, *Macromolecules* 1983, **16**, 1115.
55. S. Prager, T. Adolf and M. Tirrell, *J. Chem. Phys.* 1983, **78**, 7015.
56. E.N. Haward, H.E. Daniels and L.R.G. Treloar, *J. Polymer Sci. Phys.* 1978, **16**, 1169.
57. F. Brochard, J. Jouffroy and P. Levinson, CNRS report 542, Collège de France, Paris, 1983.
58. P.G. de Gennes, *Physics Today,* June 1983, 33.
59. E.J. Kramer, P. Green and Ch.J. Palmström, MSC Report # 4941, Cornell University, Ithaca, N.Y., 1983.
60. C.S. Henkee, E.J. Kramer, *J. Polymer Sci. Phys.* 1984, **22**, 721.
61. A.M. Donald, *J. Mater. Sci.* 1985, **20**, 2630.

6   Yield Behavior of Polymers

Norman Brown

Department of Materials Science and Engineering, University of Pennsylvania
Philadelphia, PA 19104, U.S.A.

6.1   Introduction

By definition, the *yield point* of a material is the highest stress that it can endure without manifesting a permanent strain upon unloading. Thus, the yield point is an important parameter for designing a component whose shape should not change in service. The phenomenon of yielding occurs in crystalline and glassy polymers. Yielding is associated with the stress level necessary to produce the initial permanent strain, the so called plastic strain, but the definition is not unique in that under some conditions polymers will manifest a strain after unloading which may persist only for a certain period of time. Thus, the definition of what is a permanent strain, from a practical standpoint, is arbitrary. For those materials whose stress-strain curves are monotonic the 0.2% offset method is often used to determine the yield stress. When the material exhibits a maximum in the stress-strain curve just beyond the elastic region, this maximum stress is generally called the yield stress. Most polymers, except at temperatures near T_g exhibit a maximum in the stress-strain curve, and it is this maximum stress that is usually referred to as the yield point in this paper.

In considering the yielding of polymers, a major distinction must be made between shear yielding and craze yielding because they are different phenomena from the microstructural viewpoint. *Shear yielding* basically involves the shear flow of the material with little or no change in density up to the yield point, whereas, *craze yielding* or *dilatational yielding* is highly localized in the form of thin crazes which consists of very porous fibrilla regions. Also, the macroscopic aspects of shear and craze yielding are different.

Since many of the aspects of craze yielding without an environmental effect are similar to those in gaseous environments as presented in Chapter 15 and in other environments as presented in Chapter 17, the emphasis in this Chapter will be on shear yielding.

Yielding will be presented from three viewpoints: (1) the phenomenological and macroscopic aspect, (2) the relationship between yielding and morphology, and (3) the fundamental approach in terms of intermolecular mechanisms.

6.2   Shear Yielding – Macroscopic Aspects

The most important macroscopic piece of information about the yielding of a polymer lies in its uniaxial *stress-strain curve* under a constant strain rate. From this curve one measures the yield stress, σ_y, yield strain, ε_y, and the elastic modulus. Next, it is important to know the effects of temperature, T, strain rate, $\dot{\varepsilon}$, pressure, P, and multiaxial stresses. The effects of multiaxial stressing will be considered in the section on yield criteria. In presenting the stress-strain curves, the focus will be on that portion between zero strain and the strain at the drop in the stress. The large strain region where most of the orientation is produced will not be covered from a dynamic viewpoint. However, the effect of orientation will be considered, since it is one of the most important morphological influences on yielding.

6.3 Stress-Strain Curve of Isotropic Polymers

A typical tensile stress-strain curve for an unoriented polymer is shown in Fig. 6.3.1.[1] The initial part of the stress-strain curve appears linear, but the linear range depends on temperature and the precision with which the strain is measured. The linear range decreases with increasing temperature. The sensitivity of the strain measuring device is an important factor in determining what is thought to be the linear range. Many polymers are inherently nonlinear at a very low stress. The slope of the stress-strain curve continually decreases until the stress reaches a maximum value after which the stress drops. This drop in stress is partly a geometric effect for a tensile test, since the cross section of the specimen is continually decreasing. However, as pointed out by Ward and Brown,[2] the true stress-true strain curve in tension may show a yield drop and the yield drop is also observed in shear (Fig. 6.3.2) and compressive loading (Fig. 6.3.3). Therefore, it is concluded that the yielding of polymers may involve an intrinsic drop in stress which arises from the nature of the molecular motion that occurs during yielding.

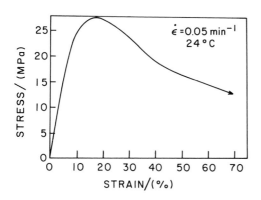

Fig. 6.3.1. Stress-strain behavior in tension of linear polyethylene, $\varrho = 0.9730$ g/cm^3 (Ref. 1). Room temperature.

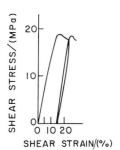

Fig. 6.3.2. Stress-strain behavior in shear of isotropic polyethylene terephthalate (Ref. 2). Room temperature.

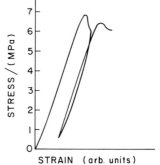

Fig. 6.3.3. Same as Fig. 6.3.2 deformed in compression (Ref. 2).

The yield strain, ε_y, is generally in the range of 5–15%. However, the yield stress depends on the type of polymer, the temperature, and the strain rate. In general, the yield point varies from 0 at T_g or T_m, depending on whether the polymer is amorphous or crystalline, to a maximum value which is determined by the interruption of the yield process by brittle fracture at the ductile-brittle transition temperature.

Polymers do not behave like the elastic-plastic solid which is approximated by many metals. The elastic-plastic material is linear elastic up to the yield point and then produces a time-independent permanent strain after unloading from above the yield point. Polymers display elastic, permanent, and time-dependent strains. The time-dependent strain may be called a persistent strain whose lifetime may vary from milliseconds to centuries depending on the type polymer and the test temperature. The complex

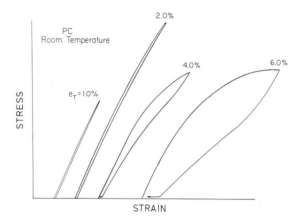

Fig. 6.3.4. Stress-strain behavior of PC after loading and unloading to successively higher stress amplitudes. Room temperatures.

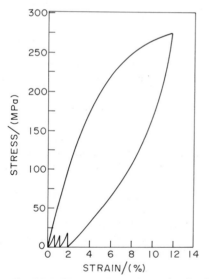

Fig. 6.3.5. Shows recovery in strain after loading PC in compression at 77 K. The last 2% strain took a few minutes to recover (Ref. 3).

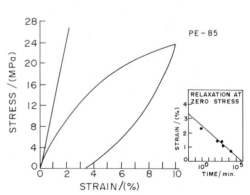

Fig. 6.3.6. Tensile stress-strain behavior of polyethylene at room temperature. Insert shows kinetics of strain recovery after unloading to zero stress (Ref. 1).

behavior of polymers is exhibited in Fig. 6.3.4 where the specimen is loaded and unloaded from successively higher stresses. The amount of persistent strain increases up
to the yield point and permanent strain does not appear until the yield point is exceeded.
Figure 6.3.5[3] shows that time-dependent recoverable strain can occur even at 77 K.
Figure 6.3.6[1] shows the kinetics of the recoverable strain in polyethylene. These observations indicate that, when the polymer is stressed up to the yield point, the total strain
that is produced consists of an anelastic, a plastic, and a time-dependent component
which is partly reversible. Thus polymers, below T_g or T_m are viscoelastic-plastic solids.
One of the questions that will be discussed is "To what extent does the yielding of
a polymer involve time when compared with an ideal elastic-plastic solid at one extreme
and a viscous liquid at the other?".

6.4   Yield Criteria and Pressure Effects

The yielding of a solid under multiaxial stresses raises the question of what is the
criterion for yielding. One of the simplest *yield criterion* was proposed by Tresca in
1864 which states that yielding occurs when the maximum shear stress exceeds a critical
value. In terms of the principal stresses, the Tresca criterion states

$$(\sigma_1 - \sigma_3)/2 \geqq \tau_y \tag{6.4.1}$$

where σ_1 is the greatest principal stress and σ_3 the least, and where negative stress
is always less than a positive one. τ_y is the shear yield point. Since the criterion holds
for all possible values of the principal stresses, it also holds for a uniaxial stress field.
Thus, the criterion reduces to

$$(\sigma_1 - \sigma_3) = \sigma_y \tag{6.4.2}$$

where σ_y is the uniaxial yield point. The Tresca criterion works well for polycrystalline
metals, but experiments with polycrystalline metals showed that a criterion proposed
by von Mises in 1913 is somewhat better. The von Mises criterion is

$$[(\sigma_1 - \sigma_2)^2 + (\sigma_2 - \sigma_3)^2 + (\sigma_3 - \sigma_1)^2]^{1/2} = \sqrt{2}\,\sigma_y \tag{6.4.3}$$

The Tresca criterion predicts that $\sigma_y = 2\tau_y$ and the von Mises criterion predicts $\sigma_y = \sqrt{3}\,\tau_y$. Experiments with multiaxial stresses on polycrystalline metals such as those
by Taylor and Quinney,[3a] generally show that the von Mises criterion is somewhat
better. However the difference between the two criterion is at most 15%. Therefore,
the Tresca criterion is often used because its form seems simpler and because for
engineering applications it gives a more conservative prediction for shear failure. The
von Mises criterion is, however, easier to handle from an analytical viewpoint because
the question of which is the greatest and least of the principal stresses does not have
to be taken into account.

It should be noted that both the Tresca and von Mises criteria are not affected
by the addition of hydrostatic pressure because hydrostatic pressure simply adds a
constant to each of the principal stresses and these criteria depend only on the difference
between principal stresses. However, many experiments with multiaxial stresses show
that the yield criteria for polymers must include the hydrostatic component of the
stress tensor, P, where

$$P = (\sigma_1 + \sigma_2 + \sigma_3)/3 \tag{6.4.4}$$

P affects the yield stress as well as the shape of the stress-strain curve. This is reflected
by the difference between the uniaxial tension and compression stress-strain curves

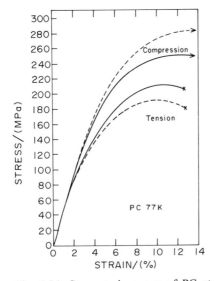

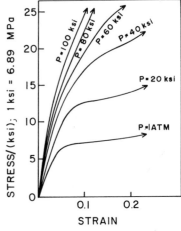

Fig. 6.4.1. Stress-strain curves of PC at 77 K in tension and compression ($- - -$) engineering curves, (———) true stress-strain curves in inert He environment (Ref. 3).

Fig. 6.4.2. Nominal compressive stress-strain curves of PCTFE at various pressures (Ref. 8).

in Fig. 6.4.1; a pure shear curve lies between them. Mears and Pae,[4, 6] Mears e.a.[5] and Radcliffe[7] have extensively measured the yield point as a function of hydrostatic pressure; Fig. 6.4.2[8] shows the effect.

Bowden and Jukes[9] reviewed the effect of P and multiaxial stresses on a number of polymers and showed that the Tresca and von Mises criteria should be modified as follows

$$(\sigma_1 - \sigma_2)/2 = \tau_y - \mu_s P \qquad \text{– modified} \qquad (6.4.5)$$
$$\text{Tresca}$$

or

$$\frac{1}{\sqrt{6}} [(\sigma_1 - \sigma_2)^2 + (\sigma_2 - \sigma_3)^2 + (\sigma_3 - \sigma_1)^2]^{1/2} = \tau_y - \mu_s P \qquad \text{– modified} \qquad (6.4.6)$$
$$\text{von Mises}$$

where μ_s is the coefficient that represents the sensitivity of yielding to the hydrostatic component, P. The uniaxial yield point, σ_y, is related to τ_y as follows

$$\tau_y = \left(\frac{1 \pm 2\mu_s/3}{2}\right) \sigma_y \qquad \text{– modified Tresca} \qquad (6.4.7)$$

$$\tau_y = \left(\frac{1 \pm \mu_s/\sqrt{3}}{\sqrt{3}}\right) \sigma_y \qquad \text{– modified von Mises} \qquad (6.4.8)$$

The positive sign is for tension and the negative for compression.

Table 6.4.1[10] shows how τ_y, σ_y, and μ_s vary for different polymers. These data are based on the reference temperature of 0 K and were obtained by a graphical extrapo-

Table 6.4.1. Yield Points at 0 K by Extrapolation

Polymer	Source	Test	$\sigma_y/(GP_a)$	Yield[9] Criterion	$\mu_s^{[9]}$	$\tau_y/(GP_a)$
PS	(11)	compression	0.30	Tresca	0.25	0.13
PS	(12)	compression	0.27	Tresca	0.25	0.11
PMMA	(13)	compression	0.82	von Mises	0.158	0.43
PMMA	(14)	compression	0.87	von Mises	0.158	0.45
PMMA	(15)	compression	0.73	von Mises	0.158	0.38
PC	(16)	shear	0.18	–	–	0.18
PC	(17)	tension	0.26	von Mises	0.12	0.16
PC	(18)	tension and compression	0.30 (av)	von Mises	–	0.17
PET	(17)	tension	0.24	von Mises	0.09	0.15
PCTFE	(19)	tension	0.25	von Mises	0.12	0.15
PE (High density)	(20)	tension	0.20	von Mises	0.038 (24)	0.12
PP (Isotactic, quenched)	(21)	tension	0.11	von Mises	0.12	0.068

lation of the experimental data. If a μ_s value was not available, the average value was used. In order to compare the yield strength of different polymers at a fundamental level it is important to use the same temperature. The value at 0 K is interesting because the effects of time and temperature on yielding are eliminated. It is noted that PMMA has the highest intrinsic shear yield strength and PP the lowest for the polymers listed in Table 6.4.1. These differences are caused by the differences in the chemistry of the molecules and in the molecular morphology.

6.5   Effect of Temperature on Yielding

The time-dependent aspects of yielding become more evident with increasing temperature. The typical effects of temperature on the tensile stress-strain behavior are shown in Fig. 6.5.1[19] for rapidly cooled PCTFE where both yield point and elastic modulus increase as the temperature decreases, and the yield strain is in the range of about 5–10%. In general, the yield stress of all polymers increases with decreasing T as does the modulus; whereas the yield strain is relatively insensitive to temperature. In general, the yield strain for *all* polymers varies from about 5–15%. The shapes of tensile stress-strain curves for a semicrystalline polymer are shown in Fig. 6.5.2.[20] These curves are similar to those of the amorphous polymers in Fig. 6.5.1.

The change in yield point with temperature is shown in Fig. 6.5.3 for polychlorotrifluoroethylene. It is seen that σ_y is 0 at T_g and increases monotonically with decreasing T until interrupted by brittle fracture. The slope of the fracture stress, σ_F, versus temperature falls below that of the extrapolated slope of the σ_y versus T curve. In common with all solids the change in σ_y with T is greater than the changes in σ_F with T. The yield and fracture stresses versus temperature for different types of PE are shown in Fig. 6.5.4[22] where the general behavior is similar to that of the amorphous polymers except that σ_y approaches zero at the melting point instead of at T_g. The differences between the various types of PE will be discussed in a later section on the effects of morphology in semicrystalline polymers.

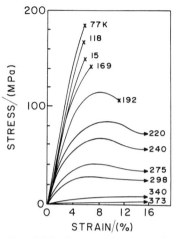

Fig. 6.5.1. Stress-strain curves in tension of quenched PCTFE at various temperatures in inert He environment (Ref. 19).

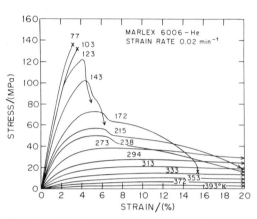

Fig. 6.5.2. Same as Fig. 6.5.1 for linear PE, $\varrho = 0.964$ g/cm^3 (Ref. 20).

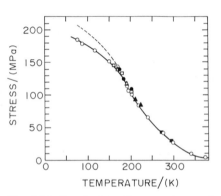

Fig. 6.5.3. Tensile strength versus temperature for quenched PCTFE. Fracture stress below 165 K and yield stress above 165 K (– – –) erxtrapolated values of yield stress (Ref. 19).

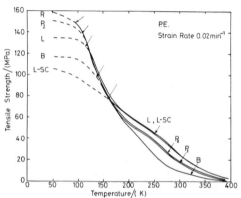

Fig. 6.5.4. (———) Yield, (– – –) fracture stress in tension for various types of PE versus temperature P_1, high molecular weight ethylene-hexene copolymer; P_2, ethylene-butene-octene copolymer; L, same as Fig. 6.5.2; B, low density branched; LSC, slow cooled, low molecular weight linear (Ref. 22).

6.6   Effect of Strain Rate

The relationship between strain rate and yield point for polymers is the same as for other solids in that to a first approximation

$$\sigma_y = B_1 + B_2 \ln \dot{\varepsilon} \qquad (6.6.1)$$

where B_1 and B_2 depend on the polymer and temperature. Figure 6.6.1[3] is an example of how the yield point of polycarbonate varies with strain rate in tension and compres-

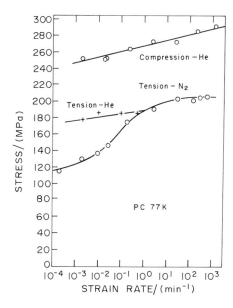

Fig. 6.6.1. Ultimate strength in compression and tension versus strain rate for PC at 77 K. Compression strength was taken at 5% strain. Curve in N_2 shows large effect of environmental crazing on the strain rate sensitivity. He represents an inert environment (Ref. 3).

sion at 78° K in an inert environment. In an active environment (N_2), where crazing occurs, the response is quite different.

A more fundamental understanding of the effect of strain rate on the yield point stress from the theory of thermal activation was originally developed by Eyring. For the moment, it is suffice to say that strain-rate sensitivity is given by

$$\frac{d\sigma_y}{d\ln\dot{\varepsilon}} = \frac{kT}{v_s} \tag{6.6.2}$$

where v_s is the shear-activation volume. Thus, the strain-rate sensitivity decreases with decreasing temperature as is generally observed.

6.7   Influence of Thermal Activation

The general effects of T, $\dot{\varepsilon}$, and P on the yield point can be succinctly described by the Eyring [1940] formalism where

$$\dot{\varepsilon} = A\, e^{-F^*/kT} \tag{6.7.1}$$

where A is a preexponential factor. The free energy for activation, F^*, may be given the following form as presented by Ward.[A]

$$F^* = Q^* - \sigma v_s - Pv_p - TS^* \tag{6.7.2}$$

where Q^* is the activation energy, v_s is the shear-activation volume, v_p is the pressure sensitive activation volume, and S^* is the entropy of activation. Therefore,

$$\dot{\varepsilon} = \dot{\varepsilon}_0 \exp[-(Q^* - \sigma_y v_s - Pv_p)/kT] \tag{6.7.3}$$

where the preexponential factor now equals $A\, e^{S^*/k}$. Consequently the yield point is given by

$$\sigma_y = 1/v_s[Q^* - Pv_p + kT\ln\dot{\varepsilon}/\dot{\varepsilon}_0)] \tag{6.7.4}$$

Thus, σ_y decreases with increasing T since $\dot{\varepsilon} < \dot{\varepsilon}_0$, increases with increasing $\ln \dot{\varepsilon}$ and increases with pressure since a hydrostatic pressure is by convention, a negative stress.

v_s, Q^*, v_p, and $\dot{\varepsilon}_0$ are physical parameters that depend on the structure of the polymer and the micromechanism of deformation which occurs during the thermally activated process. Other parameters being equal, σ_y increases as v_s decreases. Pampillo and Davis[23] with polyethylene, and Wu and Turner[16] with polycarbonate, observed that v_s decreased as the stress is increased by decreasing the temperature. Bauwens-Crowet[13] observed that v_s decreased as $\dot{\varepsilon}$ increased for PMMA. Davis and Pampillo[24] defined the strain-rate sensitivity as $\Delta \ln \sigma_y / \Delta \ln \dot{\varepsilon}$, and found that this parameter increases with P. As shown by Eq. (6.7.4) v_s and v_p are related by $v_p = v_s (d\sigma_y/dP)_T$, $= \mu_s v_s$. The values of μ_s is given in Table 6.4.1 show that v_p is generally less than v_s. All experimental data indicate that all the applied variables that increase σ_y also tend to decrease v_s at the same time.

Measurements of σ_y over a wide range of T, P, and $\dot{\varepsilon}$ show that in general v_s, Q^*, and $\dot{\varepsilon}_0$ are not constant. Thus, the Eyring formalism is generally more complex than the simple representation based on Eq. (6.7.2) because v_s, v_p, Q^*, and $\dot{\varepsilon}_0$ are functions of T, P, and $\dot{\varepsilon}$. Also Q^* is not necessarily a linear function of σ and P. The Eyring formalism is a useful phenomenological description of yielding and for giving values of Q^*, v_s, v_p, and $\dot{\varepsilon}_0$ which may be used to develop and interpret models of yielding based on molecular processes.

6.8   Relationship Between σ_y and the Elastic Modulus

The correlation between σ_y and elastic modulus has been pointed out by Buchdahl,[25] Robertson,[26] and Brown.[10, 27] It is generally observed that T and P, which increase σ_y, increase G_s and E, the shear and Young's modulae. In addition, as σ_y increases with plastic strain which produces an increase in orientation, E also increases.

Figure 6.8.1[27] shows σ_y versus E for a variety of polymers. Figure 6.8.2 shows σ_y versus E data for PC as obtained by Robertson[26] who varied the draw ratio. Using data from Boening[28] Fig. 6.8.3 shows how the density of polyethylene affects both σ_y and E. The effects of pressure on σ_y and E are shown in Fig. 6.8.4 for PE and PP as taken from the work of Mears and Pae[4] and Mears e.a.[5] Mears and Pae[6] varied P and found a linear relationship between σ_y and E for polycarbonate. The effects of the direction of orientation on σ_y and E as measured by Parrish and Brown[29] are shown in Fig. 6.8.5 where the largest value of σ_y is for the stress parallel to the draw direction and the least perpendicular. Of course, generally as T decreases both σ_y and E increase. As will be pointed out later, part of the decrease of σ_y with increasing T cannot be simply explained by the process of thermal activation as described above, but is directly connected to the change in the shear modulus with T.

The connection between σ_y and G_s is emphasized rather than with E because we are dealing with shear yielding. Poisson's ratio, v which occurs in the equation $G_s = \dfrac{E}{2(1+v)}$, does not show the large variation with T that is exhibited by G_s.

Not only does σ_y vary directly with G_s, but at low temperatures where the effect of thermal activation is small, the ratio of σ_y/G_s for all polymers is very high as pointed out by Brown.[10] Values of the ratio of shear yield point to shear modulus as obtained by extrapolating data to 0 K are shown in Table 6.8.1.[10] The average value from all known data is

$$\tau_y/G_s = 0.076 \pm 0.03 \tag{6.8.1}$$

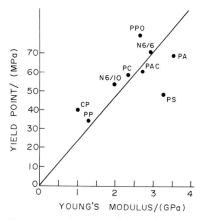

Fig. 6.8.1. Tensile yield point versus Young's modulus for various polymers (Ref. 27).

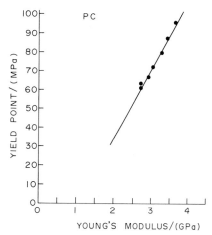

Fig. 6.8.2. Tensile yield point versus Young's modulus in PC for various draw ratios (Ref. 27). Taken from data by Robertson (26).

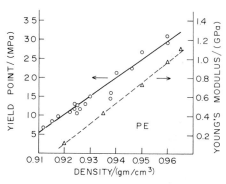

Fig. 6.8.3. Yield point and Young's modulus versus density of PE (Ref. 27). Taken from data by Boening (28).

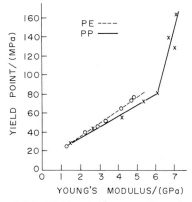

Fig. 6.8.4. Yield points versus Young's modulus for various hydrostatic pressures (Ref. 4, 5).

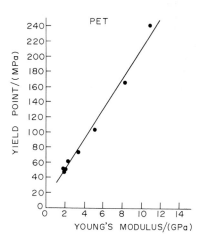

Fig. 6.8.5. Yield point versus Young's modulus for uniaxially oriented PET as produced by varying the angle between the orientation direction and the direction of the applied stress (Ref. 27). Taken from data from Parish and Brown (29).

Table 6.8.1. Ratio of shear yield point over shear modulus at 0 °K by extrapolation

Polymer	τ_y/G_s (Experimental) at 0 °K
PS	0.069
PMMA	0.133
PC	0.050
PET	0.094
PET	0.088
PCTFE	0.065
PE (high density)	0.027
PP (isotactic-quenched)	0.030
Average (excluding PE)	0.076 ± 0.03

This value is about equal to the ideal strength that should be expected for all solids. The theory by Brown,[10] based on the 6–12 Mie potential for the van der Waals force between molecules, contains the assumption that practically homogeneous deformation occurs up to the point of yielding. The theory predicts a value of $\tau_y/G_s = < 0.064 - 0.092$. A theory by Argon,[30] based on the motion and generation of defects called disinclinations, predicts a value of $\tau_y/G_s = 0.12$. Thus, the predictions and the experimental observations are in close agreement. The fact that the yield strain for all polymers is in the range of 5–15% is in accordance with these theories. It must be emphasized that because the yield point is such a large fraction of the shear modulus, linear polymers in service exhibit nearly their ideal yield strength.

6.9   Effects of Morphology and Molecular Structure

Morphology is defined here as the packing arrangement of the molecules. The influence of morphology on yielding will be presented in the simplest terms based on those aspects of morphology which are most evident. Thus, this section is divided into two domains, one covers the extreme case of amorphous packing as represented by PS and PMMA and the rapidly quenched states of polymers such as PC and PET, and the other covers the semicrystalline state as represented by PE. Only the isotropic state will be discussed here. Essentially linear homopolymers are being considered.

6.10   Amorphous Polymers

For amorphous polymers the yield point approaches zero as the temperature approaches T_g. The yield point approaches zero at T_g simply because the intermolecular forces are practically zero at T_g. The effects of molecular structure on T_g have received much discussion through the years; the cohesive energy density and chain stiffness are considered to be the primary factors. Since the yield point depends directly on the modulus, all the factors that increase the modulus, as T goes below T_g, also increase the yield point. As the temperature decreases and passes through a relaxation there is a corresponding increase in the elastic modulus and a corresponding increase in

σ_y in addition to the effect of thermal activation. In general, one can say that the greater the bonding energy, the greater the modulus and consequently the greater the yield point.

If the packing density is reduced by rapid quenching, then the bonding energy is reduced as is the modulus and the yield point. If the bonding energy is increased by very slow cooling from T_g, there is an increase in modulus and consequently of the yield point. For some polymers, the increase in yield point by slow cooling is sufficiently great that the polymer will fracture brittlely before yielding.

6.11 Semicrystalline Polymers

Figure 6.5.2 shows the stress-strain curves of a typical linear PE at various temperatures. The general change in shape of these curves with T is about the same as that for an amorphous polymer as illustrated in Fig. 6.5.1. However, an extreme type of stress-strain behavior can be exhibited by polyethylene with a low molecular weight $(M_n = 6100, M_n = 101,000)$, whose density equals 0.973 gm/cm^3, and which was slowly cooled from the melt (6° C/hr) is shown in Fig. 6.11.1.[31] This material is comparatively brittle up to about 300 K and does not exhibit necking and a drop in the yield point as exhibited by the other polymers.

The effect of crystallinity on σ_y depends on the temperature of measurement. Figure 6.5.4 shows σ_y versus T for various types of PE. At room temperature, σ_y increases with crystallinity as measured by density as shown by Fig. 6.5.4 and 6.8.3. The effect of crystallinity on σ_y is most pronounced in the temperature range around 300 K because there the crystalline region is relatively stiff and the amorphous region is comparatively weak. When the amorphous region becomes glassy at about 164 K, it is most interesting to observe that σ_y becomes nearly independent of morphology as shown in Fig. 6.5.4. The effect of crystallinity on σ_y at very low temperatures is not known for PE when brittle fracture intervenes.

The crystalline regions become soft as the temperature is increased into the α transition region and thus σ_y decreases toward zero for all types of PE as the melting point

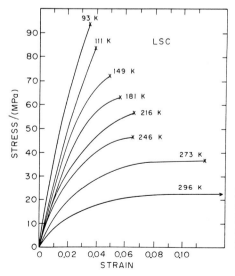

Fig. 6.11.1. Stress-strain curves in tension from 93 to 296 K, strain rate = 0.11 s^{-1} for slow cooled linear PE, $\varrho = 0.9725$ gm/cm^3 $M_n = 6.1 \times 10^3$, $M_w = 101 \times 10^3$ in an inert environment (Ref. 31).

is approached. The least crystalline material softens more rapidly than the more crystalline states, since it has the lowest melting point.

Polytetrafluoroethylene shows the same general type of dependence of σ_y on T and crystallinity except at 77 K PTFE remains ductile. The stress-strain curves of PTFE of various crystallines and at three temperatures 77 K, 200 K, and 300 K are shown in Fig. 6.11.2.[32] It is noticed that at 77 K the least crystalline material is strongest, whereas at the intermediate temperature the most crystalline material is strongest. At 300 K the effect of crystallinity is small.

The behavior of PTFE is readily understandable in terms of its morphology. At the intermediate temperature of 200 K which is above the β transition, the amorphous region is weak and the crystalline region is relatively strong, therefore σ_y increases with the percent crystallinity.

In the very low temperature range, the amorphous region is stronger than the crystalline region. The amorphous glass is stronger than the crystalline region because its yield point is about the theoretical value, whereas the crystalline region probably shears by a dislocation mechanism. The yield point is governed by the crystalline region and σ_y increases with decreasing lamella thickness[32] which varies directly with crystallinity. This effect is comparable to the effect of grain size on the yield point of a polycrystalline metal.

At some temperature interval between 78 and 200 K it is expected that σ_y would be independent of crystallinity in PTFE as is observed in PE at above 164 K. At room temperature the crystalline region becomes as weak as the amorphous region. Also the strengthening from the grain size disappears because the grain boundary is no longer rigid. Consequently, as observed, there is very little effect of the crystallinity on σ_y at higher temperatures.

In general, it can be concluded that in the temperature range where the amorphous region is weak and the crystalline region is still very strong, σ_y increases with crystallinity. In the very low temperature regions where the amorphous regions are glassy, σ_y increases with decreasing crystallinity. In the high temperature region where the crystals are very soft and of course the amorphous region is rubbery, crystallinity has a very

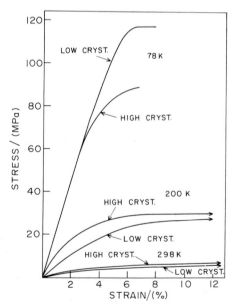

Fig. 6.11.2. Stress-strain curves for PTFE where low crystallinity = 49% and high crystallinity = 66% at 78 K, 200 K, and 298 K in an inert environment (Ref. 32).

small effect on σ_y. At some low intermediate temperature range where the yield point of the amorphous and crystalline regions are equal, differences in crystallinity will not be manifested by σ_y.

The effects of molecular weight and short chain branching on σ_y are probably indirect in that the primary effect stems from the fact that molecular weight and branching affects the crystallinity during crystallization from the melt. For very low molecular weight material where the end groups constitute more than 1% of the volume, the yield point is expected to decrease with molecular weight, if σ_y can be observed prior to brittle fracture.

6.12   Effects of Orientation

The two major effects of orientation are: (1) the effect of draw ratio and (2) the effect of the direction of a uniaxial stress relative to the draw direction. As the draw ratio increases, σ_y parallel to the draw direction increases and that perpendicular to the draw direction changes slightly. Figure 6.12.1[29] shows the effect of the direction of the applied stress relative to the draw direction where $\sigma_y(0°)/\sigma_y(90°)$ is about 10 for that particular draw ratio in PET. It should also be emphasized that it is not draw ratio per se, but the degree of orientation that controls σ_y.

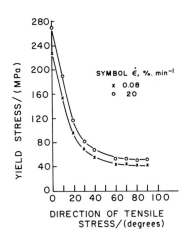

Fig. 6.12.1. The effect of the direction of stress relative to the orientation direction on the yield point of PET (Ref. 29).

6.13   Yield Criterion for a Uniaxially Oriented Polymer

Hill[33] modified the von Mises criterion for an isotropic solid as follows

$$B_9(\sigma_2-\sigma_3)^2 + B_{10}(\sigma_3-\sigma_1)^2 + B_{11}(\sigma_1-\sigma_2)^2 + \\ + 2\,L\tau_{23}^2 + 2\,M\,\tau_{31}^2 + 2\,N\tau_{12}^2 = 1 \tag{6.13.1}$$

Brown, Duckett and Ward[34] found that the large difference between the yield points in tension or compression with respect to the orientation direction required a Bauschinger term, σ_i, where σ_i is the difference between the yield point in tension and compression in the direction of orientation. In addition, it is required that the effect of the hydrostatic tension be added as was done in the pressure modified von Mises equation. Thus, the following equation for a uniaxially oriented polymer was suggested

$$
B_9(\sigma_2-\sigma_3)^2 + B_{10}(\sigma_3-\sigma_1+\sigma_i)^2 + B_{11}(\sigma_1-\sigma_i-\sigma_2)^2 +
$$
$$
+2\,L\tau_{23}^2 + 2\,M\tau_{31}^2 + 2\,N\tau_{12}^2 = (1-QP)^2 \qquad (6.13.2)
$$

where $P=(\sigma_1+\sigma_2+\sigma_3)/3$, the hydrostatic component of stress. Thus, eight parameters are required to describe completely the yield stress of a uniaxially oriented polymer.

6.14   Theory of Yielding From the Molecular Viewpoint

Many theories of yielding of amorphous polymers describe the phenomenon in continuum terms invoking concepts of free volume, springs and dashpots, and viscosity. There are theories that take a molecular viewpoint. Robertson[35] considered chain bending as the fundamental mechanism for yielding. He assumed that the yield point corresponded to that state where the amount of bends produced by the stress were equivalent to the number of bends that occur at T_g.

Yannas and Lunn[36] calculated the change in van der Waals bonding when a chain was twisted. They found that the major resistance to twisting came from the intermolecular bonds, and that a minor role was played by intramolecular bonds. Their value of σ_y/G_s was about 0.02 without thermal activation.

Bowden and Raha[37] proposed a theory based on the stress to nucleate a patch of slip. Their theory used a continuum model of a dislocation. The stress to nucleate slip without thermal activation was arbitrarily taken as $\tau_y/G_s=1/6$ which corresponds to the early Frenkel theory.[38] The effect of thermal activation was introduced by means of the stress required to nucleate a dislocation loop.

Argon[30] used a theory based on the generation and motion of kinks in the molecule as modeled by a wedge disinclination. He calculated the thermally activated stress required to generate and separate two wedge disinclinations. Argon predicted a stress of $\tau_y/G_s=0.77/1-v$ at 0 K where v is Poisson's ratio.

Brown[10] proposed a model of homogeneous yielding based on the Mie (also called Lennard-Jones) potential. Brown suggested that in an amorphous polymer there are three basic types of molecular motion up to the shear yield strain: (1) shearons which consist of the motion of molecular segments whose covalent bonds lie in the plane of shear, (2) rotons which are comparable to shearons except the covalent bond makes an angle with the plane of shear, and (3) tubons which require a force parallel to the covalent bond that allows the molecular segment to move along the shear plane. The macroscopic yield point was calculated by assuming cooperative and nearly homogeneous motion of all molecular segments. The value of the yield point was calculated from the fraction and shear resistance of each type of molecular motion produced by the shearons, rotons, and tubons in acordance with the following euqation

$$
\tau_y/G_s = f_S(\tau_{yS}/G_s) + f_R(\tau_{yR}/G_s) + f_T(\tau_{yT}/G_s) \qquad (6.14.1)
$$

where f_S, f_R, and f_T, are the fractions of the total motion from shearons, rotons, and tubons respectively and τ_{yS}, τ_{yR}, and τ_{yT} are their respective shear resistances. An analysis of the molecular motions indicated that $f_S=0.8$–0.9, $f_R=f_T=0.05$–0.10. τ_{yS}/G_s was taken as less than 0.061 which is the ideal strength of a face center cubic van der Waals bonded crystal as determined by Tyson.[39] τ_{yT}/G_s was taken to be between 0.122 based on Argon's[30] value for disinclinations and 0.37 from Kausch and Becht's[40] calculations for the stress to pull a single molecular chain from a crystal. τ_{yS} was assumed to be equal to τ_{yR}. Thus, the predicted value of the yield stress by Brown[10] was $\tau_y/G_s < 0.064$–0.92 which agrees well with the average value of 0.076 from experiments.

Those theories that predict the temperature dependence of the yield point such as Robertson, Bowden and Raha, and Argon use the Eyring formalism of thermal activation. However, as pointed out by Argon[30] and Brown,[10] temperature plays two roles. It reduces the yield point in accordance with the Eyring theory, but equally important is the direct connection of between yield point and elastic modulus as presented in the above theories. The temperature variation of the modulus is directly proportional to the variation of the intermolecular bond strength with temperature. The following is an approximate but generally useful equation for the yield point.

$$\tau_y = \alpha \, G_s(P, T) + \frac{kT}{v_s} \ln(\dot{\varepsilon}/\dot{\varepsilon}_0) \qquad (6.14.2)$$

where the average value of α is 0.076 for amorphous linear polymers. $G_s(P, T)$, the shear modulus, depends on P and T and v_s and $\dot{\varepsilon}_0$ are material parameters. Values of α for particular polymers are given in Table 6.8.1.

It is to be noticed in Table 6.8.1 that the semicrystalline polymers and isotactic PP have significantly lower values of τ_y/G_s than the amorphous polymers at 0 K. The crystalline regions become more dominant at low temperatures relative to the amorphous regions in that they permit easy slip at low temperatures probably via a dislocation mechanism. However, there are as yet no direct observations of dislocation motion in semicrystalline polymers. One can view the increase in yield point with decreasing crystallinity at 78 K in PTFE as a grain size effect. As the crystallinity is reduced the lamella size decreases. Thus, according to the Hall-Petch relationship the yield point should vary inversely as the square root of the lamella thickness. Most likely the direction of easy slip is parallel to the chain direction. Since the chains are parallel to the thin dimension of the crystal, it suggests that the lamella thickness should directly affect the yield point at low temperatures as was observed.[32]

6.15   Craze Yielding

A maximum in the stress-strain curve may be observed during craze yielding. However, the source of this maximum stress is entirely different from shear yielding. In shear yielding, the material is essentially elastic or viscoelastic up to the yield point. Then at the critical stress, a permanent strain is produced with essentially no change

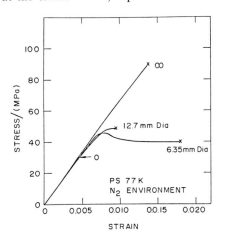

Fig. 6.15.1. Stress-strain curves of PS in N_2 at 77 K for various size specimens. Arow shows stress to initiate crazing (Ref. 42).

in volume. In craze yielding, the first craze is initiated at a stress which is well below the maximum stress as shown in Fig. 6.15.1.[41] The stress to initiate a craze depends on surface imperfections which concentrate the stress, whereas the shear yield point is relatively insensitive to surface imperfections. The very first craze produces permanent strain by a dilatation process that causes fibrillation so that plastic deformation occurs below the maximum stress which is produced by the dynamic yield mechanism as described below.[42, 43]

6.16   Dynamic Yielding by Crazing

In the conventional stress-strain test the head speed of the machine is kept constant. Thus, the total strain rate, $\dot{\varepsilon}_T$, is kept constant and equals the sum of the elastic strain rate, $\dot{\varepsilon}_E$, and the strain rate produced by crazing, $\dot{\varepsilon}_c$

$$\dot{\varepsilon}_T = \dot{\varepsilon}_E + \dot{\varepsilon}_c \tag{6.16.1}$$

For stresses below the craze initiation stress, all the strain is elastic. Beyond the craze initiation stress, σ_i, the number of crazes increase as σ increases. Also, the velocity of the crazes increase monotonically with the stress. Each craze grows with time so that even when the craze front moves with a given velocity, it generates more craze strain per unit time as it gets larger. Thus, the craze strain rate can be described by the following equation[45]

$$\dot{\varepsilon}_c = C_c \varrho_c l_c^2 v_c \tag{6.16.2}$$

where ϱ_c is the density of crazes, l_c is the size of the craze, v_c is the craze velocity, and C_c is a parameter for the shape of the craze. Thus, as the stress increases beyond σ_i, $\dot{\varepsilon}_c$ increases until $\dot{\varepsilon}_T = \dot{\varepsilon}_c$; then, $\dot{\varepsilon}_E$ becomes zero and the stress, which is related to the elastic strain, reaches its maximum value. The crazes continue to grow, but since l_c is always increasing, $\dot{\varepsilon}_c$ continues to increase. The specimen cannot extend faster than the machine so $\dot{\varepsilon}_E$ becomes negative and the stress falls. As the stress falls, v_c decreases because it varies directly with the stress; ϱ_c remains constant since it reached its maximum value when σ became a maximum or ϱ_c may even increase with time. Thus, in order to completely describe the stress-strain curve from crazing, the important parameters are σ_i, ϱ_c and v_c as a function of σ, and the craze opening displacement which produces the strain. All these parameters have been put together in a theoretical calculation[43] of the stress-strain curve. It turns out that the maximum stress is very sensitive to the applied strain rate. From the same fundamental theory, creep[45] and stress relaxation curves[44] have also been calculated.

6.17   Effect of Specimen Dimensions

Since crazes generally initiate at the surface for a homopolymer, the density of crazes depends on the cross section of the sample. Consequently the stress-strain curve depends on the size of the specimen. The significant parameter is the ratio of volume to surface area. Wu and Brown[41] measured the effect of specimen size on the stress-strain curve as shown in Fig. 6.15.1.

The ultimate strength versus volume per surface area is shown in Fig. 6.17.1. The asymptotic value of the ultimate strength, as the volume per surface area approaches zero, $\sigma_c(0)$, represents the stress to initiate the first craze since the first craze in an

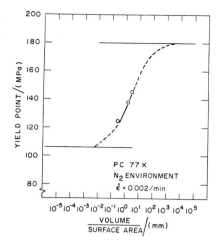

Fig. 6.17.1. Craze yield point versus ratio of volume to surface area of test specimen of PC (Ref. 42).

Table 6.17.1. Maximum variation in tensile strength with size

Polymer	$\sigma_c(0)$/(MPA)	$\sigma_c(\infty)$/(MPa)	$[\sigma_c(\infty)-\sigma_c(0)]$/(MPa)
PS	30	90	60
PMMA	82	156	74
PTFE	60	110	50
PC	105	180	75

infinitely thin specimen would be its ultimate strength. The asymptotic value of the ultimate strength, as the volume per surface area approaches infinity, $\sigma_c(\infty)$, corresponds to the shear yield point or fracture stress since in an infinitely large specimen shear yielding or fracture would occur before the crazes make a significant contribution to the strain. The maximum variation in tensile strength with respect to the extremes in volume per unit area of the specimen are given in Table 6.17.1 [41] for several polymers in liquid nitrogen where $\sigma_c(0)$ is the stress to initiate the first craze and $\sigma_c(\infty)$ is the ultimate strength in the absence of crazing.

6.18 Stress Criteria for Craze Initiation

There have been attempts to put forth a stress criterion for crazing just as the von Mises and Tresca criteria have been put forth for shear yielding. Sternstein and co-workers [46–48] used the maximum in the stress-strain curve as the critical condition for crazing. Based on multiaxial stress fields of tension-tension and tension-torsion, they proposed the criterion

$$(\sigma_1 - \sigma_3) = B_3 + B_4/P \quad \text{for } P > 0 \tag{6.18.1}$$

where B_3 and B_4 are material constants. Based on the stress to initiate the first craze, Oxborough and Bowden [49] using a tension-compression stress field, proposed the criterion

$$\varepsilon_1 = B_5 + B_6/P \quad P > 0 \tag{6.18.2}$$

where ε_1 is the maximum principal strain. Based on the work of Matsushige, Radcliffe and Baer, Moet, Palley and Baer[50] proposed the following criterion for craze initiation where the stress field was tension plus a hydrostatic pressure

$$\sigma_1 = B_7 - B_8\,P \qquad P < 0 \tag{6.18.3}$$

They[50] showed that their constant B_8 depended on whether the specimen was shielded or exposed to the fluid which transmitted the pressure. The influence of exposure was not associated with an environmental interaction between fluid and polymer, but was associated with a difference in the local stress field at a point of stress concentration. For a given hydrostatic pressure it takes a lower applied tensile stress to initiate a craze on an exposed specimen as compared to the case where the specimen is protected from the fluid. Moet, Palley and Baer interpreted their results in terms of a single criterion according to which crazing is initiated when the local maximum principal stress at the tip of a flaw reaches a critical value. Argon and Hanoosh[51] proposed the following criterion based on the stress to produce 10^4 crazes per cm^2 in $100\,s$ using a tension-torsion stress field

$$\frac{1}{\sqrt{6}}\sqrt{(\sigma_1 - \sigma_2)^2 - (\sigma_3 - \sigma_2)^2 + (\sigma_2 - \sigma_1)^2} = B_9/(B_{10} + B_{11}\,P) \qquad P < 0 \tag{6.18.4}$$

where B_9, B_{10} and B_{11} are constants. The differences in the above criteria stem partly from the use of a different critical condition for crazing. There is another major difference because, unlike shear yielding which relates directly to the applied stresses, crazing depends on the local stress at points of stress concentration. These points of stress concentration most likely have a variable character. Even homopolymers with the same chemistry may be very different because crazing depends on the surface condition of the particular specimen. Finally, another source of a difference among the criteria stems from the fact that different stress field domains most likely give a different criterion. This situation is similar to the history of stress criteria for fracture in metals where a single criterion, which governs all the domains of the stress tensor, has yet to be found. A universal criterion should not be expected because crazing, like fracture, depends strongly on the character of the surface flaws and not simply on the bulk properties of the material.

Both crazing and shear flow may occur together. PMMA[52] at room temperature can simultaneously neck and craze under a tensile stress. Thus, crazing and shear yielding are not mutually exclusive phenomena. A polymer which simultaneously shear flows and crazes involves interactions which would make the characterization of the stress-strain behavior more complex.

Acknowledgments

The support of the Gas Research Institute under Grant No. 5080-363-0382 and the use of the Central Facilities in the Materials Research Laboratory which are supported by the National Science Foundation under Grant No. DMR-8216718 are appreciated.

General References

A. I.M. Ward, *Mechanical Properties of Solid Polymers,* 2nd ed., John Wiley, New York 1983.

B. J.A. Sauer and K.D. Pae, Chapter 7 in *Introduction to Polymer Science and Technology*, edited by H.S. Kaufman and J.J. Falcetta, John Wiley, New York 1977.
C. International Conferences on *Deformation, Yield and Fracture of Polymers,* Churchill College 1970/73/76/79/82, Plastics and Rubber Institute, 11 Hobart Place, London.

Specific References

1. S. Rabinowitz and N. Brown, *J. Polymer Sci. A-2* 1967, **5**, 143.
2. N. Brown and I.M. Ward, *J. Polymer Sci. Part* 1968, **6**, 607.
3. Y. Imai and N. Brown, *J. Polymer Sci. Phys.* 1976, **14**, 723.
3a. G.I. Taylor and H. Quinney, *Philos. Trans. Royal Soc. London,* 1931, A 230, 323.
4. K.D. Mears and D.R. Pae, *J. Polymer Sci. B* 1968, **6**, 269.
5. K.D. Mears, D.R. Pae and J.A. Sauer, *J. Polymer Sci. B* 1968, 773.
6. K.D. Mears and D.R. Pae, *J. Polymer Sci. B* 1969, **7**, 349.
7. S.V. Radcliffe, in *Deformation and Fracture of High Polymers*, edited by H.H. Kausch, J.A. Hassel and R.I. Jaffe, p. 191, Plenum, New York 1974.
8. A.A. Silano and K.D. Pae, in *Advances in Polymer Science and Engineering*, edited by K.D. Pae, D.R. Morrow and Y. Chen, p. 131, Plenum Press, New York 1972.
9. P.B. Bowden and J.A. Jukes, *J. Mater. Sci.* 1972, **7**, 52.
10. N. Brown, *J. Mater. Sci.* 1983, **18**, 2241.
11. P.B. Bowden and S. Raha, *Phil. Mag.* 1970, **22**, 463.
12. J.P. Cavrot, J. Haussy, J.M. Lefebvre and B. Escaig, *Mater. Sci. Eng.* 1978, **36**, 95.
13. C. Bauwens-Crowet, *J. Mater. Sci.* 1973, **8**, 968.
14. J. Haussy, J.P. Cavrot, B. Escaig and J.M. Lefebvre, *J. Polymer Sci.* 1980, **18**, 311.
15. P. Beardmore, *Phil. Mag.* 1969, **19**, 389.
16. W. Wu and A.P.L. Turner, *J. Polymer Sci. Phys.* 1975, **13**, 19.
17. J.R. Kastelic and E. Baer, *J. Macromol. Sci. Phys.* 1973, **B 7**(4), 679.
18. C. Bauwens-Crowet, J.C. Bauwens and G. Holmes, *J. Mater. Sci.* 1972, **7**, 176.
19. Y. Imai and N. Brown, *Polymer* 1977, **18**, 298.
20. E. Kamei and N. Brown, *J. Polymer Sci. Phys.* 1983, **21.**
21. H.G. Olf and A. Peterlin, *J. Polymer Sci. Phys.* 1974, **12**, 2209.
22. N. Brown, E. Kamei and I.M. Ward, Int. Conf. Gas Research Institute, London, 1983.
23. C.A. Pampillo and L.A. Davis, *J. Appl. Phys.* 1972, **43**, 4277.
24. L.A. Davis and C.A. Pampillo, *J. Appl. Phys.* 1971, **42**, 4659.
25. R. Buchdahl, *J. Polymer Sci. A* 1958, **28**, 239.
26. R.E. Robertson, General Electric Rept. No. 64-RL (3580C) 1964.
27. N. Brown, *Mater. Sci. Eng.* 1971, **8**, 69.
28. H.V. Boening, *Polyolefins,* Elsevier, New York 1966.
29. M.F. Parrish and N. Brown, *J. Macromol. Sci. Phys.* 1970, **B 4**(3), 649.
30. A.S. Argon, *Phil. Mag.* 1973, **28**, 839.
31. N. Brown and I. Ward, *J. Mater. Sci.* 1983, **18**, 1405.
32. S. Fischer and N. Brown, *J. Appl. Phys.* 1973, **44**, 4322.
33. R. Hill, *The Mathematical Theory of Plasticity,* p. 317. Oxford University Press 1950.
34. N. Brown, R.A. Duckett and I.M. Ward, *Phil. Mag.* 1968, **18**, 483.
35. R.E. Robertson, *J. Chem. Phys.* 1966, **44**, 3950.
36. I.V. Yannas and A.C. Lunn, *Am. Chem. Soc., Polymer Prepr.* 1975, **16**, 564.
37. P.B. Bowden and R. Raha, 1974, **29**, 149.
38. J. Frenkel, *Z. Physik* 1926, **37**, 572.
39. W.R. Tyson, *Phil. Mag.* 1966, **14**, 925.
40. H.H. Kausch and J. Becht, *Deformation and Fracture of High Polymers*, edited by H.H. Kausch, J.A. Hassel and R.J. Jaffe, p. 317. Plenum Press, New York 1973.
41. J.B.C. Wu and N. Brown, *J. Mater Sci.* 1977, **12**, 1527.
42. J. Hoare and D. Hull, *Phil. Mag.* 1972, **26**, 443.

43. N. Brown, *Phil. Mag.* 1975, **32**, 1041.
44. J.B.C. Wu and N. Brown, *J. Rheology* 1979, **23**, 231.
45. N. Brown, B.D. Metzger and Y. Imai, *J. Polymer Sci. Phys.* 1978, **16**, 1085.
46. S.S. Sternstein, L. Ongchin and A. Silverman, *Appl. Polym. Symp.* 1968, **7**, 175.
47. S.S. Sternstein and L. Ongchin, *Amer. Chem. Soc. Polymer Preprints* 1969, **10**, 1117.
48. S.S. Sternstein and F.A. Myers, *J. Macromol. Sci. Phys.* 1974, **B8**, 539.
49. R.J. Oxborough and P.B. Bowden, *Phil. Mag.* 1973, **28**, 547.
50. A. Moet, I. Palley and E. Baer, *Amer. Chem. Soc. Organic Coatings and Plastics.* Preprints 1979, **41**, 424.
51. A.S. Argon and J.G. Hanoosh, *Phil. Mag.* 1977, **36**, 1195.
52. Y. Imai and N. Brown, *J. Mater. Sci.* 1976, **11**, 425.

7   General Fracture Mechanics

Kenneth J. Pascoe

Department of Engineering, University of Cambridge
Cambridge, CB 2 IPZ, Great Britain

7.1   Introduction

This volume deals with polymeric materials, but metallurgy is much older than polymer science. A systematic study of mechanical behavior, including fracture, was developed for metals first, and the concepts were later applied to polymer behavior.

With the increasing use of metals from the time of the industrial revolution, there was an increase in the number of failures of structure and machinery, often catastrophic and with loss of life. While some could be traced to poor design, others were due to deficiencies in the materials. Materials, which in a simple tensile test would behave in a reasonable ductile manner and exhibit sufficient strength, would nevertheless under certain circumstances fail dramatically at lower loading and often in a far from ductile manner.

One contributory factor, which was recognized as early as 1892[1] was the presence of preexisting crack-like flaws. These could be impurities or cavities in the material or even defects in the surface, such as scratches.

Fractures would start from these flaws. The understanding of the relationship between the size, shape, and nature of the flaws and the stress levels at which failure can occur has developed over the last two to three decades and has become known as *fracture mechanics.*

The basic concept of the subject is the imbalance between the energy needed to continue enlargement of a crack and the energy available from the work of external forces and internal elastic strain energy. This basic concept and related equations were developed by Griffith and published in 1921,[2] with some amendments three years later.[3] He considered materials which behaved in a linear-elastic manner up to or almost up to the point of fracture. Later workers modified his theory to deal with more ductile materials, which nevertheless behaved elastically except when very close to the growing crack. Within this limitation, the subject is referred to as *linear-elastic-fracture mechanics* (LEFM). When dealing with more ductile materials, such as many polymers, the basic criteria have to be modified. So at a later stage the methods of *elastoplastic-fracture mechanics* were developed.

This new subject has developed an extensive terminology of its own which will be introduced progressively through this chapter. Its application to polymers and typical results are discussed in Chapter 8 and also in Chapter 18.

7.2   Cohesive Strength

To cause failure of a material, the forces applied have to be sufficient to overcome the bonding forces between atoms or molecules. If all atomic or molecular bonds in the direction of the force share the loads equally, so that all reach their limit simultaneously, then it is possible to derive a value for the theoretical cohesive strength of the material and this can be shown to be about $E/15$ for most materials, where E is Young's modulus of the material.

The actual strengths are generally of a far lower order of magnitude, although it is possible to make almost flaw-free specimens which approach this value. An example is freshly drawn glass fiber, but exposure to the atmosphere, which has a mild corrosive effect, or even a very shallow scratch will reduce the strength dramatically.

The reason is that any scratch or crack-like flaw acts as a stress concentrator, since the stress at the tip or the point of sharpest curvature is higher than the average stress. The ratio of the maximum stress to the average stress away from the flaw is known as the *stress-concentration factor*, K_t. Its value at the end of the major axis of an elliptical hole in a plane sheet of material which is stressed elastically in the direction of the minor axis has been shown by Inglis[4] to be

$$K_t = 1 + 2\sqrt{\left(\frac{h}{l}\right)} \tag{7.2.1}$$

where $2h$ is the length of the major axis and l is the radius of curvature at each end of that axis (Fig. 7.2.1 a). Similarly a semielliptical notch of depth h extending into the material from a free surface will have the same value for K_t (Fig. 7.2.1 b). A sharp crack may be considered as an approximation to an ellipse in which $h \gg l$ (Fig. 7.2.1 c) so that

$$K_t \approx 2\sqrt{\left(\frac{h}{l}\right)} \tag{7.2.2}$$

Hence in a sheet of material with such a crack, where the mean stress away from the crack is σ, the stress at the tip of the crack will be

$$\begin{aligned}\sigma_{\text{tip}} &= K_t \sigma \\ &= 2\sigma\sqrt{\left(\frac{h}{l}\right)}\end{aligned} \tag{7.2.3}$$

It would be expected that when the stress reaches the cohesive stress, the atomic or molecular bonds would be broken, increasing the length of the crack. Hence the expected strength of the material, as measured by the average stress σ necessary to cause fracture, would be given by

$$2\sigma\sqrt{\left(\frac{h}{l}\right)} \approx \frac{E}{15}$$

or

$$\sigma \approx \frac{E}{30}\sqrt{\left(\frac{l}{h}\right)} \tag{7.2.4}$$

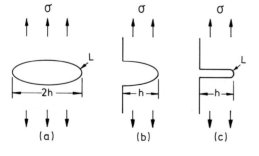

Fig. 7.2.1. Stress concentrations in a unidirectional stress field due to (a) an elliptical hole, (b) a semielliptical edge notch, (c) a sharp edge crack.

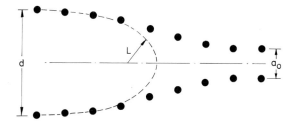

Fig. 7.2.2. Atoms at tip of a growing crack.

For a machined crack it would be possible to attach a value to *l*, but once a crack is growing, the configuration of atoms at the crack tip can be envisaged as in Fig. 7.2.2. As the atoms separate more from their equilibrium spacing the interatomic forces will first increase to a maximum and then decrease until at a separation *d* the interatomic forces are effectively negligible. The value of *l* can be taken as approximately $d/2$ which for a material with a simple atomic structure is approximately $2.5\,a_0$, where a_0 is the equilibrium atomic spacing.

In Eq. (7.2.4) the crack length appears in the denominator, so that if the crack-tip radius remains constant in a growing crack, then the stress necessary to cause propagation will decrease as the crack extends. Hence in an ideal brittle material, once the magnitude of stress necessary to start propagation has been reached, then, if that stress is maintained, the fracture would go to completion.

If however, there were some plastic flow in the highly stressed region near the crack tip, the crack might become blunted and for a constant value of σ the crack might not extend.

A relation between the stress-concentration factor, impact transition temperature, and free volume is discussed in Chapter 10.

7.3   Griffith Criterion

In the previous section, Eq. (7.2.4) was given as a condition for the spread of a crack from a flaw. This is a necessary condition, but by itself is not sufficient. The other condition was the one derived by Griffith[2,3] from the theorem of minimum energy.

Consider a body of material subjected to a series of external forces, P_i (Fig. 7.3.1), which also contains an internal flaw. The body is stressed elastically and therefore

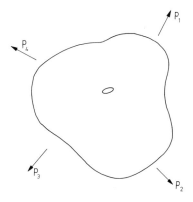

Fig. 7.3.1. Stressed body containing a flaw.

contains elastic strain energy. If there is an increase in the size of the flaw, then the body may deform overall so that the point of application of each external force moves a distance δ_i and there will be a change in the stored elastic strain energy.

Griffith postulated that under a given loading the crack extension could continue only if by doing so the total energy of the system decreases. There are three factors which contribute to this total energy:

(i) The potential energy of the external forces, F. If the body deforms, then each external force does work $P_i \cdot \delta_i$, so that the total potential energy of the system changes by $-F$ where

$$F = \sum P_i \cdot \delta_i$$

(ii) The stored elastic strain energy, U. The elastic strain energy is that portion of the work done on the body which deforms it and which can be recovered when the deforming forces are removed. For example, if a unit volume of the material is subjected to a uniaxial stress of magnitude σ, the elastic strain energy is $\sigma^2/2E$.

(iii) The work done against the cohesive forces as the new crack surfaces are formed, W. In a linear-elastic material this will equal the surface energies of the newly formed surfaces. If there is plastic flow associated with the crack growth, then the work to cause that plastic flow must also be considered. Depending upon the size of the plastic region, some of this work may be absorbed at a distance from the crack surfaces.

Suppose that for a small increase in crack size, these energies change by $-\delta F$, δU and δW respectively. Then the crack will propagate if, and only if, the sum of these three energy terms is equal to or less than zero, i.e.,

$$\delta U + \delta W - \delta F \leqq 0$$

Effectively, this is saying that the work done by the applied forces is distributed between the elastic strain energy and the crack surface energy. So fracture can occur only if

$$\delta F > \delta U + \delta W \tag{7.3.1}$$

7.4   Case of the Cracked Plate

We will apply the criterion of Griffith to the case of a large plate, of thickness t, of a material which behaves in a linear-elastic manner, and which is subjected to a uniaxial stress. Consider a through-thickness crack of length $2h$ which is perpendicular to the stress direction (Fig. 7.4.1).

With a crack of constant length $2h$ present in the plate, the load-displacement relationship would be linear as shown in Fig. 7.4.2. The slope of the straight line OA, dP/du, is known as the *stiffness* of the plate and its reciprocal, du/dP, is the *compliance*.

If the crack had been of a somewhat greater length $2(h+\delta h)$ the plate would have a greater compliance, and the load-displacement curve would be OB.

Two cases of crack growth will be considered. In the first case a constant load P is applied along two opposite edges of the plate (Fig. 7.4.1 a) which causes a displacement u of one edge relative to the other; the edges are free to move relatively as the crack extends. In the second, the edges are held fixed as the crack extends.

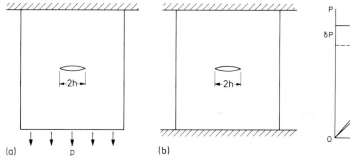
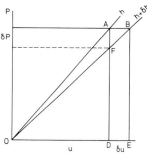

Fig. 7.4.1. Stressed plate with crack under (a) constant load, (b) constant displacement.

Fig. 7.4.2. Variation of compliance with length.

Constant Load

Now consider that the crack extends from its original length $2h$ to a new length $2(h+\delta h)$ while the plate is under the constant load P. The plate extends by δu so that the work done by the external forces is

$$\delta F = P\,\delta u \tag{7.4.1}$$

The stored elastic strain energy at load P when the crack length is $2h$ is given by

$$\int_{0}^{u} P\,du = \tfrac{1}{2}Pu \tag{7.4.2}$$

which is the area OAD under the load-extension curve. Similarly, the stored elastic strain energy at load P and crack length $2(h+\delta h)$ is the area OBE under the other load-extension curve, i.e., $\tfrac{1}{2}P(u+\delta u)$. If, therefore, the crack grows from $2h$ to $2(h+\delta h)$ while the load is constant at P, the increase in stored elastic strain energy is

$$
\begin{aligned}
\delta U &= \text{area } OBE - \text{area } OAD \\
&= \text{area } OAB \\
&= \tfrac{1}{2}P\,\delta u
\end{aligned}
\tag{7.4.3}
$$

i.e.,

$$\delta F = 2\,\delta U \tag{7.4.4}$$

Only one-half of the work done by the external forces has become stored elastic strain energy. Griffith[2] showed that this was a general result for any elastic body subject to external forces and containing a growing crack.

By substitution from Eq. (7.4.4) into Eq. (7.3.1) we find the condition for the crack to propagate is that

$$-\delta U + \delta W < 0 \tag{7.4.5}$$

We now need to derive these two quantities in terms of the known variables. Griffith used a solution previously derived by Inglis[4] for the distribution of stresses around an elliptical hole of major axis $2h$ and much smaller minor axis. He showed that in a plate of thickness t the presence of the hole reduced the elastic strain energy, compared to a plane plate by

$$U = -\frac{\pi h^2 t \sigma^2}{E} \tag{7.4.6}$$

The area of crack surface is $4ht$ and the surface energy is e_s per unit area so that

$$W = 4ht\,e_s \tag{7.4.7}$$

Differentiating Eqs. (7.4.6) and (7.4.7) with respect to h and substituting into Eq. (7.4.5) we get as the condition for crack extension

$$\frac{d}{dh}(-U+W) = \frac{d}{dh}\left(-\frac{\pi h^2 t\sigma^2}{E} + 4ht\,e_s\right) \leq 0$$

i.e.

$$-\frac{2\pi ht\sigma^2}{E} + 4t\,e_s \leq 0$$

or

$$h \geq \frac{2E\,e_s}{\pi\sigma^2} \tag{7.4.8}$$

This is the critical crack length, i.e. any crack of length equal to or greater than this value will propagate spontaneously at the stress level σ.

Thus, if there is an internal crack of length $2h$ or a surface crack of depth h in a very large sheet of a brittle material, the maximum stress that the sheet can sustain without the crack growing is

$$\sigma = \sqrt{\left(\frac{2E\,e_s}{\pi h}\right)} \tag{7.4.9}$$

This is generally known as the Griffith equation.

Constant Displacement

The other case concerns a similar plate (Fig. 7.4.1 b) stressed due to a load P and the displacement of the edges then held constant while the crack extends in length from $2h$ to $2(h+\delta h)$. The loading point will move from A to F in Fig. 7.4.2 and the load will be reduced by δP. As the applied load does not move during the crack growth no work is done by it, i.e., $\delta F = 0$.

There is a decrease in the stored elastic strain energy equal to area OAF. For a very small change in crack length

$$\text{area } OAF \approx \text{area } OAB$$

so that δU can be found by differentiating Eq. (7.4.6)

$$\delta U = \frac{2\pi ht\sigma^2}{E}\,\delta h \tag{7.4.10}$$

Hence from Eq. (7.4.5)

$$-\frac{2\pi ht\sigma^2}{E}\,\delta h + 4e_s\,t\,\delta h \leq 0$$

giving

$$\sigma \leq \sqrt{\left(\frac{2E\,e_s}{\pi h}\right)}$$

as before for the safe-stress regime.

Thus, we have shown that the Griffith equation is identical for both constant load and constant displacement conditions and we may infer that it would also be applicable for any intermediate conditions.

The two variables are σ and h, whereas E and e_s are properties of the material. The equation can be rearranged to separate them

$$\sigma \sqrt{(\pi h)} = \sqrt{(2 E e_s)} \qquad (7.4.11)$$

In the above treatment, we considered an extension of δh at each end of the crack. For an extension at one end of the crack, the strain energy release rate is $\frac{1}{2} dU/dh$ in a constant displacement condition, which equals $\pi \sigma^2 h/E$ per unit thickness of plate. This is commonly replaced by the term G_r which is known as the *elastic energy release rate*. It is also called the crack driving force. Hence

$$G_r = \pi \sigma^2 h/E \qquad (7.4.12)$$

The rate at which energy is consumed in propagating one end of the crack, $\frac{1}{2} dW/dh$, is called the *crack resistance, R*. In the case considered, R is constant and equal to $2e_s$ per unit thickness.

Both G_r and R are expressed in units of energy per unit plate width per unit crack extension.

The Griffith condition therefore states that G_r must be at least as large as R before a crack propagates. So for a case where R is constant the condition is that G_r exceeds a critical value G_c, the *critical energy release rate*.

Griffith tested his theory by experiments on thin-walled glass spheres and cylinders containing cracks of known dimensions. He pressurized these internally until they fractured by the cracks spreading, and from the pressure reached in each case he calculated the value of $\sigma \sqrt{h}$. This quantity was reasonably constant, thus confirming the general nature of the theory, even though its average value was some 80% greater than that expected from known values of E and e_s.

For materials which undergo some plastic deformation before fracture, the Griffith equation will not be applicable because some energy will be involved in plastic deformation. In a later section it will be shown how the Griffith equation has been modified to cater to this. First, however, we must study the stress distribution near the crack tip.

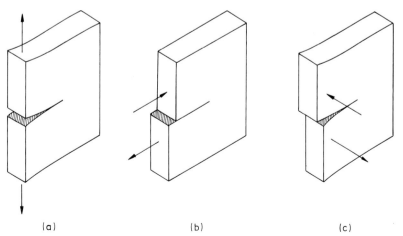

(a)   (b)   (c)

Fig. 7.5.1. The three modes of crack extension. (a) Mode I, opening or tensile mode; (b) Mode II, in-plane shear or sliding mode; and (c) Mode III, antiplane shear or tearing mode.

7.5   Crack Opening Modes

A crack in a solid can be deformed in more than one way. We distinguish three modes which are shown in Fig. 7.5.1. Cracks that we have already considered are loaded in the normal or opening mode and are referred to as Mode I. Shear within the plane of the material gives the Mode II or sliding mode, while out-of-plane shear gives Mode III or the tearing mode. A general case of cracking may include superposition of all three modes. Of the three, Mode I is the only one that will receive detailed consideration in this chapter.

7.6   Stresses Near a Crack Tip

Consider a sheet of material that behaves in a linear-elastic manner and which contains a crack of length $2h$ with infinitely sharp ends (Fig. 7.6.1). Let the sheet remote from the crack be subject to a uniform tensile stress σ acting in a direction perpendicular to the plane of the crack, that is, subject to Mode I loading. Take an origin at one crack tip with x and y axes parallel and perpendicular to the plane of the crack and within the plane of the sheet. Any position in the sheet can be described in terms of its polar co-ordinates $\bar{r}$ and θ, and the stresses on a small element at that point are normal stresses σ_x and σ_y and shear stress τ_{xy}.

The stress distribution has been solved by Westergaard[5]; his results are

$$\sigma_x = \sigma \sqrt{\frac{h}{2\bar{r}}} \cos \frac{\theta}{2} \left(1 - \sin \frac{\theta}{2} \sin \frac{3\theta}{2} \right)$$

$$\sigma_y = \sigma \sqrt{\frac{h}{2\bar{r}}} \cos \frac{\theta}{2} \left(1 + \sin \frac{\theta}{2} \sin \frac{3\theta}{2} \right)$$

$$\tau_{xy} = \sigma \sqrt{\frac{h}{2\bar{r}}} \sin \frac{\theta}{2} \cos \frac{\theta}{2} \cos \frac{3\theta}{2} \qquad (7.6.1)$$

In addition there are some constant terms, but these become relatively insignificant as $\bar{r}$ gets smaller.

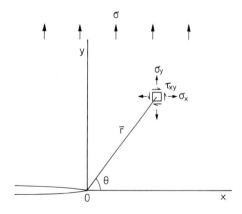

Fig. 7.6.1. Co-ordinate system in vicinity of crack tip.

In more general terms, these equations can be summarized as

$$\sigma_{ij} = \sigma \sqrt{\left(\frac{h}{2\bar{r}}\right)} f_{ij}(\theta) + C_{ij} \tag{7.6.2}$$

Concentrating first on the stress for $\theta = 0$, i.e. in a line ahead of the crack, we have

$$\sigma_x = \sigma \sqrt{\left(\frac{h}{2\bar{r}}\right)} + C_x \tag{7.6.3}$$

$$\sigma_y = \sigma \sqrt{\left(\frac{h}{2\bar{r}}\right)} + C_y \tag{7.6.4}$$

and

$$\tau_{xy} = 0 \tag{7.6.5}$$

We can write Eq. (7.6.4) as

$$\sigma_y \sqrt{(2\pi x)} = \sigma \sqrt{(\pi h)} + \sqrt{(2\pi x)}\, C_y \tag{7.6.6}$$

As x gets smaller, the term $\sqrt{(2\pi x)}\, C_y$ will become negligible compared with $\sigma \sqrt{(\pi h)}$ so that in the limit

$$\lim_{x \to 0} \sigma_y \sqrt{(2\pi x)} = \sigma \sqrt{(\pi h)} \tag{7.6.7}$$

even though the stresses tend to become infinite.

The quantity $\sigma \sqrt{(\pi h)}$ has been called the *stress-intensity factor* and is denoted by K_I *. The subscript refers to the Mode I loading condition.

The stress-intensity factor is a measure of the stress singularity at the crack tip, i.e., it characterizes the stress distribution near the tip. The dimensions of K are stress $\times$ length$^{1/2}$. Also its value depends upon both crack size and the general stress level.

For cracks of other shapes or cracks in objects of other geometries, the stress distribution will differ, but K is still defined as

$$\lim_{x \to 0} \sigma_y \sqrt{(2\pi x)}$$

This limit has been evaluated for many cases (see Section 7.7) and is usually expressed as

$$K_I = \alpha^* \, \sigma \sqrt{(\pi h)} \tag{7.6.8}$$

where α^* is a geometrical factor appropriate to the particular crack and component shape (see Section 7.12).

So a single term K_I gives all the information about the distribution of stress near a crack tip, and Eqs. (7.6.1) can be rewritten as

$$\sigma_x = \frac{K_I}{\sqrt{(2\pi \bar{r})}} \cos \frac{\theta}{2} \left(1 - \sin \frac{\theta}{2} \sin \frac{3\theta}{2}\right)$$

$$\sigma_y = \frac{K_I}{\sqrt{(2\pi \bar{r})}} \cos \frac{\theta}{2} \left(1 + \sin \frac{\theta}{2} \sin \frac{3\theta}{2}\right)$$

---

* K_I should not be confused with the stress-concentration factor (see Section 7.2); both have somewhat similar names and conventionally the same basic symbol.

and

$$\tau_{xy} = \frac{K_1}{\sqrt{(2\pi \bar{r})}} \sin \frac{\theta}{2} \cos \frac{\theta}{2} \cos \frac{3\theta}{2} \qquad (7.6.9)$$

A further consideration for a Mode I crack is whether in the direction parallel to the line of the crack tip, i.e., the z-direction, the conditions are such that plane stress or plane strain occurs. In the former case, the stress in the transverse direction, σ_z, is zero and there is a strain in that direction given by

$$\varepsilon_z = -v(\sigma_x + \sigma_y) \qquad (7.6.10)$$

where v is Poisson's ratio. Obviously, nearer the crack tip, the values of σ_x and σ_y are larger, so that the transverse strain increases and dimples form on the surfaces at the crack tip. If, however, the plate is thick relative to the crack size, such transverse strains would be limited by shear forces from the adjacent less strained regions. For a sufficiently thick plate, there will be no transverse strain near the mid-thickness, i.e., plane-strain conditions apply. In that case

$$\varepsilon_z = 0$$

and

$$\sigma_z = v(\sigma_x + \sigma_y) \qquad (7.6.11)$$

7.7 Geometrical Factors

The stress distributions given in Eqs. (7.6.9) apply strictly to cracks in infinitely large plates. In plates of finite size there are boundary conditions which have to be satisfied. In general, the stress distribution and therefore, the stress-intensity factor can be derived only by approximate methods: the stress-intensity factor becomes a function of geometry as well as $\sigma \sqrt{(\pi h)}$. An example is the case of a transverse crack of length $2h$ in a plate of width W, Fig. 7.7.1. The spread of the total load around

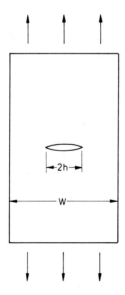

Fig. 7.7.1. Plate of finite width containing a crack.

the crack is restricted by the proximity of the sides, and so, there is a greater concentration of stress compared with a plate of infinite width. Hence the stress intensity is higher. Feddersen[6] showed that K_I is very closely approximated by

$$K_I = \sigma \sqrt{(\pi h)} \sqrt{(\sec \pi h / W)} \tag{7.7.1}$$

so that $\sqrt{(\sec \pi h / W)}$ is the geometrical factor α^* introduced in Eq. (7.6.8).

Simple analytical expressions are often not appropriate. It is common practice to express α^* as a polynomial in (h/W) where h is as before the semicrack length and W is a typical specimen dimension. Examples are given in Section 7.12.

Values of α^* have been calculated for many situations using a finite element or other methods and are listed in various works.[7,8]

7.8 Plastic Zone Size

Equations (7.6.9), which are an elastic solution would give stresses that become infinite at the crack tip. If, however, the material is elastic plastic with a yield stress σ_{ys}, then all the material where the elastic stress would exceed σ_{ys} will yield. The extent r'_p of the crack tip plastic zone in the direction away from the crack tip (Fig. 7.8.1) would therefore be given by

$$\sigma_{ys} = \frac{K_I}{\sqrt{(2\pi r'_p)}} \tag{7.8.1}$$

or

$$r'_p = \frac{K_I^2}{2\pi\sigma_{ys}^2} = \frac{h}{2}\left(\frac{\sigma}{\sigma_{ys}}\right)^2 \tag{7.8.2}$$

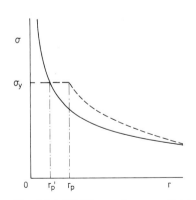

Fig. 7.8.1. Yielding and stress redistribution near crack tip, ——— elastic solution, — — — redistribution after plastic yielding.

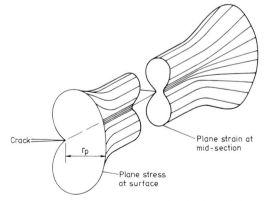

Fig. 7.8.2. Shape of plastic zone ahead of crack in a thick plate.

The yielded material does not support as much load as it would if it had remained elastic. Thus, there is a redistribution of stress and the plastic zone will be of a larger size than that given by Eq. (7.8.2), approximately

$$r_p = \frac{K_I^2}{\pi \sigma_{ys}^2} \tag{7.8.3}$$

In the plane-strain plastic zone, the transverse stress, σ_z, is positive for a positive σ_x and σ_y so that a state of triaxial tension exists. This causes the effective yield stress to be larger than the uniaxial value. The ratio of the maximum stress that can exist to the uniaxial stress is known as *the plastic constraint factor*. Its value can be as high as 3. As a consequence, the plastic zone size in the plane-strain region will be smaller than that in the plane-stress region and its variation of size through the specimen thickness can be envisaged as in Fig. 7.8.2.

7.9   Modified Griffith Equation

The Griffith treatment applied to a material which behaved elastically up to the point of fracture. This is not true of most materials, particularly most metals and polymers. If the plastic zone size is small relative to the total bulk of the material, then the bulk will still behave elastically and the stored elastic strain energy will be almost the same as in the strict Griffith treatment. Also, the compliance of the whole sheet or body will be almost the same and so the work done by the external forces would be the same.

The work necessary to form the crack surfaces will, however, be different; the work of plastic deformation has to be considered as well as the surface energy. As an outcome of the study of fracture of steel structures, in particular of ships, Irwin[9] and Orowan[10] independently produced a modification of the Griffith Eq. (7.4.9) to

$$\sigma = \sqrt{\left(\frac{2(e_s + e_p)E}{\pi h} \right)} \tag{7.9.1}$$

where e_s is the surface energy, and e_p is the work done in causing plastic deformation per unit area of crack surface. Both these terms are values of work against external forces, the first in the breaking of atomic bonds and the second in overcoming the forces opposing plastic flow.

e_p will greatly exceed e_s in plastic materials and will be dependent upon strain rate, temperature, and geometry. The ratio of e_p to e_s varies from material to material, but for different classes of materials, it is of the following order of magnitude:

metals	$10^4 - 10^6$
polymers	$10^2 - 10^4$
ceramics	10

Returning to the terminology introduced in Section 7.4, the crack resistance is $2(e_s + e_p)E$ and the critical energy release rate G_c (or G_{Ic} in plane strain) must exceed this value for the fracture to proceed. It is not practical or necessary to determine the values of e_s and e_p separately; it is sufficient to measure G_c.

7.10   Fracture Toughness

We see from Eq. (7.9.1) that fracture occurs when the stress-intensity factor $K(=\sigma\sqrt{(\pi h)})$ exceeds a critical value given by $\sqrt{(G_c E)}$. This is termed the *critical stress-intensity factor*, K_c. The value obviously depends upon the amount of plastic work that occurs in the growth of the crack, which in turn will depend upon the state of stress in that region.

7.11   Plane-Stress and Plane-Strain Conditions

It has already been said that in a thin sheet there will be plane-stress conditions, while in a thick sheet the condition approximates to plane strain. The ratio of stress to strain for elastic behavior, i.e., the apparent modulus, is greater in plane strain than in plane stress, the former, E', being given by

$$E' = E/(1-v^2) \tag{7.11.1}$$

Also as described in Section 7.8, the size of plastic zone differs and hence the amount of plastic work will also differ.

Typically, in a thin sheet, failure occurs by shear through the material on a 45° plane (Fig. 7.11.1 a). Even if a crack starts perpendicular to the plane of the sheet, it tends to rotate to the 45° configuration and fails by Mode III shearing. The force per unit length of crack to cause failure must have a component in the plane of the crack equal to the shear stress in the material times the width of crack surface. The strain to failure is also the width of crack surface. Hence, the work done to extend the crack by unit length will be proportional to the square of the thickness, or the work per unit length per unit thickness is proportional to thickness. Hence the apparent value of K_c, should increase with increasing sheet thickness.

As thicker plates are considered, the plane-strain region will occupy a progressively greater fraction of the total thickness. In this region, the plastic zone is smaller in extent and the fracture surface remains perpendicular to the plane of the plate. Near the edges, which will be more nearly plane stress, there will be 45° shear lips (Fig. 7.11.1 b and c). In the plane-strain region, less plastic flow occurs and therefore, less plastic work is expended in forming the crack surfaces. K_c, which initially rose in a linear manner with thickness, will increase at a decreasing rate as the plate thickness is in-

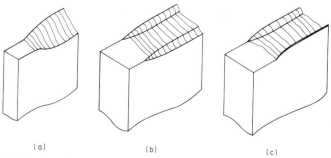

(a)   (b)   (c)

Fig. 7.11.1. Shapes of cracks in thin and thick plates. (a) From 90° starter, crack rotates to 45° plane; (b) and (c) crack mainly perpendicular to surfaces with narrow shear lips at sides.

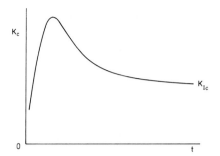

Fig. 7.11.2. Typical variation of K_c with plate thickness.

creased, pass through a maximum, and then fall asymptotically to the *plane-strain-fracture toughness* K_{Ic}, as in Fig. 7.11.2.

Whereas K_c is a function of material and plate thickness, K_{Ic} is solely a material property.

In design considerations, such as calculating the maximum size flaw that could be permitted in a structure or component without risk of fracture, it is usual to use the value of K_{Ic} for the material, because it is the lowest one, and hence the safest.

7.12   Determination of K_{Ic}

The fracture toughness of a material is measured by a test on a specimen of finite size, containing a machined notch which has been extended by controlled fatigue loading so that the end of the notch should be as sharp as possible. The specimen is then loaded in a prescribed manner until the crack propagates, this being determined acoustically or by recording a change in the opening of the notch. Standardized specimens and test procedures have been published by the British Standards Institution[11] and the American Society for Testing and Materials.[12]

Two forms of specimens used are the three-point bend and the compact tension specimens. The dimensions have to be within certain ratios and the procedures for loading, precracking, testing, and observing are prescribed. Expressions for the geometrical factor α^* in Eq. (7.6.8) are quoted in polynomial form, from which the plane-strain-fracture toughness can be calculated. Thus, for the three-point-bend specimen (Fig. 7.12.1).

$$K=\frac{PS'}{BW^{3/2}}\left[2.9\left(\frac{h}{W}\right)^{1/2}-4.6\left(\frac{h}{W}\right)^{3/2}+21.8\left(\frac{h}{W}\right)^{5/2}\right.$$

$$\left.-37.6\left(\frac{h}{W}\right)^{7/2}+38.7\left(\frac{h}{W}\right)^{9/2}\right]$$

$$(7.12.1)$$

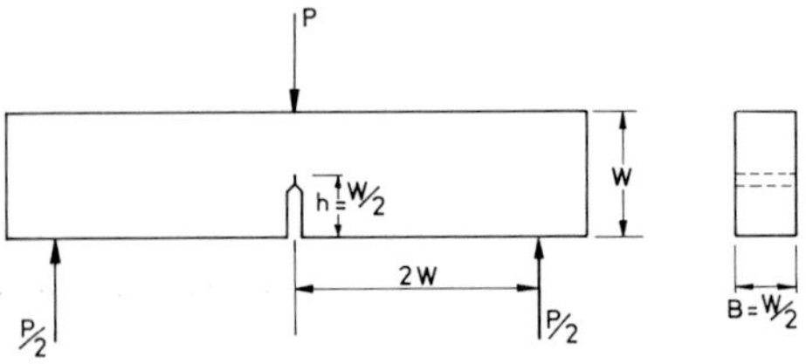

Fig. 7.12.1. Standard three-point-bend specimen for K_{Ic} determination.

where S' is the span between the supports, and for the compact tension specimen
(Fig. 7.12.2)

$$K = \frac{P}{BW^{1/2}} \left[29.6 \left(\frac{h}{W}\right)^{1/2} - 185.5 \left(\frac{h}{W}\right)^{3/2} + 653.7 \left(\frac{h}{W}\right)^{5/2} \right.$$

$$\left. - 1017 \left(\frac{h}{W}\right)^{7/2} + 6389 \left(\frac{h}{W}\right)^{9/2} \right] \qquad (7.12.2)$$

The depth h of the precracked notch can be ascertained from the final appearance
of the fracture surface, since the fatigue and fast fracture areas are clearly distinguish-
able.

The requirements on specimen size are such that plane strain should exist at the
crack tip. As stated earlier, a requirement for this is that the plastic zone size shall
be small relative to the total volume. Since there is always a plane-stress region at
the surface, the thickness has to be large if this region is to be relatively small. It
was shown in Eq. (7.8.2) that the plastic zone size is proportional to (K_{Ic}^2/σ_{ys}^2) so therefore the thickness B should be not less then $\alpha K_{Ic}^2/\sigma_{ys}^2$ where α is a value to be determined.
It has been found that if $\alpha > 2.5$ then consistent values of K_{Ic} are obtained. Another
requirement is that the crack length h should not be less than this value.

Eq. (7.8.3) gave the plastic zone size as

$$r_p = \frac{1}{\pi} \left(\frac{K_{Ic}^2}{\sigma_{ys}^2} \right)$$

Under plane-strain conditions, it is smaller by a factor of 3. Hence the requirement
is that both B and h should be about 25 times the plastic zone size with other dimensions
in proportion as in Fig. 7.12.1 and Fig. 7.12.2.

To select a specimen of sufficient size, it is, therefore, necessary to estimate the
K_{Ic} of the material. Having carried out a test and determined a K value at fracture
using the appropriate equation, this value is known as K_Q, the candidate fracture
toughness. Checks must then be made as to whether both h and B are larger than
$2.5 K_Q^2/\sigma_{ys}^2$. If all requirements are satisfied, then K_Q is taken to be the required K_{Ic}.

The opening of the crack during a K-test is monitored by a suitable clip gauge
and is plotted against load on an $X-Y$ recorder. The load versus crack opening displacement (COD) curve can have various shapes ranging from linear behavior to sudden
fracture; through the case of sudden crack extension, called pop-in; to nonlinear curves
which do not show maxima. The Standards give directions for interpreting these curves
and deciding whether a valid test result has been obtained.

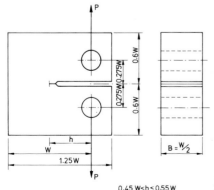

0.45 W < h < 0.55 W

Fig. 7.12.2. Standard compact tension specimen.

7.13   Applicability of Test Results

If the test result is not valid, then K_Q will be larger than the plane-strain-fracture toughness and cannot be quoted if a K_{1c} result is called for. However, if the K_Q value was obtained with a specimen of the same thickness as the plate to be used in a construction, then its value would be applicable for fracture in that structure, where plane-strain conditions would also not apply.

Materials with high toughness and low yield strength would require large values of h and B for valid tests. For example, metals such as low-strength carbon steels with yield strengths of about 220 MPa and fracture toughnesses of about 140 MN m$^{-3/2}$ would require a specimen one meter thick. Tests on specimens of such size would not be practical. Apart from the prohibitively large testing machine necessary, the properties would be atypical of material manufactured in smaller sections; in addition materials are unlikely to be used in such thick sections.

The limits of LEFM are then reached and different fracture mechanics concepts have to be used. These are discussed in Sections 7.19 and 7.20. The effect of specimen dimensions on the validity of K_c results for polymers, is discussed further in Section 8.4.

7.14   Compliance Calibration

It is possible to determine G_r and K values for a specimen by experiment without the need for calculated expressions of the type quoted in Eq. (7.12.1) and (7.12.2).

In a cracked specimen with no crack growth, the compliance D is given by

$$u = DP \tag{7.14.1}$$

where u is the displacement due to a load P, and the stored elastic strain energy is

$$U = \tfrac{1}{2} Pu = \tfrac{1}{2} DP^2 \tag{7.14.2}$$

Also by definition $G_r = -dU/dH$ for unit thickness. Hence in material of thickness B, we will have

$$G_r = \frac{P^2}{2B} \frac{\partial D}{\partial h} \tag{7.14.3}$$

If one can make a series of compliance calibrations for a given specimen geometry for different crack lengths, then from a plot of D-versus h the slope $\partial D/\partial h$ can be determined for any value of h. Specimens can be prepared with various crack or slit lengths, or the slit in a single specimen can be extended between calibrations by means of a fine saw. Alternatively, in an extension-controlled test, once a load is reached at which crack growth occurs, the load will fall off and crack growth will be limited. The specimen can be unloaded, another measurement of crack length taken and a fresh calibration obtained. Then, knowing the particular load that will cause growth for a given crack length, G_{1c} can be determined, and hence K_{1c} as

$$K_{1c} = \sqrt{(EG_{1c})} \tag{7.14.4}$$

in plane stress, and

$$K_{1c} = \sqrt{\left(\frac{EG_{1c}}{(1-v^2)}\right)} \tag{7.14.5}$$

in plane strain.

Examples of some frequently used specimens which employ this principle and which are often used for testing polymers are given in the sections that follow.

7.15   Double Cantilever Beam

This specimen (Fig. 7.15.1) is a rectangular specimen containing a crack spreading from one end. The two halves of the specimen at the cracked end are loaded either by externally applied forces or by some form of wedging action.

Provided that the material is behaving in an elastic manner, the relative displacement of the two ends by a load P will be

$$u = 2\,\frac{P h^3}{3 E I} \tag{7.15.1}$$

where I is the second moment of area of either half beam and is $\dfrac{B W^3}{2^3 \cdot 12}$. This ignores the small contributions due to shear and to the ends not being rigidly fixed at the plane of the crack tip.

Within the limitation of these factors, the compliance is

$$D = \frac{u}{P} = \frac{8 h^3}{E (W/2)^3\, B} \tag{7.15.2}$$

which substituted in Eq. (7.14.3) gives

$$G_r = \frac{12\, P^2 h^2}{E (W/2)^3\, B^2} \tag{7.15.3}$$

and from Eq. (7.14.4)

$$K = 2\,(3)^{1/2}\,\frac{P h}{B (W/2)^{3/2}} \tag{7.15.4}$$

Testing would be carried out as outlined in Section 7.14.

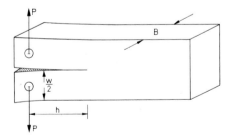

Fig. 7.15.1. Double cantilever beam (DCB) specimen.

7.16   Tapered Cantilever Beam

This specimen works on a principle similar to the double cantilever beam, but the depth increases with distance along the specimen (Fig. 7.16.1). Over a limited range of crack sizes, this specimen exhibits an almost constant K_I and constant G_I for a constant load. Hence, in a succession of loading tests, crack extension should always occur at the same value of load on each reloading.

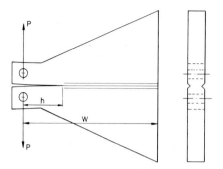

Fig. 7.16.1. Tapered cantilever beam specimen.

To prevent the crack deviating from the line of symmetry, which would upset the K calibration, as often happens in flat-sided specimens, shallow grooves are machined along the sides to direct the crack. The compliance and its variation with crack length can be obtained by calibration.

7.17   Double Torsion Specimen

Another form of specimen which gives a constant-K calibration is the double torsion specimen (Fig. 7.17.1) developed by Kies and Clark.[13] The specimen is a rectangular plate supported on two parallel rollers and the load is applied by two hemispheres near one end so that the two halves of the specimen are in torsion. The specimen may be grooved along its lower face to ensure that the crack propagates along the midplane. For a sufficiently brittle material the crack propagates in Mode I. As the crack proceeds along the specimen, the value of K remains constant for a constant load. This specimen has been used to study crack-growth rates under constant-K conditions.

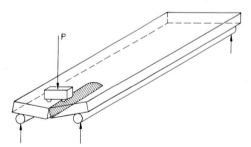

Fig. 7.17.1. Double torsion specimen.

7.18   Crack-Growth-Resistance Curve

In a sheet material containing a crack and loaded in tension, plane-stress conditions will apply and there will be no plastic constraint factor. Hence plastic deformation ahead of the crack tip will be extensive except in brittle materials and the plastic zone size will increase with increasing strain. Failure, i.e., crack extension, will be by thinning of the sheet and by growth and coalescence of microvoids ahead of the crack front. The energy required to form new crack surfaces will, therefore, depend

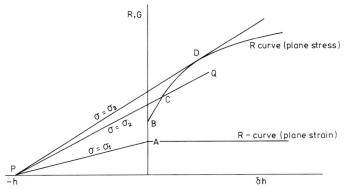

Fig. 7.18.1. G_r and R curves for plane stress.

upon the growth of the crack; that is, the crack resistance R, introduced in Section 7.4, will not be constant, but will increase from the plane-strain value (i. e. $2e_s$) with increase of crack length, probably attaining a fairly constant value for extensive crack growth. A curve of R versus crack length, i. e. a crack-growth-resistance curve, commonly called an R-curve, an idea introduced by Irwin[14] and discussed in depth later by Kraft, Sullivan and Boyle,[15] may be represented as curve BCD in Fig. 7.18.1. Suppose the plastic zone size is sufficiently small, relative to the overall sheet size, for the energy release rate to be given as in Section 7.4 by $\pi \sigma^2 h/E$. Then for a crack of length h the variation of G_r, with the length of the growing crack would be a straight line through the point $-h$ on the crack growth axis and of slope $\pi \sigma^2/E$.

Suppose the crack were loaded to a stress σ_1, with the G_r value given by line PA in Fig. 7.18.1. The available energy release rate is less than R for plane strain conditions and so no crack extension could occur. At a higher stress σ_2, the G_r value would be given by line PC, and if the crack would grow under this constant stress, energy release according to CQ would be available. However, this is lower than the R value for crack extension and, therefore, crack growth would cease at C. As the stress was increased, the crack could grow as determined by the R-curve; the condition at any stage is that

$$G_r = R$$

Hence, there would be slow crack growth controlled by the stress level. If the stress increased until the G_r-curve were tangential to the R-curve, at D, $\sigma = \sigma_3$, then rapid fracture can occur, as for any further crack growth G_r would be greater than R. The condition for the onset of failure is, therefore, that

$$\frac{\partial G_r}{\partial h} = \frac{\partial R}{\partial h}$$

If, therefore, an R-curve for a material can be derived, either experimentally[16] or analytically, then failure under conditions of plane stress can be predicted.

7.19 Crack Tip Opening Displacement

In cases where linear-elastic-fracture mechanics is not applicable (see Section 7.13) and significant plasticity occurs, the fracture process is controlled primarily by the extent of plastic deformation close to the crack tip and the separation of the crack

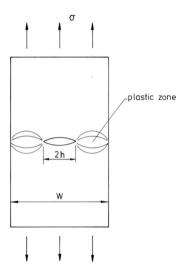

Fig. 7.19.1. Plastic yielding in tension.

faces is presumably a measure of this deformation. If the material has a high fracture toughness, the stress that would be predicted to cause fracture may be greater than the yield stress, and general yielding will occur before fracture can take place, so that no measurement of K_{Ic} is possible.

Consider a plate specimen of width W containing an internal crack of length $2h$. As the load is increased, plastic flow will spread from the crack tip across the entire remaining section before fracture starts (Fig. 7.19.1). The net stress on the uncracked portion will be

$$\sigma_{net} = \frac{W}{W-2h}\,\sigma > \sigma_{ys} \tag{7.19.1}$$

Plastic flow can occur freely at the crack tip.

A measure of the plastic strain at the crack tip is known as the crack-tip-opening displacement (CTOD). Wells[17] proposed that once a critical CTOD is exceeded, fracture will occur. It can be calculated that under linear-elastic conditions

$$\text{CTOD} = \frac{4}{\pi}\frac{K_I^2}{E\sigma_{ys}} \tag{7.19.2}$$

In the case of LEFM, fracture occurs if K_I equals K_{Ic} which implies a constant value for the critical CTOD.

For a simpler analysis, if the crack opens by a distance δ before failure, then to a first approximation the work done will be $\sigma_{ys}\,\delta$ per unit area of crack surface. Wells[17] pointed out, therefore, that we could expect

$$G_r = \sigma_{ys}\,\delta \tag{7.19.3}$$

whereas from equation 7.19.2 we would get

$$G_r = \frac{\pi}{4}\,\sigma_{ys}\,\delta \tag{7.19.4}$$

More refined models give values for the numerical factor which differ slightly.

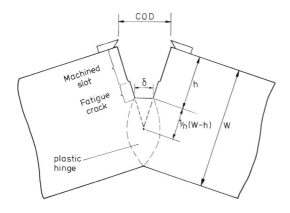

Fig. 7.19.2. Determination of CTOD, δ, in a three-point-bend test.

In general it can be shown that

$$\text{CTOD} = \frac{G_{\text{I}}}{\lambda_p \sigma_{ys}} = \frac{K_{\text{I}}^2}{E \lambda_p \sigma_{ys}} \tag{7.19.5}$$

for plane stress, or

$$\text{CTOD} = \frac{K_{\text{I}}^2 (1 - v^2)}{E \lambda_p \sigma_{ys}} \tag{7.19.6}$$

for plane strain.

The value of the constant λ_p depends upon the plastic constraint at the crack tip. Various authors give values between 1 and 2.3. Careful experimental measurements give values near unity.

In general, it is not possible to measure CTOD, but COD can be determined at the outer end of the notch with a suitable clip gauge. For example, with a notched three-point-bend specimen (Fig. 7.19.2), a plastic hinge can form so that there is relative rotation of the two ends. If the center of rotation is known, the CTOD can be calculated from the measured COD. An appropriate value of the factor, n, which locates the center of rotation can be determined by calibration tests or by calculation. A Standard has been published.[18]

7.20   *J*-integral

If there is extensive plasticity, G_r cannot be calculated from the elastic stress field because of the relatively large size of the crack-tip plastic zone. Solutions for elastic-plastic behavior are not readily available, if at all, but a quantity called the *J*-integral provides a means of determining the energy release rate for such cases.

Various contour integrals have been defined by Eshelby[19] which are path independent by the theorems of energy conservation. The two-dimensional form of one of these, the one known as the *J*-integral, is

$$J = \int_\Gamma \left\{ Z \delta y - T \frac{\partial u}{\partial x} ds \right\} \tag{7.20.1}$$

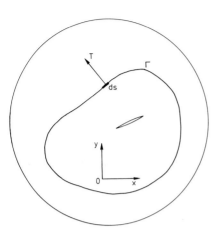

Fig. 7.20.1. Contour for definition of *J*-integral.   Fig. 7.20.2. Contour surrounding crack tip.

where: Γ is a closed contour in a stressed solid (Fig. 7.20.1),
        T is the tension vector perpendicular to the contour in an outward direction,
        u is the component of displacement of the contour in the x direction,
        ds is an element of the contour Γ, and
        Z is the strain energy (including plastic as well as elastic strain energy) per
unit volume.

Eq. 7.20.1 has been displayed for the first time as Eq. (2.6.9). This function can be shown to equal zero along any closed contour.

The integral was applied to crack problems by Cherepynov[20] and Rice.[21] A closed contour is taken as two curves surrounding the tip of the crack, one *DEF* inside the other *ABC* (see Fig. 7.20.2) which are joined by two portions of the crack surface *AF* and *CD*. The integral around the contour is zero. Along the parts *AF* and *CD* which lie parallel to the x axis and which have no normal stress on them, $T = 0$ and $\delta y = 0$. Therefore, the integral along *ABC* is equal and opposite in sign to that along *DEF*. For outward directed vectors, T, therefore, the integral is path independent.

Rice[21] has shown that the *J*-integral, as defined along a contour such as *ABC*, is the rate of change of potential energy U with extension of the crack, that is,

$$J = \frac{\partial U}{\partial h}$$

For a linear-elastic material we have seen that

$$G_r = \left(-\frac{\partial U}{\partial h}\right)_u$$

under constant displacement, and

$$G_r = \left(\frac{\partial U}{\partial h}\right)_P$$

under constant load. Hence for a linear-elastic material

$$J = G_r$$

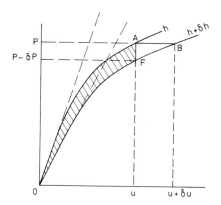

Fig. 7.20.3. Determination of *J*-integral.

In the nonlinear case, with E_γ as the total strain energy,

$$J = \left(\frac{\partial E_\gamma}{\partial h}\right)_P = \int_0^P \left(\frac{\partial u}{\partial h}\right)_P dP \qquad (7.20.2)$$

and

$$J = -\left(\frac{\partial E_\gamma}{\partial h}\right)_u = -\int_0^u \left(\frac{\partial P}{\partial h}\right)_u du \qquad (7.20.3)$$

for the constant load and constant displacement cases respectively. In general the $P-h$-curves will not be linear. The equivalent of Fig. 7.4.2 is shown in Fig. 7.20.3.

The energy release rate is represented by the shaded area which, apart from a second order effect (area *ABF*), is the same for constant load and constant displacement.

Use of these concepts implies that on unloading, the material returns along the loading curve. This is not strictly true; also, the *J*-integral is not applicable in such a case because the constant deformation theorem cannot be used for cases of unloading.

Nevertheless, where crack growth and force are associated with considerable plastic deformation, *J* has been shown to be a useful criterion.

As already shown, in the linear-elastic case

$$J = G_r$$

and therefore

$$J = K^2/E$$

So we postulate that crack growth or fracture can occur if *J* exceeds a critical value J_{1c} analogous to G_{1c} and equal to it if the material behaves in a linear-elastic manner.

In practice, *J* can be determined from changes in the load displacement diagram with changes in crack length[22] just as one determines G_r from changes in the compliance.

Having obtained an experimental value of J_{1c}, its application to the prediction of failure in a particular structure would involve the calculation of *J* values for successive increases in loading. This could involve extensive elastoplastic calculations using finite element methods.

7.21   Fracture Mechanics Applied to Fatigue Crack Growth

The discussion so far in this chapter has been limited to cases leading to monotonic crack growth and fracture, purely by the application of increasing loads. Cracks that do not propagate under a static load may nevertheless grow under an alternating load of the same magnitude. If they grow to a critical length for that load such that K_I reaches K_{Ic} then fracture occurs. Failure after a number, N, of cycles of loading in this manner is termed *fatigue failure*.

It was found by Paris and Erdogan [23] that the growth of a crack on each loading cycle was a function of the range of stress-intensity factor during a loading cycle, i.e.

$$\Delta K_I = K_{I\,max} - K_{I\,min} \tag{7.21.1}$$

The function over a wide range of K_I is of the form

$$\frac{dh}{dN} = C_0 (\Delta K_I)^m \tag{7.21.2}$$

where C_0 and m are constants; m is usually of the order of 4 for most materials. This equation is known as Paris's law.

Expressing the stress-intensity factor in terms of stress range

$$K_I = C\,\sigma\sqrt{h} \tag{7.21.3}$$

where C is another constant, and substituting in Eq. (7.21.2) we get

$$\frac{dh}{dN} = C_0\, C^m\, \sigma^m h^{m/2} \tag{7.21.4}$$

C will vary slightly with crack length. Otherwise for a constant stress range, the number of cycles N_c for a crack to grow from an initial length h_i to its final length h_f would be given by integrating Eq. (7.21.4), with the result

$$N_c = \frac{2}{(m-2)\, C_0\, C^m\, \sigma^m}\, \{ h_i^{(2-m)/2} - h_f^{(2-m)/2} \} \tag{7.21.5}$$

The behavior of polymers under fatigue loading and the agreement of the results with Paris's law is discussed in Section 18.5.

Bibliography

J.F. Knott, *Fundamentals of Fracture Mechanics,* Butterworth, London 1973.
D. Broek, *Elementary Engineering Fracture Mechanics,* Martinus Nijhoff, The Hague 1982.
D.G.H. Latzko, *Post-Yield Fracture Mechanics,* Applied Science Publishers, London 1979.

References

1. J.J. Larmor, *Phil. Mag.* 1892, **33**, 70.
2. A.A. Griffith, *Phil. Trans. Royal Soc.* 1921, **A 211**, 163.
3. A.A. Griffith, *Proc. 1st Int. Congress Applied Mech.,* p. 55, Delft 1924.
4. C.E. Inglis, *Trans. Inst. Naval Archit.* 1913, **55**, 219.
5. H.M. Westergaard, *J. Appl. Mech.* 1939, **6**, A 49.
6. C.E. Feddersen, Discussion in *Plane strain crack toughness testing of high strength metallic materials,* by W.F. Brown and J.E. Srawley, ASTM STP 410, p. 77, Amer. Soc. for Testing and Materials, Philadelphia 1967.

7. G.C. Sih, *Handbook of Stress Intensity Factors for Researchers and Engineers.* Institute of Fracture and Solid Mechanics, Lehigh University, Bethelehem, PA, 1973.
8. D.P. Rooke and D.J. Cartwright, *Compendium of Stress Intensity Factors,* HMSO, London 1976.
9. G.R. Irwin, *Trans. Amer. Soc. Metals* 1948, **40**, 147.
10. E. Orowan, *Rep. Progr. Phys.* 1948, **12**, 185.
11. B.S. 5447, *Methods of test for plane strain fracture toughness (K_{1c}) of metallic materials.* British Standards Institution, London 1977.
12. E 399-81. *Standard method of test for plane strain fracture toughness of metallic materials.* Annual ASTM Standards Part 10, p. 592, Amer. Soc. for Testing and Materials, Philadelphia 1982.
13. J.A. Kies and A.B.J. Clark, *Proc. 2nd Int. Conf. on Fracture,* Brighton, 1969; Paper 42.
14. G.R. Irwin, *Amer. Naval Research Laboratory,* Washington, D.C., *Report* 5486, July, 1960.
15. J.W. Kraft, A.M. Sullivan and R.W. Boyle, *Proc. Symposium on Crack Propagation,* p. 8, Cranfield 1961.
16. E 561-81. *Standard practice for R-curve determination.* Annual ASTM Standards Part 10, p. 680. Amer. Soc. for Testing and Materials, Philadelphia 1982.
17. A.A. Wells, *Proc. Symposium on Crack Propagation,* p. 210, Cranfield 1961.
18. B.S. 5762, *Methods for crack opening displacement (COD) testing,* British Standards Institution, London 1979.
19. J.D. Eshelby, *A continuum theory of lattice defects, Progr. Solid State Phys.* **3**, 79, Academic Press, New York 1956.
20. C.P. Cherepynov, *Appl. Math. Mech.* 1967, **31**, 503.
21. J.R. Rice, *J. Appl. Mech.* 1968, **35**, 379.
22. E 813-81. *Standard test method for J_{1c}, a measure of fracture toughness,* Annual ASTM Standards Part 10, p. 822, Amer. Soc. for Testing and Materials, Philadelphia 1982.
23. P.C. Paris and F. Erdogan, *Trans. ASME, J. Basic Eng.* 1963, **85**, 528.

8 Materials Characterization by Instrumented Impact Testing

George C. Adams
T.K. Wu, Wilmington

Polymer Products Department, Experimental Station, E.I. du Pont de Nemours & Co. Inc.
Wilmington, DE 19898, U.S.A.

8.1 Introduction

The development of toughened plastics over the years has placed increased demands on evaluation of fracture resistance and energy-absorption measurements. For various reasons, the industry has had to cope with a wide variety of fracture and failure problems. Indeed, many otherwise satisfactory plastics are rejected for engineering applications because of their frequent tendency toward brittle fracture on high-speed loading, whereas many of these materials behave in a ductile manner at low-strain rates. Unexpected brittle fracture has been attributed to stress concentrations which come about either as a design feature or in the form of an accidental scratch or other defect. The plastics industry has countered the problem by development of new and improved thermoplastics for specific properties, and the capability of plastic materials to absorb energy during shock loading is frequently considered the deciding factor in material selection. Therefore, the need is very great for a laboratory test procedure which defines the intrinsic material-toughness parameter. The parameter should be able to quantify the resistance to abuse during end use by providing a true material property that is independent of the test conditions including overall sample dimensions.

The energy-absorption standards of the plastics industry are based on tensile strength, elongation to failure, impact values, and environmental tests all of which reflect the situation in metals testing some 30 years ago. The metals industry proceeded to develop the concept of fracture mechanics and associated test methods and design procedures to deal with brittle fracture.[1] It is now realized that this body of knowledge is largely transferable to plastics. To do so involves an understanding of fracture mechanics and application of its testing methods in order to lead to the needed material parameter of toughness.

Numerous empirical impact tests to measure energy absorption in plastics have been devised. A review by Lever and Rhys[2] identifies no fewer than 15 separate methods. This large number of tests serves to emphasize the difficulty of correlating so-called "standard test results" with in-use performance. While all tests supposedly measure the energy absorbed in breaking a standard specimen, their greatest virtue is ease and rapidity and, they are, therefore, very widely quoted. Since the tests are not intrinsic measures of toughness, they do not correlate well with service performance.[3]

The problem is particularly acute in plastics because energy absorption on fracture is a combination of energy terms which represent the various stages of the fracture process. On high-speed loading, energy is absorbed in elastic and viscoelastic deformation during the initial stages, as well as local yielding and deformation of the material around the tip of the propagating crack before fracture. Variations in test type and geometry alter the separate contributions of these terms, and different aspects of the failure process are emphasized. For instance, while impact of notched specimens measures principally energy of crack propagation, impact of unnotched specimens requires greater energy for crack initiation. Both unnotched and notched impact tests are commonly used with little awareness that nothing more than a simple ranking of materials could be expected.[3]

The most common standard impact tests are the Charpy and Izod methods, which measure excess energy. These methods are included in official ASTM, BS, DIN, and ISO Standards and are, therefore, the most frequently used tests.[4] A single energy value results from these tests, so that fracture type can be identified only through inspection of the fracture surface. Material stiffness, strength, and evidence of yielding are not assessed. Consequently, most practitioners agree there is need for more information regarding force and energy transfer during high-rate loading.

The deficiencies of the standard tests cited above are largely overcome with instrumented impact testers. Present-day instrumentation employs modern data acquisition capability, which overcomes most of the drawbacks of single-energy-value tests. Today, the potential is great for the analysis of a wide selection of load forms including brittle, semibrittle, semiductile, and ductile fracture.

This chapter offers a description of the impact event through use of instrumentation. Much of the material is based on fracture-mechanics principles and, therefore, depends on analysis of force and energy values during impact. The numerous publications on the fracture mechanics of polymers will not be extensively reviewed. Likewise, an extensive citing of instrumentation is not intended. However, the authors feel that this chapter represents a much needed blend of these two subjects and is an illustration of complementary usage. The stress-intensity factor and fracture energy are both employed as underlying criteria for failure.

8.2 Historical

While instrumentation of standard impact tests would seem to be a natural process, it is not as easily done as with standard low-speed mechanical tests. There is need for separate sensors for force and displacement, with consequent complexity. In addition, many artifacts are known to occur, including inertial loading of the system, high-frequency noise, and electronic aspects. Many of these problems are peculiar only to metals and are considered by Saxton[5] in an ASTM Special Technical Publication devoted to instrumented impact of such materials.

The earliest description of an instrumented Izod test is attributed to Wolstenholme.[6] From the oscilloscope traces of force versus time, the energy absorbed upon impact was determined, and a good correlation was found between such displays and Izod impact strengths for several thermoplastics.

The earliest description of an instrumented falling weight impact test is credited to Fujyoka.[7] Arends,[8] using an accelerometer to measure deceleration and a stress gauge to measure load versus time, studied the elastic, plastic, and propagation phases of impact. Extensive calibration was necessary, because no convenient single sensor gave all the desired quantities. A step toward simplification was provided by Cessna e. a.[9] who attached an accelerometer to the falling weight. The applied force was calculated from the mass of the drop weight and both velocity and displacement were obtained by multiple integration, obviating the need for a second device to measure the latter. Today, the simplicity introduced with the accelerometer is appropriate to high-speed integration, which is the forte of modern computer systems. This simplicity clearly outweighs the higher sensitivity of accelerometer systems to a wide range of frequencies with the incumbent need for signal smoothing.

The shortcomings of early designs were also overcome by an improved accelerometer sensing system recently reported by Wnuk and co-workers.[10] These workers used a built-in integrator for obtaining force, velocity, and displacement data from the accelerometer signal as well as electronic filtering for elimination of high-frequency noise. However, the integration feature of the attached computer was described, but not employed, and the data were presented as acceleration versus time traces.

Our instrumented impact device encompasses all the desirable features in one assembly. As described previously,[11] it employs a single sensor, an accelerometer, which simplifies both the detection of signal and subsequent calculation of desired quantities. Values of force, displacement, velocity, and energy are all obtained as a function of time. This is made possible by instantaneous digitization and storage of impact data, followed by computation by a dedicated tabletop computer. The system is fully described below.

Today, the advantage of the accelerometer system seems to be overlooked, since extensive use is made of the load transducer and an optical or light beam assembly for monitoring displacement. One commercial load transducer apparatus employs software and a mathematical relation between average velocity and a load-time derived energy to obtain displacement data.[12] The absence of any raw data smoothing is an advantage in studying brittle materials.

The systems described above are based on gravity to supply the known excess of impact energy over that required to fracture the test material. Also, hydraulically driven types of apparatus which supply a wide range of test velocities and permit identification of ductile-brittle transitions are widely used.

8.3   Experimental Methods

8.3.1   Equipment

An overview of the apparatus is given in Fig. 8.3.1. The design originated with Zoller.[13] A drop tower was used as an energy device because of the ease of adjusting drop height and impact energy to any desired value within the limits of available

OVERVIEW OF INSTRUMENTED IMPACT TEST APPARATUS

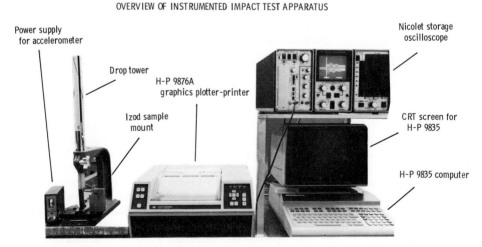

Fig. 8.3.1. Overview of instrumented impact apparatus.

tup weight and maximum drop height, a convenience not possible with a pendulum device. It has been determined that breaking energies in drop tower and pendulum devices are the same. To prevent unnecessary energy loss during impact, as suggested by Bluhm,[14] the Charpy sample support area is extensively fortified with a 100-lb steel block to which the sample supports are rigidly attached. To provide access to a wide range of impact energies, a one-meter-high drop tower is employed and drop weights range from 300 to 2800 g, resulting in impact energies of 0.01 to 27 J.

With Charpy and Izod geometries it is not necessary to impact using a pendulum energy device. These geometries have been studied with a drop tower device, which allows more complete control over energy at the start of the impact, i.e., the impact energy. The Charpy sample dimensions, given by ASTM standards,[15] include a 95.3 mm sample support span and sample dimensions of $127 \times 12.7 \times 6.3$ mm and a notch of any desired depth cut in the narrow face. Widths (B) of 3.18 mm were also used. In the notch-depth studies, notching was done with a single blade flywheel cutter using a radius of 0.25 mm. Notch depths of 1.3, 1.6, 2.5, 3.8 and 5.1 mm were used. The effect of notching and of cracks has been discussed in terms of a stress-concentration factor in Section 10.3.

An accelerometer mounted in the falling weight permits direct observation of the deceleration of the drop weight as it fractures the specimen. Use of an accelerometer as a single sensing device simplifies both the detection and the subsequent calculation of force, displacement, velocity, and energy. Because the accelerometer is calibrated, all quantities are absolute. In order to check the reliability of the instrument calibration and computer integration, high-speed cinematography was utilized to monitor an impact. The good agreement between the specimen displacement determined from the photographs with the computer results provides further support to the quantitative results of this technique (see Fig. 8.3.2).

The output voltage of the accelerometer is proportional to acceleration through an accelerometer constant supplied by the manufacturer (PCB Piezotronics). After signal filtering through a variable filter (discussed below), the signal is digitized and stored by a Nicolet Explorer III oscilloscope as 4096 discrete data points. Impacts generally last from 0.5 to 50 ms and a selectable time interval between points controls resolution. Storage of up to eight impacts is possible on a single floppy disc. Upon command from an interfaced H-P 9835 microcomputer, multiple integrations are performed to obtain velocity, displacement, energy, and force as a function of time. The results

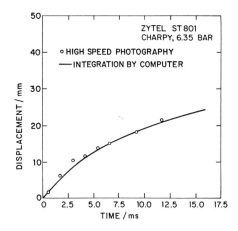

Fig. 8.3.2. Specimen displacement as recorded by instrumented impact using multiple integration and high-speed photography.

are printed or plotted on a Hewlett-Packard 7245 printer-plotter. Accessibility to programming software during experimentation makes this design particularly manageable and attractive.

Use of an accelerometer to instrument the impact event is based on Newton's laws of force and acceleration. As long as the force is not zero, the velocity and displacement of the sample are equal to that of the tup. The latter quantities are obtained from the accelerometer by successive integration as discussed by Zoller.[13] In the repeated impact work, in which the energy returned to the tup by the sample is determined, loss of contact is acceptable, because the desired quantity of the energy content of the tup is recorded. The difference between this and the impact energy is the sample retained energy, a value found to agree well with that obtained from integration of the force-displacement data.

The signal obtained when using an accelerometer requires data smoothing. This is accomplished by 1) electronic elimination of signal above a preselected frequency and 2) data averaging by software programming. An example of the effect of filtering with a Krohn-Hite variable filter is given in Fig. 8.3.3 in which signals comprising frequencies above 5000, 3000, or 2000 are eliminated. The considerable noise at 3000 Hz and above is effectively eliminated by filtering at 2000 Hz. The energy-to-break values at each of these frequencies is not affected as shown by the values quoted in the figure. In each of the curves of Fig. 8.3.3 additional smoothing is performed by use of a running average which reduces to one-eleventh the original number of data points. Care is taken to assure no signal change occurs.

The impact curves of Fig. 8.3.3 show a dampened noise which is greatest at the beginning of the impact event. The source of this noise includes effects from mechanical bending, electronic and test system ringing, and inertial acceleration loads. The separation of mechanical bending is difficult. In a simplified study Williams[16] has treated the measured signal as load oscillations by the sample at the contact points of the moving tup. The solution of the equation of motion, provided by Williams, gives very similar plots to those of Fig. 8.3.3.

Usually at least 100–200 data points are retained which describe the impact. On numerous occasions a crosscheck of the accuracy of force, displacement, or energy, was possible, and no evidence of inaccuracy was detected. As an example, data from the accelerometer based impact apparatus are found to agree with those from nine other laboratories in an ASTM round robin test using polypropylene plaques. Four of these laboratories used load and displacement sensors with drop tower energy devices. Five used hydraulically driven energy systems. Table 8.3.1 shows that good agreement is obtained between load transducer and the accelerometer system of our design.

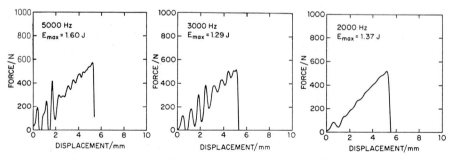

Fig. 8.3.3. Effect of upper cut-off frequency due to electronic filtering on signal noise; nylon 66.

Table 8.3.1. Impact Properties Using Load Transducer and Accelerometer Systems

Quantity from Force-Displacement Data	Average of Eight Labs in Round Robin[a]	Accelerometer System[b]
Max force/N	2375	2326
Displacement at max load/mm	15.2	17.0
Energy at max load/J	17.8	25.6
Total energy, J	32.8	33.8

[a] Impact velocity of 2.1 m/s
    Plaque thickness equals 3.17 mm (1/8 in.)
[b] The author's design.

8.3.2   Calculations

Impact energy depends on tup mass, and drop height, $(U_0 = m g h_0)$, and impact velocity depends on drop height $(v_0 = \sqrt{(2g\,h_0)})$. While these values are known at the instant of impact as U_0 and v_0, it is necessary to determine the desired quantity during impact.

Acceleration, $a(t)$, in g's is calculated from the digitized voltage-time information $j(t)$ and the sensitivity of the accelerometer, s, in mv/g (supplied by the manufacturer), as

$$a(t) = 1 - (j(t)/s) \tag{8.3.2.1}$$

This allows calculation of velocity, $v(t)$, and displacement, $u(t)$, as

$$v(t) = v_0 + \int a(t)\, dt \tag{8.3.2.2}$$
$$u(t) = \int v(t)\, dt \tag{8.3.2.3}$$

The absorbed energy, $U(t)$, is obtained from the change in kinetic and potential energy of the falling weight as

$$U(t) = m g h_0 - 1/2 \times [v_0^2 - v^2(t)] \tag{8.3.2.4}$$

where m is tup mass and h_0 is drop height. The force exerted on the sample is given as

$$F(t) = m(9.81 - a(t)) \tag{8.3.2.5}$$

The energy-to-break may also be determined by integration of the force-displacement results. Good agreement is obtained with the preferred method provided by Eq. (8.3.2.4). As stated above, in the impact-fatigue work, the sample retained energy during bouncing was determined in this manner.

The computer program relies largely on prompts. Printout occurs in the form of numerical values for force, energy, displacement, time, and velocity at points of maximum force and maximum displacement, and at the end of the run. The force-displacement plot follows.

8.4   Fracture Mechanics

8.4.1   General

Rubber toughening of polymers is a science that arose primarily to overcome brittle fracture in the parent polymer. Terms such as "high impact", "toughened" and "super tough" evolved to dramatize the effect of toughener addition. The initial introduction of a second phase imparts greater energy-absorption capability and, over the years, the ratio of toughness between improved product and unmodified matrix has risen sharply.[17] The ability to quantify the change has not kept pace with its development. As a result, component testing is the only reliable alternative when failures are costly.

When these difficulties arose for metals many years ago, a considerable interest developed in fracture mechanics with the aim to provide a more fundamental understanding of the fracture process.[18] While it is still dominated by applications to metals, an increasing portion of the fracture-mechanics literature is devoted to plastics and reinforced composites. A keener awareness of stress concentrators, critical flaw size, and the separate phenomena of crack initiation and propagation is emerging. A major achievement will be the wider use of these procedures for ranking polymers according to fracture in service.

Fracture mechanics provides a basis for expressing toughness in terms of a material parameter. The reader is also referred to Section 10.2 for discussion of the ductile-brittle transition temperature (DBTT), which is indicated by experimental results to be a material parameter.

8.4.2   Fracture Toughness

Fracture toughness or the stress-intensity factor was first introduced in Section 7.2. It is based on stress at failure. Fracture-mechanics fundamentals require the assumption that all fracture results from the presence of one or more flaws. This worst-case situation says that failure will occur when the stress-intensity factor, K, reaches a critical value, K_c, due to failure at a defect of critical length a_c. In this case

$$K_c^2 = \sigma_c^2 \, a_c \, Y^2 \tag{8.4.2.1}$$

where K_c is the fracture toughness, and σ_c is the failure stress. Y is dependent on geometry, and is equal to π for a large plate with a central flaw, or a polynomial in notch depth (flaw size) for more complex geometries.[19, 20] See Section 7.21 for the form of Y. However, K_c determination is not necessarily a straightforward procedure. The lowest value of K_c is the plane-strain-fracture toughness, K_{Ic}, and its determination must be independent of specimen dimensions and done under conditions to give brittle fracture. These necessary conditions are not always attained. The problem of obtaining the lowest or plane-strain value arises when small specimen sizes tested under laboratory conditions give ductile rather than brittle fracture. Conditions to insure brittle fracture include 1) use of small root radius and increased notch depth, as produced by precracking by fatigue, 2) thicker specimens, and 3) layer sandwiching to provide transverse constraint which raises the yield stress and suppresses crazing. A common but not necessarily advisable practice in reporting fracture toughness is to assume that brittle fracture occurs. These points and a valid test procedure for K_{Ic} are discussed in Section 7.12.

An excellent treatment, which considers the effect of specimen thickness for high impact polystyrene (HIPS), is that of Yap, Mai and Cotterell.[21] Single-edge notched (SEN) specimens measuring 70 by 200 mm were varied in thickness from 4.7 to 90 mm. Only above 40 mm was K_c constant giving plane-strain fracture toughness of $K_{Ic} = 1.46 \, \mathrm{MN/m^{3/2}}$. The minimum thickness (B) required by the ASTM Standard is calculated from $B > 2.5 \, (K_{Ic}/\sigma_y)^2$. For SEN HIPS specimens and using $K_{Ic} = 1.46 \, \mathrm{MN/m^{3/2}}$ a minimum B of approximately one third of 40 mm is obtained. For surface-notched (SN) K_{Ic} was less in error. If specimens are very thin, the plastic zone size ahead of the crack tip, calculated from

$$r_p = 1/2\pi (K_{Ic}/\sigma_{ys})^2 \tag{8.4.2.2}$$

is large compared to thickness and a plane strain fracture toughness value is not achieved; see Section 7.12. Accordingly, the ASTM criterion for minimum thickness is in need of review.[22]

The effect of notch depth must be considered as well. Nipkur and Williams[23] have modeled the fracture toughness as comprised of plane stress and plane-strain contributions for HIPS. Their data on 50 and 100 mm SEN specimens shows that, in addition to minimum thickness of 40 mm, a ratio of notch depth to total depth of at least 0.18 is required, which gives $K_{Ic} = 2.5$ MN/m$^{3/2}$. This value is considerably above that reported by Yap e.a.[21].

Instrumented impact testing with the Charpy geometry makes fracture-toughness measurements possible at impact test rates where brittle fracture is more likely to occur. The three-point bend or Charpy geometry is fundamental in that the stress-intensity calibration factor, Y, is available.[20] In the work that follows, stress-intensity factors are not to be taken as obtained under plane-strain conditions, but one intended to show how K_c varies with polymer type and reinforcement.

8.4.3   Fracture Energy

Impact strength is the most commonly quoted fracture parameter. It is absorbed energy divided by the fracture area of the specimen. While it resembles a material parameter, impact strength is a misnomer. As pointed out by Williams, its determination is too highly simplified. The method proposed by Plati and Williams[24, 25] is much preferred due to its foundation in linear-elastic-fracture mechanics (LEFM), but it is specific to brittle fracture. The work necessary to form crack surfaces was first introduced in Section 7.9.

When fracture occurs at load, P, the critical strain-energy release rate G_c is given as

$$G_c = (P^2/2B)/(dC/(d(h))) \qquad (8.4.3.1)$$

where C is the compliance, h, is the crack length, and B is specimen thickness (it should be noted that the symbol h_0 was used in a different context, namely as the drop height, in Section 8.3). Expressed in terms of energy absorbed at fracture rather than load at fracture gives

$$U = G_c BDC/(dC/d(h/D)) \qquad (8.4.3.2)$$

where D equals specimen depth.

Plati and Williams[24] defined Φ as equal to $C/(dC/d(h/D))$. Therefore, a useful equation for energy becomes

$$U = G_c BD\Phi \qquad (8.4.3.3)$$

The analysis predicts a linear relation between total energy absorbed at fracture, U, and the factor $BD\Phi$. The slope is the critical strain-energy release rate, G_c, for impact conditions. The compliance C, and therefore Φ, are determined from a series of quasi-static loading tests on notched bars for a range of h/D values as discussed in Section 7.14.

The strain-energy release rate is synonymous with fracture energy. Fracture toughness and fracture energy are interrelated through the strain-rate dependent Young's modulus, E, as

$$K_c^2 = EG_c \quad \text{(plane stress)} \qquad (8.4.3.4)$$

Equation (8.4.3.4) has not been tested by independent determination of the three quantities, because experiments to determine K_c are not performed during simultaneous determination of G_c, although instrumented Charpy impact offers this opportunity through use of Eqs. (8.4.2.1) and (8.4.3.3).

Brittle fracture, by definition, occurs by elastic loading when the sample does not exhibit local plasticity at the crack tip. LEFM is applicable. However, if there is plasticity resulting in partial ductile fracture, plastic energy is absorbed, and this can result in noticeable yielding in the high-load portion of the loading curve. In addition, the fracture energy involves an additional energy term or energy to unload, which is low in brittle fracture and considerable in ductile fracture. The combination of yielding and unloading energy associated with ductile fracture requires analysis in a more general manner. A fracture criterion based on ligament area rather than one modified by compliance is required, as discussed by Plati and Williams.[24] In this case fracture energy, U, and fracture area, A, are found linear so that the G_c notation becomes J_c where

$$U = A \cdot J_c \tag{8.4.3.5}$$

In the linear elastic case $J = G$.

These concepts are demonstrated below for brittle and ductile materials. A more extensive treatment for ductile fracture is given in Section 7.20 through use of the J-integral.

8.5   Impact Event

8.5.1   Force Displacement

The impact event is displayed as the force-displacement curve, and a typical impact involves initiation and propagation phases as illustrated in the idealized diagram shown in Fig. 8.5.1. The area to any point under this curve is the energy acquired by the specimen, and energies of initiation and propagation can be computed on the basis of crack initiation at maximum force.[26]

Typical force-displacement curves for brittle and ductile fracture are seen in Fig. 8.5.2. In agreement with the nature of brittle fracture, a linear increase in load is clearly seen for nylon 66. At the maximum load, crack initiation occurs and the load reduces catastrophically to zero in less than 5 mm total displacement, consuming about 25% of the total absorbed energy in the unloading process. The sum of initiation (0.40 J) and propagation energies (0.18 J) is 0.58 J for nylon 66, with a standard error of estimate of 1.0% on five specimens. As expected, there is no evidence for yielding,

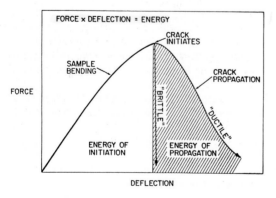

Fig. 8.5.1. Schematic representation of force-displacement curves for brittle and ductile fracture.

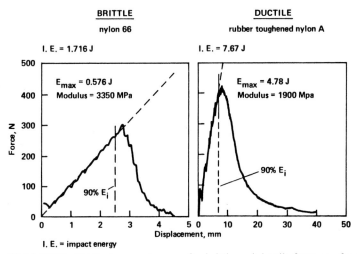

Fig. 8.5.2. Force-displacement curves for brittle and ductile fracture of nylons.

and application of equations for three-point bending gives a dynamic flex modulus of 3350 MPa. A flex modulus of 2800 MPa is obtained in slow-speed testing.

The impact profile for the toughened nylon* differs greatly. A slightly higher maximum load is achieved at about 10 mm displacement with absorption of 1.9 J of initiation energy and 2.9 J of propagation energy. Energy-to-break value is 4.8 J indicating that considerable toughening has been achieved. Nonlinear loading is evident and a yield point can not be determined accurately, so that an approximate modulus is given as 1900 MPa. Unloading consumes 60% of the total absorbed energy, with a total displacement of 40 mm.

In the toughened nylon, high-speed cinematography shows that stress whitening occurs slightly prior to maximum force and crack initiation at about the maximum load. Therefore, the assignment of initiation energy is considered valid. The total area under these curves is the conventionally determined "impact strength", and in the toughened nylon over one-half of this energy arises from the propagation phase of the impact. Depending on the application, failure may be considered to have occurred at force maximum or at the end of the propagation phase. A "toughness index" defined as the ratio of propagation to energy-to-break energies, is not found helpful in distinguishing brittle from ductile fracture.[26]

The qualitative manner in which nylon 66 is influenced by mineral fillers, glass fibers, and other additives is shown in Fig. 8.5.3. The change in stiffness due to the addition of mineral and toughener is evident. The addition of 30% glass fibers results in both higher modulus and total energy absorption. The impact strength may be calculated as the sum of initiation and propagation energies for the 3.175 mm thick specimens and extrapolated to a thickness of 1 in. in the conventional manner. Table 8.5.1 gives a summary of values for the nylons discussed.

Force-displacement curves using load transducers have been reported in both qualitative and/or quantitative detail in the literature.[10, 27–30] Often, only the qualitative features are emphasized, with the result that a thorough quantitative interpretation is often overlooked. A more complete utilization is provided by the simplicity of accelerometer signal and extensive analysis by a dedicated computer.

---

* Referred to below as Super Tough nylon, which is ZYTEL ST801.

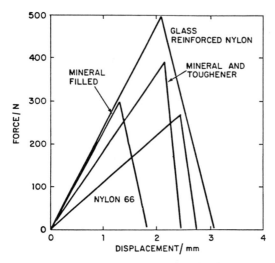

Fig. 8.5.3. Force-displacement curves for mineral and glass-reinforced nylon 66.

Table 8.5.1. Impact Properties for Toughened Nylons

Material	Designation	E_i/J	E_p/J	E_{max}/J	Modulus ——— MPa	Impact Strength ——— J/M (ft-lb/in.)
Nylon 66	ZYTEL 101	0.39	0.04	0.43	3880	69.4 (1.3)
Mineral filled	MINLON 11C40	0.21	0.07	0.48	6270	43.2 (0.81)
Mineral and toughener	MINLON 12T	0.66	0.05	0.71	6155	106 (2.0)
Glass reinforced	ZYTEL 70G33	0.54	0.24	0.78	6300	123 (2.3)

E_i the initiation energy or energy to force max.
E_p the propagation energy or energy to the right of force max.
E_{max} the sum of initiation and propagation energies.
Modulus is calculated from equations for three-point bend.

8.5.2 Impact Velocity

With the notched Charpy geometry it is possible to impact using two impact velocities that are in the ratio of 10^5 to 1. To do this, identical Charpy mounting geometries are used 1) in the drop tower impact test (2 m/s) and 2) in slow-speed Instron flex testing (2×10^{-5} m/s). The 2.54 mm (0.1-in.) notch is mounted opposite the point of impact in each case. The comparison set of force-displacement curves for ZYTEL* 408, seen in Fig. 8.5.4, shows brittle and ductile mode of fracture in high- and slow-speed impact, respectively. While the modulus values are similar, four times greater energy is absorbed in slow-speed testing. In contrast, Super Tough nylon shows little effect of impact velocity. Table 8.5.2 summarizes comparisons of impact velocity for several other polymers. It is seen that the propagation energy is low in all cases except for

---

* ZYTEL is the registered trademark of E.I. du Pont de Nemours & Co., Inc. for its nylon molding and extrusion resins.

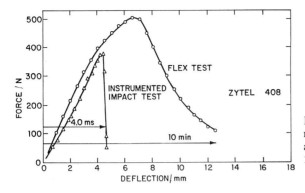

Fig. 8.5.4.   Force-displacement curves for ZYTEL 408 at impact speeds in ratio of 100,000:1.

Table 8.5.2. Effect of Test Rate on Impact Properties of Polymers

| Material | Thickness B | Test Type | Velocity | E_{max} | Ratio of E_{max} Flex/ITT |
	mm		m/s	J	
ABS	6.35	Flex	2.0×10^{-5}	0.63	1.0
	6.35	IIT	2.0		
Lexan 101	3.17	Flex	2.0×10^{-5}	2.71	1.4
	3.17	IIT	2.0	1.99	
Lexan 101	6.35	Flex	2.0×10^{-5}	not run	–
	6.35	IIT	2.0	0.96	
Super Tough Nylon	6.35	Flex	2.0×10^{-5}	5.78	1.2
	6.35	IIT	2.0	4.97	
High Impact Polystyrene	6.35	Flex	2.0×10^{-5}	0.63	1.7
	6.35	IIT	2.0	0.31	
Nylon 66	6.35	Flex	2.0×10^{-5}	1.43	3.5
	6.35	IIT	2.0	0.40	

IIT Instrumented impact technique.

glass-reinforced nylon which requires energy to break the glass fibers. While the table shows that the one brittle material (nylon 66) is the most sensitive to impact velocity, it is inadvisable to generalize from these limited results.

The question of the effect of impact velocity on impact properties often arises.[29] The effect observed for ZYTEL 408 should not be used to generalize for all polymers. Because an impact pulse corresponds to a high-frequency measurement, it effectively raises the glass transition to a higher temperature. If this occurs, some of the impact energy is absorbed by the generation of heat,[31] and higher toughness is achieved. Higher impact velocities and temperature of testing in close proximity to the glass transition temperature can result in a change in fracture mode as observed for ZYTEL 408. The effect of strain rate on toughness has been the subject of several excellent papers.[31-34]

The drop weight in a drop tower impact of necessity experiences a decrease in velocity during the impact, the amount of which depends on the velocity at the start of the impact. In an experiment to explore this effect, 6.35 mm Charpy bars of polyoxymethylene were impacted with impact energy of 0.80, 1.5, and 2.4 Joules which was greater than that required to fracture the material. In these cases, an energy-to-break

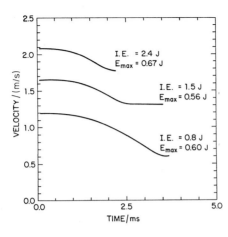

Fig. 8.5.5. Velocity change during impact using different impact energies.

Table 8.5.3. Effect of Impact Velocity on Energy-to-Break

Impact Energy	E_{max} [a]	Impact Energy	Velocity (m/s)		
J	J	J	V_0	V_f	% Δ VEL
2.4	0.67	3.6	2.1	1.8	14
1.5	0.56	2.7	1.6	1.3	18
0.8	0.60	1.3	1.2	0.6	50

[a] Energy-to-break.

value of about 0.60 J was obtained. The impact velocities associated with these impact energies are given in Table 8.5.3 as v and range from 1.2 to 2.1 m/s. The velocity at the end of impact, V_f, shows, as expected, greater decrease in velocity for the impact of lower impact energy. The significant feature of these results is that the same energy-to-break is obtained whether there is a 14% or a 50% decrease in velocity during the impact. The velocity-time curves, given in Fig. 8.5.5, are readily obtained when the accelerometer is used as sensor.

8.6   Material Toughness

8.6.1   Nylon 66 Fracture Energy

A method for analysis of impact energy in terms of molecular relaxation and bond breakage is given in Section 10.4 and the following sections. In that section an impact energy, U_0, is transmitted to the specimen during impact. In this section impact energy is given a similar meaning and is the energy content of the falling weight at the instant of impact.

The conventional toughness method obtains an Izod impact number, also called impact strength. A specimen of given notch depth is fractured according to the representation in Fig. 8.6.1. According to ASTM D 256, an energy-to-break value is determined by swinging pendulum device using either 3.17 or 6.35 mm wide and 2.54 mm notch depth samples. The procedure leads to an impact strength expressed as J/m (ft-lb/in.)

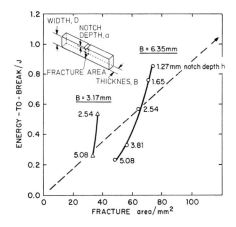

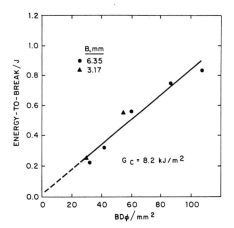

Fig. 8.6.1. Energy-to-break at several specimen fracture areas, obtained by notching to different depths for nylon 66. Sample thickness, *B*, is as indicated. Notch depth is as indicated, expressed as mm.

Fig. 8.6.2. Energy-to-break values for nylon 66 in which the fracture area has been modified by a fracture-mechanics derived compliance correction factor. Sample thickness, *B*, is as indicated.

of sample thickness. The extrapolation to 1 in. of thickness that occurs in this procedure is illustrated in the figure for nylon 66. Energy-to-break versus fracture area (obtained from use of different notch depths) is shown for 3.175 mm and 6.35 mm thickness specimens. The deficiency of the conventional single-point extrapolation procedure is apparent, because the energy-to-break is neither linear with fracture area nor extrapolates through the origin.

Application of linear-elastic fracture mechanics is necessary to obtain accurate energy-fracture area results from impact testing. Use of a geometrical correction factor based on LEFM is necessary as proposed by Plati and Williams.[24] The parameter Φ is a geometrical correction factor which depends on notch depth *h*, sample depth *D*, and support span *L*, and is available from tabulations for both Izod and Charpy geometries. The dimension *D* is sample depth. The parameter Φ is applicable to brittle fracture only. To obtain a linear relation between energy and fracture area, the unnotched cross section, *BD*, must be modified through the factor Φ as shown in Fig. 8.6.2. Five specimens of nylon 66 were impacted at each of four notch depths. The Φ values range from 0.323 to 1.329. Regardless of sample thickness, all data points comprise a straight line with a high correlation coefficient. This indicates that LEFM is applicable for the brittle fracture of nylon 66. This study on nylon 66 has been presented previously.[35]

Extrapolation performed to a fracture area represented in conventional terms by 1 in. of sample thickness, a notch depth of 2.54 mm and the associated Φ value, produced an impact strength of 76.3 J/m (1.3 ft-lb/in.) for nylon 66. Here 1 J/m equals 0.0187 ft-lb/in. This value is arbitrary, since it depends on the arbitrary selection of a 2.54 mm-in. notch depth and the associated Φ value for 1 in. of thickness.

Conventional impact testing provides an apparent fracture energy, but the analysis does not consider sufficient variation in area nor an area modified according to fracture mechanics.[15] Making this modification through use of *BD*Φ provides the correct surface energy from the slope of energy versus *BD*Φ. A critical strain-energy release rate, G_c, in brittle fracture of nylon 66 is discussed above. Ductile fracture, using the

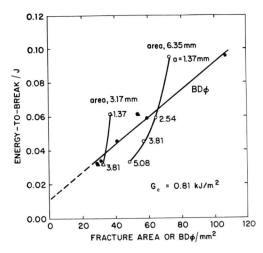

Fig. 8.6.3. Energy-to-break values for poly(methyl methacrylate) showing unmodified and modified fracture area. Notch depth is as indicated, expressed as mm.

toughness parameter J_c, will be illustrated below for ABS*. The toughness parameter so obtained, called fracture energy, is independent of sample geometry.

Linear energy-$BD\Phi$ plots are obtained from material which undergoes brittle fracture. As another example, PMMA follows LEFM as shown in Fig. 8.6.3. Polyoxymethylene and glass-reinforced polyester, behave in a similar manner.

A kinetic energy or intercept term is obtained from Figs. 8.6.2 and 8.6.3. This value is commonly taken as the energy required by the tup to remove the broken test pieces from its path.[24] Because of the catastrophic nature of brittle fracture, the value is low, but the intercept increases considerably with increasing ductility, thereby emphasizing the need for multiple-point energy analysis.

Literature reports of the use of impact test rates and the Charpy geometry to obtain G_c are very limited. Two examples will be cited. Marshall e.a.[36] obtained a G_c of PMMA of 1 kJ/m² using a fracture area approach which employes the Φ parameter. This work constitutes the first use of Φ treatment to fracture area. Plati and Williams[37] made use of a conventional pendulum apparatus in which the fracture energy, G_c, was obtained for several polymers by the variation in fracture area. The work shows that a carefully constructed device which is not instrumented is adequate.

8.6.2   ABS Fracture Energy

Brittle fracture results from elastic loading and stored energy prior to catastrophic failure, in which case LEFM is applicable. As stated, a combination of yielding during loading and considerable energy absorption during unloading requires analysis in terms of ligament fracture area.[24] The nonlinear nature at around 300 N of the force-displacement curve for ABS, seen in Fig. 8.6.4, identifies considerable yielding during loading. The high propagation energy of 51% of the total fracture energy is compared to 31% for nylon 66. The linear energy-area plot of Fig. 8.6.5 for ABS indicates that this is the preferred basis for analysis. The positive intercept on the energy axis indicates a higher kinetic energy term (0.27 J) than for nylon 66 (0.013 J) and results from lower stored energy at fracture for ABS. Subtracting the intercept gives an extrapolated impact strength of 218 J/m (4.1 ft.-lb/in.).

---

* Monsanto Lustran 640

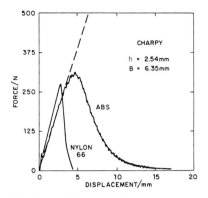

Fig. 8.6.4. Force-displacement curves for brittle fracture of nylon 66 and ductile fracture of ABS.

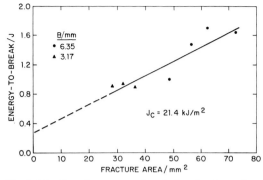

Fig. 8.6.5. Use of energy-to-break for the determination of fracture energy, G_c, for ABS.

For both brittle and ductile polymers, the slope of the energy-fracture area curve represents an improved measure of toughness. A study of sample dimensions provides the basis of its determination and a value of fracture energy of ABS given by the more general notation for nonlinear fracture, J_c, is 21.4 kJ/m^2. A value of 8.2 kJ/m^2 for nylon 66 indicates that about 2.6 times greater energy is required to fracture ABS than nylon 66.

High-impact polystyrene*, ABS and Super Tough nylon are examples of ductile polymers, and the toughness parameters for these materials are given in Table 8.6.1. The impact strengths, also included and calculated from multiple-point area analysis, show a good correlation with G_c. Since single-point extrapolation neglects the positive intercept on the energy axis, multiple-point extrapolation will give lower impact strengths than those obtained by single-point extrapolation.

While improved energy values are obtained through use of instrumentation, a conventional pendulum energy device can be used in an energy-fracture area analysis. If the hammer weight selection does not provide for low-impact energies, some materials such as PMMA may be impacted with a large excess of impact energy by this device. This is less desirable than the use of a wide selection of energy available from a drop tower.

Table 8.6.1. Fracture Energy of Brittle and Ductile Polymers

Material	Type of Fracture	Impact Strength[a]		Fracture Energy
		J/m (ft-lb/in.)		G_c/(kJ/m^2)[b]
PMMA	Brittle	9.6(0.18)		1.0
Nylon 66	Brittle	77(1.4)		8.2
HIPS	Ductile	75(1.4)		7.5
ABS	Ductile	218(4.1)		21.4
Super Tough Nylon	Ductile	795(15)		75.7

[a] Obtained from extrapolation of energy-fracture area data.
[b] From slope of energy-fracture area data.

---

* Monsanto Lustrex 4300.

8.6.3   Fiber-Reinforced PET

Toughness of fiber-reinforced composites has received considerable attention. A variety of test methods has been employed including slow-speed deformation of flat plaques and of three-point-bend specimens. A high-speed test that employs a fundamental geometry is pertinent to the present work. Studies of this type include epoxy-matrix composites,[26] polyester glass,[38] and epoxy glass.[39] The latter study takes into consideration an experimental compliance procedure which accounts for the specific reinforcement geometry in use. In the present work the applicability of LEFM to composite materials is demonstrated.

A comparison of two materials of very similar toughness serves as an example of the improved applicability of fracture mechanics and multiple-point over single-point fracture area analysis. Glass-fiber-reinforced PET composites were prepared from two glass fibers of different tensile strengths. The conventional or single-point Charpy impact strength by Izod is given in Table 8.6.2. It appears that the composite prepared from the *higher* tensile strength glass fibers is the more energy absorbing. Instrumented impact strength values by single-point analysis give the same results. However, several fracture areas and energy values determined from instrument impact give linear energy-$BD\Phi$ plots. Because of the considerable propagation energy involved in fracture of the glass fibers, it is appropriate to use energy to maximum force, i.e., initiation energy, in this analysis. Both the extrapolated energy values from fracture area and the fracture energy, G_c, from the slope indicate that composites prepared from *lower* rather than higher tensile strength fibers possess higher energy absorption or toughness upon impact.

A further comparison illustrates additional utility of the fracture mechanics approach for these materials. As previously shown, fundamental equations exist for the Charpy or three-point-bend geometry, which enable the calculation of the stress at fracture, σ_c. The fracture toughness, K_c, may be calculated from load at fracture, P_c, notch depth h and a $Y(h/D)$ calibration given by Brown and Srawley[20] using Eq. (8.4.2.1) as described previously. The fracture toughness values for an h/D of 0.4 are given in Table 8.6.2. It is not surprising that the reinforcing fibers of *higher* tensile strength impart greater toughness based on strength or fracture toughness. This contrasts with results shown from the energy analysis of impact that toughness based on energy absorption is obtained with the *lower* tensile strength glass fibers. The difference in the properties of the reinforcement can be assessed through appropriate use of multiple-point analysis and the principles of fracture mechanics.

Table 8.6.2. Tabulation on Impact Values for Glass-Reinforced Poly(ethylene terephthalate)

Material	% Glass/type[b]	Conventional[a] Izod Impact Strength	Impact Strength from E_i-$BD\Phi$ Data	G_c, (INIT) kJ/m^2	K_c MN/m$^{3/2}$
		J/m(ft-in.)	J/m(ft-lb/in.)		
A	30/E	107(2.0)	74.5(1.40)	8.2	5.8
B	32/S	187(3.5)	65.7(1.23)	6.5	7.2
C	45/E	130(2.4)	42.2(0.79)	4.4	6.4
D	43/S	198(3.7)	41.1(0.77)	4.1	7.6

[a] Based on single-point extrapolation.
[b] Tensile strength: E-glass 50 kpsi; S-glass 65 kpsi
    Modulus: E-glass 12.4 kpsi; S-glass 13.5 kpsi

8.6.4   Epoxy Composites

A primary attribute of the instrumented impact device is that it permits analysis of energy transferred into and out of the material. The high stiffness of advanced composites and the absence of yielding make analysis by LEFM appropriate. It has been shown that properties of the reinforcement have a direct effect on the composite toughness.[26] Therefore, it is of interest to study the converse, that is, to vary matrix toughness at constant reinforcement geometry. Toughened and untoughened epoxy resins were used, and impacts were made on Charpy bars cut from 1.27 mm thick plaques. The construction was 12 layers of a quasi-isotropic graphite fabric. Fracture area was varied through use of several notch depths. Variations in notch depth enabled the energy-$BD\Phi$ analysis given in Fig. 8.6.6. The good correlation with a fracture-mechanics derived fracture area gives confidence that differences in energy absorption reflect the G_c values for these epoxy composites. The application of fracture mechanics to determine G_c shows that 40% greater energy absorption occurs for the toughened matrix composite.

The notch depth study also enables calculation of fracture toughness, K_c, for each notch depth based on a crack length, h, equal to the notch depth. Table 8.6.3 shows nearly constant K_c at each notch depth and also that the two epoxy composites have essentially the same fracture toughness. In this limited study an increase in energy absorption with toughening, but no change in fracture toughness, was observed.

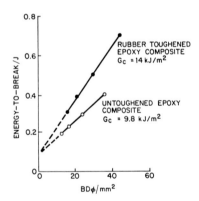

Fig. 8.6.6. Energy-to-break data for epoxy/graphite composite showing the determination of fracture energy, G_c.

Table 8.6.3. Toughness Parameters for Epoxy/graphite Composites

Material	Fracture Toughness, $K_c/(\mathrm{MN/m^{3/2}})$						Fracture Energy, G_c
	Notch Depth/mm						$\mathrm{kJ/m^2}$
	1.27	1.65	2.54	3.81	5.08	AVE	
Rubber toughened epoxy/graphite composite	16.0	17.3	17.0	16.7	17.1	16.8	14
Untoughened epoxy/graphite composite	15.5	15.5	15.5	15.9	16.8	15.8	9.8

8.6.5   Charpy Versus Compact Tension Geometry

The validity of a single parameter, K_c, to define a strength basis of crack resistance independent of geometry was investigated. Polyester reinforced with 0–50% glass fibers was prepared. Plaques of the compact-tension geometry were subjected to slow-speed Instron testing by Friedrich[40] in a coordinated effort with the authors. A test speed of 0.2 mm/min was used to give nearly constant K_c values over an (h/D) range of 0.35 to 0.75 for each composition. Charpy specimens were also prepared from the plaques providing the same longitudinal fracture. Charpy bars were notched to 2.54 mm ($h/D = 0.2$). The two geometries are sketched in Fig. 8.6.7. Using the maximum force to calculate K_c from Eq. (8.4.2.1) and the appropriate polynominal, $Y(h/D)$, gave the results of Fig. 8.6.8. The higher values for the slow testing rate are reasonable. The similarity in values for two diverse geometries demonstrates the utility of the geometry correction $Y(h/D)$ in Eq. (8.4.2.1). A reasonable accurate K_c value for longitudinal fracture is obtained and the stress-intensity factor is shown to be a universal parameter related to the material toughness.

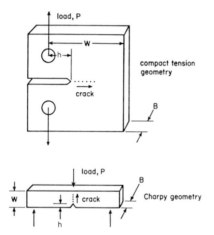

Fig. 8.6.7. Representation of specimen geometry used for slow- and high-speed testing of glass-reinforced poly(ethylene terephthalate).

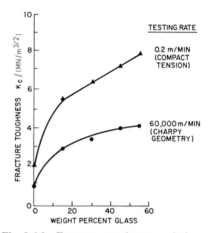

Fig. 8.6.8. Fracture toughness variation with glass content for glass-reinforced poly(ethylene terephthalate).

8.7   Impact Fatigue

8.7.1   Introduction

Much of the analysis for toughness of polymers centers on monotonic stress, often to bring about rapid failure. These tests have permitted the advancement of a considerable array of toughened materials. Less attention has been directed to the behavior of polymers under fatigue or conditions of repeated stress. Yet, polymers are today being used increasingly in such load bearing purposes as in gears, hinges, springs, and automated arms in which a low and periodic stress is experienced.

Studies on impact fatigue of polymers which use high-impact energy are very few. By impact fatigue is meant loading under conditions of constant impact energy as opposed to fatigue-crack growth or loading under conditions of constant load. The latter is discussed in Section 7.21. Bhateja e.a.[41] studied the failure of ultrahigh molecular weight polyethylene in less than 20 cycles using notched three-point-bend specimens. Ohishi e.a.[42] studied polycarbonate under repeated impacting and constructed S-N curves. Response to impact fatigue by high-molecular weight polyethylene was studied by Bhateja e.a.[43] Takemori[44] studied biaxial stresses and fatigue-crack-propagation curves in biaxially oriented polycarbonate film. These limited repeated stressing studies, which destroy the sample in few impacts, indicate a need to employ a highly instrumented technique to this important area of mechanical property behavior.

It is possible that fatigue loading using high-impact energy can provide knowledge of a phenomenological nature in a manner not possible with single impact to break tests. One might suppose that fatigue loading has the advantage of generating material responses which are not detected during the more conventional monotonic deformation. For instance, the addition of adventitious flaws to elevate the local stress and crystallinity, to name but two, may have a significant effect on the toughness or fatigue performance over and above that detected by single-blow impact. In this initial fatigue study, testing has been performed on notched specimens using less than 50% of the impact energy necessary to cause fracture in a single impact and less than 200 impacts or cycles to failure. The deformation is clearly not linear, which distinguishes it from the previously established dynamic response or fatigue-crack propagation by constant load testing.

A sizable body of knowledge of polymeric fatigue exists for constant loads much lower than those mentioned above. Thus, dynamic loading giving linear response has resulted in generation of a large body of fatigue-crack-propagation data. The elements of fracture mechanics have been used to obtain the stress-intensity factor, K, which characterizes the severity of the stress field at the tip of the crack. In these load control studies, values of K, coupled with fatigue-crack-propagation (FCP) rates, permit a comparison between materials. The foundation of this work is the Paris equation, which is discussed in more detail in Section 7.21. The relationship is given as

$$dh/dN = A(\Delta K)^n \qquad (8.7.1.1)$$

where A and n are material constants. A large number of polymers have been characterized by Eq. (8.7.1.1) for FCP rates by Hertzberg and Manson[45] and others.[46,47] With a compact-tension geometry under constant- and low-load control, up to 1×10^6 cycles may be used in a single experiment at frequencies of 1–10 Hz. Thus, this conventional or stress-derived approach to fatigue, which is embodied in the stress-intensity factor, entails monitoring both load and crack length over the duration of the test.

It turns out that the experimental details of such fatiguing by low and dynamic loading are found directly applicable to a study in repeated impacting at higher loads with constant-impact energy using the instrumented-impact technique.

8.7.2 Constant-Impact-Energy Fatigue

Advantage is taken of the extensive energy description available from the instrumented-impact event. Repeated impacting with constant-impact energy to the point of failure is accomplished with a falling weight impact tester and a fundamental or Charpy specimen geometry. By systematic study of impact energies less than that obtained at maximum force in a single impact to break, i.e., less than the initiation energy, impacting is repeated until, for ductile materials, the natural crack is gradually moved through the sample cross section. A constant impact-energy of 30% to 95% of the initiation energy was used. From 1 to 300 impacts were required at the impact energy depending on the material. As might be expected, ductile materials fail by yielding and crack propagation, whereas brittle materials fail catastrophically and without warning or change in physical appearance. The former will be discussed first. The work has been presented previously.[48-50]

The ductile materials under study have initiation and energy-to-break values as listed in Table 8.7.1. A given force-displacement curve for the ductile material Super Tough Nylon under repeated impacting is given in Fig. 8.7.1. During each cycle a nearly linear increase to maximum force is observed, followed by a decrease in load and displacement. Therefore, the force-displacement curve traces out a loop. With increasing number of impacts, the curve deteriorates continually to exhibit a decrease in maximum force but an increase in the area enclosed by the loop in the manner shown in Fig. 8.7.1.

Brittle materials behave much differently upon repeated impact. The force-displacement curves for repeated impacting of nylon 66 are given in Fig. 8.7.2 using 0.8 J of impact energy which is 98% of the crack-initiation energy. Upon repeated impacting there is no visible crack advance, no reduction in maximum force, and no permanent displacement. The elastic nature of the impact results in no yielding. The curve essentially retraces itself up to the 145th impact, which precedes by one impact the failure impact. The area inside the force-displacement loop is small, indicating little permanent

Table 8.7.1. Energy Values for Polymers

Material	E_i/J[a]	E_{max}/J[b]
Nylon 66	0.88	0.91
HIPS	0.51	0.58
ABS	1.2	1.5
Super Tough Nylon	1.6	5.4

[a] Initiation energy or energy to force max.
[b] Energy-to-break.

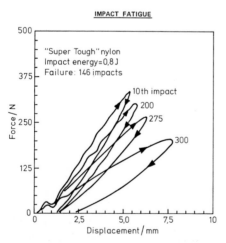

Fig. 8.7.1. Force-displacement results for impact fatigue study of Super Tough nylon.

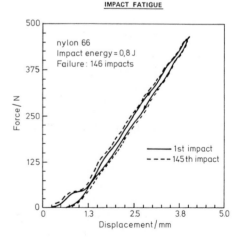

Fig. 8.7.2. Force-displacement results for impact-fatigue study of nylon 66.

sample damage. The occurrence of failure is quite sensitive to impact energy, since use of a slightly higher value such as 0.9 J results in failure in a single impact.

As impact energy is decreased, a greater number of impacts is required to bring about failure. A relatively simple fatigue curve may be constructed from the data. The fatigue curve is taken as the number of impacts to failure versus impact energy as shown in Fig. 8.7.3 for nylon 66, rubber toughened nylon A, HIPS and ABS. Rubber toughened nylon A is an experimental toughened nylon. As the impact energy decreases, there is gradual convergence of the ductile or toughened nylon failure curve with that of nylon 66, the matrix polymer. For the matrix polymer, a strong dependence on impact energy is observed.

An impact energy may be identified in which there are more than 100–200 impacts to failure such as two to three orders of magnitude greater. This transition to a large number of impacts has some semblance to a vertical line in Fig. 8.7.3. No implication is made that failure does not occur for impacts below this transition impact energy; rather, a transition impact energy is implied by the vertical line. Impact energies below this value result in long-term fatigue performance. Impact energies above this value result in short-term failure. A region of classical or constant-load control fatigue testing, i.e., fatigue-crack-propagation testing, exists below the transition impact energy. Above this energy, the newly introduced concept of impact fatigue by the instrumented impact technique applies, which employs constant-impact energy.

The results show the value of varying impact energy in fatigue studies. Failure occurs over a wide range of impact energy for ductile polymers, and over a narrow range for brittle polymers. These impact-energy values are accurately identified by the impact-fatigue technique. The data suggest that polymer crystallinity improves fatigue performance. For crystalline nylon, the point of convergence of the fatigue curve occurs at an impact energy of 0.8 J, while the point of convergence of the polymers HIPS and ABS is considerably lower at 0.2 J. In their studies using the constant-load

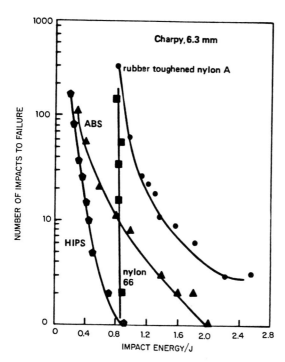

Fig. 8.7.3. Fatigue curves for brittle and ductile polymers.

technique, Hertzberg and Manson[45] and Ramirez e. a.[51] found that crystallinity plays a dominant role in determining the fatigue behavior of a polymer. These workers point out that polymers most resistant to fatigue-crack propagation are nylon 66, nylon 6, poly(vinylidene fluoride), and polyacetal.[45] On a limited number of polymer types, the constant-impact-energy fatigue technique further illustrates the importance of crystallinity in fatigue performance.

Clearly, impacts of low-impact energy are as informative as impacts of high-impact energy. Most applications correspond to low-energy impact, whereas most toughness testing is relevant to high-impact energy. The understanding of toughness that has been built on a base of single-blow impacts does not identify all the important parameters affecting toughness. The present study shows that low-energy impacts identify crystallinity as useful in toughness.

8.7.3 Constant-Load Fatigue

The constant-load fatigue testing of Hertzberg and Manson[45] requires monitoring load and crack length during cycling. Since these parameters are readily identified in constant-impact-energy fatigue testing, it was also of interest to construct fatigue-crack-propagation curves from such data in order to investigate the applicability of the Paris Law, Eq. (8.7.1.1) using a Charpy geometry.

Linearity is observed in fatigue-crack-propagation curves of Fig. 8.7.4 for HIPS. Such linearity indicates that the manner of loading and unloading by constant-impact-energy is amenable to analysis by the Paris Law. Fatigue-crack-propagation rates are higher by a factor of 10 to 100 for constant-impact-energy fatigue, and higher stress-intensity factors are also obtained. The constant-impact-energy fatigue lines are an extension of constant-load-fatigue curves and of a higher slope. This is not unexpected and results from an increase in propagation rate at higher stress levels.

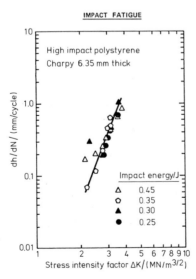

Fig. 8.7.4. Fatigue-crack-propagation plots from impact-fatigue study by instrumented impact of high-impact polystyrene.

References

1. *Fracture Toughness Testing,* ASTM 381, 1965.
2. A.E. Lever and J. Rhys, *The Properties and Testing of Plastic Materials,* Temple Press, London 1957.
3. *Plastics Design Forum,*"The Perils of Izod", May/June 1980.
4. a) BS 2782, British Standards Institute, London; b) ASTM Standards Part 27, American Society for Testing Materials, Philadelphia 1971.
5. Instrumented Impact Testing, ASTM STP 563, p. 50, 1974.
6. W.E. Wolstenholme, *J. Appl. Polymer Sci.* 1962, **6**, 332.
7. K. Fujioka, *J. Appl. Polymer Sci.* 1969, **13**, 1421.
8. C.B. Arends, *J. Appl. Polymer Sci.* 1965, **9**, 3531.
9. L.C. Cessna, J.P. Lehane, R.H. Ralston and T. Prindle, *Polymer Eng. Sci.* 1976, **16**, 419.
10. A.J. Wnuk, T.C. Ward and J.E. McGrath, *Polymer Eng. Sci.* 1981, **21**, 313.
11. G.C. Adams and T.K. Wu, *SPE ANTEC* 1981, **27**, 185.
12. DYNATUP Model 8000, *Effects Technology,* Santa Barbara, CA
13. P. Zoller, *Polymer Testing* 1983, **3**, 197.
14. J.I. Bluhm, *ASTM STP 176,* P. 84, 1955.
15. *Impact Resistance of Plastics and Electrical Insulating Materials,* ASTM D 256-81.
16. J.G. Williams, *Fracture Mechanics of Polymers,* Wiley, New York 1984.
17. C.B. Bucknall, *Toughened Plastics,* Applied Science Publ., London 1977.
18. *Fracture Toughness Testing,* ASTM STP 381, 1965.
19. S.T. Rolfe and J.M. Barsom, *Fracture and Fatigue Control In Structures,* Prentice-Hall, Englewood Cliffs, New Jersey 1977.
20. *Plane Strain Crack Toughness Testing of High Strength Metallic Materials,* ASTM STP 410, 1966.
21. O.F. Yap, Y.W. Mai and B. Cotterell, *J. Mater. Sci.* 1983, **18**, 657.
22. *Plane Strain Fracture Toughness of Metallic Materials* ASTM E399-74.
23. K. Nikpur and J.G. Williams, *J. Mater. Sci.* 1979, **14**, 467.
24. E. Plati and J.G. Williams, *Polymer Eng. Sci.* 1975, **15**, 470.
25. J.G. Williams, *Polymer Eng. Sci.* 1977, **17**, 144.
26. P.W.R. Beaumont, R.G. Riewald and C. Zweben, *ASTM STP 568,* 1974, p. 134.
27. F. Ramsteimer, *Polymer* 1979, **20**, 839.
28. A.G. Miller, P.G. Hertzberg and V.W. Rantala, *SAMPE Quart.* 1981, **12**, 36.
29. H. Gonzalez and W.J. Stowell, *J. Appl. Polymer Sci.* 1976, **20**, 1389.
30. J.C. Radon, *J. Appl. Polymer Sci.* 1978, **22**, 1569.
31. B. Hartman and G.F. Lee, *J. Appl. Polymer Sci.* 1979, **23**, 3639.
32. P.I. Vincent, *Polymer* 1974, **15**, 111.
33. M. Kisbanyi, M.W. Birch, J.M. Hodgkinson and J.G. Williams, *Polymer* 1976, **20**, 1200.
34. I.G. Zewi, W.J. Rudik and R.D. Corneliussen, *Polymer Eng. Sci.* 1980, **20**, 622.
35. G.C. Adams and T.K. Wu, *SPE ANTEC 1982,* **28**, 898.
36. G.P. Marshall, J.G. Williams and C.E. Turner, *J. Mater. Sci.* 1973, **8**, 949.
37. E. Plati and J.G. Williams, *Polymer* 1975, **16**, 915.
38. P.W.R. Beaumont and D.C. Phillips, *J. Mater. Sci.* 1972, **7**, 682.
39. H. Harel, G. Morom, S. Fischer and I. Roman, *Composites* 1980, p. 69.
40. K. Friedrich, Microstructure and Fracture of Fiber Reinforced Thermoplastic Polyethylene Terephthalate, Fortschr.-Ber. VDI-Z., Reihe 18, Nr. 12 (1982).
41. S.K. Bhateja, J.K. Rieka and E.H. Andrews, *J. Mater Sci.* 1979, **14**, 2103.
42. F. Ohishi, S. Nakamura, D. Koyama, K. Minabe and Y. Fujisawa, *J. Appl. Polymer Sci.* 1976, **20**, 79.
43. S. Bhateja, J. Rieke and E. Andrews, *Ind. Eng. Chem. Prod. Res. Dev.* 1980, **19**, 607.
44. M.T. Takemori, *J. Mater. Sci.* 1982, **17**, 164.

45. R.W. Hertzberg and J.A. Manson, *Fatigue of Engineering Plastics,* Academic Press, New York 1980.
46. M.T. Takemori, *Polymer Eng. Sci.* 1982, **22**, 937.
47. A.M. Serrano, G.E. Welsch and R. Gibala, *Polymer Eng. Sci.* 1982, **22**, 946.
48. G.C. Adams and T.K. Wu, *SPE ANTEC* 1983, **29**, 541.
49. G.C. Adams, T.K. Wu, *Fatigue in Polymers,* The Plastics and Rubber Institute, London 1983.
50. G.C. Adams, ASTM STP, Instrumented Impact Testing of Polymers and Composites, to be published.
51. A. Ramirez, J.A. Manson and R.W. Hertzberg, *Polymer Eng. Sci.* 1982, **22**, 975.

9   Deformation and Failure of Thermoplastics on Impact

Georg Menges

Institut für Kunststoffverarbeitung, RWTH Aachen
Pontstraße 49, D-5100 Aachen, West Germany

Heinz-Eberhard Boden

Barmer Maschinenfabrik AG
Leverkuser Straße 65, D-5630 Remscheid 11, West Germany

9.1   Introduction

In practice, nearly all thermoplastic moldings are frequently subjected to impact loads. Since most thermoplastics have a relatively high deformability, toughness, and mechanical damping, they are generally well suited for this type of loading.

However, even in the case of normally very ductile thermoplastics, a glass-like brittle fracture with low elongation at break can occur under certain conditions. This dangerous type of failure occurs above all at low temperatures and at very high deformation rates. It is particularly feared, because in such cases the failure occurs without much plastic deformation, in other words virtually without warning.

In recent years, it has become even more important to find a solution to this problem because technical components, i.e., high-quality parts of vital importance for the functioning of an entire piece of equipment or component, are being increasingly manufactured from polymers. Under such circumstances, failure can result in considerable financial damage and even personal injuries.

Since the prices of raw materials and production costs have risen enormously and the regulations on producer liability have become far more stringent, reliable designing is nowadays more necessary than ever before.

In recent years, the Institut für Kunststoffverarbeitung (IKV) in Aachen, West Germany, has been specifically studying the principles of designing impact-loaded polymeric moldings. Above all, the work has been concerned with establishing the behavior of thermoplastic materials under impact loading and with determining the material data that are suitable for designing calculations.[1]

The tests, which are the subject of this work, were performed on a servohydraulic testing machine over a wide range of times and temperatures. In order to obtain a clear and simple stress distribution over the cross section of the test specimen, the investigations involved uniaxial tensile tests on ideal test bars. In all experiments the force or piston position was taken as the controlled variable. The tests can be divided into the following types:

– *ultra-short-term creep test*
  (10^{-2} s $\leq t \leq 10^3$ s)
– *high-speed tensile test*
  (time to fracture approximately 10^{-2} s)
– *ramp impulse test*
  (rise time 10^{-2} s to 10^0 s; fall time approximately 3 ms)
– *ultra-short-time relaxation test*
  (10^{-2} s $\leq t \leq 10^3$ s)

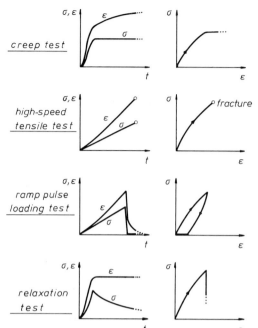

Fig. 9.1.1. Schematic representation of the test methods.

Figure 9.1.1 provides a schematic representation of the tests. With the available equipment, it was possible to achieve loading times or times to fracture down to approximately 5 ms (piston velocity maximum 2 m/s). The temperature could be varied in the range from $-190°$ C to $+100°$C.

9.2   Failure Behavior

In most cases the permissible loading should be fixed to ensure that there is a considerable safety margin to the failure threshold. So it is very important to have information about this threshold. The problem is the pronounced decrease in the elongation at break, found with all thermoplastics examined under very sharp impact loading. The values achieved in the worst case, e.g., with rigid polyvinylchloride PVC, are only approximately 1/5 of the yield strain determined in the standardized short-term test according to German standard DIN 53455.[2]

The types of failure occurring on impact loading can be divided into four types:

1. *Failure with necking.* There is a pronounced yield point, and when it is exceeded, the stress drops appreciably. In this case the yield stress, σ_S, and yield strain, ε_S, have been given as the characteristic failure data.

2. *Failure with a weakly defined yield point.* There is neither a drop in the stress nor any necking. In this case, the stress at break σ_R and the respective elongation ε_R have been taken as the characteristic failure data.

3. *Failure without yield point with simple fracture.* (Fracture is perpendicular to the direction of loading.) As is usual, the tensile strength σ_B and the elongation at break ε_B ($\triangleq$ strain at maximum force) were determined.

4. *Multiple splinter failure (brittle fracture).* A primary break generally forms perpendicular to the direction of loading, fracturing the material into tiny splinters within
a break zone. Besides this, several secondary fractures often occur, which can run
in any direction through the test specimen and have a smooth, glass-like appearance.
The fracture data for this are given as with 3.

   The speed of the unstable spreading of fracture can, with this type of failure,
reach values in the range of 200 m/s to 1000 m/s.[3,4] As is shown below, the brittle
failure starts with linear viscoelastic behavior until fracture occurs. This means that
one of the first activated flaws is transferred immediately into unstable growth
and thus triggers off the fracture.

   All the strain data in the following results refer to the definition of nominal elongation (technical elongation).

$$\varepsilon = (l - l_0)/l_0 \tag{9.2.1}$$

In the strain range under examination (up to 10%), the nominal elongation is larger
than the true elongation by a maximum factor of 1.05. The given stresses always
refer to the initial cross section, that means, they are engineering stresses.

   Since, with most plastics the Poisson ratio v is between 0.3 and 0.5,[5] the cross-
sectional area changes because strains up to 10% lie in the range of

$$0.06\,A \leq A \leq 0.1\,A \tag{9.2.2}$$

It should be pointed out with regard to the results of the creep tests, that failure
only occurred within the first 1000 s after application of the load under the condition
that the threshold of linear viscoelastic range had already been exceeded during loading.

   This means that we must assume in this case, that flaws become activated during
loading and that the growth of the flaws overlaps the subsequent creep process. In
this respect, the experiments presented here have a close connection with the fracture
mechanics which examine the conditions governing the propagation of an artificially
created macroscopic crack. In this case, however, a distribution of flaws and crazes
is created simulating one encountered in practice, and an observation is made of the
influence on the failure.

   Unless otherwise stated, the measuring curves shown below represent the mean
data from three to seven measurements. If not, the individual measurements are plotted.

9.3   Ultimate Strain

According to Boyer,[6] Kleinemeier and Boden,[7] and Menges,[8] it can be assumed
that, for complete embrittlement of plastics, a lower threshold of elongation at break
exists. Since this is a minimum value, it is safe to use this in calculations for designing
purposes.

   It must, nevertheless, be borne in mind, that in the most unfavorable case the
initial fracture can already occur if the minimum elongation at break is exceeded at
any point in the component. Unfortunately, this cannot be macroscopically measured
in every case.

   These problems affect all material testing methods, since the specimens, like moldings, always have certain defects (e.g., notches, voids, entrapped air, phase boundaries).
An attempt is therefore generally made to determine the basic data on test specimens
that have as few defects as possible, and then to make the appropriate reductions
for practical application.

High molecular weight polymethylmethacrylate (HMW-PMMA) showed a very marked decline in the elongation at break at room temperature, with the time to fracture becoming shorter than 1 s as shown in Fig. 9.3.1. The following characteristic effects can be observed:
— The scattering of the elongation at break decreases considerably in the range of complete embrittlement ($t_f \leqq 10^{-2}$ s).
— There is a transition from the simple fracture (type III) to the multiple splinter fracture (type VI).
— The threshold, ε_F, from which the crazes become optically visible, increases with decreasing time to fracture. For times to fracture below approximately 1 s, no crazes become visible up to the fracture.
— The threshold of the linear-viscoelatic range, ε_{lin}, also rises and is intersected in the range of complete embrittlement by the fracture line. The failure in this case thus occurs virtually direct from the linear-viscoelastic behavior.

This description of HMW-PMMA also applies to PA6–25% SFR (polyamide 6 with 25% short fiber reinforcement), with the exception that with this material, it is not crazes that occur but damage effects similar to the slip lines. These defects disappear totally where embrittlement is complete.

The fracture lines have been determined for several temperatures in creep tests and supplemented by results from high-speed tensile tests. The results from HMW-PMMA and PA6–25% SFR are shown in Figs. 9.3.2 and 9.3.3. For HMW-PMMA, a lower threshold of elongation at break (minimum elongation at break) of approximately 2.2% could be derived from Fig. 9.3.3.[9, 10] This figure is also confirmed by the results of other researchers.[7, 11–13]

A comparable decline in the elongation at break was also shown by PA6–25% SFR (Fig. 9.3.3). A minimum elongation at break of approximately 1.8% was extrapolated from the flat path of the curve for $-60°$ C.

According to the available results for test specimens of different cross sections, the threshold itself (the minimum elongation at break) proved to be independent of the geometry.[12, 14, 15] On the other hand, the combination of temperature and time to fracture, at which the decrease in elongation at break started, was clearly dependent on the geometry—particularly in the case of PA6–25% SFR.

Thin test bars became brittle only under sharper impact loads (lower temperature, shorter time to fracture) than thick or wide ones. But as already mentioned, the threshold itself proved to be independent of the geometry.

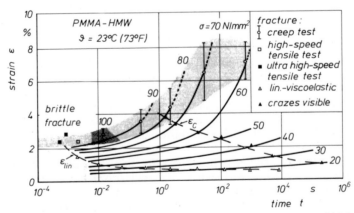

Fig. 9.3.1. Mechanical behavior of HMW-PMMA in the creep and high-speed tensile tests.

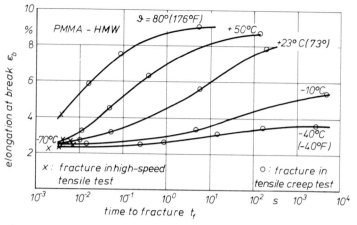

Fig. 9.3.2. Fracture or yield strain of HMW-PMMA.

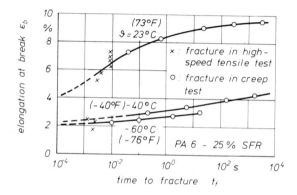

Fig. 9.3.3. Elongation at break of PA6-25% SFR.

In numerous investigations,[8, 10, 12, 15] it has been proved empirically that the time-temperature shift principle (see also Sections 2.3 and 10.5) can also be applied in the temperature range below the glass transition temperature, T_g, or the crystalline melt temperature, T_m. This is particularly the case where the material behaves linear-viscoelastically so that no irreversible flow processes take place.[8] This is true of thermoplastics in the region of embrittlement up to elongations of about 1.5–2%. It has therefore been possible, by applying the time-temperature shift principle to the fracture curves determined at low temperatures, to obtain an estimate for the elongation at break for times to fracture of less than 5 ms.

With rigid PVC and polyoxymethylene (POM) the decline in the elongation at break was also clearly recorded. It was, nevertheless, not possible to determine as pronounced a threshold for HMW-PMMA and PA6–25% SFR. The ultimate strain determined for rigid PVC is shown in Fig. 9.3.4.

At $-30°$C, a pronounced embrittlement and the onset of decline in elongation at break was recorded.[15] A further fall in the temperature to $-60°$C did not bring any significant change in the line of fracture as compared with $-30°$C.[14] However, at $-120°$C, it was possible to estimate a threshold value of 2%.[16] The material, because of its globular morphology,[17] evidently behaves in some respects not like an amorphous polymer, but like a semicrystalline plastic.[18]

Semicrystalline POM failed in creep tests at 23°C with pronounced necking. It was not possible, from the measurements, to say exactly when this process started.

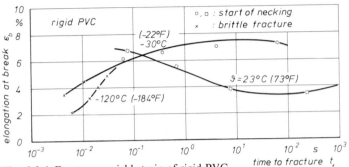

Fig. 9.3.4. Fracture or yield strain of rigid PVC.

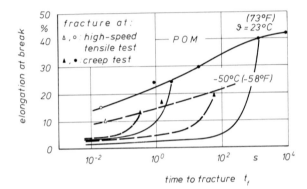

Fig. 9.3.5. Creep and fracture behavior of POM.

For this reason, the elongation at break was plotted in Fig. 9.3.5. It shows a marked decline as the load increases. In high-speed tensile tests at 23° C and at −50° C, generally for POM, no further necking was evident; there was, nevertheless, a pronounced yield point at approximately 4 to 5% elongation as shown by the stress-strain curve.

Surprisingly, despite the high elongation at break of approximately 10%, there was no measurable permanent deformation after the fracture. In other words, the two parts of the fractured test bar spontaneously returned to their original length.

The time-strain (creep) curves plotted in Fig. 9.3.5 show that after exceeding approximately 5% elongation, there is a well above-average rise in the strain. From all available results, 4% elongation was estimated as the minimum elongation at break.

The decrease in elongation at break and yield strain for POM at 23° C has also been confirmed in strain rates up to $10^4 \%/s$ ($\cong t_f \approx 1$ ms).[19] In this case, values of at least 8% were determined for ε_s. The threshold of recognizable clouding of the material up to the fracture rose with increasing strain velocity, and it can be assumed that, for $\dot{\varepsilon} > 10^5 \%/s$, a fracture is likely without any detectable cloudiness. This behavior is thus very similar to that of amorphous materials concerning their threshold of visible crazes.

For polypropylene (PP), which was not examined here, Vincent[11] determined a minimum elongation at break of approximately 1.8% at temperatures to −190° C, though he did not state the time to fracture.

Nonreinforced and short-glass fiber-reinforced polycarbonate (PC) were examined as representative of polymers with particularly high impact strength. As can be seen from Fig. 9.3.6, the elongation at break of PC 20% SFR was much higher at −50° C than at 23° C.

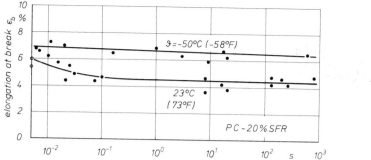

Fig. 9.3.6. Elongation at break of PC–20% SFR. *time to fracture t_f*

Table 9.3.1. Minimum elongation at break on impact loading

Material	ε_{min}
	(%)
HMW-PMMA	2.2
PA6–25% SFR	1.8
Rigid PVC	2.0
POM	≈ 4.0
PC–20% SFR	≈ 4.0
PC	≈ 6.0
PP	1.8 (Ref. 11)

In high-speed tensile tests at $-170°$ C, an elongation at break of approximately 4.5% was measured. According to the available results, the lowest value of elongation at break is about 4%. This was taken as the minimum elongation at break.

The fracture pattern of PC 20% SFR was macroscopically identical to that of the short-glass fiber-reinforced PA6.

With nonreinforced PC, only a slight decrease in the elongation at break was found, even at $-170°$ C. In the high-speed tensile tests an elongation at break of approximately 6% was estimated. Here, macroscopically brittle multiple fractures occurred, even tough the plot of force against piston displacement still clearly indicated a viscoelastic deformation component. The figure of 6% was taken as a minimum elongation at break.

In the range between $23°$ C and $-40°$ C, the fracture behavior in the creep test was identical to that of rigid PVC. The yield strain rose at $23°$ C to between 8 and 12% and the fracture occurred with necking.[20]

It is of interest to note that the fracture surfaces of PC and HMW-PMMA looked identical when necking occurred (PC at $= -40°$ C; HMW-PMMA at $80°$ C; time to fracture about 1 s). The fracture always started at an edge of the test bar. Table 9.3.1 summarizes the minimum elongation at break that was established.

9.4 Ultimate Stress and Energy Absorption

The ultimate stress increased with falling temperature over the entire range examined. It thus approached more and more the theoretically expected figure of approximately 700 N/mm², which has been estimated on the basis of the secondary valence forces (specifically: dispersion forces).[8]

However, because of the numerous defects that are always present in industrially manufactured plastics, this figure is never reached in practice.[21] Only specially oriented crystalline fibers or so-called "whiskers" come near.[22]

The increasing tensile strength in the region of embrittlement can be explained because the modulus of elasticity continues to rise with falling temperature and falling time to fracture, whereas the elongation at break moves toward the minimum threshold. This necessarily results in the increase of tensile strength in accordance with Hooke's law.

As can be seen from Figs. 9.4.1 and 9.4.2, the fracture curves in the plot showing creep rupture strength become flatter and flatter with falling temperature until they are almost parallel with the time axis. When this occurs, the range of complete embrittlement has been reached. The lowest respective stress value, σ_{min}, is used for calculating the minimum absorbable volume-specific energy, w_{min}. The data determined for σ_{min} are summarized in Table 9.4.1.

With PC, PC–20% SFR, and POM, the value for σ_{min} could only be estimated, because the creep fracture curves did not become horizontal in the temperature range examined. With rigid PVC, the lifetime could be split into two ranges for temperatures below approximately 50° C.

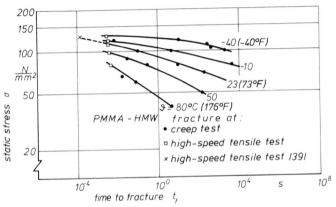

Fig. 9.4.1. Creep rupture strength curves of HMW-PMMA.

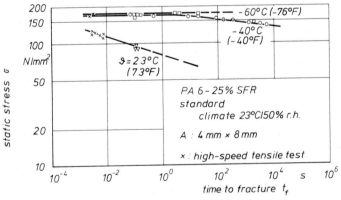

Fig. 9.4.2. Creep rupture strength curves of PA6–25% SFR.

Table 9.4.1. Stress at break with minimum volume-specific energy absorption up to fracture

Material	σ_{min}
	(N/mm²)
HMW-PMMA	135
PA6–25% SFR	175
Rigid PVC	125
POM	> 130
PC–20% SFR	> 110
PC	> 70

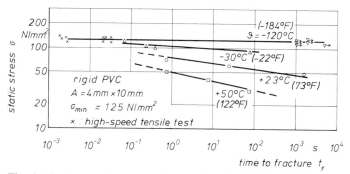

Fig. 9.4.3. Creep rupture strength curves for rigid PVC.

Either the test specimens broke immediately at the end of loading, or they attained times to fracture of more than 10^3 s.[16] This is clearly evident in Fig. 9.4.3 from measurements at $-120°$C. At this temperature, the relaxation and flow processes are so small that the plastics behave essentially elastically. The discontinuity in the load increase is probably triggered by the presence of flaws. The globular structure of the PVC and the press-molding process used here to manufacture the test specimens encourage this.

From the considerations already described on the growth of defects, an attempt was made to extrapolate the long-term behavior from the ultra-short-time creep rupture strength curves that had been determined. Döll and Könczöl[23] followed a similar procedure by calculating the long-term fracture behavior from short-term observations of the crack growth in mechanics-fracture experiments on HMW-PMMA. As Fig. 9.4.4 indicates, it is safe to carry out an extrapolation on the data determined in this work.

Conformity with respect to extrapolation, was also satisfactory for rigid PVC and PC.[24] The extrapolation can, of course, only be performed with a certain amount of reliability if it is known from other tests that no changes in the creep rupture curves occur that were caused by a change in the form of failure behavior. For example, it is not possible to employ such extrapolation for polyolefin pipes in which a change in the fracture behavior occurs.[8, 25]

Figure 9.4.5 shows the stress-strain and fracture behavior of PMMA in the stress-strain diagram. By measuring the areas under the curves, $\sigma(\varepsilon)$, the volume-specific energy is obtained. It can be seen that this initially declines in the range $\dot{\varepsilon} \geq 10^4\%$/h which is relevant for impact.

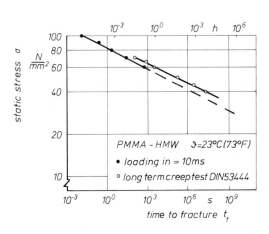

Fig. 9.4.4. Creep rupture strength curves of HMW-PMMA in long-term tests and in creep tests with a loading time of 10 ms.

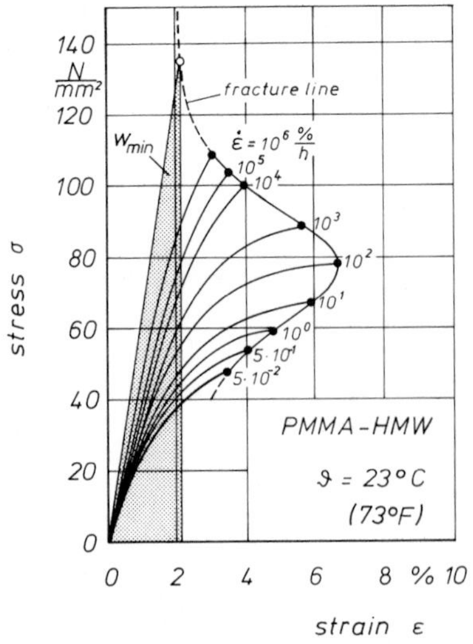

Fig. 9.4.5.   Stress-strain   behavior   of HMW-PMMA during loading under constant strain-rate.

For HMW-PMMA an elongation at break of 2.2% and a stress at break of 135 N/mm², represent a minimum of volume-specific energy because the stress at break for higher strain rates continues to increase, whereas the elongation at break remains constant.

Assuming a linear behavior, Eq. (9.4.1) can be used to estimate the minimum volume-specific energy absorption up to fracture

$$w_{min} \approx 0.5\,\sigma_{min}\,\varepsilon_{min} \tag{9.4.1}$$

With this equation and the data already determined for σ_{min} and ε_{min}, the value of w_{min} for the thermoplastics examined here can be given. These are summarized in Table 9.4.2.

If the approximate stress-strain distribution in the component is known, it is possible to estimate the minimum energy absorption capacity via w_{min}. It can be assumed here that failure occurs if w_{min} is exceeded in the part of the component subjected to the highest load.

Table 9.4.2. Minimum volume-specific energy applied until fracture

Material	w_{min}
	(Nmm/mm³)
HMW-PMMA	1.5
PA6–25% SFR	1.6
Rigid PVC	1.3
POM	>2.6
PC–20% SFR	>2.2
PC	>2.1

9.5 Estimating the Threshold of Brittle Fracture

A brittle fracture in the ultra-short-term range means that the failure occurs at a low elongation and without any macroscopically recognizable flow processes, usually as a multiple-splinter fracture. In tensile testing, the primary break normally occurs perpendicular to the direction of loading. This may be accompanied by several secondary fractures.

The cause of brittle failure is the inability of the material to quickly break down the stresses generated as a reaction to the forced deformation, via relaxation and flow processes. This occurs when the time to failure is in the order of the speed of the relaxation process which dominates the mechanical behavior in the observed range. With the thermoplastics under examination, the question of the relaxation process relevant for impact loads was answered with the aid of the results from tests with a torsion pendulum carried out according to German standard DIN 53445.[26] This can be explained using rigid PVC as an example.

The results from the torsion pendulum test with this material are shown in Fig. 9.5.1. The β-relaxation process is expressed here by a pronounced maximum of the mechanical damping factor, d, which occurs at $-55°$ C for $f = 1$ Hz. Whether a relaxation process significantly influences the mechanical properties or not can generally be recognized from the curve for the storage shear modulus G'_s, in the range of the damping maximum.

Since, in Fig. 9.5.1, a distinct modulus step occurs in this range, it can be assumed that here the β-process will also have a major influence on the mechanical behavior on impact loading. This has been confirmed by our own studies and independently.[27, 28]

It has been stated, however, that the occurrence of relaxation secondary maxima alone is not a necessary and adequate condition for impact strength.[6] This means that additional information must always be obtained as to whether a given relaxation process has a significant influence on the impact behavior. In the simplest case, this may be high-speed tensile or impact-bending tests[29] at temperatures a little below and a little above the transition temperature of the respective relaxation process.

If a correlation is found between the impact strength and a relaxation process – this was always the case with the studies presented here – then a relationship must be established between the frequency of the torsion pendulum test and the time to fracture. For this purpose, the following relationship for the time to failure, which is frequently

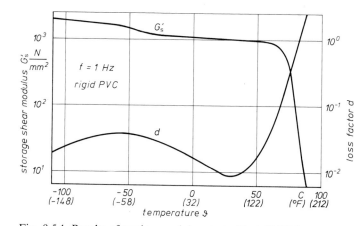

Fig. 9.5.1. Results of torsion pendulum test with rigid PVC.

given in the literature, proved suitable

$$t_f = 1/(2\pi f) \tag{9.5.1}$$

According to this, for example, the following applies

$$1\ \text{Hz} \triangleq 160\ \text{ms}$$
$$10\ \text{Hz} \triangleq \ 16\ \text{ms, etc.} \tag{9.5.2}$$

In addition, the influence of temperature and frequency on the time constants of the relaxation processes have to be given, since, with increasing frequency, the damping maxima shift toward higher temperatures parallel to the temperature axis.

This shift can be described by Eyring's theory,[30, 8] using an Arrhenius equation

$$\dot{t}_{ref} = \dot{t}_0 \cdot e^{\frac{\Delta U}{RT_{ref}}} \tag{9.5.3}$$

This equation can be used to calculate the change in the relaxation-time constants if the activation energy, U, is known for the respective process. This is true of many plastics. References to this can be found in the literature.[29, 31, 32]

The activation energy can be determined from mechanical or dielectric measurements in which the shift of the relaxation maxima under the influence of temperature and frequency is observed.

Using Eq. (9.5.4), it is possible to calculate by how many degrees the damping maximum of a relaxation process shifts per decade of frequency change.

$$\frac{\Delta T}{\text{Decade}} = \frac{1}{\dfrac{R}{\Delta U \log e} + \dfrac{1}{T_{ref}}} - T_{ref} \tag{9.5.4}$$

This process can be illustrated from the example of the shift in the β-secondary maximum for PMMA, rigid PVC, and PC which, with these materials, is linked to the embrittlement. T_{ref} is chosen as equal to the temperature at which the β-secondary maximum is measured for $f = 1$ Hz.

With T_β (1 Hz) and ΔU the figures given in Table 9.5.1 for the shift ΔT_β can be calculated.

Plotting $\log t^*$ against $1/T$, straight lines, are produced as shown in Fig. 9.5.2 for PMMA, rigid PVC, and PC. The region of embrittlement is, in each case, to the right below the determined threshold. In this area, t_f is smaller than t^*, which means that the relaxation process can no longer be significantly effective up to fracture.

$$\left(\text{For the straight lines themselves, } t_f = \frac{1}{\omega} = t^*.\right)$$

Table 9.5.1. Shift of β-relaxation maximum

Material	T_β (1 Hz)	ΔU	Source	ΔT_β
	(K)	(kJ/mol)[a]		(K/decade)
PMMA	293	71.162	Ref. 32	21.4
Rigid PVC	218	54.418	Ref. 32	15.5
PC	173	46.046	Ref. 33	11.6

[a] 1 kcal = 4.184 kJ

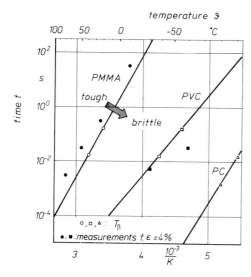

Fig. 9.5.2. Estimation of the region of brittle failure for three amorphous thermoplastics.

In order to experimentally check the agreement, it was necessary to fix an embrittlement strain. Since the minimum elongations at break that were determined lay in the range between 2 and 6%, a mean figure of 4% elongation was regarded as suitable.

The pairs of time/temperature values resulting from the points of intersection of the curves for elongation at break with $\varepsilon_B = 4\%$ are plotted in Fig. 9.5.2 for comparison with the respective estimates. For PMMA and rigid PVC, there is good agreement between the estimate and actual experiment.

For PC, the estimated embrittlement limit is so low that is difficult to check by experiment. So far, it can be said that, at $-170°$ C, $t_f = 10^{-2}$ s multiple fractures occurred and the respective force/displacement diagrams showed no further flow limit. At $-40°$ C, $t_f = 10^{-2}$ s, on the other hand, there was still pronounced necking. Both these statements are not a contradiction to the calculated embrittlement limit. Further tests are necessary to check this point.

Davis and Macosko[33] provide a simple equation for amorphous thermoplastics to estimate the shift in the β-relaxation maximum

$$\frac{\Delta T_\beta}{\text{Decade}} = 0.07 \, T_\beta \, (1 \text{ Hz}) \tag{9.5.5}$$

The values calculated from this for PMMA, rigid PVC, and PC differ by a maximum of $\pm 5\%$ from the numbers in Table 9.5.1.

With short-glass fiber reinforced thermoplastics, the relaxation steps occur at the same temperature as in the respective nonreinforced material. The modulus is merely increased and the dissipation factor reduced. The glass fiber reinforcement thus does not have any influence on the time- and temperature-dependent position of the embrittlement threshold, only on the relative brittleness. With semicrystalline polymers, the same procedure can also be adopted.

With the aid of Eq. (9.5.4) and literature values for the activation energy of the relaxation processes,[32] the values shown in Table 9.5.2 for the position of these secondary maxima on the temperature scale at 1 Hz and the expected shift per decade of frequency change can be calculated.

Table 9.5.2. Relaxation maxima of three semicrystalline thermoplastics

Material	T (1 Hz)	ΔU	ΔT
	(K)	(kJ/mol)	(K/decade)
POM (Cop)	$T_\gamma = 206$	80.79	9.6
PA6[a]	$T_\alpha = 308$	173.72	10.1
PP	$T_\beta = 273$	117.21	11.65

[a] Conditioned in standard climate 23/50

If the resultant data are plotted as previously, they produce Fig. 9.5.3. The embrittlement area is again to the right, below the threshold lines. In the experiments, a macroscopically brittle fracture occurred with POM at $-50°$ C and 10^{-2} s time to fracture. This figure is plotted in Fig. 9.5.3. In agreement with the estimated threshold, nonbrittle fractures occurred at $23°$ C up to 5 ms time to fracture, and very brittle multiple fractures at $-120°$ C, 5 ms time to fracture. In the experiments with PA6–25% SFR, there was a clear relationship between the geometry and the brittle failure threshold.

A comparison between measurement and calculation showed that the 15-mm wide specimens confirmed the estimate, while the thinner, 8-mm wide specimens only became brittle at lower temperatures. This is probably due to the anisotropy of the fiber-reinforced material and the influence of the β- or γ-relaxation secondary maxima at $-67°$ C and $-150°$ C respectively.

Fracture tests were not carried out on PP, but on applying the estimate to three different types of PP, it was possible to clearly assess their impact strength properties, while their results agreed with those from the Izod impact strength test.[34]

Since there was also an ethylene propylene diameterpolymer (EPDM) blend among the three PP types, it would seem that it is also possible to apply the estimation method to modified materials. This would be an enormous benefit as a quick decision making aid with regard to the effectiveness of modifications when developing new high-impact materials.

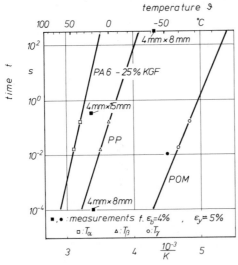

Fig. 9.5.3. Estimation of the region of brittle failure for three semicrystalline thermoplastics.

It is to be expected that the estimation for semicrystalline materials is affected to a greater extent by the morphology and thermal history of the test specimens than with amorphous materials. In the case of polyamide, it is to be expected that the moisture content plays a major role. This is clearly evident in the curves of the dissipation factor, and is expressed in a shift of the secondary maxima and the respective shear modulus data to lower temperatures with increasing moisture content.

9.6 Stress-Strain Behavior

For designing calculations, it is necessary to know not only the thresholds of deformation and load but also to have information on the relationship between stress and strain. In order to examine the latter, tests were performed with sawtooth shaped-loadings.[35] In addition, an evaluation was made of the load-increase phases of the creep tests. The loading time was varied between 10 ms and 1000 ms and the temperature between $-120°$ C and $+80°$ C. This meant that dependence on time and temperature was taken into account over a wide range.

For PMMA, for example, the stress-strain curve shown in Fig. 9.6.1,[36] was obtained which shows the following points:

— Even at low strains, the stress-strain behavior is nonlinear.
— The influences of load-increase time and temperature are large enough for them not to be ignored. Refer to the work by Lörtsch and Retting.[37]
— A change in temperature of approximately 15° C to 20° C has the same effect as a change in time by a factor of 10.
— If, in the range $\varepsilon \leq 1.5\%$, a difference between the measured value and approximated value of approximately $\pm 15\%$ can be tolerated, then it is possible to describe $\sigma(\varepsilon)$ with a secant modulus for 0.5% elongation ($E_{0.5}$).

These statements also apply to the other thermoplastics included in this study. The dependence of rigid PVC and PC on time was, however, lower than it was with PMMA. It may, therefore, be possible within exactly defined limits of validity to use the approximated time-independent modulus in calculations. Figure 9.6.2 enables a comparison to be made between PMMA and PC.

The secant moduli for $\varepsilon = 0.5\%$, which were determined from the measured stress-strain curves for rigid PVC, HMW-PMMA, and PC, are shown in Figs. 9.6.3, 9.6.4,

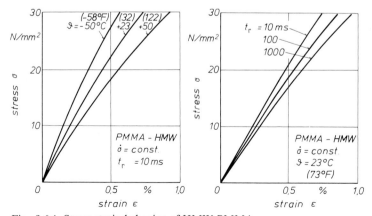

Fig. 9.6.1. Stress-strain behavior of HMW-PMMA.

and 9.6.5. The dependence of the moduli on time can also be estimated with the plot of the storage shear modulus G'_s measured in the torsion pendulum test.

In order to make the differences easier to see, a linear plotting of G'_s as in Fig. 9.6.6 is more suitable than the generally adopted method of plotting log G'_s against temperature. The steeper the $G'_s(v)$ curves, the greater the change of modulus at constant temperature if the loading time ($\cong$ frequency) changes and, as a result, the modulus curves shift approximately parallel to the temperature axis.

With the aid of Eq. (9.6.1),[29] it is possible to obtain a very good estimate of Young's modulus for nonreinforced thermoplastics from G'_s.

$$E_0 = 2G'_s(1+v) \tag{9.6.1}$$

Both G'_s and the Poisson ratio v should be plotted as a function of temperature and frequency. For the time/frequency allocation, Eq. (9.5.1) can be used. If $v(f, \vartheta)$ is unknown, then the calculation is made with $v = 0.37$ for impact load. Since v for thermoplastics lies between 0.3 and 0.5, the maximum error can then be -8.7 to $+5.4\%$.

As the nonlinearity of the stress-strain behavior shows, e.g., in Fig. 9.6.1, the fact that the real stress-strain ratios are always smaller than E_0 must be taken into account in calculations including the estimated or measured Young's modulus.

Fig. 9.6.7 shows how the estimation is applied for PMMA. By using the time-temperature shift principle, the curves can also be determined for larger or smaller

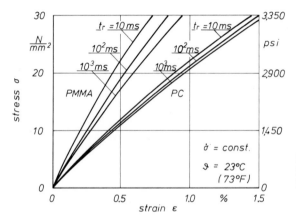

Fig. 9.6.2. Stress-strain behavior of PC and HMW-PMMA.

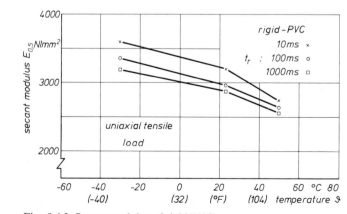

Fig. 9.6.3. Secant modulus of rigid PVC.

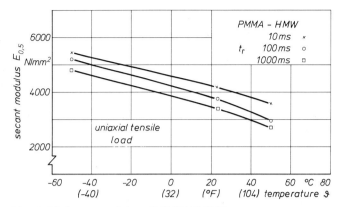

Fig. 9.6.4. Secant modulus of HMW-PMMA.

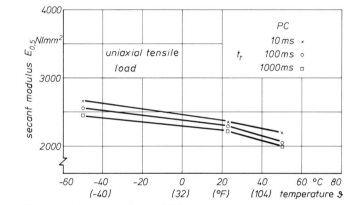

Fig. 9.6.5. Secant modulus of PC.

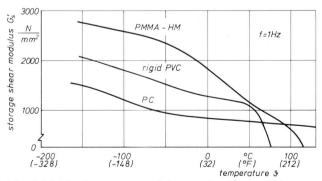

Fig. 9.6.6. Elastic component of the complex shear modulus versus temperature.

load increase times. The measured data have also been included in the plot[38] to show the usefulness of estimation.

In the range of load increase or on impulse loading for the relationship between stress and strain (secant compliance) a clear dependence on time can be recognized. Figure 9.6.8 shows the results for PMMA as an example. This shows that, at 23° C,

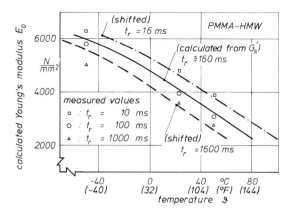

Fig. 9.6.7. Estimation of Young's modulus of HMW-PMMA.

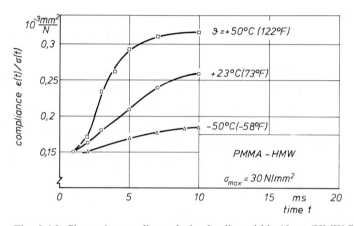

Fig. 9.6.8. Change in compliance during loading within 10 ms (HMW-PMMA).

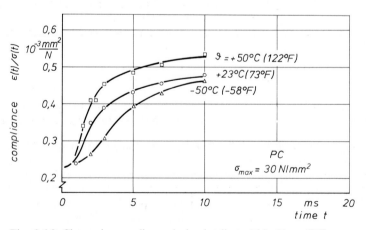

Fig. 9.6.9. Change in compliance during loading within 10 ms (PC).

Table 9.6.1. Estimation of quasi-elastic component of compliance

Material	ϑ	G	J_0
	(°C)	(N/mm²)	(mm²/N)
HMW-PMMA	−150	2600	$1.4 \cdot 10^{-4}$
Rigid PVC	−150	2020	$1.81 \cdot 10^{-4}$
PC	−150	1500	$2.43 \cdot 10^{-4}$
POM	−150	3400	$0.93 \cdot 10^{-4}$

a considerable increase in strain due to retardation is evident during loading in 10 ms. The compliance which was established at the end of the loading time and the so-called "spontaneous strain" thus describe not the purely elastic part of the deformation, but also contain viscoelastic components whose time constants are smaller than or about the same as the load-increase time. At − 50° C, PMMA approaches elastic behavior, as can be seen from Fig. 9.6.8.

As is evident from Fig. 9.6.9, PC behaves virtually identically at both 23° C and − 50° C. The compliance still changes to a relatively large extent during loading at − 50° C.

With both materials, the curves for $t \rightarrow 0$ lead toward the purely elastic part of the compliance. Although the measurements in this range showed a relatively large amount of scatter, the large number of tests provided a good guarantee for this figure. It also agrees very well with the value calculated from the storage shear modulus G'_s for very low temperatures. In this case, the purely elastic behavior is approached since the relaxation processes are frozen. These statements also apply to the other nonreinforced materials included in the study.

Table 9.6.1 provides a summary of the obtained data. J_0 was calculated with $v = 0.37$.

9.7 Notch Effect

For investigation of notch effect, test bars with drilled notches were used. The notches were drilled in the 10 mm wide measuring strip centrally with respect to the thickness direction; diameter 1 mm or 3.5 mm.

For a linear-elastic material, according to Neuber's notch stress theory for drilled notches, the form factors α_k given in Table 9.7.1 are obtained.[39]

Notch factors β_k were then determined by experiment using the thermoplastics listed in Table 9.7.2

$$\beta_k = \frac{F_{\text{max, }u}\, A_{\text{Rest}}}{F_{\text{max, }k}\, A} \tag{9.7.1}$$

The tests were carried out at $+23°$ C and $-170°$ C. The time to fracture was in each case approximately 10 ms. As can be seen from Table 9.7.2, the notch factors β_k were

Table 9.7.1. Form factor α_k (Neuber's theory)

Bore diameter	Specimen width	α_k
1 mm	10 mm	2.7
1 mm	15 mm	2.8
3.5 mm	10 mm	2.3

Table 9.7.2. Comparison between form factor α_k and experimental notch factor β_k

Material	Temperature	α_k	β_k	α_k/β_k
	°C			
PMMA	23	2.7	1.6	1.7
	23	2.3	1.5	1.5
	23	–	1.25[a]	–
	–170	2.7	2.5	1.1
Rigid PVC	23	2.7	1.55	1.7
	–60	2.7	1.7	1.6
	–170	2.7	1.75	1.5
PC–20% SFR	23	2.7	1.0	2.7
PC	23	2.7	0.9–1.0	2.7
PS	23	2.7	1.26	2.1
PS	–120	2.7	1.56	1.7
PC	–170	2.7	1.84	1.5
PC–20% SFR	–170	2.7	1.56	1.7
PA6–25% SFR	–170	2.8	1.4	2.0
POM (C 9021)	–120	2.7	1.45	1.9

[a] Cut notch.

Drilled notch, tension, break in 10 ms.

in most cases 1.5 to 2.7 times lower than the respective form factors, α_k. However, in PMMA at $-170°$ C, there was $\beta_k \approx \alpha_k$.

This means that, according to the available results, Neuber's theory can be safely applied with regard to stress at break. It must, however, be taken into account that, through the notch effect, the elongation at break generally declines very appreciably without this being expressed in the notch factor. It should also be borne in mind that even small cut notches can bring about a reduction in the stress at break of around 20%.[40] It must, therefore, be expected that small grooves in the base of the notch, at changes of cross section, at the bottom of ribs, or at parts of additional generation of internal tensile stresses on the surface of the part, may cause a considerable intensification of the expected notch effect.

Through the effect of the notch, the tensile strength and the elongation at break are, as a first approximation, equally reduced in the case of brittle failure. In this case, the strain is to be measured in an area which is not directly affected by the notch. Without the notch, it can be approximated that

$$w_B \approx 0.5 \sigma_B \, \varepsilon_B \tag{9.7.2}$$

Under the influence of the notch, the volume-specific energy absorption up to fracture, outside the notched cross section falls to

$$w_{B,k} \approx w_B/\beta_k^2 \tag{9.7.3}$$

9.8   Relationship Between the Impact Bending Test and Falling Dart Test

Since the impact bending test, as per German standard DIN 53 453,[41] is a frequently used and simple test method, an examination was made as to whether a relationship could be found with the data determined in the present study.

Table 9.8.1. Volume-specific fracture energy

Material	ϑ	w_k	w_{min}	w_n
	(°C)	(Nmm/mm³)	(Nmm/mm³)	(Nmm/mm³)
HMW-PMMA	23	1.1	1.5	2.7
Rigid PVC	−20…23	1.1…1.6	1.3	NF
PC	23	>10.8	>2.1	NF
	−40	>5.4	>2.1	NF
PC–20% SFR	23	3.8	>2.2	14.6
PA6–25% SFR	23	5.2	1.6	7.7
POM copolymer	23	2.7	>2.6	NF
POM copolymer	−40	2.2	>2.6	−

Standard climate 23/50.
NF No fracture.

For this purpose, the data for several materials given in the literature[5, 29, 42] for impact and notched impact strength, according to the above standard, were converted into a volume-specific energy absorption threshold with the aid of the K-factors given by Oberbach.[43] Table 9.8.1 provides a summary of this.

The conversion was made as follows

— The impact strengths a_n and a_k are calculated from the energy absorbed until fracture W_n and W_k

$$a_n=\frac{W_n}{A_n}, \quad a_k=\frac{W_k}{A_k} \tag{9.8.1}$$

— According to Oberbach,[43] the volume-specific energy absorption values w_k and w_n can be obtained from W_k and W_n with the aid of the respective K-factors

$$w_k=\frac{W_k}{K_k}, \quad w_n=\frac{W_n}{K_n} \tag{9.8.2}$$

— For K_n, of the three-point bending bar, Eq. 9.8.3 is applicable[43]

$$K_n=\frac{b\,h\,l}{9} \tag{9.8.3}$$

— The value of the K-factor for the small standard test bar with U-notch has been given[43] as

$$K_k=30 \text{ mm}^3 \tag{9.8.4}$$

— Thus, for the small standard test bar as per German standard DIN 53453 ($l=40$ mm)

$$K_n=107 \text{ mm}^3 \tag{9.8.5}$$

— By inserting the values for K_n or K_k and from Eq. (9.8.1) and Eq. (9.8.2), the appropriate equations are derived

$$w_k=\frac{a_n}{\text{kJ/m}^2}\,0.54 \text{ Nmm/mm}^3 \tag{9.8.6}$$

$$w_n=\frac{a_n}{\text{kJ/m}^2}\,0.225 \text{ Nmm/mm}^3 \tag{9.8.7}$$

($A_n=24$ mm², $A_k=16.2$ mm²)

A comparison between the calculated w_k values and the w_{min} values given in the present study shows that they correlate very well with the w_{min} values in regard to their magnitude and would, therefore, appear suitable for assessing the impact behavior. On the other hand, the w_n values — as it is possible to determine them — do not provide any comparable information.

One point in favor of using the w_k values is that they implicitly contain information on notch sensitivity.

Delp[44] established the following relationship for the deterioration energy using HMW-PMMA for 23° C in falling dart tests according to German standard DIN 53443.[45]

$$W_{50} = w_D \, h^2 \, R \qquad\qquad (9.8.8)$$

For the constant w_D, a figure of 1.51 mm/mm³ was determined. (Testing conditions: $\vartheta = 23°$ C; $v \approx 5$ m/s; $R = 20, 40, 130$, and 360 mm; $h = 3$ and 5 mm.) This figure corresponds exactly to the threshold of the volume-specific energy absorption w_{min} determined in this study. It is still to be tested whether this relationship is also valid for other thermoplastics in the range of the brittle fracture.

In Delpy's work,[46] a report is given on impact tests until fracture, carried out on glass-fiber reinforced PA6–35% glass fiber and PBTP–30% glass fiber polybutylene terephtalate in which Eq. (9.8.8) could also be applied with small modifications. For example, the exponent in Eq. (9.8.8) had to be chosen as 1.6 for describing the measurements on PA6–35% glass fiber. In this case, a figure of 16.5 Nmm/mm³ was determined for w_D. Comparable figures of our own are unfortunately not available.

Additionally, the relationship between the energy absorption and the square of the sheet thickness was confirmed by Dunn and Williams,[47] for PMMA. According to Fig. 2 in their work, the influence of the radius of the support is opposite to that reported by Delp[44]; for this reason we think that in this case, the radius data on the measuring curves in Fig. 2[47] are probably reversed.

If this situation is corrected, Eq. (9.8.9) can be used to describe the total energy absorption to fracture

$$W_{max} = (2.4 \text{ Nmm/mm}^3) \, h^2 \, R \qquad\qquad (9.8.9)$$

(measurements for $R = 17$ mm and $R = 40$ mm, $\vartheta = 23°$ C).

The volume-specific energy absorption up to fracture of 2.4 Nm/mm³ for 23° C is above the determined minimum of 1.5 Nmm/mm³ of this study, since in this case the deformation velocity was only, at maximum 8.33×10^{-3} m/s.

Dunn and Williams,[47] mentioned that the exponent of the thickness h for ductile plastics is likely to be nearer 1 than 2.

If we apply the considerations made by Oberbach,[43] with regard to the K-factor to the results of Delp[44] and Dunn and Williams,[47] the following K-factor can then be assumed for the falling dart test until fracture

$$K_D = h^2 \, R \qquad\qquad (9.8.10)$$

It should be borne in mind here that the stress-strain state produced in the test specimen in the falling dart test until fracture is governed very much by the following factors:

— geometry of the point of the falling dart,
— type and clamping conditions on the support,
— ratio between maximum flexure and thickness of test specimen,
— level of embrittlement and percentage of plastic deformation, and
— isotropy and morphology of the test specimen (above all with glass-fiber reinforced and semicrystalline thermoplastics).

All in all, however, this would appear to be a useful base for engineering calculations and it should be the subject of further studies.

9.9 Abbreviations

(Latin)

A	area; amplitude
a_k	notched impact strength (German standard DIN 53453)
A_k	residual cross-sectional area of the notched standard small bar (German standard DIN 53453)
a_n	impact strength (German standard DIN 53453)
A_n	cross-sectional area of the standard small bar (German standard DIN 53453)
A_{Rest}	residual cross-sectional area of the notched tensile bar
a_T	time shift factor
b	width
c_p	specific heat (constant pressure)
CFK	carbon fiber reinforced plastics
d	mechanical dissipation factor; diameter
D_1, D_2, D_3	coefficients for describing the viscoelastic and viscose component of deformation
DDM	direct digital measurement
DMS	strain gauge
E	elasticity modulus
E_0	Young's modulus (origin modulus)
$E_{0.5}$	secant modulus for 0.5% elongation
$E_{0.1}, E_{0.2}$	coefficients for describing the quasi-elastic component of deformation
EPDM	ethylene propylene dieneterpolymer
f	flexing; frequency
f_B	flexing at start of fracture
F	force
F_B	force at start of fracture
$F_{max,k}$	maximum force in notched tensile bar
$F_{max,u}$	maximum force in unnotched tensile bar
$F(f)$	spectral component
$g(t)$	function at the beginning of a system
GF	glass fiber
G'_s	real part of the complex shear modulus (storage shear modulus)
h	thickness
$h(t)$	impact reply of a system
HDPE	high-density polyethylene
HIPS	high impact polystyrene
HMW	high molecular weight
i	running index
I	angular impulse (flexure)
J_0	elastic component of compliance
k	constant
K_k	K-factor according to Oberbach[43] for the notched standard small bar
K_n	K-factor according to Oberbach[43] for three-point bending bar
K_D	K-factor for impact test until fracture
KGF	short glass fiber

l	length; distance between supports
l_0	length of the unloaded tensile bar
m	coefficient of the Findley equation
m_L	coupled mass
M_b	bending moment
M_n	number average of the molecular weight distribution
M_w	weight average of the molecular weight distribution
n	exponent in the Findley equation
PA6	polyamide 6
PBTP	polybutylene terephtalate
PC	polycarbonate
PE	polyethylene
PMMA	polymethylmethacrylate
POM	polyoxymethylene
PP	polypropylene
PVC	polyvinylchloride
R	general gas constant $(8.31 \, \mathrm{J \, mol^{-1} \, K^{-1}})$; radius of the support
$s(t)$	function at the beginning of a system
SFR	short fiber reinforcement
t	time
t_{an}	increase time of load or elongation
$\tan \delta$	mechanical dissipation factor
t_B	time to fracture
t_L	life time = time to fracture or to reach the yield point
t_{min}	minimum time to fracture
t_0	reference time of the experiment (in general, end of loading)
t^*	time constant of a relaxation process;
t^*	integration variable
t^*_{ref}	time constant of the relaxation process at reference temperature
t^*_0	time constant of the relaxation process for T
T	absolute temperature; period of time
$T_\alpha, T_\beta, T_\gamma$	absolute temperature of the position of a damping maximum
T_h	homologous temperature
T_{ref}	selected absolute reference temperature
T_m	crystalline melt temperature
T_v	delay time
U	electrical voltage
v	piston velocity
V	volume
w	volume-specific energy
w_R	volume-specific energy absorption up to fracture
$w_{B,k}$	volume-specific energy absorption for notched tensile bar
w_D	constant according to Delp[44] (impact test until fracture)
w_k	volume-specific energy absorption up to fracture, notched standard small bar
w_{min}	volume-specific energy absorption up to fracture, minimum value
w_n	volume-specific energy absorption up to fracture, standard small bar
w_{sp}	volume-specific energy, stored component
$\tilde{w}_{sp}$	volume-specific energy on sinus-shaped loading, stored component
w_v	volume-specific energy, dissipated component
$\tilde{w}_v$	volume-specific energy under sinus-shaped loading, dissipated component

w_{zu}	total volume-specific energy applied
W	energy absorption
W_b	moment of resistance (flexing)
W_B	energy absorption to fracture
WLF	Williams, Landel, Ferry (see Ref. 55)
W_k	energy absorption to fracture, notched standard small bar
W_{max}	energy absorption to fracture
W_n	energy absorption to fracture, standard small bar
W_{50}	energy absorption leading to damage in 50% of the specimens

(Greek)

α	characterization of relaxation process
α_k	form factor as per Neuber
β	characterization of relaxation process
β_k	notch factor (determined by experiment) characterization of relaxation process
δ	loss angle
ΔU	apparent activation energy
ε	engineering strain
ε_B	engineering strain, fracture value
ε_F	engineering strain, optically recognizable flow zones
ε_{lin}	engineering strain, threshold of linear-viscoelastic deformation area
ε_{min}	engineering strain, minimum value for brittle fracture
ε_R	engineering strain, fracture on exceeding the yield point
ε_S	engineering strain, yield point
ε_{vol}	engineering strain, volume strain
ε_w	true strain
ε_0	quasi-spontaneous engineering strain
$\hat{\varepsilon}$	engineering strain, outer fibers
$\dot{\varepsilon}$	engineering strain, strain velocity
ν	Poisson's ratio
ϱ	density
σ	engineering stress
σ_B	engineering stress at fracture
σ_{min}	engineering stress at minimum
σ_n	engineering stress as calculated on the residual cross section of a notched test bar
σ_0	engineering stress, upper stress, dynamic
σ_R	engineering stress, fracture on exceeding the yield point
σ_S	engineering stress, yield point
σ_{st}	engineering stress, static
σ_u	engineering stress, lower stress, dynamic
$\hat{\sigma}$	engineering stress, outer fibers
$\dot{\sigma}$	engineering stress, increase velocity
ω	angular frequency

References

1. H.E. Boden, *Das mechanische Verhalten von Thermoplasten bei stoßartiger Belastung,* Doctoral Thesis, RWTH Aachen 1983.
2. B. Carlowitz, *Plastverarbeiter* 1981, **32**, 437.
3. W. Döll, *Colloid Polymer Sci.* 1978, **256**, 904.

4. K.P. Großkurth, in *Vorabdruck zur 9. Sitzung des Arbeitskreises Bruchvorgänge,* DVM e.V., Berlin 1977.
5. K. Oberbach, *Kunststoff-Kennwerte für Konstrukteure,* Carl Hanser, München 1980.
6. R.F. Boyer, *Polymer Eng. Sci.* 1968, **8**, 161.
7. B. Kleinemeier and H.E. Boden, *Das Spannungs-Dehnungsverhalten von Thermoplasten unter Stoßbeanspruchung,* Final report of research project AIF 4094, IKV Aachen 1980.
8. G. Menges, *Werkstoffkunde der Kunststoffe,* Carl Hanser, München 1979.
9. H.E. Boden, *Stoßverhalten von Thermoplasten unter Berücksichtigung von Glasfaserverstärkung, Kerben und tiefen Temperaturen,* Final report of research project AIF-No. 4707, IKV Aachen 1982.
10. K. Esser, unpublished work, IKV Aachen 1979.
11. E.H. Andrews, *Fracture in Polymers,* Oliver & Body, London 1968.
12. E. Döring, unpublished work, IKV Aachen 1978.
13. C.M. Meysenburg and R. Henkhaus, *Z. Werkstofftech.* 1975, **6**, 306.
14. J. Rellmann, unpublished work, No. S 8243, IKV Aachen 1982.
15. A. Stelkens, unpublished work, No. S 8052, IKV Aachen 1980.
16. A. Matzke, unpublished work, IKV Aachen 1981.
17. G. Menges and D. Pütz, *GWF-Abwasser* 1977, **118**, 524.
18. H. Schlüter, unpublished work, 1982.
19. R. Henkhaus, *Das Verformungsverhalten ausgewählter Thermoplaste in Abhängigkeit von der Dehngeschwindigkeit,* Doctoral Thesis, RWTH Aachen 1980.
20. M. Paar, unpublished work, IKV Aachen, 1979.
21. A.A. Griffith, in *Physik der Hochpolymere,* Springer Verlag, Berlin 1956.
22. D.W. van Krevelen, *Properties of Polymers,* Elsevier, Amsterdam 1976.
23. W. Döll and L. Könczöl, *Kunststoffe,* 1980, **70**, 563.
24. H.E. Boden, *Bruch- und Verformungsverhalten von Thermoplasten beim Kriechen im Zeitbereich von 10^{-3} s bis 10^3 s,* Final report of research project DFG/Me 272/144, IKV Aachen 1980.
25. L.J. Zapas and J.M. Crissmann, *Polymer Eng. Sci.* 1979, **19**, 99.
26. DIN 53445 (German standard), *Torsionsschwingversuch,* 1965.
27. B. Hartmann and G.F. Lee, *J. Appl. Polymer Sci.* 1979, **23**, 3639.
28. W. Retting, *Rheol. Acta* 1969, **8**, 259.
29. G. Schreyer, *Konstruieren mit Kunststoffen,* Carl Hanser, München 1972.
30. H.H. Kausch, *Polymer Fracture,* 2nd ed., Springer, Heidelberg-New York 1986.
31. J.D. Ferry, *Viscoelastic Properties of Polymers,* Wiley, New York 1980.
32. N.G. McCrum, B.E. Read and G. Williams, *Anelastic and Dielectric Effects in Polymer Solids,* John Wiley & Sons, London 1967.
33. W.M. Davis and Ch.W. Macosko, *Polymer Eng. Sci.* 1977, **17**, 32.
34. M. Sönmez, unpublished work, 1982.
35. H.E. Boden, *Kriech- und Verformungsverhalten von Thermoplasten bei verschiedenen Lastverläufen und Temperaturen in einem Zeitbereich von 10^{-3} s bis 10^3 s,* Final report of research project DFG/Me 272/150-2, IKV Aachen 1982.
36. W. Janke, unpublished work, IKV Aachen 1981.
37. W. Lörtsch and W. Retting, *Materialprüfung* 1972, **14**, 299.
38. H.E. Boden and G. Menges, *Proc. 40th Ann. Techn. Conf. Soc. Plastics Engineers,* edited by L.B.Weisfeld, p. 7–8. Society of Plastics Engineers, Brookfield Center, CT 1982.
39. N.N., *Dubbel, Taschenbuch für den Maschinenbau − Band I,* Springer, Berlin 1974.
40. G. Hahn, unpublished work, IKV Aachen 1979.
41. DIN 53453 (German standard), *Schlagbiegeversuch,* 1975.
42. B. Carlowitz, *Thermoplastische Kunststoffe,* Zechner und Hüthig Verlag, Speyer 1980.
43. K. Oberbach, *Plaste Kautschuk* 1982, **29**, 178.
44. St. Delp, *Spektrum* 1980, **25**, 11.
45. DIN 53443 (German standard), *Fallbolzenversuch,* 1975.
46. U. Delpy, *Kunststoffe* 1983, **72**, 476.

47. C.M.R. Dunn and M.J. Williams, *Plast. Rubber* 1980, **5**, 90.
48. B. Carlowitz, *Untersuchungen zum Bruchverhalten ausgewählter thermoplastischer Kunststoffe bei Schlagbeanspruchung,* Doctoral thesis, RWTH Aachen 1983.
49. B. Carlowitz, *Kunststoffe* 1980, **7**, 405.
50. DIN 53455 (German Standard), *Zugversuch,* 1981.
51. K. Oberbach, *Ingenieur Digest* 1979, **18**, 59.
52. H.W. Paffrath and G. Wübken, *Kunststoffe* 1982, **9**, 8.
53. W. Retting, *Europ. Polymer J.* 1970, **6**, 853.
54. R.F. Schwarzl, *Die beschreibende Theorie der linearen Relaxationserscheinungen,* Course notes, RWTH Aachen 1971.
55. M.L. Williams, R.F. Landel and J.D. Ferry, *J. Amer. Chem. Soc.* 1955, **77**, 3701.

10 Impact Strength: Determination and Prediction

Witold Brostow

Department of Materials Engineering, Drexel University
Philadelphia, PA 19104, U.S.A.

At high elongations the load is borne unequally by the chains in the network; some of them, by virtue of their situations, are more susceptible to overextension than are others. When these chains are broken, others in turn are vulnerable; and so on to the eventual catastrophe: rupture of the specimen.

(Paul J. Flory in *The Science of Macromolecules*[1])

10.1 Introduction

A polymeric material is usually accompanied by a long list of test results. Lay users—and to a lesser extent polymer scientists and engineers—often find such lists bewildering. A number of questions immediately appear. Can one really predict the service behavior on the basis of parameters given? Are the quantities measured in standard tests the most pertinent ones? Would a shorter list suffice?

No single answer exists to a series of questions such as quoted above. This, of course, is one reason why we have a whole volume on the subject of mechanical failure. In general, answers depend on the type of service: under a constant load, under cyclic loads, in degrading environments, at elevated temperatures, under cyclic temperature changes, under impacts, and so on. Indeed, under specified conditions one parameter can become much more important than the remaining characteristics. As an example, if the material serves in an electric field, the breakdown strength (defined for instance in [2]) becomes the key parameter.

Although we are concerned here primarily with service, the same statement can apply to processing: a single parameter can be decisive for the behavior of the material. Thus, Santamaria, Guzmán and Peña[3] have studied degradation of poly(vinyl chloride) during extrusion in a capillary rheometer. Thermal degradation was measured in terms of the time to color and the time to black. The Spanish researchers have found that both of these parameters go symbatically with melt viscosity. In other words, viscosity represents the key measure of thermal stability.

In this Chapter we consider the failure resulting from crack propagation and/or under impact. There is one parameter which determines behavior under such conditions; it will be defined in the following Section. Most of the Chapter is concerned with the nature of this parameter and with ways of predicting it from a minimum of experimental data.

10.2 Impact-Transition Temperature: Definition

The *impact transition temperature* T_I is defined[4] as the temperature at which the response of a material changes from brittle to ductile under *high-impact* load conditions.

The italicized part of the definition is important. For metals the ductile-brittle transition temperature (DBTT) has been used for some time to characterize their mechanical behavior. At DBTT both ductile and brittle behavior is possible in samples of identical geometry. Attempts to transfer the definiton from metals to polymers in general were not very successful. Bueche[5] has discussed what he called "brittle temperature". He noted that values of this temperature depend on the experimental procedure used. Andrews[6] has provided a fairly detailed, but only qualitative, discussion of DBTT. After a survey of experimental data Andrews has concluded that what we here call DBTT is not a single temperature, but rather a transition range. The size of this range varies from one material to another, but intervals of up to 10 K have been observed.

In 1979, Zewi and Corneliussen[7] noted that impact conditions play a role in the transition. This led to the definition given in [4] with which we started this Section. In contrast to DBTT, our T_I corresponds to a single temperature which can be precisely located. From experimental results[8,9] it appears that T_I represents a basic material property.

The distinction between ductile behavior and brittle failure in impact is readily apparent. Macroscopically, the fracture surface of a brittle failure appears smooth. In contrast, in the *ductile impact* mode the specimen does not actually break but merely deforms, in response to the applied load. Under the scanning electron microscope, brittle failures in polyethylene display a "flaky" fracture surface morphology; see Fig. 10.2.1. This appearance has been attributed to microscopic crack branching.[10] It should be noted that the brittle impact phenomenon appears to be different in nature than the brittle slow crack growth phenomenon. The latter is discussed in detail by Lustiger in Chapter 16 in terms of an interesting model of behavior of tie molecules.

The impact-transition temperature T_I can be conveniently determined by performing the *Charpy test*. A specimen of definite size with a notch of known dimensions is impacted at high velocity. The geometry of the test represents three-point bending. A rectangular specimen resting against two posts is struck midway between the supports. Details of the experimental procedure have been discussed in [9]. We do not discuss any such details at this point since the entire Chapter 8 by Adams and Wu is devoted to impact testing, in fact to a highly accurate form of it called instrumented impact testing. Moreover, details of behavior of a number of thermoplastic materials under impact have been provided in the preceding Chapter by Menges and Boden.

Stress concentration in the specimen takes place at the notch. At the same time, the notch tip constitutes a constraint to plastic deformation. Therefore, we shall characterize the notch and the resulting stress concentration in the following Section.

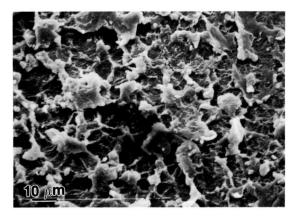

Fig. 10.2.1. Scanning electron micrographs of the fracture surface of polyethylene after undergoing brittle failure in the Charpy test at 173 K. Before taking the micrograph the surface was coated by sputtering with a 10–20 nm layer of a gold + palladium alloy.

10.3 Stress-Concentration Factor

As noted in Section 2.2, the theoretical cohesive strength of most material is about $E/15$. Here E is the Young modulus, discussed in any textbook of materials science and engineering, for instance in Chapter 12 of [2]. Due to the presence of scratches, cracks and other imperfections, a real material has a much lower strength. This is why in impact testing we study effects of notches of varying dimensions.

The detrimental effect of a scratch, crack or an artificial notch is, on the basis of work of Inglis[11] and Griffith[12, 13] characterized by

$$K_t = 1 + 2 \left(\frac{h}{l} \right)^{1/2} \tag{10.3.1}$$

Here K_t is the *stress-concentration factor*; h is the depth of the notch, or one-half the length of the major axis in an elliptical hole; l is the radius of curvature at the bottom of the notch, or at each end of the major axis of an elliptical crack. The above equation is the same as Eq. (7.2.1); further discussion of it can be found in Section 7.2. For the present it is sufficient to note that Eq. (1) corresponds well to our intuitive notions of the "evil" produced by a crack: the deeper the crack is, the higher the stress-concentration factor K_t; the blunter the crack, the smaller K_t results.

The geometry of the deepest crack is only one element in the failure story. An important question now appears: what fraction of the energy U_0 provided to the specimen during an impact test goes into bond breaking? This is in contrast with the remainder of U_0 which is dissipated by chain relaxation and is thus harmless. To be more accurate, U_0 also contains other contributions such as plastic deformation; all of them along with relaxation have a common characteristic, namely they do not bring about mechanical failure. Before proceeding further, let us note that the parameter U_0 is *not* related to another parameter represented by the same symbol and used by Zhurkow, Korsukov, Regel, Slutsker and their colleagues[14−16] in a kinetic theory of mechanical stability of solids. Properties of our impact energy U_0 will be analyzed in the following Section.

10.4 Impact Energy

As noted above, a part of the impact energy U_0, such as the part "invested" by the hammer of the Charpy machine into the specimen, certainly goes into breaking bonds in polymer molecules. Let us denote the energy so utilized by U_b, and the corresponding rate constant by c_b. At the same time, a part of the input energy goes into *relaxation of macromolecular chains*; denote this energy term by U_r, and the respective rate constant by c_r. Take both bond rupture and molecular relaxation to be first-order processes. At any time t, denote the unutilized energy by U. That is, U represents the part of input energy which did not go, as yet, into either of the two competitive processes. Thus

$$U = U_0 - U_b - U_r \tag{10.4.1}$$

All parameters featured in Eq. (1), except for U_0, are in general functions of time t. Clearly

$$U(0) = U_0 \tag{10.4.2}$$

Consider now the derivative dU/dt. Given the two first-order processes which bring about a decrease in U, then

$$\frac{dU}{dt} = -(c_b + c_r)\, U \qquad\qquad (10.4.3)$$

Integrating Eq. (3) and remembering Eq. (2), then

$$U = U_0\, e^{-(c_b + c_r)t} \qquad\qquad (10.4.4)$$

Substituting Eq. (4) into Eq. (3) results in

$$\frac{dU}{dt} = -\frac{dU_b}{dt} - \frac{dU_r}{dt} \qquad\qquad (10.4.5)$$

where

$$\frac{dU_r}{dt} = c_b\, U_0\, e^{-(c_b + c_r)t} \qquad\qquad (10.4.6a)$$

and

$$\frac{dU_r}{dt} = c_r\, U_0\, e^{-(c_b + c_r)t} \qquad\qquad (10.4.6b)$$

Integrating Eq. (6a) and again remembering Eq. (2), then

$$U_b = \frac{c_b\, U_0}{c_b + c_r}\, [1 - e^{-(c_b + c_r)t}] \qquad\qquad (10.4.7a)$$

Analogously

$$U_r = \frac{c_r\, U_0}{c_b + c_r}\, [1 - e^{-(c_b + c_r)t}] \qquad\qquad (10.4.7b)$$

Thus, for $U(\infty) = 0$ separation of the original U_0 into U_b and U_r components depends on the fraction $c_b/(c_b + c_r)$. Various consistency checks of the present set of equations can be easily made: for instance, direct differentiation of Eq. (1) produces Eq. (5); summation of Eqs. (7a) and (7b) reproduces Eq. (1).

From Eqs. (7) we have immediately

$$\frac{U_b}{U_r} = \frac{c_b}{c_r} \qquad\qquad (10.4.8)$$

The parameter c_b is necessarily related to stress concentration. We, therefore, make the simplest possible assumption

$$c_b = c_b'\, K_I \qquad\qquad (10.4.9)$$

where K_I is the stress-concentration factor defined by Eq. (10.3.1) while c_b' is a material parameter independent of temperature.

By contrast, the relaxation-rate parameter c_r is necessarily related to the thermodynamic temperature T. As noted in general by Jedlinski,[17] the usability of polymeric solids is to a large extent determined by T. Intuitively, we expect that a given crack will become less "dangerous" after a temperature increase, and the ratio displayed in Eq. (8) will then decrease. In molecular terms, we expect that an increase in free volume will promote relaxation. Therefore, in the following Section we shall consider the free volume problem; this will provide a background necessary for determination of c_r.

10.5 Free Volume

In general, we can write

$$v^* = v - v^f$$ (10.5.1)

Here we take v as the volume per molecule for nonpolymeric materials and the volume per segment for polymers; v^* is a characteristic parameter often called the hard-core volume; v^f is the remainder, that is the *free volume*, again per molecule or per one polymer segment. A reduced volume $\tilde{v}$ is also in use; it is defined by

$$\tilde{v} = \frac{v}{v^*}$$ (10.5.2)

The problem now is in the definition of v^* (or of v^f). Since the problem of free volume is particularly important for liquid phases, various approaches to polymeric solids often take over definitions of v^f from theories of liquids. Thus, the Flory statistical-mechanical theory of the liquid state[18, 19] leads to

$$\tilde{v} = \left[\frac{\alpha T}{3(1 + \alpha T)} + 1 \right]^3$$ (10.5.3)

Here α is the *isobaric expansivity* defined by

$$\alpha = \frac{1}{V} \left(\frac{\partial V}{\partial T} \right)_p$$ (10.5.4)

and V can be taken as the specific (or molar) volume of the material. Eq. (3) has prompted studies of temperature dependence of volume for liquids and polymeric solids.[20, 21]

In addition to reduced volume $\tilde{v}$, there are also reduced temperature $\tilde{T}$ and reduced pressure $\tilde{P}$ defined by analogs of Eq. (2), that is

$$\tilde{T} = \frac{T}{T^*}$$ (10.5.5a); $$\tilde{P} = \frac{P}{P^*}$$ (10.5.5b)

The Flory theory gives the following prescriptions for calculating the parameters named above for zero pressure:

$$T^* = \frac{T \tilde{v}^{4/3}}{\tilde{v}^{1/3} - 1}$$ (10.5.6)

$$P^* = T \tilde{v}^2 \frac{\alpha}{\kappa_T}$$ (10.5.7)

Here κ_T is the *isothermal compressibility*

$$\kappa_T = -\frac{1}{V} \left(\frac{\partial V}{\partial P} \right)_T$$ (10.5.8)

A modification of the Flory theory, assuming the existence of holes as independent entities in the material, has been developed by Simha and Somcynsky.[22] In the reduced form it reads

$$\frac{\tilde{P} \tilde{v}}{\tilde{T}} = [1 - 2^{-1/6} y (y \tilde{v})^{-1/3}]^{-1} + \frac{2y}{\tilde{T}(y \tilde{v})^2} \left[\frac{1.011}{(y \tilde{v})^2} - 1.2045 \right]$$ (10.5.9)

where $(1-y)$ is the fraction of holes; the reduced pressure $\tilde{P}$ and temperature $\tilde{T}$ are defined by Eqs. (5). Since Eq. (9) is hardly convenient for practical calculations, it has been represented[23] by an interpolation formula:

$$\tilde{v}=0.9299+0.4478\,\tilde{T}+37.33\,\tilde{T}^2-327.3\,\tilde{T}^3 \tag{10.5.10}$$

Coefficients in Eq. (10) are supposed to be universal.

As noted, in [24], there are a variety of other definitions of v^f. One more will be mentioned, developed by Roszkowski for polymer melts[25] and then extended to polymeric solids down to 0 K.[26] However, the free volume relations given above are sufficient for our immediate purposes.

The next step towards relating the chain relaxation rate c_r to free volume is provided by the well-known empirical Doolittle equation[27]:

$$\ln \eta = \ln A' + B\,\frac{v^*}{v^f} \tag{10.5.11}$$

where η is the viscosity while A' and B are constants. Intuitively, an increase in free volume should increase fluidity and thus lower viscosity; Eq. (11) is a quantitative statement of this fact.

Since the capability of chains to relax goes along with free volume, Eq. (11) – or rather its inverse – provides us with a measure of this capability. However, a measure of changes of relaxation capability with temperature would be more convenient. As such a measure we choose the *shift factor* a_T defined[28] as

$$a_T = \frac{\eta\,T_{\text{ref}}\,\varrho_{\text{ref}}}{\eta_{\text{ref}}\,T\varrho} \tag{10.5.12}$$

where ϱ denotes mass density and the index ref refers to any convenient reference values. The factor a_T plays an important role in time-temperature superposition, which is also called the method of reduced variables. The method is explained by Kenner in Section 2.3. By combining now Eqs. (11) and (12) and using Eq. (2) we obtain

$$\ln a_T = -\frac{B}{\tilde{v}_{\text{ref}}-1}+\ln\frac{T_{\text{ref}}\,v}{Tv_{\text{ref}}}+\frac{B}{\tilde{v}-1} \tag{10.5.13}$$

The first r.h.s. term in Eq. (13) is a constant by definition; we shall denote it by the symbol B'. The second r.h.s. term varies only slowly with the temperature; neglecting its changes with T constitutes a reasonable assumption. Accordingly, we write

$$A = B' + \ln\frac{T_{\text{ref}}\,v}{Tv_{\text{ref}}} \tag{10.5.14}$$

and we assume that

$$\frac{dA}{dT}=0 \tag{10.5.15}$$

Validity of relation (15) will be checked later. Now we are able to rewrite the shift factor Eq. (12) as

$$\ln a_T = A + \frac{B}{\tilde{v}-1} \tag{10.5.16}$$

Equation (16) constitutes a key step in our quest for a method of predicting the impact transition temperature. However, before using the equation just derived and pursuing this quest further, we shall digress to the WLF-equation in the following Section. Those not interested in the WLF relation may proceed directly to Section 10.7.

10.6   Williams-Landel-Ferry Equation

We do not need the Williams-Landel-Ferry *(WLF)-equation*[29] now. The reason we are dealing with it here is because it is still so popular; in this book it is referred to in at least four other Chapters.

The general Eq. (10.5.16) cannot be used unless a specific relation between the reduced volume $\tilde{v}$ and temperature T is assumed. When T increases, total volume also increases (water in a certain temperature range is an exception). Accordingly, $\tilde{v}$ increases also, while its reciprocal $\tilde{v}^{-1}$ decreases. As it happens, Williams, Landel, and Ferry worked with $\tilde{v}^{-1}$. They have made the simplest possible assumption, namely that of a linear relationship

$$\tilde{v}^{-1} = \tilde{v}_{ref}^{-1} - \alpha_f(T - T_{ref}) \tag{10.6.1}$$

where α_f is a proportionality constant, *not* equal to the expansivity α.

Other than the accidental choice of $\tilde{v}^{-1}$ (rather than $\tilde{v}$, for instance) and simplicity, there is no basis for Eq. (1). Substituting it into Eq. (10.5.16) leads to

$$\ln a_T = \frac{\alpha_f(T - T_{ref})\, B}{(1 - \tilde{v}_{ref}^{-1})\,[1 - \tilde{v}_{ref}^{-1} + \alpha_f(T - T_{ref})]} \tag{10.6.2}$$

We note that Eq. (2) blows up when $(1 - v_{ref}^{-1})/\alpha_f = (T_{ref} - T)$. The main advantage of this relation is the capability of calculating the shift factor in the function of temperature. It has virtually opened the whole field of time-temperature superpositions or master curves, already mentioned; see again Section 2.3. Particularly important in this flurry of activities is the study by Tschoegl and his school[30, 31] of dependence of a_T on both temperature and pressure, P. Thus, Fillers and Tschoegl[31] have provided an explicit formulation of the general shift factor $a_{T, P}$.

Since Eq. (1) was only a guess, the success of Eq. (2) clearly relies on the validity of our general formula (10.5.16). As it happens, the WLF formula (2) gives satisfactory results above the glass-transition temperature, T_g, in particular around $T_g + 50$ K. It breaks down around $T_g + 100$ K; for some materials its upper limit of applicability is even lower. The lower limit of applicability is somewhat above T_g. In the immediate vicinity of T_g and below the WLF formula is useless; we shall return to this problem briefly in Section 10.8. Ferry himself[28] notes that what we have written as Eq. (2) certainly does not hold below the glass-transition temperature.

We conclude that the WLF Eq. (2) represents one of the possible relations between the shift factor a_T and temperature. Experimental evidence shows that Eq. (1) constitutes the weak link in the otherwise seminal work of Williams, Landel, and Ferry.

10.7   Prediction of Impact-Transition Temperature

In view of Eq. (10.4.8), we continue toward the evaluation of c_r, and eventually toward relating the impact-transition temperature T_I and the stress-concentration factor. Now, however, we are in a better position than before, since we can utilize the relations from Section 10.5.

As stated earlier, more free volume should increase the relaxational ability of a material. A measure of this ability can be the fluidity or the reciprocal viscosity —related to free volume by the Doolittle equation (10.5.11). Another candidate for such a measure is the shift factor defined by Eq. (10.5.12). In view of repeatedly noted important uses of the shift factor α_T in obtaining master curves or superposition operations, the latter is a preferred candidate.

We note that when temperature increases the shift factor decreases. Therefore, we simply assume that the relaxation-rate parameter c_r is inversely proportional to the shift factor

$$c_r = \frac{c_r'}{a_T} \tag{10.7.1}$$

Here c_r' is also a material constant independent of temperature.

Now from Eqs. (10.4.9), (10.4.8) and the present Eq. (1) we obtain for the stress-concentration factor

$$K_t = \frac{c}{a_T} \cdot \frac{U_b}{U_r} \tag{10.7.2}$$

Here c is a new abbreviation, namely

$$c = \frac{c_r'}{c_b'} \tag{10.7.3}$$

A further simplification can be made. According to the definition (10.3.1), K_t depends on the crack geometry only. For a given crack geometry and at a given temperature, U_b/U_r is necessarily a constant. Introducing a new symbol

$$f = c \frac{U_b}{U_r} \tag{10.7.4}$$

we can rewrite Eq. (2) as

$$K_t = \frac{f}{a_T} \tag{10.7.5}$$

Eq. (5) can be combined with Eq. (10.5.13). By introducing the symbol

$$F = \frac{f}{e^A} \tag{10.7.6}$$

we arrive at the final formula [4]

$$K_t = F \, e^{-\frac{B}{\bar{v} - 1}} \tag{10.7.7}$$

Thus, the stress-concentration factor is an explicit function of the reduced free volume. The constant B comes from the Doolittle equation. The constant F comes partly from our problem and also from the Doolittle constant A.

Since the free volume, reduced or otherwise, is a function of temperature, Eq. (7) contains implicitly a relation between K_T and the corresponding impact-transition temperature T_I. An *explicit* form would involve a specific $\bar{v}(T)$ formula. Then the equation could be inverted to a form $T_I = T_I(K_t)$. At this time a commitment to any $\bar{v}(T)$ is not yet warranted: for this reason Eq. (7) is kept in its present general form. However, in the following Section an exemplary calculation will be provided.

10.8   Calculation of Impact-Transition Temperature: Low-Density Polyethylene

We shall now study the validity of Eq. (10.7.7) in terms of experimental data. Impact-transition temperatures corresponding to a set of stress-concentration factors (razor notches) have been determined by Zewi and Corneliussen [7] for *low-density polyethylene* using the Charpy method (see Section 10.2). The range of impact-transition temperatures found extended from 293 K down to 179 K.

As noted above, one needs a $\tilde{v}(T)$ formula to be able to use Eq. (10.7.7). Eq. (10.5.10) was taken for this purpose. This does *not* mean any general endorsement of the latter formula; a value of the hard-core temperature corresponding to it, namely $T^* = 7956$ K was available. [33]

Since Eq. (10.7.7) involves to constants, it was solved for two pairs of K_t and T_I values, namely for $T_I = 269$ K and 293 K. The results are $F = 0.1239$ and $B = 0.1053$. A slightly better agreement between calculation and experiment could conceivably have been obtained by solving an over-determined system of Eqs. (10.7.7) for all experimental points for two unknowns, F and B. Such a solution, however, would have detracted from the fact that a *pair* of experimental points is sufficient.

With the calculated pair of F and B values, we have computed stress-concentration factors corresponding to experimental impact transition temperatures. The results are listed in Table 10.8.1, together with experimental K_t values. Corresponding $\tilde{v}$ values as well as factors $(\tilde{v} - 1)^{-1}$ are listed also.

Clearly, the agreement between calculation and experiment is very good; particularly because there is some scatter of experimental data, visible at 285 K and 288 K, when we have two experimental K_t values at each. The same results are presented graphically in Fig. 10.8.1.

On our way towards Eq. (10.7.7) we made some assumptions, including Eq. (10.5.15). The latter will now be checked using the same set of experimental data. We take the hard-core parameters as the reference ones; as noted by Ferry [28] there are no limitations in the choice of the reference state. With this particular choice Eq. (10.5.14) can be rewritten as

$$A = B' + \ln \frac{\tilde{v}}{\tilde{T}} \qquad\qquad (10.8.1)$$

By using the same T^* value as before, and taking the $\tilde{v}$ values listed in the second column of Table 1, we find the r.h.s. of Eq. (1) equal to 3.748 at 179 K and to 3.282 at 293 K. By comparison, $(\tilde{v} - 1)^{-1}$ values in the same temperature interval — as seen in the third column in the table — exhibit a much larger variation. Thus, the validity of assumption (10.5.15) for the present purposes is upheld, particularly in view of

Table 10.8.1. Impact transition temperatures T_I, corresponding stress concentration factors K_t and reduced volumes from Eq. (10.5.10) for low density polyethylene. Experimental values from [7].

T_I/K	$\tilde{v}$ (Eq. (10.5.7))	$1/\tilde{v} - 1$	K_t	
			Experimental	Calculated
179	0.9552	−22.32	1.0	1.3
269	0.9751	−40.16	8.5	8.5
273	0.9760	−41.67	9.4	10.0
285	0.9787	−46.95	12.6	17.4
285	–	–	11.9	–
288	0.9795	−48.78	19.4	21.1
288	–	–	16.0	–
289	0.9797	−49.26	22.2	22.2
290	0.9799	−49.75	24.5	23.4
291	0.9803	−50.76	27.0	26.0
293	0.9807	−51.81	29.0	29.0

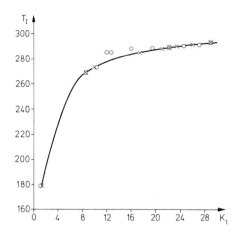

Fig. 10.8.1. Relation between the stress concentration factor K_t and the impact transition temperature T_I for low-density polyethylene. Experimental values (circles) determined by the Charpy method[7]; calculated values (crosses) obtained from Eq. (10.7.7).

limited accuracy of experimental data seen in Table 1. Of course, if at any time this accuracy will improve, then the simplification (10.5.15) can be abandoned and $\ln(\tilde{v}/\tilde{T})$ calculated easily.

An attempt has also been made to use the WLF-equation (10.6.2). The results, as discussed in [4] were a failure, and we do not provide further details here.

One more comment on the present results is in order. There is still a common belief, voiced for instance by Ferry,[28] that the constant B in the Doolittle equation (10.5.11) is close to unity, and, therefore, taking $B=1$ constitutes a reasonable assumption. We have found above B of the order of 0.1, but this is by no means the only such result. Fillers and Tschoegl[31] found values of B that varied from 0.175 for Viton B (a copolymer of vinylidene fluoride and hexafluoropropylene, lightly filled with carbon black) to 0.608 for poly(vinyl chloride). Clearly, in the general case the assumption $B=1$ has to be treated with caution; it is convenient when there is a paucity of experimental data, or when we particularly want to limit the number of independent parameters.

10.9 Future Work and Concluding Remarks

As noted in Section 10.7, a commitment to a specific relation between volume and temperature in Eq. (10.7.7) would be premature. In other words, we believe that a still better equation than (10.5.10) can be found. A promising candidate is the Guggenheim-Lu-Ruether (GLR) equation. Decades ago, Guggenheim[34] proposed a formula for the temperature dependence of the mass density ρ of argon. Lu, Ruether and their collaborators[35, 36] have generalized the Guggenheim formula. The GLR-equation can be written as

$$\tilde{\varrho} = \frac{\varrho}{\varrho_{\text{ref}}} = 1 + \varrho_\alpha \left(1 - \frac{T}{T_{\text{ref}}}\right) + \varrho_\beta \left(1 - \frac{T}{T_{\text{ref}}}\right)^{1/3} \tag{10.9.1}$$

Actually, Lu, Ruether and collaborators used the liquid-vapor critical point as the reference state; they applied the formula successfully to a number of organic liquids and their mixtures. Good results have also been obtained for isobaric expansivities (see Eq. (10.5.4)) of organic as well as inorganic liquids.[20] Of course, for polymers the critical point is hardly a convenient reference state; however, as implied by the general form of Eq. (1), other reference states are possible. In fact, good results have already been obtained by using Eq. (1) for solid polymers. They will be reported in a future paper.[21]

The problem discussed in this Chapter involves additional aspects. Behavior of a specimen under impact depends on the sample thickness. Some aspects of this have already been studied experimentally by the Izod method,[37] but clearly, further experimental and theoretical study are required. Related to this is the problem of internal distribution of stresses in a material. These stresses manifest themselves in bond rupture. Fortunately, a convenient experimental method of studying the bond rupture has been developed by Roylance.[38, 40] He uses electron spin resonance (ESR) spectroscopy to monitor covalent bond rupture during fracture. The ESR results are then translated into an atomistic fracture model which represents the internal stress distribution.

Many kinds of properties of polymeric materials (thus, not only mechanical) can be calculated with good accuracy by using additivity schemes. A number of such schemes were developed by Van Krevelen; an excellent review of his schemes as well as of those developed by other authors is provided in his book.[41] There is, as yet, no such scheme for the shift factor. Van Krevelen does note[41] difficulties involved in the use of the WLF-equation.

So far we have been talking about "pure" materials, disregarding the composition variable. Of course, this variable is unavoidable for blends, and also when fillers, dyes, flame retardants and other additives are used. Interesting here, is the work of Knauss and Kenner[42] who studied the creep compliance of poly(vinyl acetate) in shear as a function of temperature, and also as a function of the quantity of water adsorbed. They have found that water concentration affects the time scale of creep in a manner similar to temperature. This clearly implies the existence of a *concentration-dependent* shift factor. For more on this subject consult the chapter by Kenner in this book, Section 2.3 in particular.

We have just mentioned additives. There exist various classes of additives aimed at improving the impact behavior of plastics. Comminuted polymers can be added to unsaturated polyesters and other plastics.[43] Rubbery polymers are added to poly(vinyl chloride).[44, 45] Further, elastomers are being added to unsaturate polyester prepolymers to improve the impact resistance of sheet molding compounds. Moreover, short cellulose fibers derived from wood are well suited for the reinforcement of elastomers.[46] The problem here is the prevention of direct fiber-fiber interaction[47]; this can be achieved by using bonding agents to promote adhesion of the fiber to an elastomeric matrix. Apparently, short fiber + polymer composites have an advantage over continuous fiber cord composites, since the former can be mixed and processed by the methods used for nonreinforced polymers.

The volume of additives improving impact resistance and other mechanical properties of polymeric materials is growing steadily. However, our understanding of the mechanisms involved is not exactly complete, and a line of further research is clearly open.

Instead of a single artificial notch or a prevailing crack we can have a number of cracks of comparable size which result in *fragmentation* of the specimen. The fragmentation problem was studied theoretically by Sir Nevill F. Mott during the Second World War. The timing of this research was not exactly accidental, nor was the fact that the results were published only two years after the war ended.[48] A cumulative distribution followed from the work of Mott. Namely, the frequency of the occurrence of particular masses is given by

$$\frac{N_m}{N} = e^{-\left(\frac{m}{\mu}\right)^{1/2}}$$

(10.9.2)

where N_m is the number of fragments with the mass of each fragment greater or equal than m, N is the total number of fragments, while

$$\mu = \frac{\bar{m}}{2} \qquad (10.9.3)$$

and $\bar{m}$ is the average mass of the fragment. Experiments on brittle hypereutectoid steels by Weimer and Rogers[49] have shown the applicability of the Mott distribution (2). This was somewhat surprising, since brittleness has not been taken into account in the derivation of (2). More recently, however, a general formalism of the theory of information (see for instance Chapter 3 in [2]) has been applied to the fragmentation problem. No assumption concerning material brittleness – or otherwise – has been made, and the Mott distribution (2) along with Eq. (3) is re-derived.[50] In view of this, experimental studies of polymer fragmentation and comparing the results with Eq. (2) is clearly worthwhile.

Very important in processing and in subsequent service is the existence of *knit lines* (also called weld lines) in polymer products. This is why we have a whole chapter in this book on the subject. We shall note here only one conclusion of Criens and Moslé[51]: the destructive role of knit lines is weakly correlated with tensile behavior of components, but definitely correlated with impact behavior. Thus, our progress in understanding impact events can help in the evaluation of the effects of knit lines.

Finally, there is a phenomenon of *rapid crack propagation* (RCP) in plastic pipes: velocities between 100 and 430 m/s have been observed in polyethylene pipes.[52] A trend toward using pipes of larger diameters will make this phenomenon even more dangerous. We do not know enough about the nature of this kind of crack propagation, with obvious consequences for our capabilities to prevent the occurrence of RCP. One thing, however, is known: RCP in polymers is intimately related to impact behavior. This has been demonstrated by American researchers[53] as well as by Gaube and Müller in Germany.[54, 55] An equation relating the impact energy absorbed in the Charpy test to the length of the crack produced by RCP has been developed.[56] Thus, once again our work on impact behavior provides help in solving a related problem, in this case that of RCP prevention.

References and Notes

1. P.J. Flory in *Outlook for Science and Technology,* National Research Council of the U.S., p. 370, Freeman, San Francisco 1982.
2. W. Brostow, *Science of Materials,* Wiley, New York-London 1979; W. Brostow, *Introduccion a la ciencia de los materiales,* Editiorial Limusa, México, D.F. 1981; W. Brostow, *Einstieg in die moderne Werkstoffwissenschaft,* Carl Hanser, München-Wien 1985.
3. A. Santamaria, G.M. Guzmán and J.J. Peña, *Mater. Chem. Phys.* 1982, **7**, 347.
4. W. Brostow and R.D. Corneliussen, *J. Mater. Sci.* 1981, **16**, 1665.
5. F. Bueche, *Physical Properties of Polymers,* Wiley, New York-London 1962.
6. E.H. Andrews, *Fracture in Polymers,* American Elsevier, New York 1968.
7. I.G. Zewi and R.D. Corneliussen, *Amer. Chem. Soc. Polymer Papers* 1978, **20-1**, 960.
8. R. Corneliussen, E. Lind and W. Rudik, *Proc. Ann. Tech. Conf. Soc. Plastics Eng.* 1978, **36**, 283.
9. I.G. Zewi, W.J. Rudik, R.D. Corneliussen and E.V. Lind, *Polymer Eng. Sci.* 1980, **20**, 622.
10. C.G. Bragaw, *Plast. Rubber Process. Appl.* 1979, **1**, 145.
11. C.E. Inglis, *Trans. Inst. Naval Archit.* 1913, **55**, 219.
12. A.A. Griffith, *Phil. Trans. Royal Soc. A* 1921, **211**, 163.
13. A.A. Griffith, *Int. Congr. Appl. Mech. Delft* 1924, **1**, 55 A.
14. S.N. Zhurkov and V.E. Korsukov, *Fiz. Tverd. Tela* 1973, **15**, 2071.

15. V.R. Regel, A.I. Slutsker and E.E. Tomashevskii, *Kineticheskaya priroda prochnosti tverdykh tel*, Nauka, Moskva 1974.
16. M.G. Zaitsev and I.V. Razumovskaya, *Vysokomol. soed. B* 1980, **22**, 198.
17. Z. Jedlinski, *Thermal Stability of Polymers Containing Naphtalene Units in the Chains*, Polish Scientific Publishers, Warsaw 1977.
18. P.J. Flory, *J. Amer. Chem. Soc.* 1965, **87**, 1833.
19. P.J. Flory, *Disc. Faraday Soc.* 1970, **49**, 7.
20. W. Brostow, E.M. del R. Enriquez and A.L. Espiritu, *Mater. Chem. Phys.* 1983, **8**, 541.
21. W. Brostow, M.A. Macip and J.A. Ruether, paper in preparation.
22. R. Simha and T. Somcynsky, *Macromolecules* 1969, **2**, 342.
23. R. Simha and R.K. Jain, *J. Polymer Sci. Phys.* 1978, **16**, 1471.
24. W. Brostow, *Polymer* 1980, **21**, 1410.
25. Z. Roszkowski, *Mater. Chem. Phys.* 1981, **6**, 455.
26. Z. Roszkowski, *Makromol. Chem.* 1984, **185**, 1767.
27. A.K. Doolittle, *J. Appl. Phys.* 1951, **22**, 1741.
28. J.D. Ferry, *Viscoelastic Properties of Polymers*, 3rd Ed., Wiley, New York 1980.
29. M.L. Williams, R.F. Landel and J.D. Ferry, *J. Amer. Chem. Soc.* 1955, **77**, 3701.
30. S.C. Sharda and N.W. Tschoegl, *Trans. Soc. Rheol.* 1976, **20**, 361.
31. R.W. Fillers and N.W. Tschoegl, *Trans. Soc. Rheol.* 1977, **21**, 51.
32. J.J. Aklonis and W.J. MacKnight, *Introduction to Polymer Viscoelasticity*, 2nd ed., Wiley, New York 1983.
33. R. Simha and R.K. Jain, *J. Polymer Sci. Phys.* 1978, **16**, 1471.
34. E.A. Guggenheim, *J. Chem. Phys.* 1945, **13**, 253.
35. B.C.-Y. Lu, J.A. Ruether, C. Hsi and C.-H. Hsiu, *J. Chem. Eng. Data* 1973, **18**, 241.
36. C.-H. Hsiu, C. Hsi, J.A. Ruether and B.C.-Y. Lu, *Can. J. Chem. Eng.* 1973, **51**, 751.
37. W. Brostow, R.D. Corneliussen and T.A. Frederick, paper presented at the November 1982 Meeting of the American Physical Society, Philadelphia, November 3–5, 1982.
38. D.K. Roylance in *Applications of Polymer Spectroscopy*, Chapter 13; Academic Press, New York-London 1978.
39. R. Popli and D.K. Roylance, *Polymer Eng. Sci.* 1982, **22**, 1046.
40. D. Roylance, *Int. J. Fracture* 1983, **21**, 107.
41. D.W. Van Krevelen, *Properties of Polymers – Their Estimation and Correlation with Chemical Structure*, 2nd Edition, Elsevier, Amsterdam-Oxford-New York 1976.
42. W.G. Knauss and V.H. Kenner, *J. Appl. Phys.* 1980, **51**, 5131.
43. R.B. Seymour, in *Additives for Plastics*, ed. R.B. Seymour, Vol. 1, Chapter 1, Academic Press, New York 1978.
44. M.W. Riley, *Plast. Technol.* 1977, **23** (8), 91.
45. R.W. Seymour, in *Additives for Plastics*, edited by R.B. Seymour, Vol. 1, Chapter 8, Academic Press, New York 1978.
46. A.Y. Coran, K. Boustany and P. Hamed, *Rubber Chem. Technol.* 1974, **47**, 396.
47. P. Hamed and A.Y. Coran, in *Additives for Plastics*, edited by R.B. Seymour, Vol. 1, Chapter 3, Academic Press, New York 1978.
48. N.F. Mott, *Proc. Royal Soc. A* 1947, **189**, 300.
49. R.J. Weimer and H.C. Rogers in *Proc. 2nd Int. Conf. Mech. Behav. Materials*, p. 1473, Amer. Soc. Metals, Metals Park, OH 1976.
50. W. Brostow and H.C. Rogers, *Mater. Chem. Phys.* 1985, **12**, 499.
51. R.M. Criens and H.-G. Moslé, Chapter 21 in this book.
52. Anonym, *Plastic Pipe Line* 1982, **3**, (2), 1.
53. K. Prabhat, W.A. Maxey, L.E. Hulbert and M.M. Mamoun, *Plastic Fuel Gas Pipe Sym. Proc.* 1983, **8**, 128.
54. E. Gaube and W.F. Müller, *Kunststoffe* 1980, **70**, 72.
55. E. Gaube and W.F. Müller, *Plastic Fuel Gas Pipe Symp. Proc.* 1983, **8**, 137.
56. W. Brostow and W.F. Müller, *Polymer* 1986, **27**, 76.

11 Physical Aging: Influence on the Deformation Behavior of Amorphous Polymers

Leendert C.E. Struik

Plastics and Rubber Research Institute of TNO
P.O. Box 71, 2600 AB Delft, The Netherlands

11.1 Introduction

Physical aging* has been known for many years as an inherent property of glassy materials.[1-8] It occurs because a non-equilibrium state is frozen-in during the cooling of the material from above to below glass temperature T_g. Above T_g, *retardation times* are short enough to enable material properties such as volume and entropy to follow the changes in temperature, but below T_g the retardation times become too long. To a first approximation, the state existing at T_g is frozen-in and the material properties at temperature $T < T_g$ deviate from those in the equilibrium state at T. The consequence is a slow structural relaxation process (aging) that induces changes in many material properties; volume and entropy decrease, stiffness, yield stress, and viscosity increase, and creep- and stress-relaxation rates decrease.

This simple isothermal aging after rapid cooling from above to below T_g is briefly reviewed in Section 11.2.a. The next two subsections, 11.2.b and 11.2.c, deal with a number of facts, known for some time, but either not generally recognized or not understood. The implications of aging in predicting the long-term behavior are summarized in 11.2.d.

Section 11.3 deals with the more general and more realistic case of a glass, subjected to changes in temperature (whereby T remains below T_g) or to high mechanical stresses. Under such conditions aging is no longer a monotonic process, always running in one and the same direction (increasing stiffness and viscosity, decreasing volume). Deformation or a slight increase in temperature or a brief, temporary exposition to vapors[9, 10-19] may revert the process and (partially) de-age the glass. This behavior strongly resembles that of thixotropic materials such as margarine. When left at rest and at constant temperature (overnight in the refrigerator), the margarine stiffens due to the gradual formation and perfection of a network of fat crystals. As shown,[20] this aging strongly resembles that of polymeric glasses. However, when the aged margarine is slightly heated or plastically deformed (kneading during breakfast), it softens (de-ages) because of the partial destruction of the network of crystals. This thixotropic nature of glasses will be outlined in Section 15.3.

Some consequences for the practice of measuring mechanical properties are reviewed in 11.3.a; some deductions concerning nonlinear viscoelasticity, shear banding, and crazing are discussed in 11.3.b. The central thesis of this section is that the whole nonlinear viscoelastic behavior of glassy polymers can be viewed as a competition between "structure formation" due to aging and "structure destruction" due to deformation. Aging is further shown to influence the localization of strain in shear bands and crazes and so to affect fracture properties.

---

* The adjective "physical" is hereafter omitted in this Chapter.

The main emphasis of this chapter is not on a comprehensive decription of the aging phenomenon, but on the conceptual difficulties which still appear to exist in incorporating the facts about aging into the framework of polymer physics. In describing these difficulties, we will often refer to Struik[8] for detailed information.

11.2 The Simple Case: Isothermal Aging After Quenching

11.2.1 General Picture

The general views on aging can be deduced from the ideas about *time-temperature superposition* which were developed in the forties[21] and which cumulated in the famous papers of Williams[22] and Williams, Landel, and Ferry.[23] In a limited temperature range above T_g, temperature affects mechanical or dielectric response curves by merely shifting them along the logarithmic time/frequency scale without changes in shape. Such behavior is called *thermorheological simplicity*[24] and forms the basis of the time-temperature superposition procedures. It implies that the mechanical response of a certain material can be characterized by a single time parameter t_R, for instance arbitrarily chosen as the time at the midpoint of the relaxational transition; all effects of temperature can be described by the function $t_R(T)$. Later, the same was shown to hold for volume relaxation (volumetric response upon a (small) stepwise change in temperature or pressure)[25, 26] and it appeared[27] that, in the glass transition range, the t_R's were roughly the same for relaxation in volume and shear. We therefore neglect the differences in t_R here and suppose that the whole relaxational behavior of the material is characterized by one single parameter, t_R.

Williams e.a.[23] proposed that the changes of t_R with T are not caused by thermal activation, but by thermal expansion, i.e., by the temperature dependence of free volume v^f. Using Doolittle's formula,[28] which connects viscosity with v^f, the authors could derive the famous WLF-equation which is equivalent to the Vogel-Fulcher-Tamman equation found thirty years earlier.[29−31]

For the sequel it is important to note that, according to the WLF treatment, $\ln t_R$ changes with v^f (and therefore also with specific volume v) in a more or less linear way. This follows from the Doolittle equation

$$\ln t_R = B_1 + \frac{B_2}{v^f} \tag{11.2.1}$$

where B_1 and B_2 are constants.

For $v^f \to \infty$ (extremely high temperatures) we have $\ln t_R = B_1$ and writing $B_1 = \ln t_{R_0}$, we find

$$\ln \frac{t_R}{t_{R_0}} = \frac{B_2}{v^f} \tag{11.2.2}$$

Experimental data[23] show that the limiting value t_{R_0} is many orders of magnitude smaller than t_R at T_g (factor of 10^{10} or so).

Differentiation of (11.2.2) yields

$$\frac{d \ln t_R}{d v^f} = -\frac{B_2}{v^{f2}} = -\frac{1}{B_2}\left(\frac{B_2}{v^f}\right)^2 = -\frac{1}{B_2}\left(\ln \frac{t_R}{t_{R_0}}\right)^2 \tag{11.2.3}$$

Since around T_g, t_R is many orders of magnitude larger than t_{R_0}, $(\ln t_R/t_{R_0})^2$ will vary only slowly with t_R. For example, with $t_{R_0} = 10^{-10}$ s, a change in t_R from 1 to 1000 s induces a change in $d \ln t_R/d v^f$ of only a factor of $(\ln 10^{13}/\ln 10^{10})^2 = 1.7$. Consequently the slope of the curve of $\ln t_R$ versus v^f (or v) will change substantially only if t_R varies over many decades. So much for this mathematical excursion, the results of which will be used later on.

In the early sixties, the idea of applying the WLF treatment in the temperature range *below* T_g was initiated. In this range, the (free) volume exceeds its equilibrium value and slowly *decreases* with time (*volume relaxation* [4-7]). It is logical to expect that this volume contraction will be accompanied by an increase in t_R, i.e. by horizontal shifts of the mechanical and dielectric response curves along log time scale. In other words, it is expected that during isothermal volume contraction below T_g, the same will happen as during cooling in the temperature range (just) above T_g.

In fact the above assumption was a rather bold one. In spite of its empirical success, the WLF treatment had not at all proved the physical validity of the free volume concept; this concept served only as an elegant means to interpret experimental facts. Since the material is in thermodynamic equilibrium above T_g, v and t_R are unique functions of T and without theoretical justification it is impossible to decide which quantity (T or v) is the cause of the variations in t_R. Another reason for scepticism about the applicability of the WLF ideas below T_g was the following: their model was based entirely on measurements in thermodynamic equilibrium and a theoretical basis for the model had been offered by Cohen and Turnbull [32, 33] from considerations about the equilibrium size distribution of holes. However, below T_g the material is out of equilibrium and the distribution of hole sizes will be out of equilibrium too; therefore, the theoretical basis for the free volume concept no longer exists below T_g.

Various authors [27, 34-40] performed pioneering experiments in order to test the aging hypothesis mentioned above. Glassy materials were quenched from above to below T_g, next aged at some temperature $T < T_g$, and tested after various values of the time t_e elapsed at T. The tests comprised creep, [27, 39, 40] dynamic mechanical, [34-36] and dielectric measurements [37, 38]; the materials were amorphous glassy polymers and low molecular weight organic glasses. [36, 39] All authors found the expected phenomenon, namely a decrease in relaxation rate with increasing t_e; in this way the earler results on inorganic glasses [2] were generalized to organic glasses.

A dramatic illustration of the aging effect is shown in Fig. 11.2.1. It shows that aging induces horizontal shifts of the creep curves and that the creep rate decreases by many orders of magnitude, namely by approximately a factor of 10 for each tenfold increase in t_e. The last point, which will be discussed in detail later, implies that t_R increases in proportion to t_e.

Following these early studies, a comprehensive investigation of the effect of aging on small-strain creep was done by the author. [8] About forty materials were investigated; most of them were synthetic amorphous polymers, but some were semicrystalline polymers, natural polymers, or low molecular weight glasses; one epoxy resin was also tested. In all cases, aging effects similar to those of Fig. 11.2.1 were observed. Similar effects for stress-relaxation were reported by Sternstein [41-43] and Cizmecioglu e.a. [44] An extension of the work to cured epoxy resins and fiber reinforced resins was reported by Kong [45] and Chang e.a. [46] All of this experimental evidence strongly supported the aging hypothesis mentioned earlier as well as the assumed applicability of the free volume concept below T_g.

The most critical check of the applicability of the free volume concept at temperatures below T_g was made by the author in 1968. The results are described in Section 9.1 of Struik [8] and will be reviewed here briefly. We started from Kovacs's observation [4, 7] that the development of the volume relaxation below T_g critically depends on the preceding thermal history. As shown in Fig. 11.2.2, a quench from an equilibrium state at temperature T_0 to final temperature T_∞ (85° C in Fig. 11.2.2) results in a monotonic contraction process (curve 7 in Fig. 11.2.2). Similarly, an upquench from

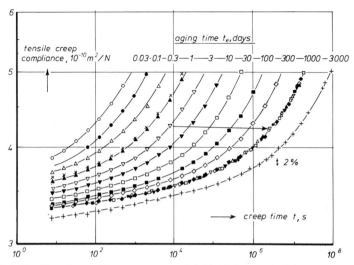

Fig. 11.2.1. Small-strain tensile creep curves of rigid PVC quenched from 90° C (i.e. about 10° C above T_g) to 40° C and kept at $40\pm0.1°$ C for a period of ten years. The different curves were measured for various values of the time t_e elapsed after the quench. The master curve at $t_e=$ 1000 days gives the result of a superposition by shifts which were almost horizontal; the shifting direction is indicated by the arrow. The crosses refer to another sample quenched in the same way, but only measured for creep at a t_e of one day. For details see Chapter 3 of [8]; Reproduced with permission.[8]

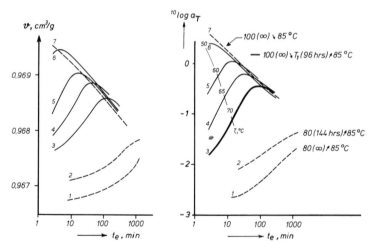

Fig. 11.2.2. Isothermal volume relaxation (left) and simultaneous changes in the mechanical creep properties (right) of PS heated or cooled to the final temperature T_∞ of 85° C in various ways. $^{10}\log a_T$ is the shift of a creep curve relative to the one measured 25 minutes after a simple quench from 100 to 85° C. In the two diagrams, curves with the same test number (1–7) refer to the same experiment. The notation $100(\infty) \searrow T_1$ (96 hours) $\nearrow$ 85° C means: quench from thermodynamic equilibrium at 100° C to intermediate temperature T_1; waiting time at T_1 of 96 hours, followed by an upquench to 85° C; t_e denotes the time elapsed at the final temperature of 85° C (to prevent overcrowding, the original datapoints have been omitted). Reproduced with permission.[8]

equilibrium at $T_0 < T_\infty$ results in a monotonic dilatation process (curve 1 in Fig. 11.2.2). However, maxima in the volume-time curve will be observed when the sample is heated from a nonequilibrium state at $T_1 < T_\infty$ (curves 2–6 in Fig. 11.2.2). The explanation of those volume relaxation maxima is discussed in Section 9.2 of [8].

On the basis of the free volume concept, applied to $T < T_g$, one would expect the following:

– After a quench from equilibrium at T_0 to final temperature T_∞, the monotonic volume contraction will be accompanied by a monotonic shifting of the creep curve to the *right*. This was already shown in Fig. 11.2.1; another example, for PS, is given in Fig. 11.2.3; it refers to the quench $100(\infty) \searrow 88°$ C. In Fig. 11.2.2 the monotonic shifts to the right for the quench $100(\infty) \searrow 85°$ C are given by curve 7.

– After an upquench from equilibrium, the monotonic dilatation will be accompanied by a monotonic shifting of the creep curves to the *left*. This behavior is displayed by the curves in Fig. 11.2.2; actual creep curves which show shifts to shorter times are given in Fig. 11.2.3.

– After an upquench from a nonequilibrium state, the volume maximum will be accompanied by a shift to the left during the dilatation period and by a shift to the right during the contraction period. An example of this behavior is shown in Fig. 11.2.4; the full set of data is given in the right hand diagram of Fig. 11.2.2 (curves 3–6). It is obvious that the shifts log a in the mechanical creep curves precisely

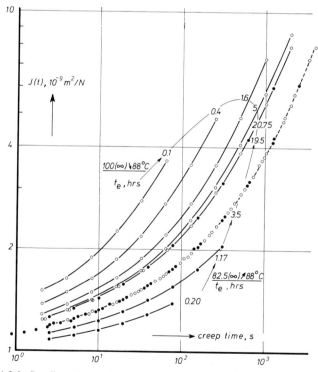

Fig. 11.2.3. Small-strain torsional creep curves at 88° C for PS at various times t_e elapsed after temperature jumps from equilibrium states at 100° C (o) and 82.5° C (●); the master curve at 3.5 hours after the upquench was obtained by horizontal superposition of all data (downquench and upquench). Reproduced with permission.[8]

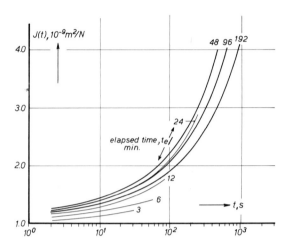

Fig. 11.2.4. Small-strain torsional creep curves at 85° C for PS at various times t_e elapsed after 100° C (∞) ↘ 65 (96 hours) ↗ 85° C; to prevent overcrowding, measuring points have been omitted. Reproduced with permission.[8]

follow the changes in specific volume v. This is further demonstrated in Fig. 11.2.5 in which $^{10}\log a$ from Fig. 11.2.2 (right) is crossplotted versus the value of v (Fig. 11.2.2 (left)) measured at the same moment (t_e) in the same experiment. The data clearly show that the position of the creep curve on (log) time scale is uniquely determined by the momentary value of v. Similar results were obtained for a number of temperatures below T_g for PS, PVC (see Fig. 11.3.1), and PC. These results definitely show that the *free volume concept can indeed be applied in the temperature range below T_g*. Some problems, arising with this concept at temperatures far below T_g, are discussed in point 11.2.6.

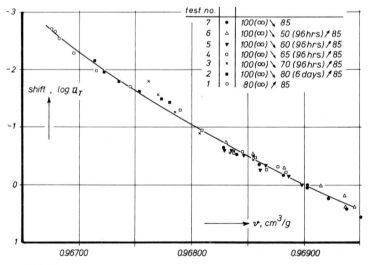

Fig. 11.2.5. Shift log a_T from the right hand diagram of Fig. 11.2.2 versus the value of specific volume v, measured at the same time t_e in the same experiment (left hand diagram of Fig. 11.2.2). Reproduced with permission.[8]

11.2.2 Two Well-Understood Peculiarities

To find some of the aspects of aging, we consider the volume-relaxation process. Formally we can write

$$\frac{dv^f}{dt_e} = -\frac{F}{t_R} \tag{11.2.4}$$

where v^f denotes the free volume and F the driving force for the relaxation.

If the process could be described by a single parameter, F would be proportional to the distance $v^f - v_\infty^f$ from equilibrium. This theory has been worked out by Kovacs.[6, 7] However, data as those from Fig. 11.2.2 clearly show that the process cannot be described in this way (memory effects; $dv^f/dt_e = 0$ at different values of v!). More elaborate theories have been worked out by Narayanaswamy and Gardon,[47, 48] de Bolt e.a.,[49] Kovacs and co-workers,[50, 51] and by the author.[8] As shown by the author[8] and by Chow and Prest,[52] all these treatments are basically identical.

For transparency, we will not use the elaborate theory here (see Appendix D of [8]) but the following simplified reasoning. As shown in Section 11.2.a, $\ln t_R$ will vary more or less linearly with v^f (see also Fig. 11.2.5). Therefore, dv^f/dt_e in Eq. (11.2.4) can be replaced by

$$\frac{dv^f}{dt_e} = -\frac{1}{\gamma}\frac{d\ln t_R}{dt_e} = -\frac{F}{t_R} \tag{11.2.5}$$

where

$$\gamma = -\frac{d\ln t_R}{dv^f} \sim \text{constant} \tag{11.2.6}$$

Equation (11.2.5) yields

$$\frac{dt_R}{dt_e} = \gamma F \tag{11.2.7}$$

At temperatures not too close to T_g, specific volume v remains far from its equilibrium value (see Fig. 11.2.6) and the relative variations in $v - v_\infty$ are only moderate. For simplicity, we therefore also assume that the driving force F will vary only slightly with t_e, which results in

$$\frac{dt_R}{dt_e} \sim k_1 = \text{constant} \tag{11.2.8}$$

or

$$t_R = k_1 t_e + t_R (t_e = 0) \tag{11.2.9}$$

During aging, t_R varies orders of magnitude (Fig. 11.2.1), so for large values of t_e, t_R will be much larger than $t_R (t_e = 0)$. Consequently

$$t_R \sim k_1 t_e \tag{11.2.10}$$

$$\frac{d\log t_R}{d\log t_e} \cong 1 \tag{11.2.11}$$

The same result is obtained by applying the more elaborate theory (Appendix D of [8]). It implies that during aging the relaxation time t_R will increase proportionally to aging time t_e. For small-strain creep this has been found by the author[8]; for dielectric relaxation it was found by Kästner,[37, 38] Matsuoka and co-workers,[53] and by Uchidoi e.a.[54, 55]; for the creep of a low-molecular weight organic glass by Plazek and Magill[39];

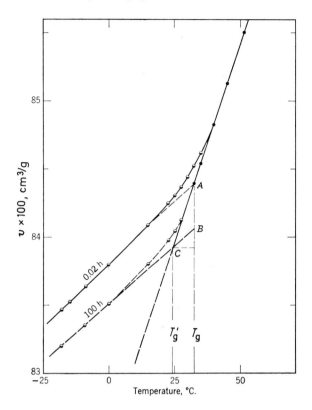

Fig. 11.2.6. Specific volume versus temperature for polyvinylacetate, measured after cooling quickly from well above T_g. Black points, equilibrium values; half blacks, values measured 0.02 hours and 100 hours after quenching the sample. Reproduced with permission.[5]

for the changes in the viscosity of inorganic glasses by many authors, e.g., Prod'homme[56]; and for the aging of amorphous metals by many authors.[57, 58]

Remarkably, Eq. (11.2.11) is also fulfilled for the aging of wool fibers[59] which occurs after wetting (similar to heating to above T_g), followed by rapid drying (similar to quenching from above to below T_g). This aging of wool has an unexpected practical consequence.[59] Just after wetting (which occurs in dry cleaning and pressing), the fibers are "young" and show considerable stress relaxation; this explains the poor wrinkling recovery. The wrinkling recovery becomes much better when, after dry cleaning, the wool suit is hung out (aged) for some time in a cupboard. Since new suits have always been stored (aged) for some time, their wrinkling resistance will be better than that of freshly pressed/cleaned suits.

After this brief excursion into textile technology, we now derive another peculiar aspect of aging. Below T_g the distance from the equilibrium volume v_∞ is given by

$$\frac{v-v_\infty}{v}=\Delta\alpha(T_g-T) \qquad (11.2.12)$$

where $\Delta\alpha$ denotes the jump in cubic isobaric expansivity α at T_g (compare Fig. 11.2.6).

Assuming as above that the driving force F for volume relaxation is proportional to $(v-v_\infty)/v$, we find that constant $k_1=\gamma F$ in (11.2.8) and (11.2.10) is proportional to $\Delta\alpha(T_g-T)$. Substituting this in Eq. (11.2.10) we find

$$t_R\sim k_3\, t_e\,(T_g-T) \qquad (11.2.13)$$

in which k_3 is some constant.

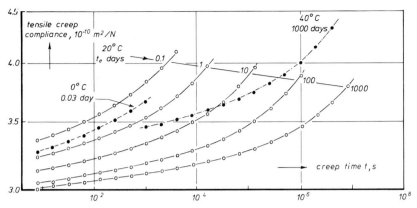

Fig. 11.2.7. Small-strain tensile creep of rigid PVC quenched from 90° C to 0, 20, and 40° C and tested at the aging temperature after various values of t_e. For 20° C the data are given for t_e's of 0.1–1000 days, for 40° C for a t_e of 1000 days only, and for 0° C for a t_e of 0.03 day only. Notice that at 40° C, creep may be slower than at 0° C. Reproduced with permission.[8]

Again, this equation can be derived in a more sophisticated way (compare Section 4.10 of [8]). In fact it shows that *in contrast with the usual thinking, t_R does not change exponentially with T (or $1/T$) but linearly.* This has been confirmed by experiment (Section 4.10 of Struik[8]).

It should be noted that in the above derivation, we have not abandoned the possibility that t_R is thermally activated and, e.g., given by

$$t_R = t_{R_0} \phi (v^f) \exp \left(\frac{\Delta H}{RT} \right) \qquad (11.2.14)$$

The point, however, is that thermal activation influences both the rate of volume relaxation (aging process) and the rate of mechanical relaxation. The two effects cancel out when the relation between t_R, t_e, and T is considered. If, e.g., by thermal activation t_R is shortened by a certain factor, the volume-relaxation process is speeded up and at a certain t_e, v^f is just such an amount smaller that the decrease in v^f compensates for the thermal activation effect. Consequently, *in comparing states at different temperatures but equal values of t_e, the whole concept of thermal activation does not apply.* This implies that care must be taken with T-jump experiments, as proposed by e.g., McCrum and co-workers,[60, 61] because the interpretation of these tests is based completely on thermal activation concepts. Other problems with such tests are discussed in Section 11.3.

A dramatic illustration of the dominating role of t_e over T is given in Fig. 11.2.7. By changing the aging time t_e, we can produce a faster creep rate at 0° C than at 40° C. This immediately *implies that creep data in which t_e is not specified are almost useless.*

11.2.3 Two Poorly Understood Peculiarities

A number of peculiarities about aging or related to aging has not been explained so far. The first problem, described in detail in Section 13.2 of [8] concerns the relation (11.2.3) between $d \ln t_R/dv$ and $\ln t_R$ which was derived in Section 11.2.a from the WLF-Doolittle model (Eq. 11.2.1). Both quantities can be measured: $\ln t_R$ follows from the position of the creep curve on the time scale, $d \ln t_R/dv$ from the volume-relaxation rate $dv/d\ln t_e$, and the aging rate $\mu = d\ln t_R/d\ln t_e$.

$$\frac{d\ln t_R}{dv} = \frac{d\ln t_R/d\ln t_e}{dv/d\ln t_e} \tag{11.2.15}$$

As discussed above, $\ln t_R$ varies only moderately with temperature below T_g, therefore, Eq. (11.2.3) predicts that there will be only slight variations in $d\ln t_R/dv$ with temperature below T_g. This is in disagreement with the experiment; between T_g and 100–200° C below T_g, $d\ln t_R/dv$ increases by a factor of about 10. The empirical reason is that below T_g, $d\ln t_R/d\ln t_e$ remains about unity over a temperature range where $dv/d\ln t_e$ falls steeply (compare Fig. 128 of [8]). These results clearly show that *there are limitations in the applicability of the WLF-Doolittle model below T_g.* Even when this model is generalized (power law dependence between $\ln t_R$ and v^f and an additional thermal activation term in Eq. (11.2.1)) the problems do not disappear.[8]

The second peculiarity does not concern the aging phenomenon itself, but the shape of the creep curves in the temperature range where aging is important. It was found[8] that in this temperature range, creep can be described by

$$J(t) = J_0 \exp(t/t_0)^m \tag{11.2.16}$$

where $J(t)$ denotes the small-strain creep compliance, J_0 the limiting compliance at $t \rightarrow 0$, and t_0 a time parameter related to t_R. J_0 and t_0 depend on temperature, aging time, and type of material; but for a wide class of materials, m turned out to be remarkably constant ($m \sim 1/3$). The equation is restricted to creep tests which are short in comparison to the previous period of aging ($t \ll t_e$); for longer creep tests ($t > t_e$), the creep process becomes affected by *simultaneous aging* and the creep curves become flatter (Section 11.2.d). The equation is also invalid in the nonlinear region of high stresses (Chapter 8 of [8]); moreover, exponent m is no longer equal to 1/3 for plasticized polymers (this has been found for plasticized PVC and for a double-base rocket propellant[62]).

The constancy of m implies that in a double-logarithmic diagram, the shape of the creep curve is invariant, i.e., creep curves measured at different temperatures after different thermal histories and on different polymers can be superimposed by horizontal and vertical shifts, which account for the changes in the parameters t_0 and J_0 (compare Figs. 30 to 34 of [8]).

As said before, Eq. (11.2.16) with $m \sim 1/3$ only applies to the temperature range below T_g, where physical aging dominates the small-strain mechanical behavior. Consequently, the equation appears to apply to creep processes which are due to the onset of the glass-rubber transition, i.e., to slow diffusional micro-Brownian motions (compare Fig. 21 of [8], as well as [63]). At low temperatures, where the creep is determined by local relaxation modes,[64] Eq. (11.2.16) is no longer valid, and the aging phenomena disappear (Section 4.5 of [8]).

We believe that our understanding of the mechanical and aging behavior of glassy polymers would be greatly increased when Eq. (11.2.16) can be explained theoretically. To understand this, we start with the ascertainment that the equation has a long history in material science. More than a century ago it was used by Kohlrausch (1866) to describe the creep of glass, and at one time Kohlrausch considered the equation to be a universal expression for creep.[65] In 1923, Pierce[66] showed that the stress relaxation of cotton and many other fibrous materials can be described by an approximate counterpart of Eq. (11.2.16), namely

$$E(t) = E_0 \exp(-(t/t_0)^m) \tag{11.2.17}$$

in which $E(t)$ denotes the stress-relaxation modulus. Remarkably, Pierce also found a value of $m = 1/3$ for many materials.

Other examples of the use of Eqs. (11.2.16) and (11.2.17) are:

1. Kurkjian[67] and De Bast and Gilard[68] applied Eq. (11.2.17) to the stress relaxation of inorganic glasses. It appears than m is no longer a constant, but varies between 1/3 at low temperatures to 1/2 or even higher at temperatures around T_g (see also Douglas[69]). Later, the equation was also used to describe the results of optical correlation spectroscopy on inorganic and organic glasses.[70-72]

2. Narayanaswamy[48] and De Bolt e.a.[49] applied Eq. (11.2.17) to the volume recovery of *inorganic glasses,* quenched from above to below T_g. A similar application was possible for organic glasses,[51, 73] not only for volume relaxation, but also for enthalpy relaxation.[52, 74, 75]

3. Tobolsky and co-workers[76, 77] and Djiauw and Gent[78] applied Eq. (11.2.17) to the stress relaxation and elastic recovery of unvulcanized rubbers at temperatures above T_g. As mentioned in points 1 and 2, parameter m does not appear to be constant.

4. Williams and Watts[79, 81] used Eq. (11.2.17) for the dielectric relaxation of many amorphous polymers and found m-values ranging from 0.38 to 0.6.

5. The present author applied Eq. (11.2.16) with $m=1/3$ to the small-strain mechanical creep of polycrystalline metals such as lead and to an organic low-molecular weight glass such as amorphous sugar (compare Figs. 33 and 34 of [8]). In addition, van 't Spijker[82] used Eq. (11.2.17) to describe the stress relaxation of metallic glasses ($m \neq 1/3$).

Moreover, it should be noted that Eq. (11.2.16) reduces to the famous Andrade creep equation for $t \ll t_0$ (Plazek[39]) and that Eq. (11.2.17) seems to be connected to the "universal" stress-relaxation law, found by Kubát[83, 84] (compare the discussion in Section 14.2 of [8]; see also Chapter 4).

An interesting question is why do Eqs. (11.2.16) and (11.2.17) have such a wide applicability. Of course, it may be argued that we deal with very versatile curve fitting equations; variations in the relaxation strength can be accounted for by changing E_0, variations in the mean relaxation time by changing t_0, and variations in the width of the *"relaxation spectrum"* by changing m.

On the other hand, the finding that $m=1/3$ applies to the small-strain creep of many amorphous glassy polymers as well as to lead and amorphous sugar, has a physical meaning; there must be some reason for the invariance in the shape of the small-strain creep curves. Moreover, there must be some reason for the increase of m with temperature, as found for inorganic glasses (point 1. above) as well as for the fact that in both mechanical and dielectric relaxation the lower limit of m appears to be approximately 1/3 (compare [8] and points 1., 4., and 5. above).

The wide applicability of Eqs. (11.2.16) and (11.2.17) indicates that the relaxation processes considered above have something in common. Let us start with the reminder that, if a wide range of materials would show simple *Debye relaxation* (one relaxation time, $m=1$), the common factor would be that in all materials the relaxation would be due to the orientation of *isolated* electrical or mechanical dipoles in a viscous environment. The common factor behind Eqs. (11.2.16) and (11.2.17) may well be that in the materials considered, the relaxation is basically not of the Debye type and that the deviations from the Debye relaxation are due to interactions among dipoles (cooperative effects). A similar conclusion about the non-Debye character of relaxations in solids has been drawn by Jonscher,[85] Ngai,[86] and Dissado and Hill[88]; attempts to explain Eq. (11.2.17) have theoretically been made by e.g. Gény and Monnerie[89] and Shore and Zwanzig.[90] If the non-Debye character of relaxation can be established firmly, the conclusion follows that the spectral representation of mechanical and dielectric response functions[21] by a series of exponentials, is merely a mathematical formalism, without any physical meaning. One then should be very careful with giving, explicitly or implicitly, a physical meaning to the springs and dashpots of the models, representing a spectrum of retardation times, or to the internal parameters $\xi_1, \xi_2, \ldots, \xi_n$ in the multiparameter theory for structural relaxation.[51] The situation becomes similar to that of heat diffusion; also that process can be described by an infinite series of exponential functions, each with a different *"relaxation time"* (solution of the Fourier equation). Nobody, however, would connect different physical entities to the different exponential functions.

The merit of the above remarks is quite simple: *we have no (molecular) theory of mechanical relaxation* and this hampers the development of our understanding of mechanical properties.

11.2.4 Predictions of Long-Term Behavior from Short-Time Tests

Because of aging, a distinction must be made between long-term and short-term tests. A *short-time test* is defined as an experiment in which the testing time, t, remains short compared with the aging time t_e at the beginning of the test. An example is a 100 s creep test, started at a t_e of 10 hours. During such a test aging hardly proceeds, t_R remains approximately constant and we take, so to speak, a "snapshot" of the properties of the aging material. Such properties are called *momentary properties;* all creep data shown in Figs. 11.2.1 to 11.2.4 and Fig. 11.2.7 were obtained with such short-time measurements. In a *long-term test*, testing time t becomes large compared to the t_e at the beginning of the test; an example is a creep test of 100 hours, started 10 minutes after quenching. During such a long-term test the aging process proceeds and the mechanical properties *change during testing*. The resulting behavior is called *long-term behavior;* in view of the changes in properties during testing, there is some analogy with tests under nonisothermal conditions.

It should be noted that the distinction between short- and long-term tests is not based on the absolute duration of the test, but on the ratio t/t_e. It should be noted further that laboratory tests are often short-time tests while practical loading situations often have a long-term character.

Fundamental differences exist between short- and long-term properties. As shown in Chapters 10 to 12 of[8], the short-time small-strain creep behavior of amorphous polymers below T_g obeys the linear-viscoelastic theory (Boltzmann's superposition principle) as well as the time-temperature superposition principle. Long-term creep does *not* obey the linear-viscoelastic theory (the properties explicitly change with time, which conflicts with Boltzmann's principle), and time-temperature superposition does not apply (see below).

Some basic aspects of the long-term behavior can be understood from the simple example of a material with a Newtonian viscosity η that increases by aging in proportion to t_R. (This behavior is studied in detail for inorganic and metallic glasses.[56-58]) If stress σ is applied at elapsed time t_e, the creep strain $\varepsilon(t)$ develops according to

$$\frac{d\varepsilon}{dt} = \frac{\sigma}{\eta(t)} \tag{11.2.18}$$

At loading time t, the aging time equals $t_e + t$, so from Eq. (11.2.13) and using the assumption that η is proportional to t_R, we find

$$\eta(t) = \eta(t_e) \frac{t_e + t}{t_e} = k_4(t_e + t)(T_g - T) \tag{11.2.19}$$

where $k_4 = k_3 \eta / t_R = $ constant.

Substituting this in (11.2.18) we find

$$\frac{d\varepsilon}{dt}=\frac{\sigma}{k_4(T_g-T)}\frac{1}{t_e+t} \tag{11.2.20}$$

or

$$\varepsilon(t)=\frac{\sigma}{k_4(T_g-T)}\ln(1+t/t_e) \tag{11.2.21}$$

which reduces to

$$\varepsilon(t)=\frac{\sigma}{k_4(T_g-T)}\ln t/t_e \quad \text{for } t\gg t_e \tag{11.2.22}$$

We thus conclude that when aging and creep occur simultaneously (long-term behavior), the creep curve becomes straight on log time (t) scale. Further, the logarithmic creep rate $d\varepsilon/d\ln t$ becomes an increasing function of temperature, so that time-temperature superposition is no longer possible. Finally, the long-term creep curve (ε versus log t) will shift to longer times when t_e increases (just as do the short-term creep curves, Fig. 11.2.1).

The same results were obtained by the much more sophisticated treatment given in Chapters 10 to 12 of [8]; instead of the viscous creep law (11.2.18) we used the empirical Eq. (11.2.16) and instead of the simple aging law (11.2.13) we used the generalized formula $t_R \propto t_e^\mu$ ($0\leq\mu\leq1$). On the basis of this treatment, methods could be designed for predicting long-term creep from tests of short duration. They apply to creep, stress relaxation (Section 11.4 of [8]) and to dimensional instability phenomena such as skrinkage (Section 11.6 of [8]). *Dimensional instability* is considered as creep under the action of internal stresses. Such stresses already start to act during the (final stages of the) processing cycle; the relevant t_e will be very short and, therefore, dimensional instability is an outstanding example of long-term behavior. The prediction methods mentioned above were cast in the form of simple recipes (Chapter 11 of [8]), and they allow for extrapolations on a time scale of over two to three decades; moreover, the maximum errors of the extrapolation could be derived theoretically. Altogether it can be concluded that *the problem of predicting long-term creep at small strains (no crazing!) has been solved.*

An interesting point, discussed in Chapter 12 of [8] is, that attempts to accelerate creep by performing tests at higher temperatures, are generally useless for predicting

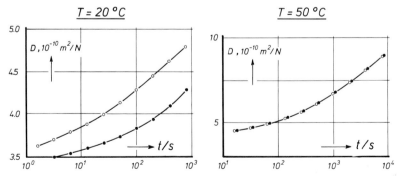

Fig. 11.2.8. Small-strain tensile creep curves of rigid PVC quenched from 90 to 20° C and aged at 20° C for 1/3 hours (o, sample 1) or six hours (●, sample 2). After these two aging histories at 20° C, the samples were tested at 20° C (left) and at 50° C (right). The equilibration time at 50° C was one-half hour. Reproduced with permission.[8]

the long-term creep at the lower working temperature. Firstly, the assumed time-temperature superposition does not apply (see above), and second, heating changes the state of the material; the properties measured at the higher temperature are not representative of those at the working temperature (see Fig. 11.2.8). How misleading such tests at higher temperatures can be, is shown in Fig. 126 of [8].

11.3 General Case: Aging Under Changing Conditions; Thixotropic Nature of Glassy Polymers

11.3.1 Evidence for and Practical Consequences of Partial De-Aging By Heating or High Stresses

That heating to temperatures still below T_g can de-age the material is illustrated by Fig. 11.2.8. Two samples are compared; at 20° C their aging times differed, by a factor of 18. When (creep) tested at 20° C, the creep properties differed in the usual way. However, when the two samples, aged at 20° C were heated to and tested at 50° C, their creep properties were identical. Thus, the heating to 50° C erased the differences in aging history at 20° C.

The origin of this erasing effect is clear from Fig. 11.2.2. On heating from T_1 to final temperature T_∞ (85° C in Fig. 11.2.2), part of the volume contraction that occurred at T_1 is removed (dilatation period). At longer times, i.e., right from the volume maximum, the $v - t_e$ curves approximately merge with the curve measured after a direct quench from above T_g to T_∞ (curve 7 in Fig. 11.2.2). This behavior has been discussed in great detail by Kovacs.[4,7] It implies that right from the maximum the aging history at T_1 has been erased. This not only applies to specific volume but also to the creep properties (Fig. 11.2.2, right). These volume-relaxation maxima completely agree with the effects seen in Fig. 11.2.8. PVC samples are aged at 20° C for 1/3 and six hours respectively; the volume-relaxation maxima at 50° C occur within a t_e of much less than one-half hour (see Fig. 11.3.1). Consequently, the differences between the two samples will be erased when the creep tests of Fig. 11.2.8 are started at a t_e at 50° C of one-half hour.

The de-aging effect described above has also been reported by Wright[91] (creep properties of PVC and PMMA) while the volume-relaxation peaks have been discussed by Kovacs,[4,7] Hozumi,[26] Udichoi e.a.,[54,55] Corsaro,[92] Adachi and Kotaka,[93] and the auhor.[8] The peaks can be explained from a distribution in glass-transition tempera-

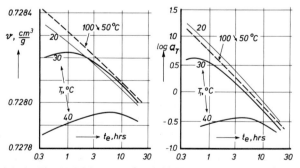

Fig. 11.3.1. As Fig. 11.2.2 but now for rigid PVC at $T_\infty = 50°$ C after 100 ↘ T_1 (120 hours) ↗ 50° C or after a direct quench from 100 to 50° C. Log a_T is the shift of each creep curve with respect to the one at 5.33 hours after the quench from 100 to 50° C. Reproduced with permission.[8]

tures in the material and can be described by the multiparameter theory[47-51]; for a detailed discussion about the origin and consequences of the effects, see Sections 9.2, 9.3, and 9.5 of [8].

De-aging is also well-known from DSC studies on organic[9, 74, 75] or metallic glasses.[95] If a glass is aged at a temperature well below T_g (sub-sub T_g range according to Chen[95]) a DSC heating run shows an endothermic peak below T_g. The area under the peak and the peak temperature increase with increased aging time and only when the aging temperature approaches T_g sufficiently, the endothermic sub-T_g peak changes into the well-known endothermic overshoot peak at T_g. Again, this behavior can be described with the multiparameter theory.[74, 75]

A summary of the de-aging behavior is given in Fig. 11.3.2. The picture strongly resembles that of partial melting and recrystallization in fast cooled semicrystalline polymers.[96, 97] For such materials, the crystalline structure is not stable at temperatures above T_g. If the material is annealed at $T_a > T_g$, it recrystallizes and upon heating a melt peak is found at 10–30°C above T_a; thus, the new structure developed at T_a is again unstable and the material melts and recrystallizes at 10–30°C above T_a.

De-aging has important practical consequences. First, the data of Fig. 11.2.8 illustrate why attempts to predict long-term creep at, say, 20°C by accelerated tests at higher temperatures, are unsuccessful. The most important parameter for the long-term creep at the working temperature of 20°C is the aging time t_e at 20°C; by heating to 50°C the state of aging is completely changed (compare the discussion at the end of Section 11.2.d).

A second example (Fig. 11.3.3) shows how confusing the *de-aging effects* may work out. The data refer to an experiment in which the effect of annealing at 60°C on the yield stress at 35.5°C was investigated. In contrast to usual thinking, annealing first causes a decrease of σ_y; only after some time σ_y begins to increase. The origin of this effect is the de-aging at 60°C after aging at room temperature for several months. This is clearly demonstrated by the DSC diagrams. The original endotherm,

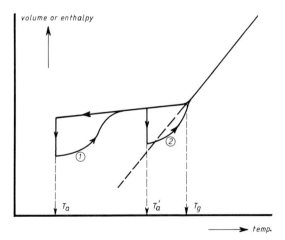

Fig. 11.3.2. De-aging during heating after aging at temperature T_a or $T_a' > T_a$ respectively. The schematic course of volume or enthalpy versus temperature is plotted; data are available.[8, 9, 74, 75] After aging at T_a, complete de-aging occurs *below* T_g (full recovery) and plots of thermal expansivity (dv/dT) or heat capacity (dH/dT) show peaks *below* T_g (curve 1). If the aging temperature approaches $T_g(T_a')$, the endothermic C_p-peak changes into the well-known overshoot peak at T_g (curve 2).

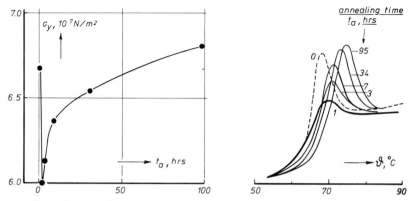

Fig. 11.3.3. Effect of annealing (aging) at 60° C for various times t_a on the tensile yield stress σ_y at 35.5° C (strain rate 10% per minute) and on the DSC diagrams measured at a heating rate of 32° C per minute. The material was a sample, machined from a commercial (extruded) sheet of rigid PVC that had been stored under room temperature conditions for several months. Reproduced with permission.[8]

due to the aging at room temperature, disappears, and only after some time a new endothermic peak, related to the aging at 60° C, developes.

The third example, given in Fig. 11.3.4, concerns the dynamic mechanical loss, tan δ, of a quenched, a slowly cooled, and an as-received sample (*slow* cooling has the same effect as isothermal aging). The as-received sample has aged for some months at room temperature, but it is de-aged when during the torsional pendulum measurements, the temperature is raised considerably above 20° C. This explains why the as-received sample behaves as an aged material below 20° C and as a quenched material above 20° C (crosses in Fig. 11.3.4). Other examples of this behavior are given in Section 9.4 of Struik.[8]

De-aging should also be considered in thermally stimulated discharge TSD[98] or thermally stimulated creep TSC[99] measurements. These methods are entirely based on thermal activation theory and should be reconsidered for the temperature range where aging and de-aging effects are important. A nice example is Chai and McCrum's[99] experiment on PP. The sample was loaded at 60° C for 4000 s, cooled to 15° C, unloaded and recovered and aged at 15° C for various times t_e. Next, the continuation of the recovery was measured during heating (TSC). Since in this tempera-

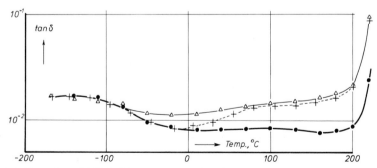

Fig. 11.3.4. Damping of PPPO (Tenax) at 1 Hz (torsional pendulum) measured during stepwise heating from −170 to +220° C at an average rate of 0.8° C per minute. ----+---- material as-received, ——△—— after a quench from 230 to −170° C, and ——●—— after cooling at 0.5° C per hour from 220 to 25° C. Reproduced with permission.[8]

ture range the aging of the semicrystalline PP is similar to that of amorphous polymers,[100] we expect that during the heating run, de-aging will occur, so that the recovery
at temperatures considerably above 15° C is in no way influenced by the aging/recovery
time at 15° C. This was indeed found, but since the de-aging effects were disregarded,
the results were interpreted differently. Similar results were obtained for PMMA and
PC.[101]

De-aging can also be brought about by *high mechanical stresses*. An example is
given in Fig. 11.3.5. After 17 hours of aging at 20° C, a high stress is applied for
a short time. Before and after the pulse, the small-strain creep properties were measured
as a function of age. Large changes were observed between 1/4 and four hours after
the pulse, i.e. between t_e's of 17.75 and 21.5 hours after quenching. In a normal aging
test (no high-stress pulse) this increase in t_e would have produced a shift of the creep
curves by at most $^{10}\log(21.5/17.75) = 0.08$ decade. The actual shifts are 10 times larger
and in addition the material shows a faster creep (decreased value of t_R) after the
pulse than before. This shows that the high stress has de-aged the sample and that
the aging process is reactivated after removal of the stress, a behavior schematically
outlined in Fig. 11.3.6.

Other examples of the de-aging effect of high stresses are discussed in detail in
Section 8.6 of [8], a nice experiment is the one in which the (cross) effect of mechanical
stresses on a frozen-in electrical polarization was studied.[102] It was shown that the
froozen-in polarization nearly disappeared by the decrease in t_R during *cold drawing*.

Other authors have reported the same effect. Matsuoka e.a.[103] found that the endothermic C_p peak at T_g of an aged sample is removed by cold drawing, Chang e.a.[46]
showed the same for cold rolling of a cured epoxy resin. Berens and Hodge[9] studied
the effect of annealing (aging) on the sub-T_g endothermic peaks for PVC samples
with different pretreatments (prior to the aging test). A pretreatment strongly affects
the magnitude of the annealing effects, as found with the DSC. Actually, the annealing
effects increase in the order (of pretreatment): hot pressing ~ normal cooling < liquid
nitrogen quenching < cold drawing ~ pressing. It could be shown[74] that the increase
in the magnitude of the annealing effects is due to an increase in the initial frozen-in

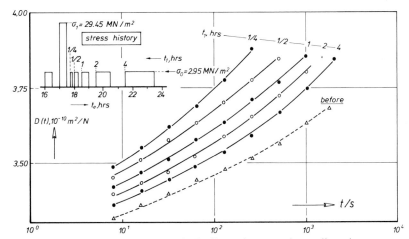

Fig. 11.3.5. Effect of a high-tensile stress "pulse" of intensity σ_1 on the small-strain creep properties
of rigid PVC. The material was quenched from 90 to 20° C, and σ_1 was applied at 17 and
removed at 17.5 hours after quenching. Time t_e is the time elapsed after quenching; time t_1
the time after removal of stress σ_1. The exact stress history, including the small-strain tests done
under a stress σ_0, is given in the insert. Reproduced with permission.[8]

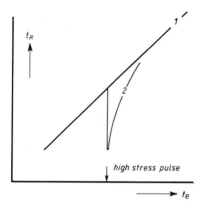

Fig. 11.3.6. Effect of high stresses on aging. When high stresses are absent, the spontaneous aging follows curve 1. When a high-stress pulse is applied, the material is first de-aged and next (after the pulse) the aging process is reactivated. Obviously, a high-stress pulse has a similar effect as a brief heating period.

enthalpy (i.e., after the pretreatments). Consequently, cold drawing or pressing increases the frozen-in enthalpy to a higher level than quenching in liquid nitrogen.

In fact, the de-aging effect of high stresses is not unexpected. Most glassy polymers can be brought to yield, particularly when brittle fracture is prevented by using shear strains or uniaxial compressions. Yielding implies that the mobility has been increased to a level, comparable with that at T_g; since the work of Lazurkin[104] and Vincent[105] we have known that the mobility is increased by the stress. So, just as by heating, high stresses can transform the mechanical response from *glasslike to rubberlike*. Also the well-known fact that the yield stress σ_y increases with age[106-109] (also see Section 8.2 of Struik[8]) fits into this picture. With increasing age, t_R increases, and more mechanical work is required to "draw" the material out of the rigid glassy state (see also [106]).

Phenomenologically, the behavior discussed above (de-aging by heating or high stresses) strongly resembles that of thixotropic materials such as, e.g., margarine. Its aging (*stiffening*) at rest and de-aging by heating or kneading has been described in the introduction; similar phenomena occur in many substances, e.g., in waxy crude oils, in cement pastas, and in many metals. The general picture that can be drawn is shown in Fig. 11.3.7. Mobile defects exist in the solid (dislocations, free volume packages) which are responsible for creep or anelastic behavior in general. The number of defects is increased by deformation, i.e. large deformations generate defects. This is well-known for dislocations in metals. For polymers this picture has been discussed by the author (Section 8.6 of [8]), by Bauwens,[110] Bultel e.a.[63] and by Robertson.[111] The author's tentative proposition was that deformation processes involving segmental motion produce free volume, irrespective of whether the deformation is a tension, a shear, or a *uniaxial compression*. The idea was born from the well-known observation that compacted ("aged"!) powders (e.g., sand) dilate when they are sheared.[112] This idea is not related at all to the idea of Ferry and Stratton[113] and many others (see for instance Rusch and Beck[114]) that the mechanical dilatation $(1-2\nu)\,\varepsilon$, which accom-

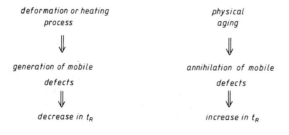

Fig. 11.3.7. Schematic illustration of the two competitive phenomena occurring in a deformed glassy polymer; for details see text.

panies a tensile strain ε ($v =$ Poisson's ratio) produces the increase in free volume. As shown in Section 13.3 of [8], the latter idea is largely incorrect, as is the reverse one that the volume compaction produced by cooling under a high hydrostatic pressure through the T_g-range, would decrease the free volume.[115-119]

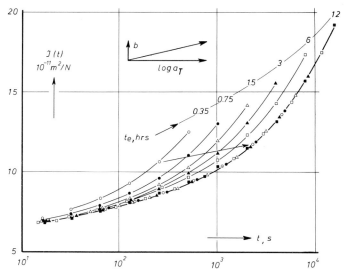

Fig. 11.3.8. Small-strain torsional creep (measured on strips of 2 mm thickness, $\tau_{max} = 0.56$ MPa, $\gamma_{max} < 1.1 \times 10^{-4}$) of duraluminum 51S-T at 350° C; the creep tests were started at various values of the time, t_e, elapsed at 350° C after heating the sample from room temperature. The master curve was found by using the horizontal and vertical shifts as indicated.

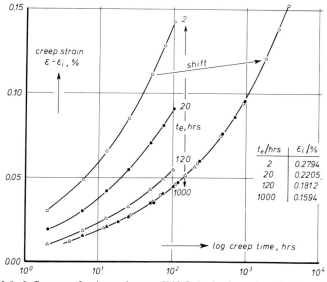

Fig. 11.3.9. Influence of prior aging at 593° C (t_e is the aging time) on the creep behavior at 593° C at a stress level of 86.2 MPa. Type 304 stainless steel; replotted from data of Cho and Findley.[120, 127] The mastercurve is given at $t_e = 1000$ hours. The initial strains ε_i are given in the insert.

The increase of the number of defects by deformation processes is, however, not the only effect. In the absence of stresses, the material ages because the number of mobile defects is larger than at thermodynamic equilibrium. This annihilation of mobile defects by aging is just as general as the generation of defects by deformation; in fact aging phenomena similar to those in glassy polymers have been found in polycrystalline metals such as lead and tin,[8] duraluminum (Fig. 11.3.8), and stainless steel (Fig. 11.3.9).[120, 127]

So we conclude that there are *two competitive processes* in a strained solid: first, the gradual decrease in the number of defects by aging, second, the counteracting effect of the production of defects by the deformation process. If the stresses are low, the production of defects can be neglected and the material behaves in the linear-viscoelastic fashion. At high stresses the behavior becomes nonlinear and determined by the competition between production and annihilation of defects. (We restrict ourselves here to nonlinearities caused by the effects of high stresses on the rate constants.) This is the reason why aging phenomena are so important for our understanding of the behavior at high stresses; some examples are discussed in Section 11.3.b. The above most probably also implies that the multiple integral theory[121] for nonlinear-viscoelastic behavior is not applicable because in this theory it seems to be assumed that the properties of the material are invariant with respect to time.

11.3.2 Some Deductions Concerning Yielding, Shear Banding, and Crazing

The picture on yielding, developed in Section 11.3.a, is that yielding occurs when the mechanical deformation process has regenerated the defects that were annihilated during cooling through the T_g-range and subsequent aging below T_g. This picture at least qualitatively explains why yield stress σ_y increases with age.[8, 106–109] We can understand further that the upper yield stress σ_y (see Fig. 11.3.10, 11.3.12, and 11.3.13) increases much more with t_e than the (lower) drawing stress σ_d.[122–125, 133] After yield, the aging history is erased and, therefore, σ_d should be independent of previous aging.

The discussion in Section 11.3.a also suggests that a close relation exists between the yield peak seen in Fig. 11.3.10 and the well-known endothermic DSC peak at T_g. Both develop with increasing age and both peaks are removed by yielding. With the latter statement we mean the following: when the drawing machine is stopped in point A (Fig. 11.3.10) and the sample is unloaded, we observe that

- upon restarting the drawing machine, no new yield peak is found (Fig. 11.3.10),[105, 122, 125]
- when the necked material is tested in a DSC, no endothermic peak is found at T_g.[103, 106]

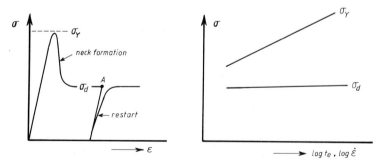

Fig. 11.3.10. Effect of aging time t_e on the upper-yield stress σ_y and on the drawing stress σ_d; for details see text.

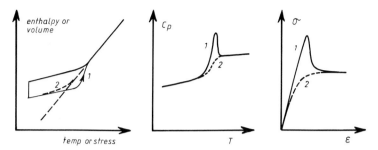

Fig. 11.3.11. Schematic illustration of the relation between DSC peaks and yield peaks. Curves 1 refer to high heating/straining rates, curves 2 to low rates.

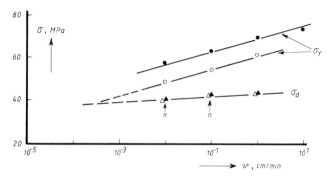

Fig. 11.3.12. Upper-yield stress σ_y (circles) and drawing stress σ_d (triangles) versus drawing rate v for rigid PVC dumbbells (prismatic part $22 \times 5 \times 3$ mm) measured at $20°$C with an Instron machine. Open symbols, at $t_e = 1/2$ hours after a quench; filled symbols, after three months of aging at $23°$C; n, necking; h, homogeneous draw (see Fig. 11.3.13).

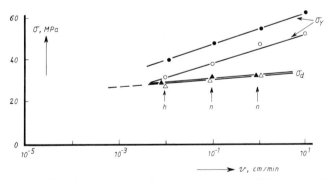

Fig. 11.3.13. As Fig. 11.3.12, but now for measurements at $40°$C (aging still at $23°$C).

Consequently, both peaks disappear simultaneously. Further, both peaks can be restored by aging.

So far for the experimental evidence. Our interpretation is shown in Fig. 11.3.11. The transition from the glassy (= aged) state to the rubbery (= de-aged) state can be induced by heating or stressing. If these processes are performed at sufficient speed, overshooting occurs, which is reflected in a DSC peak or a yield peak (curve 1). If the process is performed at a sufficiently slow rate, the structural relaxation can keep up with the changes in conditions (temperature, stress) and the overshooting effects disappear (curve 2). This is well-known for DSC peaks (effect of heating rate), but also confirmed by experiment for the yield peak. Some results for rigid PVC are shown in Figs. 11.3.12 and 11.3.13; at low-strain rates σ_y becomes equal to σ_d and the yield peak disappears. We thus conclude (see Fig. 11.3.10, right) that the yield peak will disappear at either a low-strain rate or a short-aging time.

An important consequence follows when the above results are combined with the claim of Bowden and Raha[126] that strain softening (drop form σ_y to σ_d) induces *localization of "plastic" deformation*. A first kind of strain localization is the neck formation in a tensile test. According to Bowden's criterion, necking is the more favored the larger the yield drop $\sigma_y - \sigma_d$, i.e., the higher the strain rate and the longer the aging time. Conversely, necking will disappear at very low strain rates (σ_y becomes equal to σ_d) or just after quenching. The first point has been known for a long time (Lazurkin[104]) and is observed in the 40° C PVC data of Fig. 11.3.13; data confirming the second point are unknown to the author.

The discussion in Section 11.3.a also explains the disappearance of necking close to T_g: the yield drop, responsible for necking, is due to aging and aging effects disappear at T_g. Further, it is now understood why the strain rate dependences of σ_y and σ_d are so different (compare Figs. 11.3.10, 11.3.12, and 11.3.13). Stress σ_y measures the stresses needed to de-age the sample, σ_d is the flow stress of the de-aged sample; there is no a priori reason for assuming a similarity in strain-rate dependence between these two stresses. We finally can understand why the effect of aging on σ_y is much smaller than expected from tests at very small strains (compare Fig. 64 of [8]). The de-aging process that ultimately leads to yielding starts in the earlier phases of the tensile tests. Consequently, part of the de-aging work is done before σ_y is reached and the differences which originally existed between samples of different ages are partly removed.

So far for the necking instability. It should be realized that the strain softening discussed above is only one of the factors that determine the neck formation. The other is the strain hardening due to the large rubberlike deformation in the yielded material. The molecular factors which determine this strain hardening are discussed by Haward.[128]

The above mentioned localization of "plastic" deformation may also be important for the fracture properties. As Adam e.a.[106] have shown, aging reduces the size of the plastic zone in an impact test and so reduces the impact energy.

Strain localization is not restricted to necking, but may also occur on a much smaller scale, for example in crazes or shear bands. Therefore, the question arises whether the propensity for crazing and shear banding increases with age. Wales[129] found that a freshly quenched PVC sample was less sensitive to crazing (in air) than an aged one. Various authors[130-133] have reported that the formation of shear bands (at least the coarse ones[131]) is favored when the material is aged or tested at a higher strain rate.

A, at first sight, puzzling behavior was found by Illers[134] and Petrie and co-workers.[107] These authors performed tensile tests on semiductile materials (PVC and PETP) which were previously annealed for various lengths of time. At short annealing times the samples necked and fracture occurred when a large fraction of the sample was transformed in necked material (high apparent fracture strain). With increasing annealing time the fracture point moved back to the yield point and the apparent fracture strain decreased. Since yielding de-ages the sample, one would expect no effect of aging on the post-yield fracture behavior. A possible explanation is the following. Samples with long annealing times are more likely to craze before yield. These crazes survive the necking process, but form the initiation points for cracks in the necked material.[135] The longer the aging time preceding the experiment, the more the necked material is seeded with defects (*crazes*). In this respect it is interesting to notice that Petrie and co-workers[107] reported an eventual increase in the apparent fracture strain at very long annealing times, and they noted that this increase was due to a manifold of crazes.

Our last remark concerns the possibility of a threshold stress for yielding. At small strains, the mechanical deformation process does not influence the aging process and t_R increases proportionally with t_e. We do not know the exact conditions for yielding, but it seems very unlikely that yielding can occur when t_R increases steadily. This gradual stiffening by aging must be "broken down" by a minimum of stress before yielding can take place. This suggests the existence of a *threshold stress for yielding*; below this threshold, yielding will not take place, however long the time available. Data of Narisawa e.a.[136] seem to confirm this.

References and Notes

1. F. Simon, *Z. Anorg. Allgem. Chem.* 1931, **203**, 219.
2. H.R. Lillie, *J. Amer. Ceram. Soc.* 1933, **16**, 619.
3. E. Jenckel, *Z. Elektrochem.* 1937, **43**, 769; 1939, **45**, 202.
4. A.J. Kovacs, *Thèses*, Fac. Sci., Paris 1954.
5. A.J. Kovacs, *J. Polymer. Sci.* 1958, **30**, 131.
6. A.J. Kovacs, *Trans. Soc. Rheology* 1961, **5**, 285.
7. A.J. Kovacs, *Fortschr. Hochpolym. Forsch.* 1964, **3**, 394.
8. L.C.E. Struik, *Physical Aging in Amorphous Polymers and Other Materials*, Elsevier, Amsterdam-New York 1978.
9. A.R. Berens and I.M. Hodge, *Macromolecules* 1982, **15**, 756; also see the related papers 10–19 on diffusion and sorption:
10. D.J. Enscore, H.B. Hopfenberg, V.T. Stannes and A.R. Berens, *Polymer* 1977, **18**, 105.
11. A.R. Berens, *J. Macromol. Sci. Phys.* 1977, **14**, 483.
12. A.R. Berens, *Polymer. Eng. Sci.* 1980, **20**, 95.
13. A.R. Berens and H.B. Hopfenberg, *J. Polymer Sci. Phys.* 1979, **17**, 1757.
14. A.R. Berens, *Pure Appl. Chem.* 1981, **53**, 365.
15. J.H.M. Fechter, H.B. Hopfenberg and W.J. Koros, *Polymer Eng. Sci.* 1981, **21**, 925.
16. A.H. Chan, D.R. Paul, *J. Appl. Polymer Sci.* 1980, **25**, 971.
17. J.A. Yavorsky and H.G. Spencer, *J. Appl. Polymer Sci.* 1980, **25**, 2109.
18. D.A. Blackadder and P.I. Vincent, *Polymer* 1974, **15**, 2.
19. S.P. Chen, *Polymer Eng. Sci.* 1981, **21**, 922.
20. L.C.E. Struik, *Rheol. Acta* 1980, **19**, 111.
21. A.J. Staverman and F.R. Schwarzl, in *Die Physik der Hochpolymeren*, edited by H.A. Stuart, Vol. IV, § 5, Berlin 1956.
22. M.L. Williams, *J. Phys. Chem.* 1955, **59**, 95.
23. M.L. Williams, R.F. Landel and J.D. Ferry, *J. Amer. Chem. Soc.* 1955, **77**, 3701.
24. F.R. Schwarzl and A.J. Staverman, *J. Appl. Phys.* 1952, **23**, 838.
25. G. Goldbach and G. Rehage, *Rheol. Acta* 1967, **6**, 30.
26. S. Hozumi, T. Wakabayashi and K. Sugihara, *Polymer J.* 1970, **1**, 632; 1971, **2**, 756; 1978, **10**, 161.
27. L.C.E. Struik, *Rheol. Acta* 1966, **5**, 303.
28. A.K. Doolittle, *J. Appl. Phys.* 1951, **22**, 1471.
29. H. Vogel, *Physik Z.* 1921, **22**, 645.
30. G.S. Fulcher, *J. Amer. Ceram. Soc.* 1925, **8**, 339, 789.

31. G. Tamman and G. Hesse, *Z. Anorg. Allgem. Chem.* 1926, **156**, 245.
32. M.H. Cohen and D. Turnbull, *J. Chem. Phys.* 1959, **31**, 1164.
33. D. Turnbull and M.H. Cohen, *J. Chem. Phys.* 1961, **34**, 120.
34. A.J. Kovacs, R.A. Stratton and J.D. Ferry, *J. Phys. Chem.* 1963, **67**, 152.
35. H.H. Meyer, P.M.F. Mangin and J.D. Ferry, *J. Polymer Sci.* 1965, **A3**, 1785.
36. H.H. Meyer and J.D. Ferry, *Trans. Soc. Rheol.* 1965, **9**(2), 343.
37. S. Kästner and M. Dittmer, *Kolloid-Z.* 1965, **204**, 74.
38. S. Kästner, *J. Polymer Sci.* 1968, **16**, 4121.
39. D. Plazek and J.H. Magill, *J. Chem. Phys.* 1966, **45**, 3038.
40. S. Turner, *British Plastics,* December 1964, 682.
41. S.S. Sternstein and T.C. Ho, *J. Appl. Phys.* 1972, **43**, 4370.
42. F.A. Myers and S.S. Sternstein, *Proceedings VIIth Intern. Congress on Rheology,* edited by C. Klason and J. Kubát, p. 260, Göteborg 1976.
43. S.S. Sternstein, *Amer. Chem. Soc. Polymer Prepr.* 1976, **17**, 136.
44. M. Cizmecioglu, R.F. Fedors, S.D. Hong and J. Moacanin, *Polymer Eng. Sci.* 1981, **21**, 940; *Amer. Chem. Soc. Polym. Prepr.* 1983, **24**, 31.
45. E.S.W. Kong, *J. Appl. Phys.* 1981, **52**, 5921; idem, *Amer. Chem. Soc. Org. Coatings & Plastics Prepr.* 1982, **46**, 568.
46. T.D. Chang, S.H. Carr and J.O Brittain, *Polymer Eng. Sci.* 1982, **22**, 1221, 1228.
47. R. Gardon and O.S. Narayanaswamy, *J. Amer. Ceram. Soc.* 1970, **53**, 380.
48. O.S. Narayanaswamy, *J. Amer. Ceram. Soc.* 1971, **54**, 491.
49. M.A. de Bolt, A.J. Easteal, P.B. Macedo and C.T. Moynihan, *J. Amer. Ceram. Soc.* 1976, **59**, 16.
50. J.M. Hutchinson, J.J. Aklonis and A.J. Kovacs, *Amer. Chem. Soc. Polymer Prepr.* 1975, **16**(2), 94.
51. A.J. Kovacs, J.J. Aklonis, J.M. Hutchinson and A.R. Ramos, *J. Polymer Sci. Phys.* 1979, **17**, 1097.
52. T.S. Chow, W.M. Prest, *J. Appl. Phys.* 182, **53**, 6568.
53. G.E. Johnson, S. Matsuoka, H.E. Bair, *Amer. Chem. Soc. Org. Coatings & Plastics Prepr.* 1978, **38**, 350.
54. M. Uchidoi, K. Adachi, Y. Ishida, *Reports Progr. Polymer Physics Japan* 1975, **18**, 421.
55. M. Uchidoi, K. Adachi, Y. Ishida, *Polymer J.* 1978, **10**, 161.
56. M. Prod'homme, *Rheol. Acta* 1973, **12**, 337.
57. P.M. Anderson and A.E. Lord, *Mater. Sci. Eng.* 1980, **44**, 279.
58. A.I. Taub and F. Spaepen, *J. Mater. Sci.* 1981, **16**, 3087.
59. B.M. Chapman, *J. Text. Inst.* 1973, **64**, 667, 729; *Rheol. Acta* 1975, **14**, 466; *J. Text. Inst.* 1975, **66**, 339, 343.
60. N.G. McCrum and M. Pizzoli, *J. Mater. Sci. Letters* 1977, **12**, 1920.
61. C.K. Chai and N.G. McCrum, *Polymer* 1982, **23**, 589.
62. P.J. Greidanus and L.C.E. Struik, *AIAA/SAE/ASME, 16th Joint Propulsion Conf.,* June 30–July 2, 1980, Hartford, CT, USA.
63. C. Bultel, J.M. Lefebvre and B. Escaig, *Polymer* 1983, **24**, 476.
64. J. Heijboer, *Intern. J. Polym. Mat.* 1977, **6**, 11; idem, Annals NY Acad. Sci. 1976, **279**, 104.
65. H. Leaderman, *Elastic and Creep Properties of Filamentous Materials and other High Polymers,* p. 13, The Textile Foundation, Washington DC 1944.
66. F.T. Pierce, *J. Text. Inst.* 1923, **14**, T 390.
67. C.R. Kurkjian, *Phys. Chem. Glasses* 1963, **4**, 128.
68. J. de Bast and P. Gilard, *Phys. Chem. Glasses* 1963, **4**, 117.
69. R.W. Douglas in *Amorphous Materials,* edited by R.W. Douglas and B. Ellis, p. 3, Wiley London 1972.
70. C.P. Lindsey and G.D. Patterson, *J. Chem. Phys.* 1980, **73**, 3348.
71. G.D. Patterson, P.J. Carrol and J.R. Stevens, *J. Polym. Sci. Phys.* 1983, **21**, 605, 613.

72. J.A. Bucaro, H.D. Barby, *J. Non-Crys. Solids* 1977, **24**, 121.
73. H. Sasabe and C.T. Moynihan, *J. Polym. Sci. Phys.* 1978, **16**, 1447.
74. I.M. Hodge and A.R. Berens, *Macromolecules* 1982, **15**, 762.
75. I.M. Hodge and G.S. Huvard, *Macromolecules* 1983, **16**, 371.
76. M. Markis, I.L. Hopkins and A.V. Tobolsky, *Polymer Eng. Sci.* 1970, **10**, 66.
77. W.F. Knoff, I.L. Hopkins and A.V. Tobolsky, *Macromolecules* 1971, **4**, 750.
78. L.K. Djiauw and A.N. Gent, *Amer. Chem. Soc. Polymer Prepr.* 1973, **14**, 62.
79. G. Williams and D.C. Watts, *Trans. Faraday Soc.* 1970, **66**, 80.
80. G. Williams, D.C. Watts, S.B. Dev and A.M. North, *Trans. Faraday Soc.* 1971, **67**, 1323.
81. G. Williams and D.C. Watts in *Dielectric Properties of Polymers,* edited by F.E. Karasz, p. 17, Plenum, New York 1972.
82. F. van 't Spijker, *Proc. Conf. Metallic Glasses,* Budapest, June 30–July 4, 1980.
83. J. Kubát, *Nature* 1965, **204**, 378.
84. J. Kubát, *Thesis,* Stockholm 1965.
85. A.K. Jonscher, *Colloid Polymer Sci.* 1975, **253**, 231; *Nature,* 1975, **253**, 717; *J. Mat. Sci.* 1981, **16**, 2037.
86. K.L. Ngai, *Comments Solid State Phys.* 1979, **9**, 127.
87. R.M. Hill, *J. Mater. Sci.* 1982, **17**, 3630.
88. L.A. Dissado and R.M. Hill, *J. Mater. Sci.* 1981, **16**, 638.
89. F. Gény and L. Monnerie, *J. Polymer Sci. Phys.* 1977, **15**, 1.
90. J.E. Shore and R. Zwanzig, *J. Chem. Phys.* 1975, **63**, 5445.
91. D.C. Wright, *Polymer* 1976, **17**, 77.
92. R. Corsaro, *J. Amer. Ceram. Soc.* 1976, **59**, 115.
93. K. Adachi and T. Kotaka, *Polymer J.* 1982, **14**, 959.
94. H.S. Chen and T.T. Wang, *J. Appl. Phys.* 1981, **52**, 5898.
95. H.S. Chen, *J. Non-Cryst. Solids* 1981, **46**, 289; idem, *J. Appl. Phys.* 1981, **52**, 1868.
96. S. Sakaguchi, L. Mandelkern, J. Maxfield, *J. Polymer Sci. Phys.* 1976, **14**, 2137.
97. D.T.F. Pals, *Proceedings VIIth Intern. Congr. on Rheology,* edited by C. Klason and J. Kubát, p. 588, Göteborg 1976.
98. J. van Turnhout, *Thermally Stimulated Discharge of Polymer Electrets,* Elsevier, Amsterdam 1975.
99. C.K. Chai and N.G. McCrum, *Polymer* 1980, **21**, 706; Polymer 1982, **23**, 589.
100. L.C.E. Struik, *Plastics and Rubber Processing and Applications,* 1982, **2**, 41.
101. L. Guerdoux and E. Marchal, *Polymer* 1981, **22**, 1199.
102. L.C.E. Struik, P.Th.A. Klaase, P.H. Ong, D.T.F. Pals and J. van Turnhout, *1976 Annual Report on the NAS-NRC Conference on Electrical Insulation and Dielectric Phenomena,* Buck-Hill Falls, PA, USA, October 1976; see also *Proceedings Workshop on Thermally Stimulated Processes,* Montpellier, 1976, edited by G.C. Gibbings, in *J. Electrostatics* 1977, **3**, 171.
103. S. Matsuoka, C.J. Aloisio and H.E. Bair, *J. Appl. Phys.* 1973, **44**, 4265.
104. J.S. Lazurkin, *J. Polymer Sci.* 1958, **30**, 595.
105. P.I. Vincent, *Polymer* 1970, **1**, 7.
106. G.A. Adam, A. Cross and R.N. Haward, *J. Mater. Sci.* 1975, **10**, 1582.
107. R.M. Minnini, R.S. Moore, J.R. Flick and S.E.P. Petrie, *J. Macromol. Sci. B* 1973, **8**, 343.
108. D.H. Ender, *J. Macromol. Sci. B* 1970, **4**, 635.
109. J.H. Golden, B.L. Hammont, E.A. Hazell, *J. Appl. Polymer Sci.* 1967, **11**, 1571.
110. J.C. Bauwens, *Polymer* 1980, **21**, 699; idem, Chapter 12 in this volume.
111. R.E. Robertson, *J. Chem. Phys.* 1966, **44**, 3950; *Appl. Polymer Symp.* 1968, **7**, 201.
112. R.L. Brown and J.C. Richards, *Principles of Powder Mechanics,* p. 91–92, Pergamon, 1970.
113. J.D. Ferry and R.A. Stratton, *Kolloïd-Z.* 1960, **171**, 107.
114. K.C. Rusch and R.H. Beck, *J. Macromol. Sci. Phys.* 1969, **B 3**, 365; 1970, **B 4**, 621.
115. H.W. Bree, J. Heijboer, L.C.E. Struik and A.G.M. Tak, *J. Polymer Sci. A-2,* 1974, **12**, 1857.

116. K. Tanaka, T. Nose, T. Hata, *Rep. Progr. Polym. Phys. Japan* 1972, **15**, 229.
117. S. Matsuoka and H.E. Bair, *J. Appl. Phys.* 1977, **48**, 4058.
118. S. Matsuoka, *Polymer. Eng. Sci.* 1981, **21**, 907.
119. T.T. Wang, S. Matsuoka, *J. Polymer Sci.* 1980, **18**, 593.
120. U.W. Cho and W.N. Findley, Eng. Mat. Res. Lab. Report EMRL-75, Brown University, Providence, RI, USA, July 1980.
121. F.J. Lockett, *Non-linear Viscoelastic Solids,* Academic Press, London-New York 1972.
122. R.C. Richards and E.J. Kramer, *J. Macromol. Sci. Phys.* 1972, **B 6**, 229.
123. M.F. Milagin and N.I. Shishkin, *Mekhanika Polimerov* 1976, **1**, 8.
124. G. Titomanglio and G. Rizzo, *Polymer* 1980, **21**, 461.
125. D.T.F. Pals, Internal TNO Report, CL79/85 dated 1979-08-03.
126. P.H. Bowden and S. Raha, *Phil. Mag.* 1970, **22**, 463.
127. E. Krempl, *The role of aging in modeling of elevated temperature deformation,* Int. Conf. Creep and Fracture of Eng. Materials and Structures, University of Swansea, UK, March 24–27, 1981.
128. R.N. Haward, *Colloid. Polym. Sci.* 1980, **258**, 643.
129. J.L.S. Wales, *Polymer* 1980, **21**, 684.
130. T.E. Brady and G.S.Y. Yee, *J. Appl. Phys.* 1971, **42**, 4622.
131. J.B.C. Wu and J.C.M. Li, *J. Mater. Sci.* 1976, **11**, 434; 1979, **14**, 1593.
132. N. Walkes, *Polymer* 1980, **21**, 857.
133. K. Ito, *Trans. Soc. Rheol.* 1971, **15**, 389.
134. K.H. Illers, *Angew. Makromol. Chem.* 1969, **8**, 87.
135. P.L. Cornes and R.N. Haward, *Polymer* 1974, **15**, 149.
136. I. Narisawa, M. Ishikawa, H. Ogawa, *Rep. Progr. Polym.Phys. Japan* 1978, **21**, 227.

12 Physical Aging: Relation between free Volume and Plastic Deformation

Jean-Claude Bauwens

Service de Physique des Matériaux de Synthèse, Université Libre de Bruxelles
Avenue Adolphe Buyl, 87, 1050 Bruxelles, Belgium

12.1 Introduction

It has been known for a long time that mineral and organic glasses are in a metastable state, more or less far from thermodynamic equilibrium which they tend to approach.

We will call this metastable state of glasses which is characterized by thermodynamic parameters such as entropy and enthalpy "structural state". Many physical properties are linked to this structural state: mechanical and dielectric losses, glass-transition temperature, creep and relaxation, yield behavior,[1] and toughness.[2]

The evolution toward equilibrium, called "physical aging" by Struik,[3] produces a densification of the glass, the kinetics of which was investigated by Kovacs[4] some years ago.

All the physical properties we quoted are conditioned by molecular mobility; yet, it is generally accepted that this mobility depends mainly on the configurational free volume which decreases as a function of physical aging and characterizes a given structural state. Time and temperature affect the structural state, so do thermal and mechanical histories.

The effects of plastic deformation are opposite to those of physical aging. For instance, plastic deformation produces an increase of enthalpy and mechanical losses and a decrease of the yield stress. Therefore, at a given time the physical behavior and especially the mechanical behavior of a glassy material are linked to thermal and mechanical histories through the related structural state which controls the kinetics of evolution.

In order to study plastic deformation, we ought first to analyze other physical properties connected to the structural state, because they will help to identify the parameters included in the equations of behavior which we shall use.

12.2 Effect of Thermal History on Physical Properties

12.2.1 Specific Volume

Kovacs[5] studied the effects of thermal treatments on specific volume in detail, especially isothermal contraction.

When a polymer is suddenly cooled from the glass transition temperature T_g, to a given temperature T_1, it densifies as a function of time, t, as shown in Fig. 12.2.1.a. Let us assume that this polymer reaches equilibrium at $T_2 < T_1$, and that later is reheated to T_1. At T_1, it undergoes an increase of volume until this reaches the equilibrium value corresponding to that which was attained when it was cooled from T_g to T_1 (Fig. 12.2.1.b). Therefore, the polymer tends to reach an equilibrium state which only

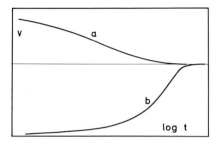

Fig. 12.2.1. Isothermal volume relaxation of a glassy polymer
a after quenching from T_g to T_1
b after heating to T_1 a specimen in equilibrium at $T_2 < T_1$.

depends on temperature, while the kinetics depends on both temperature and initial state.

In Kovacs's treatment, the kinetics, at a given time, depend on temperature, free volume, and the distribution of free volume within the glass which is considered as a set of cells each characterized by a time constant. This treatment takes into account a relaxation spectrum and allows description of complex phenomena such as dilatation followed by contraction which is observed when a glass is reheated after quenching (see de-aging in Chapter 11). Neglecting, for a first approximation, the contribution of a relaxation spectrum, the basic equation derived from Kovacs's treatment may be written

$$\frac{dv}{dt} = (v_\infty - v) \, v(T) f(v^f) \tag{12.2.1}$$

where v denotes the specific volume at a given time t, v_∞ the equilibrium volume, $v(T)$ a frequency factor depending on temperature T, and v^f of the configurational free volume from which variations lead to volume changes in the sample.

Assuming that the activation processes law holds, $v(T)$ may be expressed as a function of temperature by

$$v(T) = J_0 \exp\left(-\frac{\Delta G^*}{RT}\right) \tag{12.2.2}$$

where ΔG^* denotes the activation energy and J_0 a frequency factor. Only $f(v^f)$ may still be defined.

According to Doolittle,[6] the molecular mobility of a fluid depends mainly on free volume; the following relation was proposed

$$f(v^f) = \exp\left(-\frac{v_0}{v^f}\right) \tag{12.2.3}$$

where v_0 is a constant.

Considering that above T_g, the formation of free volume produces an increase of thermal expansion and assuming that below T_g, a free volume equal to v_g^f remains frozen-in Williams, Landel, and Ferry[7] proposed an expression of v^f

$$v^f = v_g^f + \Delta\alpha(T - T_g) \tag{12.2.4}$$

where $\Delta\alpha = \alpha_l - \alpha_g$ is the difference between thermal expansion related to liquid and glass respectively.

Taking into account Eq. (12.2.4), Doolittle's relation becomes

$$f(v^f) = \exp\left(-\frac{v_0}{v_g^f + \Delta\alpha(T - T_g)}\right) \tag{12.2.5}$$

which gives, with the universal mean values of the parameters $\dfrac{v^f}{v_0}$ and $\dfrac{\Delta\alpha}{v_0}$

$$-\log a_T = \ln \frac{f(v^f)}{f(v_g^f)} = \frac{17.44\,(T-T_g)}{51.6+(T-T_g)} \tag{12.2.6}$$

It is the well-known WLF-equation, often used to describe the kinetics of the relaxation processes of high polymers above T_g. See, however, Section 10.6. Below T_g, relation (12.2.4) is no longer valid; the free volume v_g^f remains in a metastable state which is susceptible of evolution following relation (12.2.1), as a first approximation. Therefore, any process having kinetics connected with v^f, will probably depend on the thermal history of the polymer.

12.2. Enthalpy Relaxation

At the glass transition, high polymers exhibit a specific heat increment Δc_p. When they are subjected to annealing treatments, at temperatures near but below T_g, an endothermic peak appears on the curve giving the specific heat as a function of temperature, measured in DSC experiments[8] (Fig. 11.3.3). The area enclosed between traces related to annealed and unannealed samples gives the increase of enthalpy ΔH, which is found to depend on temperature and duration of the thermal treatment.

Petrie[9] measured the kinetics of the enthalpy relaxation for several polymers; an example is given on Fig. 12.2.2. The data presented are in agreement with a set of parallel straight lines. Their horizontal shifts allow calculation of an apparent activation energy reaching about 200 to 400 kJ/mol.

This approach ceases to be valid in the following limiting cases when:
1) thermal treatments are performed very near T_g; then ΔH tends asymptotically to an equilibrium value ΔH_∞ increasing with $T_g - T$.
2) annealing is conducted at relatively low temperatures; in this case, the slope is lower and increases slowly.

This behavior schematically presented in Fig. 12.2.3 agrees with the proposition of Kovacs e.a.[10] who consider that enthalpy relaxation may be treated in the same manner as volume relaxation, provided one substitutes $\Delta H - \Delta H_\infty$ and Δc_p for $v - v_\infty$ and $\Delta\alpha$, respectively. In this case, Eq. (12.2.1) becomes

$$\frac{d\Delta H}{dt} = (\Delta H_\infty - \Delta H)\,v(T)\,f(v^f) \tag{12.2.7}$$

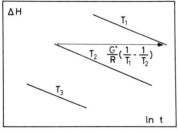

Fig. 12.2.2. Enthalpy relaxation of a glassy polymer at various temperatures.

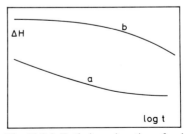

Fig. 12.2.3. Enthalpy relaxation of a glassy polymer
a just below T_g
b far below T_g.

Indeed if ΔH_∞ and v_∞ are the equilibrium enthalpy and the equilibrium volume at a given temperature T and, if a given structural state is characterized by given values of ΔH and v and a temperature θ, one must have

$$\frac{\Delta H_\infty - \Delta H}{\Delta c_p} = \frac{v_\infty - v}{\Delta \alpha} = T - \theta \tag{12.2.8}$$

In fact, all this amounts to a consideration that any given structural metastable state may be associated to a temperature θ at which the glass would be in equilibrium: therefore θ obeys the following relations

$$\frac{d\theta}{dt} = (\theta_\infty - \theta)\, v(T)\, f(\theta) \tag{12.2.9}$$

$$dv = \Delta \alpha\, d\theta \tag{12.2.10}$$

$$d\Delta H = \Delta c_p\, d\theta \tag{12.2.11}$$

12.2.3 Dynamic Shear Relaxations

Loss tangent curves of glassy polymers exhibit typical peaks as a function of temperature (Fig. 12.2.4).[11] The more prominent one, denoted α is related to the glass transition. Secondary peaks, called β, γ, ... are associated with local motions in the polymer chain, for instance, torsion of the main chain or rotations of side groups. These peaks may be considered to be insensitive to thermal treatments.

This is not the case of α', located near but below T_g.[12] This peak disappears from the loss curve related to samples slowly cooled or annealed below T_g, but on the contrary, is enhanced for samples quenched from temperatures above T_g. It is not easy to investigate the evolution of α' as a function of thermal histories, because it appears at temperatures where kinetics are rapid and the sample is subjected, during the measurements, to an unintentional thermal treatment, difficult to estimate. However, why the α' peak seems to vanish for annealed samples may qualitatively be explained as follows.[13] For a first approximation, it can be assumed that it lies at frequencies linked to the shear modulus G_s and the viscosity η by

$$v_0 = \frac{G_s}{2\pi\eta} \tag{12.2.12}$$

The viscosity η depends on molecular mobility explicit in relations (12.2.2) and (12.2.3) and therefore may be expressed by

$$\eta = \frac{\eta_0}{v(T)\, f(v^f)} = \frac{\eta_0}{J_0 \exp(-G^*/RT)\, (\exp - v_0/v^f)} \tag{12.2.13}$$

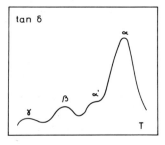

Fig. 12.2.4. Mechanical losses of glassy polymer versus temperature.

which leads to

$$\frac{v_0}{v^f} + \frac{G^*}{RT} = \ln \frac{J_0 G_s}{2\pi \eta_0} = \text{constant} \qquad (12.2.14)$$

where η_0 is a constant.

As a consequence of (12.2.14) a decrease of the free volume produces a shift of α' towards higher temperatures (and vice versa). This shift happens on annealing a sample; in this case α' is displaced beyond T_g. It should be noted that the opinion of Struik concerning the α' peak is quite different; see the preceding Chapter.

12.2.4   Electrical Properties

Dielectric losses and depolarization current are also linked to molecular motions and therefore, depend on the structural state.[14] Figure 12.2.5 shows depolarization current of PVC samples polarized at 60° C, unaged and aged 120 h at 60° C respectively, prior to testing. The aging treatment induces a strong decrease of depolarization current by slowing the kinetics.

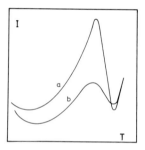

Fig. 12.2.5. Thermostimulated depolarization current versus temperature of quenched (a) and annealed (b) PVC samples.

12.2.5   Creep at Low Stresses

The effect of aging on creep at low stresses has been intensively studied by Struik [3, 15] for a number of materials, polymers in particular (see also Chapter 11). According to Struik "it is obvious that the shifts $\log a_T$ in the mechanical creep curves precisely follow the changes in specific volume v". This can bee seen in the curve of $\log a_T$ versus v in Fig. 11.2.5; the curve gives an acceptable fit to the WLF-equation.

12.2.6   Conclusions

Free volume seems to be the main measurable physical quantity linked to the structural state. It remains in a metastable state below T_g and slow decreases following kinetics which depend on both the temperature and its actual value. Other physical properties linked to the structural state, exhibit kinetics similar to that of free volume.

12.3   Effect of Mechanical History on Physical Properties

The structural state of a glassy polymer is modified when it is submitted to stress. Measurements performed in tensile tests have shown that the level of such active stresses is far below the yield point. For example, if a sample is loaded and unloaded repeatedly in tensile tests, below the yield point and in such a way that each cycle corresponds to an increase in deformation (Fig. 12.3.1), the resulting tensile curve deviates more and more from Hooke's law and is located far below the standard tensile curve (dashed line on the graph). On the other hand, a test piece deformed beyond the yield stress in alternated bending, finally reshaped to its original dimensions, and then submitted to a tensile test, exhibits a tensile curve with a vanishing yield point and a shorter linear part than a control specimen tested under the same conditions (Fig. 12.3.2).[16] Therefore, plastic deformation causes a strain softening. Struik[17] suggests that this effect reflects the destruction of aging which occurs below T_g, just the reverse of an annealing treatment. Cross e. a.[18] obtained a result similar to strain softening by quenching a sample prior to the tensile test. Again, the yield point disappeared from the tensile curve of such a sample. Adam e.a.[19] compared compression curves related to samples of PC, unannealed and annealed below T_g (Fig. 12.3.3). The work needed to deform the annealed sample was higher. The authors compared the increase in work to the energy related to the DSC endothermal peak which appeared at T_g as a consequence of the annealing treatment; they measured the same value: 2.26 J gr^{-1}, a result confirmed by data obtained on other polymers (PET and PVC).[20]

It can, therefore, be suggested that structural changes produced by plastic deformation may be described by using the same parameters as those used to characterize the structural state induced by annealing — mainly the free volume.

Indeed, plastic deformation tends to remove the structure from its thermodynamic equilibrium while duration produces the inverse effect. It follows that physical aging proceeds more rapidly when deformation brings the glass to a state far from equilibrium; this may cause a strain hardening during deformation when both aging and deformation kinetics are characterized by time constants of the same magnitude. Probably, this mechanism produces the transitory phenomena which occurs in tensile tests with sudden changes of strain rate during the neck propagation. The aspect of the related tensile curves is given in Fig. 12.3.4. An increase of the strain rate requires the destruction of the structure corresponding to a lower strain rate in the necking zone (stress overshot). On the contrary, if the strain rate is lowered, the structure in equilibrium with the reduced value requires some time to set in (stress undershot).

Sternstein and Ho[21] and Struik[22] have measured an opposite effect to strain softening when a glassy polymer is submitted to stresses far lower than the yield stress. In this case, the kinetics of physical aging are accelerated by the action of the stress. Thus, one realizes that the effect of thermomechanical history is rather complex and difficult to be modelized.

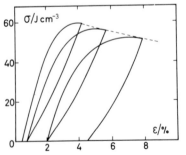

Fig. 12.3.1. Tensile curve of a PVC sample loaded and unloaded repeatedly.

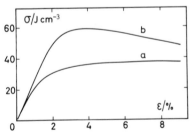

Fig. 12.3.2. Tensile curves of PVC samples
a the initial structure had been destroyed in bending before the test
b control specimen.

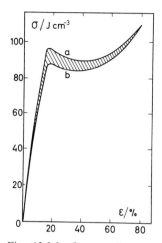

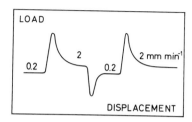

Fig. 12.3.3. Compressive curves of PC
samples
a untreated
b quenched.

Fig. 12.3.4. Part of the tensile curve related
to the neck propagation when sudden
changes of strain rate are imposed.

12.4   Nature of the Glassy State

Many authors, including Gibbs and DiMarzio,[23] Nose,[24–27] Simha and Som-cynsky,[28, 29] and Rehage[30] developed models able to account for the glass transition and particularly the influence of hydrostatic pressure p on T_g and $\Delta\alpha$. We shall now describe experimental conclusions concerning the nature of the glass transition.

When a polymer is cooled from a temperature above T_g, under the hydrostatic pressure p,[31] the following behavior may be observed: α_l, α_g, and $\Delta\alpha$ decrease when p increases while T_g increases with p (Fig. 12.4.1).

When pressure is applied to a polymer in the glassy state, if it is heated above T_g and then cooled, the curve of Fig. 12.4.2 is obtained, which gives the specific volume

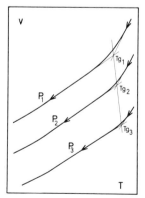

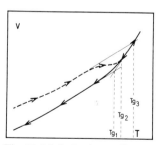

Fig. 12.4.1. Isobaric specific volume versus
temperature of samples cooled from temperatures above T_g.

Fig. 12.4.2. Isobaric specific volume versus
temperature related to a sample compressed in the glassy state, heated to temperatures above T_g and then cooled.

as a function of temperature. The polymer in the glassy state is densified, this effect remains if the pressure is suppressed. The two temperatures, T_{g1} and T_{g2}, may then be defined as the intersections of the straight lines related to v_l and v_g, respectively; even a third temperature T_{g3}, where both curves merge, may be considered.

A contradiction arises when T_g is seen as the temperature at which molecular mobility becomes too slow, accounting for the cooling rate, to allow the equilibrium of free volume with temperature. In this case, the higher the free volume, the higher T_g, yet the free volume theories imply the inverse situation.

Using the method described above, Bree e.a.[32] densified four polymers and measured the following mechanical properties: torsional creep compliance, dynamic modulus and mechanical losses, Charpy impact strength, thermal expansion below T_g, and torsional yield stress. The mechanical behavior differs little from that of control specimens not densified although the density increase reaches from 0.4 to 0.6% which corresponds to 20% of the WLF free volume frozen-in below T_g. Such a densification ought to slow molecular mobility by more than three decades. Therefore, the glassy structural state cannot be defined by free volume alone.

Ehrenfest's relations hold if only one order parameter is implied in the expression of the volume as a function of temperature and pressure. Thus, we must have

$$\frac{dT_g}{dp} = \frac{\Delta B}{\Delta \alpha} = \frac{T\Delta \alpha}{\Delta c_p} \qquad (12.4.1)$$

In the case of PS, the second relation is almost satisfied but the first one is, not.

According to Gee[33] this means that two glasses, densified and not densified, have roughly the same entropy and enthalpy, but differ by their free volume.

Also, the experiments performed by Brown e.a.[34] cast some light on the nature of free volume and of densification under pressure. These authors have measured the DSC trace and the effect of dilatometry on samples of PS, densified prior to testing (Figs. 12.4.3 and 12.4.4, respectively). The densified sample exhibits an endothermal peak followed by an exothermal one far below T_g; moreover, the specific volume increases more rapidly above 330 K than the volume of a control sample does.

As a consequence, one may assume that molecular rearrangements may take place which raise the specific volume even far below T_g. This may, therefore, not be connected with an excess of free volume in a metastable state which at first sight disagrees with the WLF-equation. In order to suppress these contradictions, one may assume that:
— holes compose free volume (Nose,[24-27] Gibbs and DiMarzio[23]),
— the size of these holes decreases when a pressure is applied,

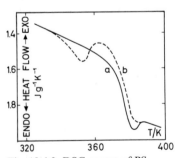

Fig. 12.4.3. DSC curves of PS
a untreated
b densified under pressure above T_g.

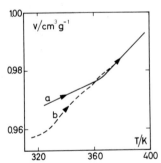

Fig. 12.4.4. Specific volume versus temperature
a untreated
b densified under pressure above T_g.

– the fraction of holes remains the same below T_g, for equal cooling rates,
– the influence of free volume on molecular mobility depends on the fraction of holes and not on their size.

This may mean that the entropic part of the free energy which influences molecular mobility depends on the fraction of holes; this is not necessarily the case of enthalpy which is thought to be linked both to the size of the holes and to the pressure. On the other hand, Goldbach and Rehage,[35, 36] conclude that the effect of the size of the holes on kinetics is insignificant. The experiments of these authors concern the dilatation of PS, near T_g, which occurs when pressure is suddenly changed. They measured the kinetics of this dilatation.

Let p_a be atmospheric pressure; $\Delta p = p - p_a$ the increase of pressure; v, v_p, and v_a the volume at t and the equilibrium volumes at p and p_a, respectively. The curves giving $(v - v_a)$ versus $\log t$ are represented on Fig. 12.4.5 for p_1 and p_2, two different values of p. These curves superpose well, for any pressure, if the reduced variable $x = \dfrac{v - v_a}{v_p - v_a}$ is plotted as a function of $\log t$, as in Fig. 12.4.6. This observation implies that this phenomenon is linear, and that at constant time kinetics are independent of volume.

Let us apply the same treatment as that proposed by Kovacs to account for isothermal contraction after quenching. From Eq. (12.2.1), we must have

$$\frac{dx}{d \log t} = - x\, t\, v(T) \exp\left(-\frac{v_0}{v_1^f + x(v_a - v_p)} \right) \qquad (12.4.2)$$

where v_1^f is the free volume in equilibrium at p_a and $v(T)$ a frequency factor depending on temperature.

At constant time and when dilatation begins (i.e., x nearly equals unity), it follows from (12.4.2) that

$$\frac{dx}{d \log t} \sim \exp\left(-\frac{v_0}{v_1^f + (v - v_a)} \right) \qquad (12.4.3)$$

Now, in this case, it follows from the WLF-equation that v_0 is close to unity and that v_1^f equals 2.5%.

As $(v_{p1} - v_{p2})$ reaches about 7×10^{-4}, the ratio of the slopes of the curves given in Fig. 12.4.5 might be about three in the range of short times (around a few minutes). This ratio disagrees with the data because a fairly good superposition is observed.

The following explanation of this discrepancy may be suggested: enthalpy, as well as entropy, depends only on the pressure at which experiments are conducted and is not influenced by the size of the holes. Until now, all the described measurements have been performed on samples densified at different pressures prior to being tested

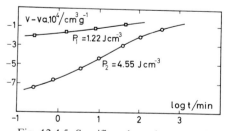

Fig. 12.4.5. Specific volume increase of PS versus log time when an applied pressure is suddenly removed.

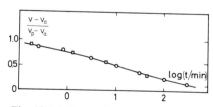

Fig. 12.4.6. Data of Fig. 12.4.5 expressed in reduced variables.

at p_a. Data obtained under the action of a pressure p are more difficult to interpret, because p raises T_g and thus entropy is modified. Therefore, the effect of pressure on enthalpy cannot be separated from that of entropy.

Measurements conducted on polymers in the glassy state, where entropy is supposed to remain constant, mainly concern the influence of pressure on Young's modulus, on density, and especially on plasticity, the kinetics of which have been studied for a number of polymers. We will come back later to this subject.

12.5 Free Volume Theories

Free volume can be viewed in at least two different ways. It can be linked to the redistribution of the extra volume which arises from thermal dilatation, as assumed by Cohen and Turnbull.[37, 38] Free volume can also be assumed to consist of a set of sites characterized by a configuration with a higher specific volume (holes, for example); this point of view has led to a number of theories, but we will summarize here the one developed by Gibbs and DiMarzio.[23]

12.5.1 Cohen and Turnbull Model

In this model, free volume is linked to the anharmonic terms of the attractive potential $V(r)$ between molecules (*Mie potential* for example). The distance r between molecules increases with the thermal dilatation. Below the inflexion of $V(r)$, shown on Fig. 12.5.1, an inhomogeneous distribution of the volume requires some energy contribution, which equals zero when $V(r)$ varies linearly with r. A critical value of r, related to a critical value v_{cr}, may be distinguished above which volume fluctuates between different sites without extra energy contribution. This type of volume is denoted v_p. Let $\Delta \bar{v}$ be the increase as a function of temperature: it can be expressed by

$$\Delta \bar{v} = v_p + v_{cr} \tag{12.5.1}$$

If a critical temperature T_c is associated with v_{cr}, it follows that

$$v^f = \alpha(T - T_c) \tag{12.5.2}$$

In addition, this treatment assumes that v^f consists of a set of discrete cells distributed at random. Using Lagrangian multipliers, the probability P_{vf} that the free volume of a cell reaches a value higher than v_0, can be expressed by

$$P_{vf} = \exp\left(-\frac{v_0}{v^f}\right) \tag{12.5.3}$$

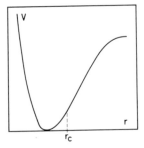

Fig. 12.5.1. Potential energy versus distance.

As a function of temperature, the probability P_T may be written

$$P_T = \exp\left(-\frac{v_0}{\alpha(T-T_c)}\right) \tag{12.5.4}$$

And therefore, from (12.5.4), we obtain the WLF-equation again, by

$$\log \frac{P_T}{P_{T_g}} = \frac{v_0}{\alpha(T-T_g)\,2.303} \cdot \frac{T-T_g}{(T-T_g)+(T_g-T_c)} \tag{12.5.5}$$

provided

$$\frac{v_0}{2.303\,\alpha(T_g-T_c)} = 17.44 \quad \text{and} \quad T_g-T_c = 51.6 \tag{12.5.6}$$

Let us point out that this treatment is similar to that leading to the Maxwell-Boltzmann distribution.

12.5.2 Gibbs and DiMarzio Model

This model is based on Flory's theory[39] of macromolecular configuration in dilute solutions. The links of the macromolecular chains are considered as points of a quasi-lattice of which the coordination number equals Z (for a carbon backbone chain, $Z=4$).

If a given site is occupied by a given link i, the link $(i+1)$ may be located on any of the $(Z-1)$ other sites; one of them corresponds to a nonflexed position generally more stable (Fig. 12.5.2). The $(Z-2)$ other sites related to flexed bonds require an extra energy, ΔE_f. Moreover, this model assumes the existence of vacant sites (i.e., holes created due to an extra energy, ΔE_c). Let P_{cf} be the probability of a configuration associated to a fraction, f, of flexed bonds and, c, of holes. The Gibbs Function ΔG may be expressed by

$$\Delta G = c\,\Delta E_c + f\,\Delta E_f - kT \ln P_{cf} \tag{12.5.7}$$

where k is the Boltzmann constant. This energy is minimized as a function of c and f to obtain the state of equilibrium realized for given values $c(T)$ and $f(T)$. Now, P_{cf} decreases with c and f and reaches zero for the couple c_0 and f_0. Therefore, some holes and flexed bonds characterize the amorphous structure; they are related to a temperature T_0 below which c and f remain constant. This temperature is lower than T_g because additional holes do subsist in a metastable state.

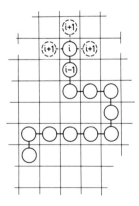

Fig. 12.5.2. Two dimensions quasi-lattice model.

Adam and Gibbs[40] developed a theory of viscosity wherein they assumed that deformation of a liquid implies the local rearrangement of a given number of molecules (or of segments of macromolecules if a polymer is considered). They calculated the related probability and obtained a relation similar to the WLF-equation

$$\ln \frac{P_T}{P_{cf} T_g} = \frac{C_1 (T - T_S)}{C_2 + (T - T_S)} \tag{12.5.8}$$

where C_1 and C_2 depend on T_0, defined above, and T_S denotes a characteristic temperature.

Gee[33] and Nose[24-27] have proposed theories analogous to that of Gibbs and Di-Marzio. Gee's theory consists of a thermodynamic treatment involving an order parameter which remains constant in the glassy state but is temperature and pressure dependent above T_g. Nose's contribution is a *"hole theory"* in which the glassy state is distinguished from the liquid state by freezing the holes, and the glass transition occurs at an isoconfigurational entropy.

12.5.3 Conclusions

The free volume theories may be classified in two different groups according to how glass transition is thought to occur: at constant volume or at constant entropy S.

The latter conception does not disagree with the WLF-equation; it may, for instance, mean that the fraction of holes has to be put in the place of free volume in Doolittle's relation (12.2.3), or more generally, that $1/k[S(T) - S(T_g)]$ fulfills the WLF-equation above T_g but keeps a constant value below T_g. The glassy state is then characterized by an excess of entropy in a metastable state, susceptible to decrease with increasing time or on annealing.

12.6 Theories of Plastic Deformation in Relation with Free Volume

We now consider data and theories dealing with general yield behavior, we will discuss models which take into account the structural state in the glassy range which allows the inclusion of physical aging.

12.6.1 Robertson Theory[41, 42]

Robertson assumes glasses are very viscous liquids (10^{15} poises or more) and that the effect of an applied stress induces a more fluid structure which allows viscous deformation to occur. A good approximation to the yielding state is one that has the structure of the melt but the vibrational temperature of the glass. He thinks that the diversity of sequences along the chains which describes the configurational entropy generally correlates with the free volume, and that the flexed bond fraction may be used instead of the free volume to calculate the viscosity of a polymer. Further, he assumes that the unflexed state is lower in energy than the flexed state by an amount denoted by ΔE. This amount also includes the change in intermolecular energy and in free volume concomitant with changes in the conformation. He sets ΔE equal to $15.9 T_g$ J mol^{-1} K^{-1} which is the value derived by Gibbs and DiMarzio. Assuming two configurations, unflexed and flexed respectively, and a Boltzmann distribution, the fraction χ_{eq} of flexed bonds, in equilibrium with the temperature, above T_g, is expressed by

$$\chi_{eq} = \frac{\exp\left(-\frac{\Delta E}{kT}\right)}{1+\exp\left(-\frac{\Delta E}{kT}\right)} \tag{12.6.1}$$

Below T_g this fraction remains frozen-in at the fraction existing at or near the glass transition.

The energy difference between the flexed and unflexed conformations and thus the equilibrium distribution of conformations will be changed by the application of a shear stress (Fig. 12.6.1) and two bonds can be promoted from the unflexed to the flexed conformation. The effect of the stress is to change the energy difference from ΔE to $\Delta E - v_{fl}\,\tau \cos\varphi$, where v_{fl} is the flexed volume (the distance moved times the area over which the stress acts per bond being flexed). τ is the local stress and φ the angle between the direction of motion and τ). Then the fraction of flexed bonds in the new equilibrium state induced by the application of the shear stress at a temperature T is given by

$$\chi_{\tau(\varphi)} = \frac{\exp\left(-\frac{\Delta E - v_{fl}\,\tau \cos\varphi}{kT}\right)}{1+\exp\left(-\frac{\Delta E - v_{fl}\,\tau \cos\varphi}{kT}\right)} \tag{12.6.2}$$

The question is now to calculate the maximum fraction of flexed bonds

$$\chi_{max} = \chi_i + \chi_\tau \tag{12.6.3}$$

where χ_i is the initial fraction of such bonds and χ_τ is the integral of $\chi_{\tau(\varphi)}$ for the values of φ such that $\chi_{(\varphi)} > \chi_i$. The result is

$$\chi_{max} = \frac{kT}{2 v_{fl}\,\tau} \left(\ln \frac{1+\exp\left(-\frac{\Delta E - v_{fl}\,\tau}{kT}\right)}{1+\exp\left(-\frac{\Delta E}{k\theta_g}\right)} \right)$$

$$+ \left(\frac{v\tau}{kT} + \frac{\Delta E}{kT} + \frac{\Delta E}{k\theta_g}\right) \frac{\exp\left(-\frac{\Delta E}{k\theta_g}\right)}{1+\exp\left(-\frac{\Delta E}{k\theta_g}\right)} \tag{12.6.4}$$

where θ_g is the temperature related to the configurational entropy when no stress is applied.

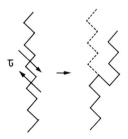

Fig. 12.6.1. Bonds flexed under the action of a shear stress (according to Robertson).

Robertson assumes that the stress induces a configurational entropy characterized by the temperature θ at which χ_{max} would be in equilibrium according to the following relation

$$\chi_{max} = \frac{\exp\left(-\frac{\Delta E}{k\theta}\right)}{1+\exp\left(-\frac{\Delta E}{k\theta}\right)} \tag{12.6.5}$$

Equating (12.6.4) and (12.6.5), one can compute θ. Thus by increasing the fraction of flexed bonds, the polymer structure becomes even more disordered than it was in the glass and because of the general relationship between order and volume[43] one can expect the volume also to increase.

The next step consists in stating that the viscosity obeys the WLF-equation. The Gibbs Function of activation at the temperature θ is

$$\Delta G(\theta) = 2.303 \frac{C_1 C_2 k\theta}{\theta - T_g + C_2} \tag{12.6.6}$$

where C_1 and C_2 are the two WLF parameters. The strain rate at yield is given by

$$\dot{\gamma} = \frac{\tau}{\eta} \tag{12.6.7}$$

and the viscosity by

$$\eta^{-1} = A \exp\left(-\frac{\Delta G(\theta)}{kT}\right) \tag{12.6.8}$$

where A is a constant.

To compare this result with the data, Robertson assignes the following numerical values to the parameters;
— C_1 and C_2 are the universal values 17.44 and 51.6, respectively,
— ΔE is the value obtained by Gibbs and DiMarzio by fitting the measured glass transition temperature,
— v is the volume of a segment containing two flexible bonds, and
— A is adjusted by assuming that the viscosity at T_g reaches 10^{13} poises. Thus, except for T_g, all the parameters are "universal" values not dependent on the considered polymer.

Although not perfect, the comparison with the data obtained on PS (Fig. 12.6.2) is surprisingly good; it is the reason why this theory is very attractive. This model, which does not take into account the normal stress components, has been modified by Robertson himself by computing the relative effect of the stress component normal to the shear plane. It has also been modified by Duckett e.a.[44] who added, to the

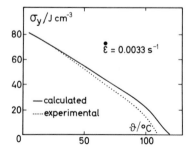

Fig. 12.6.2. Tensile yield stress of PS versus temperature.

work done by the shear stress a term proportional to the hydrostatic component of the stress p so that the difference in energy becomes

$$\Delta E - \tau_{max} \, v \cos \varphi + p\Omega$$

where Ω is a constant and τ_{max} the maximum shear component of the stress. Torsion tests under hydrostatic pressure are in good agreement with this simple treatment.

12.6.2 Ree-Eyring Theory[45, 46]

Eyring[45] considered that the jump of a segment of macromolecule from one equilibrium position to another requires an energy barrier, Q, to be overcome. This diffusion process gives rise to a macroscopically permanent deformation when some directions are favored (under the action of a shear stress τ, for example). Let v_0 be the volume undergoing the elementary shear γ_0 during one jump. When τ is applied, the energy barrier is lowered by a quantity $\tau v_0 \gamma_0$ in the direction of the stress and raised by the same quantity in the opposite direction. The shear strain rate $\dot\gamma$ is given by

$$\dot\gamma = \gamma_0 \, J_0 \exp\left(-\frac{Q}{kT}\right) \sinh \frac{\gamma_0 \, v_0 \, \tau}{kT} \tag{12.6.9}$$

where J_0 is a frequency factor containing an entropy term. At yield, where $\gamma_0 \, v_0 \, \tau \gg kT$, relation (12.6.9) can be written

$$\frac{\tau}{T} = \frac{k}{\gamma_0 \, v_0}\left(\ln \dot\gamma - \ln \frac{\gamma_0 \, J_0}{2} + \frac{Q}{kT}\right) \tag{12.6.10}$$

The relation remains linear for tensile stresses and strains (see the end of this paragraph). This behavior does not agree with the data around room temperature, except for PC, where a quite accurate linear fit is found[47] (Fig. 12.6.3). For other polymers one observes a more or less pronounced curvature (Fig. 12.6.4). This discrepancy may be overcome by assuming that more than one degree of freedom is involved in the deformation process and by applying the treatment proposed by Ree and Eyring[46] to describe the behavior of liquids. According to these authors, the applied stress is the sum of the partial stresses τ_α, τ_β, ... required to free, at the imposed strain rate, the different kinds of molecular motions implied in the deformation process.

Generally, one obtains a good fit to the data by calling on only two processes α and β.[48, 49] The α *processes* is supposed to correspond to the displacement, from

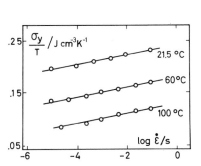

Fig. 12.6.3. Ratio of tensile yield stress of PC to temperature versus log strain rate at various temperatures.

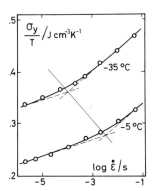

Fig. 12.6.4. Analogous plot as in Fig. 12.6.3, but now for PVC.

one equilibrium position to another, of segments belonging to the main chain; while the β *process* is related to local motions such as torsion of the main chain or rotation of side groups. Parameters related to each process are adjusted to fit the data; the value found for Q_β, the activation energy associated with the β process, has been correlated, for at least four polymers, with that of molecular motions giving rise to secondary transitions revealed by loss peaks in the glassy state.

We proposed a model able to describe both the yield and the damping behavior.[50] Relation (12.6.10), which applies to shear, has been modified for an arbitrary stress tensor, using $\sqrt{\frac{3}{2}}(\tau_{\text{oct}}+\mu p)$ instead of τ, where τ_{oct} is the octahedral shearing stress, p the isotropic stress tensor, and μ a constant.[51, 52] The modification is described below.

12.6.3 Modified Ree-Eyring Theory

At first sight, the Ree-Eyring model is unable to account for polymer structure and in particular, for physical aging.

However, this drawback can be overcome as follows.

Relation (12.6.9) must be rewritten as a function of the Gibbs function, ΔG of the yield process instead of the activation energy, such that

$$J_0 \exp\left(-\frac{Q}{RT}\right)=v_0 \exp\left(-\frac{\Delta G}{kT}\right)=v_0 \exp\left(\frac{\Delta S}{k}\right)\exp\left(-\frac{Q}{kT}\right) \tag{12.6.11}$$

where v_0 is a frequence factor and ΔS the configurational entropy which may depend on the initial structure and vary with physical aging. Thus, if we consider that ΔS follows the WLF-equation, relation (12.6.9) becomes

$$\dot{\gamma}=v_g\,\frac{C_1(\theta-T_g)}{C_2+(\theta-T_g)}\exp\left(-\frac{Q}{kT}\right)\sinh\frac{\gamma_0\,v_0\,\tau}{kT} \tag{12.6.12}$$

where v_g is a frequency factor containing the entropy at T_g and θ the configurational temperature, as in the Robertson treatment; θ can be assumed constant in the glassy state and equals T above T_g.

12.7 Comparison of Theories with the Data

Now, we have to compare these theories with the data in order to check their respective advantages and weaknesses. We shall use chiefly results obtained on PC for two reasons:
- The effect of aging on plastic deformation has been intensively studied.
- The yield behavior may be described within a wide range of temperature and strain rate, using a single deformation process (probably because the β transition occurs at temperatures which are far lower than for other glassy polymers). This allows accurate determination of the parameters and the variation of the yield stress.

12.7.1 Effect of Temperature and Strain Rate at Constant Structure

The data of Fig. 12.6.3 quite accurately agree with the relation (12.6.10) derived from the Eyring model, implying that
1) curves giving the yield stress, σ_y, as a function of T at constant strain rate, are straight lines intersecting at zero Kelvin;

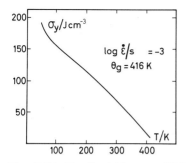

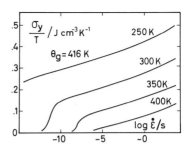

Fig. 12.7.1. Tensile yield stress of PC versus temperature computed from Robertson's model.

Fig. 12.7.2. Analogous plot as in Fig. 12.6.3, but now computed from Robertson's model.

2) isotherms giving σ_y/T versus $\log \dot{\varepsilon}$ are parallel straight lines which can be superposed by a horizontal shift accurately following an Arrhenius plot.

We have applied the Robertson relations (12.6.3–8) taking the parameters he proposed for PC[42] in order to compare his theory to the empirical conclusions recalled above.

The curve giving σ_y versus T is found to be nearly a straight line (Fig. 12.7.1).

The curves giving σ_y/T versus $\log \dot{\varepsilon}$ may be considered as parallel straight lines within the range of usual strain rates (Fig. 12.7.2); but the shift does not fit an Arrhenius plot.

12.7.2 Effect of Physical Aging on the Yield Stress

Effects of physical aging on the yield stress of PVC have been studied by Struik.[17] He compared the horizontal shift factor $\log a_T$, required to superpose:
1) two creep compliance curves under small stresses related to samples aged during different times (Fig. 12.7.3); and
2) the curves giving σ_y versus $\log \dot{\varepsilon}$ related to the same kind of samples (Fig. 12.7.4).

All these data referred to ambient temperature. Aging induces a smaller shift for yielding as does creep under small stresses. Struik concluded that the effect of physical aging is partially destroyed in a tensile test. He quoted the same results with creep tests conducted under high stresses. Bauwens-Crowet and Bauwens[13] compared for PC, the isotherms σ_y/T versus $\log \dot{\varepsilon}$ for three different physical treatments — as received, ice-quenched from above T_g, and annealed 45 hours at 120° C.

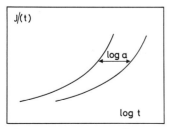

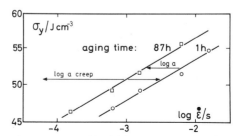

Fig. 12.7.3. Creep compliance versus log time related to PVC samples aged during two different times.

Fig. 12.7.4. Tensile yield stress versus log time related to PVC samples aged during two different times.

Results are given in Fig. 12.7.5. The straight lines drawn through the data fit the following relation

$$\frac{\sigma_y}{T} = A\left(\ln\dot\varepsilon - \ln J_0 + \frac{Q}{kT}\right) \tag{12.7.1}$$

with constant values of the parameters A and Q. Only J_0 depends on thermal treatment. It follows from (12.6.11) and (12.7.1) that J_0 contains an entropy factor and is expressed by

$$J_0 = v_0 \exp\frac{\Delta S}{k} \tag{12.7.2}$$

Thus, we can conclude that, within experimental accuracy, physical aging affects entropy much more than enthalpy.

Matsuoka and Kwei[53] compared the effects of the annealing time t_a and the strain rate on the yield stress for PC. They found that the plot of $\log\sigma_y$ versus $\log t_a$, at constant strain rate and versus $\log\dot\varepsilon$ at constant t_a respectively, gives straight lines with the same slope, s. Moreover, if E and v^f are the relaxation modulus and the fractional free volume respectively, data given for $\log E$ versus $\log t$ and $\log v^f$ versus $\log t_a$ also fit straight lines, the slope of which is compared with s; the same absolute value is found as

$$s = \frac{d\log\sigma_y}{d\log\dot\varepsilon} = \frac{d\log\sigma_y}{d\log t_a} = -\frac{d\log v^f}{d\log t_a} = -\frac{d\log E}{dt} = 0.02 \tag{12.7.3}$$

The kinetics seem to be identical and strongly dependent on free volume. Within the range of annealing times and strain rates investigated, Matsuoka pointed out that an increase, by one decade, in either the strain rate or the annealing time, will result in the same increase in the yield stress. This is not in agreement with Struik's data on PVC.[17] Bauwens-Crowet and Bauwens[54] measured the increase in the yield stress $\Delta\sigma_y$ for PC at ambient temperature and constant strain rate, as a function of annealing time t_a for treatments performed at different temperatures (Fig. 12.7.6a and b). The following theoretical model is proposed, assuming that through annealing a structural change which can be characterized by θ takes place. This temperature has the same meaning as in Robertson's theory and is defined in relation (12.2.3).

The WLF relation and an Arrhenius law are used to express $v^f(\theta)$ and $v(T)$, respectively. The relation (12.2.9) becomes

$$d\theta = -(\theta - T_a)J_0\, 10^{\frac{17.44(\theta-\theta_i)}{51.6+(\theta-\theta_i)}} \exp\left(-\frac{Q}{kT}\right) dt \tag{12.7.4}$$

where T_a is the annealing temperature and θ_i, close to T_g, the structural temperature related to the initial state.

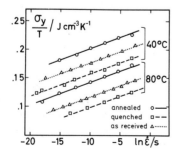

Fig. 12.7.5. Same plot as in Fig. 12.6.3 but now related to PC samples differing by their thermal history.

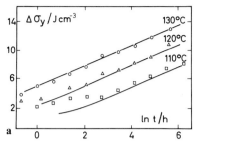 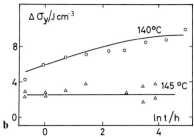

Fig. 12.7.6. Figures a and b. Increase in yield stress of PC samples produced by annealing at various temperatures, as a function of log of annealing time.

From relation (12.6.12), rewritten for uniaxial tensile tests and yield conditions, it is found that

$$\Delta\sigma_y = \frac{kT}{\varepsilon_0 v_0} \frac{2.303}{51.6 + (\theta - \theta_i)}$$ (12.7.5)

where ε_0 is constant. For sake of clarity, we have omitted a small correction arising from enthalpy relaxation.

Relations (12.7.4) and (12.7.5) allow to compute $\Delta\sigma_y$ as a function of $\log t_a$. The related theoretical curves, drawn in Fig. 12.7.6, have been calculated using the value of Q obtained from the shift of the isotherms given in Fig. 12.6.3, while J_0 is adjusted to get the best fit with the data. As an acceptable agreement is found, it can be assumed that the same molecular motions are involved in yield behavior and physical aging. According to relation (12.6.5), one must have

$$\Delta H = \Delta c_p (\theta - \theta_i)$$ (12.7.6)

From relation (12.7.4) we calculate θ as a function of t_a and T_a, and therefore theoretical values of ΔH, which were compared with the data (Fig. 12.7.7). A good fit is obtained, particularly if one considers there is no adjustable parameter left.

For comparison, we computed relation (12.6.8) of Robertson's theory for two different values of θ_g. The related curves giving σ_y/T versus $\log \dot{\varepsilon}$ are drawn in Fig. 12.7.8 where it can be seen that they cannot be superposed and, for this reason, do not agree with the data.

Until now, we have studied the consequences of aging within a range of temperatures where effects related to secondary relaxations may be neglected. Let T_β denote a temperature below which these effects must be taken into account.

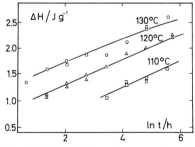 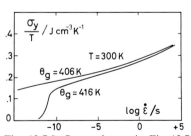

Fig. 12.7.7. Enthalpy change produced by annealing at various temperatures versus log of annealing time.

Fig. 12.7.8. Same plot as in Fig. 12.7.2, but now computed for two different structural states.

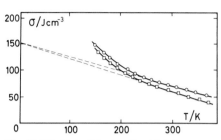

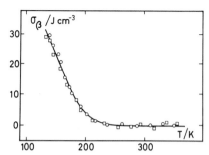

Fig. 12.7.9. Plot of the tensile yield stress versus temperature at constant strain rate for PC samples. □ and ○ refer to the quenched and the annealed (2 hours at 130° C) forms respectively.

Fig. 12.7.10. Plot of the β component of the yield stress versus temperature at constant strain rate (same PC forms as in Fig. 12.7.9).

As noted by Struik[3] shear modulus and mechanical losses are not affected by thermal history below T_β. However, according to the Ree-Eyring model, some effects must be observed on the yield stress in this range of temperatures. Relations (12.6.9) and (12.7.1) may be generalized as

$$\sigma_y = \sigma_\alpha + \sigma_\beta = A_\alpha T \left(\ln \dot{\varepsilon} - \ln J_{0\alpha} + \frac{Q_\alpha}{kT} \right) + A_\beta T \sinh^{-1} \left(\frac{\dot{\varepsilon}}{2 J_0} \exp \frac{Q_\beta}{kT} \right) \quad (12.7.7)$$

when two processes α and β are supposed to take part in the yield behavior (say below T_β).

Even if the β process, and thus σ_β are considered to be insensitive to aging, it follows from (12.7.7) that the yield stress depends on thermal history through σ_α, which is affected. This must be reflected in the plot of σ_y versus T, at constant strain rate, for different aging treatments. Such a plot is given in Fig. 12.7.9 where it can be seen that σ_y is sensitive to thermal history all over the range explored.[55] Above T_β, $\sigma_y \approx \sigma_\alpha$, according to (12.7.7), data must fit straight lines intersecting at zero Kelvin, which is accurately checked. Below T_β, the σ_β component may be obtained by subtracting from the measured yield stress the extrapolated value of σ_α. Results are plotted as a function of T in Fig. 12.7.10. No effect of thermal history can be discerned. This agrees with Struik's measurements; but the plot of Fig. 12.7.9 does not match his conclusions when he states that aging has no effect below T_β.

12.7.3 Effect of Hydrostatic Pressure on the Yield Stress and the Free Volume

As has been shown previously, cooling under pressure from temperatures above T_g produces a densification of the polymer which remains in part when the pressure is removed, but which does not influence the mechanical properties at atmospheric pressure. On the other hand, pressure applied in the glassy state affects the yield stress.

Tensile tests under pressure have been conducted widely by Matsushige e.a.[56] on PS and PMMA. Results are shown in Fig. 12.7.11 where one can see that the pressure effect is important. However, this effect does not result from the decrease of free volume as it can be deduced from the following numerical comparison. In the case of PMMA, an applied pressure of 100 J cm^{-3} produces an increase of 60% in the yield stress and a volume decrease of 3.3% which exceeds the frozen-in free volume, estimated by using the WLF relation. Let us point out that, in this range of high pressures, Raghava e.a.[57] proposed a nonlinear relationship between yield stress and pressure which was more accurate than the linear ones.[44, 51, 52]

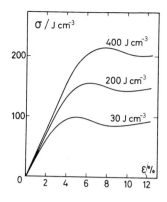

Fig. 12.7.11. Tensile curves of PMMA under different pressures.

12.7.4 Yield Behavior in the Vicinity of the Glass Transition Temperature

Near the glass transition, the yield behavior becomes more complex for several reasons. First, at temperatures just below T_g, heating causes an involuntary thermal treatment such that the yield stress varies with the time elapsed at these temperatures during the test. Second, the structural state above T_g depends on the temperature at which the test is performed. Therefore, the yield behavior may be investigated on two kinds of samples: annealed at temperatures at which the effect of further aging may be neglected, or remaining in the environmental chamber during a time long enough to reach a structure in equilibrium with temperature.

Results[58] related to PC samples annealed 48 h at 133° C and in structural equilibrium with temperature, respectively, are given in Fig. 12.7.12 where σ_y/T is plotted against $\log \dot{\varepsilon}$, for different T. The isotherms drawn throughout the data belong to a master curve built up by horizontal shifts (Fig. 12.7.13). These shifts denoted $\Delta \log a_T$ are plotted as a function of temperature in Fig. 12.7.14. Two different parts can be discerned on the curve: (a), related to a constant structural state below T_g, corresponds to a constant activation energy 318 kJ mol^{-1} and (b), related to samples having a structural state in equilibrium with temperature above T_g, fits the WLF relation

Now let us consider the master curve which may be separated in two portions. Within the range where $\sigma_y/T > 0.16$ J cm^{-3} K^{-1}, it is a straight line; below this level, σ_y/T decreases more rapidly for slower strain rates. We think that this faster decrease may be correlated with the diffusion process causing physical aging as well as fluctuation of the structural state toward values of θ higher than T. This last process allows defor-

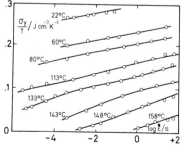

Fig. 12.7.12. As in Fig. 12.6.3, but now in a wide range of temperatures and strain rates. The curves belong to the master curve of Fig. 12.7.13.

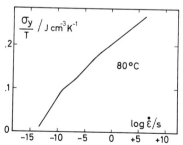

Fig. 12.7.13. Master curve, reduced to 80° C, related to the plot given in Fig. 12.7.12.

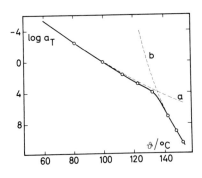

Fig. 12.7.14. Horizontal shift of the curves of Fig. 12.7.12 versus temperature
a Eyring relation
b WLF relation.

mation under smaller stresses. The frequency of the entropy fluctuation may be computed from relation (12.7.4). It roughly equals the strain rate at which σ_y/T is 50% lower than the extrapolated value of the linear part of the muster curve. Thus, the proposed assumption is acceptable.

However, an experimental fact remains unexplained; contrary to relation (12.7.4), the shift factor above T_g is found to depend on entropy only; no enthalpy effect appears.

12.8 Theories Concerning the Mechanical History

The effect of prior mechanical treatment on the yield behavior has been less intensively investigated than physical aging. However, since the early work of Vincent,[59] some models have been proposed. Kramer[60] used an asymmetric *Eyring type model* to account for the decrease of the neck propagation rate during creep tests performed on polyamides.

G'Sell and Jonas[61] explained the transient effects observed during a tensile test, when the strain rate is suddenly modified (see Fig. 12.3.4) by a density change of molecular misfits.

Yee and Detorres[62] suggested that a fast strain rate accumulates fewer flexed bonds and more elastic energy than a slow one; they showed qualitatively that the Robertson theory may explain transient effects. Struik[17] assumes that "mechanical deformations connected with segmental motions generate free voume, irrespective of whether they are tensile deformation, uniaxial compressions, or shear deformations"; see Chapter 11.

We proposed a crude treatment to account for a structural change during a tensile test.[63] We characterized the initial structural state by a spectrum of relaxation times and postulated that time constants are shifted towards shorter values when macromolecular segments undergo a plastic strain (this is equivalent to a local increase of entropy). Thus, at a given strain, two sorts of segments coexist, the elastically deformed one, distributed into the initial spectrum and the plastically deformed one, characterized by a spectrum with shorter relaxation times. The number of the elements belonging to the first spectrum decreases with increasing strain while that related to the second spectrum increases. A suitable choice of both spectra allows us to describe the whole tensile curve and the yield drop in particular.

The above treatment has several weakness: the choice of the spectra is empirical, it does not take into account the aging process which occurs when the initial structure has been destroyed, and it does not link plastic deformation and increase of entropy. Therefore, much has to be done in this field.

12.9 Concluding Remarks

The structural state strongly influences the relaxation times of many physical properties and the yield process in particular. It follows that embrittlement depends on the structural state also, because the stress concentration at a crack tip is closely related to the plastic behavior. As shown by Brostow and Corneliussen,[64] the impact transition temperature depends on relations similar to those describing the structural state. However, the structural state and the kinetics are not yet perfectly understood and the proposed models invoke more or less empirical relations.

References

1. J.H. Golden, B.L. Hammant and E.A. Hazell, *J. Appl. Polymer Sci.* 1967, **11**, 1571.
2. G. Allen, D.C.W. Morley and T. Williams, *J. Mater. Sci.* 1973, **8**, 1447.
3. L.C.E. Struik, *Physical aging in amorphous polymers and other materials'*, Elsevier, Amsterdam 1973.
4. A. Kovacs, *J. Polymer Sci.* 1958, **30**, 131.
5. A. Kovacs, *Fortsch. Hochpolym.-Forsch.* 1963, **3**, 394.
6. A.K. Doolittle, *J. Appl. Phys.* 1951, **22**, 1471.
7. M.L. Williams, R.F. Landel and J.D. Ferry, *J. Amer. Chem. Soc.* 1955, **77**, 3701.
8. M.S. Ali and R.P. Sheldon, *J. Appl. Polymer Sci.* 1970, **14**, 2619.
9. S.E.B. Petrie, *J. Polym. Sci. A 2* 1972, **10**, 1255.
10. A. Kovacs, J.J. Aklonis, J.M. Hutchinson and A.R. Ramos, *J. Polymer Sci. Phys.* 1979, **17**, 1097.
11. R.F. Boyer, *Polymer* 1976, **17**, 996.
12. K.H. Illers and H. Breuer, *J. Colloid. Sci.* 1963, **18**, 1.
13. C. Bauwens-Crowet and J.C. Bauwens, *J. Mater. Sci.* 1979, **14**, 1817.
14. J. Van Turnhout, *Thermally stimulated discharge of polymer electrets*, Elsevier, Amsterdam (1975).
15. L.C.E. Struik, *Rheol. Acta* 1966, **5**, 303.
16. C. Bauwens-Crowet and J.C. Bauwens, *J. Macromol. Sci. Phys.* 1977, B 14(**2**), 265.
17. L.C.E. Struik, *Physical. aging in amorphous polymers and other materials*, Chapter 8, Elsevier, Amsterdam 1973.
18. A.R. Cross, R.N. Haward and N.J. Mills, *Polymer* 1979, **20**, 288.
19. G.A. Adam, A. Cross and R.N. Haward, *J. Mater. Sci.* 1975, **10**, 1582.
20. R.N. Haward and Co-workers, *Colloid Polymer Sci.* 1980, **258**, 643.
21. S.S. Sternstein and T.C. Ho, *J. Appl. Phys.* 1972, **43**, 4370.
22. L.C.E. Struik *Polymer* 1980, **21**, 962.
23. J.H. Gibbs and E.A. DiMarzio, *J. Chem. Phys.* 1958, **28**, 373.
24. T. Nose, *Polymer J.* 1971, **2**, 124.
25. Ibid. p. 428.
26. Ibid. p. 437.
27. Ibid. p. 445.
28. R. Simha and T. Somcynsky, *Macromolecutes* 1969, **2**, 342.
29. T. Somcynsky and R. Simha, *J. Appl. Phys.* 1971, **42**, 4545.
30. G. Rehage, *J. Macromol. Sci., Phys.* 1980, B 18(**3**), 423.
31. P. Heydemann and H.D. Guicking, *Kolloid Z. Z. Polymer.* 1964, **193**, 16.
32. H.W. Bree, J. Heijboer, L.C.E. Struik and A.G. Tak, *J. Polymer Sci. Phys.* 1974, **12**, 1857.
33. G. Gee, *Polymer* 1966, **7**, 179.
34. I.G. Brown, R.E. Wetton, M.I. Richarson and N.G. Savill, *Polymer* 1978, **19**, 659.
35. G. Goldbach and G. Rehage, *Rheol. Acta.* 1967, **6**, 30.
36. G. Goldbach and G. Rehage, *J. Polymer Sci.* 1967, C **16**, 2289.
37. M.H. Cohen and D. Turnbull, *J. Chem. Phys.* 1959, **31**, 1164.

38. D. Turnbull and M.H. Cohen. *J. Chem. Phys.* 1961, **34**, 120.
39. P.J. Flory, *Proc. Royal Soc. A* 1956, **234**, 60.
40. G. Adam and J.H. Gibbs, *J. Chem. Phys.* 1965, **43**, 139.
41. R.E. Robertson, *J. Chem. Phys.* 1966, **44**, 3950.
42. R.E. Robertson, *Appl. Polymer Symp.* 1968, **7**, 201.
43. R.E. Robertson, *J. Phys. Chem.* 1965, **69**, 1575.
44. R.A. Duckett, S. Rabinowitz and I.M. Ward, *J. Mater. Sci.* 1970, **5**, 909.
45. H. Eyring, *J. Chem. Phys.* 1936, **4**, 283.
46. T. Ree and H. Eyring, in *Rheology,* vol. II, edited by Fr. Eirich, Chapter 3, Academic press, New York 1958.
47. C. Bauwens-Crowet, J.C. Bauwens and G. Homès, *J. Polymer Sci., Part A-2* 1969, **7**, 735.
48. C. Bauwens-Crowet and G.A. Homès, *C.R. Acad. Sci. Paris* 1964, **259**, 3434.
49. J.A. Roetling, *Polymer* 1965, **6**, 311.
50. J.-C. Bauwens, *J. Polymer Sci. C.* 1971, **33**, 123.
51. J.-C. Bauwens, *Mémoires scientifiques Rev. Métallurg,* LXV, 1958, **4**, 355.
52. S.S. Sternstein, L. Ongchin and A. Silverman, *Appl. Polymer Symp.* 1969, **7**, 175.
53. S. Matsuoka and T.K. Kwei, in *Macromolecules,* edited by F.A. Bovey and F.H. Winslow, Chapter 6, Academic Press, New York 1979.
54. C. Bauwens-Crowet and J.-C. Bauwens, *Polymer* 1982, **23**, 1599.
55. C. Bauwens-Crowet and J.-C. Bauwens, *Polymer* 1983, **24**, 921.
56. K. Matsushige, S.V. Radcliffe and E. Baer, *J. Polymer Sci. Phys.* 1976, **14**, 703.
57. R. Raghava, R.M. Caddell and G.S.Y. Yeh, *J. Mater. Sci.* 1973, **8**, 225.
58. J.-C. Bauwens, *Polymer* 1984, **25**, 1523.
59. P.I. Vincent, *Polymer* 1960, **1**, 7.
60. E.J. Kramer, *J. Appl. Phys.* 1970, **41**, 4327.
61. C. G'Sell and I.J. Jonas, *J. Mater. Sci.* 1981, **16**, 1956.
62. A.F. Yee and P.D. Detorres, *Polymer Eng. Sci.* 1974, **14**, 691.
63. J.-C. Bauwens, *J. Mater. Sci.* 1978, **13**, 1443.
64. W. Brostow and R.D. Corneliussen, *J. Mater. Sci.* 1981, **16**, 1665; W. Brostow, Chapter 10 in this volume.

13 Estimation of Long-Term Behavior from Short-Term Tests

Georg Menges
Rainer Knausenberger
Ernst Schmachtenberg

Institut für Kunststoffverarbeitung, RWTH Aachen
Pontstraße 49, D-5100 Aachen, West Germany

13.1 Introduction

Mechanical loading capacity is evaluated according to relatively simple tensile tests such as the German standard DIN 53455. Unfortunately, however, the measured data for origin modulus, yield point and tensile strength are not suitable as characteristic data for designing building components. The deviation of both the long-term load bearing capacity and the impact strength from this data is so great, that so far it has not been possible to generate suitable correlation functions. This unsatisfactory state necessitates expensive trials, e.g., for determining long-term data. A further problem, although polymers also share this with other materials, is the uncertainty as to how multiaxial loads are best considered.[1]

In Chapter 3 it is shown how a short-term test can be modified in such a way as to provide data which could be used for dimensioning purposes.[2,3] Therefore, it has been assumed that slightly more expensive instrumentation could be used if it saved time.

13.2 Uniaxial Tensile Tests with Controlled Strain Rate

The standardized tensile tests, e.g., German standard DIN 53455, involve the specimen clamps moving at constant speeds (see Fig. 13.2.1). It is then generally assumed that the strain rate would also be constant. To prove this assumption, we carried out such tests and measured the strain rate in the middle of the sample.[2] Figure 13.2.2 shows the strain measurement. We see, that contrary to the assumption, the strain rate is not constant. There are various causes for this, e.g., the positioning of the specimen clamps. Without investigating these causes, we decided to regulate the strain rates of the specimens to make them constant by taking the strain measured on the measured section of the specimen as the basis for the speed control of the machine. The strain occurring on the specimen during the tests is measured continuously and compared with a highly accurate linear set value profile plotted against time and produced by a function generator. A control circuit on the machine ensures a minimum difference between the set value and the actual value.[2]

Test results are normally subject to a varying degree of scatter between measurements. This makes it necessary to perform a relatively large number of identical tests in order to ensure statistical accuracy. When testing with constant strain rates, the results show that scattering of the measurements is reduced to a remarkable extent. This is explained by the fact that it is now possible to ensure that every test specimen

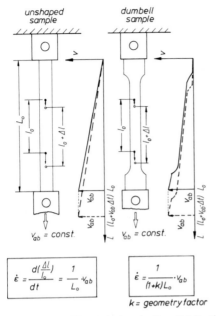

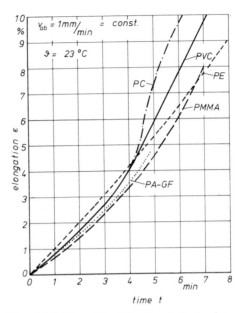

Fig. 13.2.1. Theoretical velocity distribution within sample during short-term-tensile test.

Fig. 13.2.2. Elongation behavior of plastics during short-term testing.

is subjected to the same deformation profile since possible differences in the clamping conditions or the geometry of the specimens no longer have any effect.

The stress-strain behavior at various constant strain rates at room temperature is shown in Fig. 13.2.3 for PMMA. There is a clear relationship between the mechanical

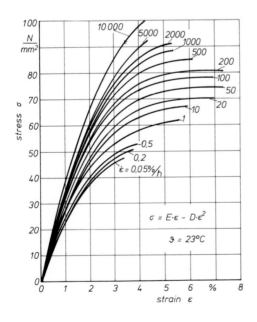

Fig. 13.2.3. Stress-strain behavior of PMMA at constant strain rates.

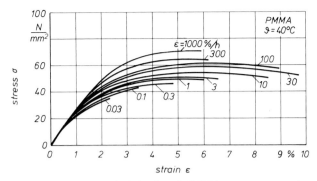

Fig. 13.2.4. Stress-strain behavior of PMMA at constant strain rates at 40° C.

properties and the strain rate, which correspond approximately to isochronous stress-strain diagrams. The time and the strain rate have, as expected, opposite effects on the behavior of the material. To determine the relationship between temperature and the mechanical properties at constant strain rates, appropriate tests at elevated temperatures have also been carried out (Fig. 13.2.4).

13.3   Functional Description of the Mechanical Behavior

The stress-strain profiles in Figs. 13.2.3 and 13.2.4 can be regarded as branches of a parabola, whose vertex lies at high stresses and strains and where the branches pass through the origin of co-ordinates. The analytic equation in Cartesian co-ordinates for this is

$$y = a_1 \cdot x - a_2 \cdot x^2 \tag{13.3.1}$$

This relationship, assumed purely on the basis of experience, signifies a nonlinear relationship between stress and strain which can consequently be described by

$$\sigma = E \cdot \varepsilon - D \cdot \varepsilon^2 \tag{13.3.2a}$$

$$\sigma = \varepsilon \cdot (E - D \cdot \varepsilon) \tag{13.3.2b}$$

Since, however, the tests were carried out at constant strain rates, for small deformations ($\varepsilon < 0.04$) it is also true that

$$\varepsilon = \dot{\varepsilon} \cdot t \tag{13.3.3}$$

Inserting Eq. (13.3.3) into Eq. (13.3.2a) and (13.3.2b) and taking account of the fact that different parameters are associated with every strain rate

$$\sigma = E \cdot \dot{\varepsilon} \cdot t - D \cdot \dot{\varepsilon}^2 \cdot t^2 \tag{13.3.4a}$$

$$\sigma = \dot{\varepsilon} \cdot t \cdot (E - D \cdot \dot{\varepsilon} \cdot t) \tag{13.3.4b}$$

This form of relationship was also selected because, for small strains, the second order quadratic term becomes small and thus extends into Hooke's law. Consequently, the factor E corresponds to a strain-rate-dependent modulus of elasticity, whereas D represents a measure of the nonlinearity of the material behavior – a damping effect.

As is evident from Eq. (8.3.4), the stress at constant temperature depends only on the strain rate and the time (in tests at constant strain rate). Figure 13.3.1 shows the relationship between the origin modulus and the strain rate at various temperatures.

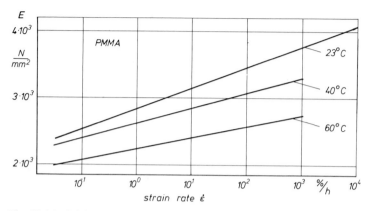

Fig. 13.3.1. Origin modulus E versus strain rate at different temperatures.

The decrease of E with increasing temperature is in accord with experience concerning the behavior of the elasticity modulus. Furthermore, the plot of E against the strain rate and the temperature is suitable for applying the time-temperature shift principle. The time-temperature shift principle and shift factor are explicitly dealt with by Brostow in Section 10.5 and by Kenner in Section 2.3. If the origin-modulus curves are shifted using the factors shown in Fig. 13.3.2, a single curve is obtained which, for the strain-rate-dependent elasticity modulus at room temperature (taken here as the reference temperature) covers the strain rate

$$10^{-5}\%/\mathrm{h} < \dot{\varepsilon} < 10^{4}\%/\mathrm{h}$$

The same shift factors applied to the damping D produce the result shown in Fig. 13.3.4. Here, too, a master curve is obtained, from which it is evident that the nonlinear influence decreases considerably for very small strain rates and thus also does so in a static creep test over very long times. It follows from the two master curves in Figs. 13.3.3 and 13.3.4 that, with tests carried out at constant strain rates, a pair of values E and D can be exactly attributed to every strain rate at constant temperature.

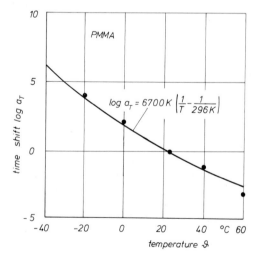

Fig. 13.3.2. Time-temperature shifting comparison: measurement calculation.

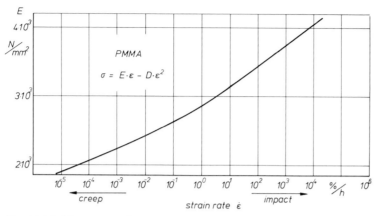

Fig. 13.3.3. Master curve for the origin modulus E at 23° C.

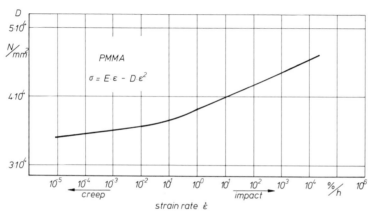

Fig. 13.3.4. Master curve for the coefficient D at 23° C.

This implies that the stress in Eq. (13.3.4) must also represent an explicit function of the strain rate and of time. Thus for the following functional relationship

$$\sigma = f(\dot{\varepsilon}, t, \vartheta) = \text{const.} \qquad (13.3.5)$$

there exists an inverse function of the form

$$\dot{\varepsilon} = g(\sigma, t, \vartheta) = \text{const.} \qquad (13.3.6)$$

With this inverse function, it must be possible, at a given stress and time, to calculate the strain for a test with an unknown, but constant strain rate. However, the functions $E(\dot{\varepsilon})$ and $D(\dot{\varepsilon})$ contain the exponential function (sections of straight lines in semilogarithmic representation). This is the reason why an analytical formation of the inverse function is not possible and the problem can only be solved by iteration.

Using such a method of calculation, curves describing the strain as a function of time for a constant load have been calculated. These curves were obtained by determining the strain at a given stress and time from the curves for the tests carried out at constant strain rate. These curves show good agreement in the deformation range $\varepsilon < 2\%$ with long-term creep tests for static loads (Fig. 13.3.5). The strains above this deformation range, which were found in the calculation process to be too small, can

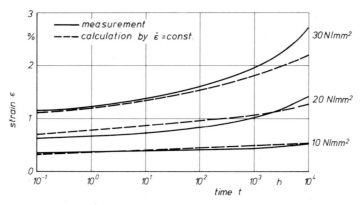

Fig. 13.3.5. Comparison between measured and calculated creep curves.

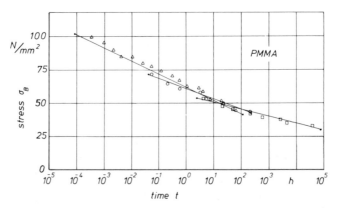

Fig. 13.3.6. Master curve of ultimate strength at constant strain rates.

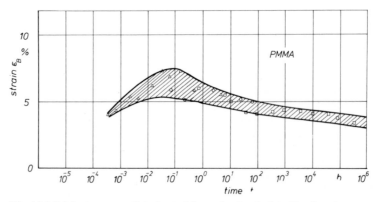

Fig. 13.3.7. Master curve of strain at ultimate load, calculated by time-temperature shifting principle.

be explained by the onset of damage in the material. The similarity between the two curves over a wide time range testifies to the correctness of the procedure.

Finally, a further empirically determined relationship for the uniaxial stress state can be derived. Applying the same shift factors to the times to fracture belonging to the tensile strengths

$$t_B = \varepsilon_B / \dot{\varepsilon} \tag{13.3.7}$$

to tests carried out at constant strain rates at the various temperatures, we get the tensile strength at room temperature shown in Fig. 13.3.6. This extends over a period up to 5×10^4 h (5.7 years) and conforms very well indeed with our experimental results. Similar agreement is obtained for the strain at ultimate load, which becomes the elongation at break, when the values determined at elevated temperatures are shifted to longer times using the same shift factors. This curve is shown in Fig. 13.3.7. The fact that this procedure is also successful with materials other than PMMA has been established in tests which are still continuing at the moment. In these tests, similar results have already been obtained with the semicrystalline HDPE and other materials.

Viewed as a whole, it is clear that tests carried out at a constant strain rate can be interpreted to provide information on long-term properties.

13.4 Multiaxial Short-Term Stress-Strain Tests Under Constant Strain Rates

13.4.1 Test setup

It was assumed that the strain rate with multiaxial loads has a similar significant influence as in uniaxial tensile tests.

For this reason, a suitable tensile testing machine was re-designed so that pipe specimens with variable, regulated, axial, and tangential strain rates could be loaded. The magnitude of the two individual deformations could be adjusted independently of one another by altering the position of the cross-head or the internal pressure. Measurement was again made of the longitudinal and the circumferential strain; both, along with the thickness strain, were also fed back as measurements to the control system. Figure 13.4.1 shows a schematic diagram of the test equipment and Fig. 13.4.2 illustrates the measuring sensors for the circumferential strain and for measuring the thickness of the pipe wall.

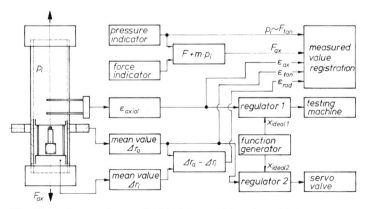

Fig. 13.4.1. Test equipment for biaxial material testing with constant strain rates.

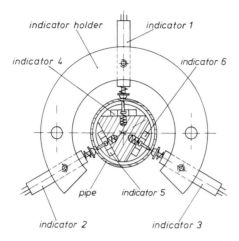

Fig. 13.4.2. Measurement of circumferential strain and wall thickness of pipes.

Table 13.4.1. Test Parameters

Type of test	$x/\%$	$y/\%$	$x_2/(\text{N}\cdot\text{mm}^{-2})$	$y_2/(\text{N}\cdot\text{mm}^{-2})$
a	0.50	0	—	—
b	0	0.50	—	—
c	0.25	0.50	—	—
d	0.50	0.25	—	—
e	0.50	0.50	—	—
f	0.50	—	—	0
g	—	0.50	0	—

x tangential.
y axial.

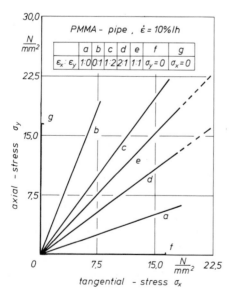

Fig. 13.4.3. Biaxial stressing for different strain ratios.

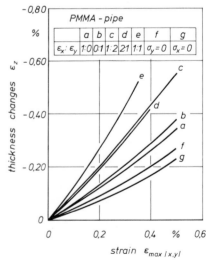

Fig. 13.4.4. Thickness changes at different strains.

With this machine, it is thus possible to test pipe specimens under various stress states. The strain rates in two directions or one strain rate and the stress in the other direction were taken as test parameters, giving certain stress and strain ratios similar to Table 13.4.1.

The results of these tests on extruded PMMA are shown in Figs. 13.4.3 and 13.4.4.

13.4.2 Results of Quasi-Uniaxial Tensile Tests on Pipe Specimens of PMMA

The tests *f* and *g* (Table 13.4.1) are the equivalent of an uniaxial tensile test. In each case, stresses occur here only in the axial or tangential direction (neglecting the low radial compressive strength in *f* from the internal pressure). They are thus suitable for providing information about the anisotropy of the material. Figure 13.4.5 shows the elasticity modulus for these two directions, calculated from $E_x = \sigma_x/\varepsilon_x$; $E_y = \sigma_y/\varepsilon_y$. As was expected, the modulus of elasticity in the axial direction, i.e., in the extrusion direction of the pipe, is higher. This can be explained by a greater orientation in this direction. Extrusion material was of lower molecular weight than the material in Section 13.3. It is noticeable that the plot for E_y is a curve, whereas E_x inclines linearly. This difference is also evident in the other tests. The modulus of elasticity in the thickness direction cannot be determined directly. Because the dependence on the direction is altogether lower (maximum difference only 3.6%), a virtually isotropic material behavior can be assumed. A further point in favor of this is that the *u*-values (Lamé constant = shear modulus) have, for all the tests, a mean deviation of only 6.2% calculated on the smallest of these three values. This scattering already contains the measuring inaccuracy.

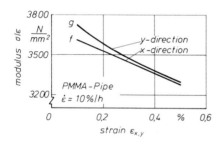

Fig. 13.4.5. Modulus in relationship to the direction in effective uniaxially loaded pipes.

13.4.3 Determining the Material Data by a Multiaxial Pipe Test

In these tests, the stresses and strains were determined in all three spatial directions. Under the condition of isotropy, however, it is basically sufficient to measure five of these stress or strain values in order to determine the material data such as elasticity modulus, transverse contraction (Poisson's ratio), and the Lamé constants. In the biaxial or uniaxial stress state, the number of values to be measured is reduced to four and three respectively, since the boundary conditions provide information about further values. However, since measurements never provide exact values for the stress of deformation state, additional measurement of the other values allows information about the quality of the calculated material data to be determined. In addition, it has been found that it is very important which measurements are used to determine the material data. This can be explained with the example of a biaxial stress state ($\sigma_z = 0$): Here

$$v = \frac{\sigma_x^2 - \sigma_y^2}{\varepsilon_x \cdot \sigma_x - \varepsilon_y \cdot \sigma_y} \tag{13.4.1}$$

or $$v = \frac{\varepsilon_x \cdot \sigma_y - \varepsilon_y \cdot \sigma_x}{\varepsilon_x \cdot \sigma_x - \varepsilon_y \cdot \sigma_y}$$ (13.4.2)

and the characteristic data can be determined without knowing the strain in the z-direction. For stress states, $\sigma_x \approx \sigma_y$, the denominator of these expressions nevertheless approaches zero and even small measuring errors can result in major changes in the characteristic value.

For these types of tests, values for the strain measurement in the wall-thickness direction (z-direction) have to be used for determining the material data. On observing the measuring systems and their resolution, however, it is found that the accuracy of the radial strain measurement is poorer by a factor of around 15–25 than that of the longitudinal or circumferential strain measurement.

This means that, for the evaluation, it is necessary to weigh these two aspects against one another in order to obtain the maximum probability for the calculated characteristic data. The procedure adopted was to calculate the data from the strains in the circumferential and axial directions. Using the characteristic data obtained in this way (points in Figs. 13.4.6 and 13.4.7), the strain state in the radial direction was determined.

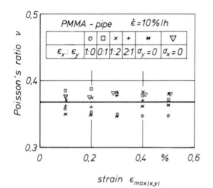

Fig. 13.4.6. Poisson's ratio for different strain levels.

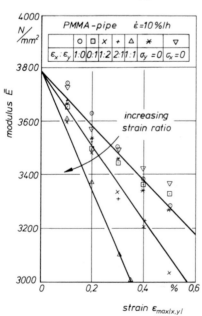

Fig. 13.4.7. Modulus versus strain for different strain ratios.

A comparison between measurement and calculation then provided information about the quality of the previously determined characteristic value. Where the deviations were excessive, these measurements were not taken into account for determining the characteristic data.

Hence the following conclusions can be drawn from the characteristic data.

13.4.4 Poisson Ratio

The Poisson ratio is constant for tests with constant strain rate, in other words it is dependent neither on the degree of multiaxiality nor on the strain.[4] For the present measurements, a value of $v=0.37$ has been determined (Fig. 13.4.6).

Besides this, the Poisson ratio has been determined in an uniaxial tensile test (under constant strain rate). It was also found here that v is independent of the strain and dependent only on the strain rate or on the temperature. Figure 13.4.8 shows the Poisson ratio for various temperatures and strain rates. It is interesting to note that the time-temperature shift principle proved to be usable with the same shift factors as for the modulus. Conformity with the values obtained from the multiaxial tests is excellent.

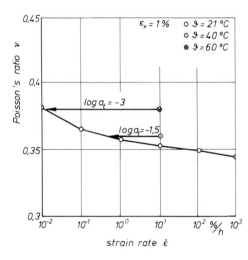

Fig. 13.4.8. Poisson's ratio in relationship to strain rate (uniaxial tension).

13.4.5 Modulus of Elasticity

In Hooke's law, the modulus of elasticity is defined as the quotient of stress and deformation. In ideal-elastic materials, this quotient is independent of the load. In plastics, however, this quotient decreases with increasing strain considering the uniaxial loading of PMMA. From the equation

$$\sigma = E\cdot\varepsilon - D\cdot\varepsilon^2 \qquad (13.3.2a)$$

the quotient of stress and deformation can be formed

$$\frac{\sigma}{\varepsilon} = E - D\cdot\varepsilon \qquad (13.4.3)$$

For sufficiently small deformations, origin modulus can thus be obtained

$$\lim_{\varepsilon\to 0}\frac{\sigma}{\varepsilon} = E \qquad (13.4.4)$$

In addition, Eq. (13.4.3) describes the path of the secant modulus as is determined from the stress-strain diagram for a given strain. For the statement describing the stress-strain behavior in line with Eq. (13.3.2a), a linearly inclined function is obtained for the quotient $\sigma/\varepsilon = E$ (referred to from this point as the modulus).

An analysis of Fig. 13.4.7 shows that here too, a linear function exists for a multiaxial stress state. A comparison with the origin modulus measured in the uniaxial tensile test on dumbbell specimens shows that the origin modulus of the pipe specimens was about 14° higher. This difference can be explained by the varying types of material (cast or extruded PMMA). Tensile specimens produced from the pipe material showed conformity in their elasticity modulus with the quasi-uniaxial tensile tests performed on the pipe specimen.

The second interesting property of the modulus function is its gradient, which is determined by the coefficient D. It was found that, with increasing multiaxiality, these gradients increase. This property indicates that greater relaxation processes occur with multiaxiality. The multiaxial deformation states calculated with characteristic data from uniaxiality are therefore also lower than the measured deformations. An attempt was made to record the higher decline in the modulus in the multiaxial stress state.

One obvious possibility is to plot the modulus against the sum of forced strains (Fig. 13.4.9), in order to take into account the higher stress, say of a pipe test, for example in the 1:1 strain state. The forced strain is defined as those strains which lie directly in the direction of the external application of force.

It was found that the curves for various biaxial stresses coincide with the uniaxial ones. In accordance with the previous considerations, the creep modulus could, in analogy to Eq. (13.4.3) be defined as

$$E_c = E_{(t,\,\vartheta)} - D_{(\varepsilon,\,t,\,\vartheta)} \cdot \varepsilon \tag{13.4.5}$$

This means that the dependence on the multiaxial load is reflected only in the coefficient D for the attenuation. This is a question of practicability which will become evident on further use.

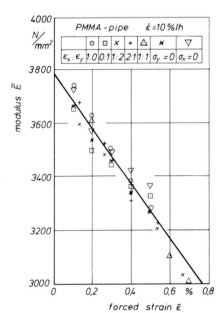

Fig. 13.4.9. Modulus for different strain ratios versus forced strain.

13.4.6   Concept of Forced Strain

From the previous considerations, it is possible to develop a concept with which multiaxial stress-strain states can also be calculated.

Basically, an extension of Hooke's law is used

$$\varepsilon_x = \frac{1}{\bar{E}}(\sigma_x - v \cdot (\sigma_y + \sigma_z)) \tag{13.4.6a}$$

$$\varepsilon_y = \frac{1}{\bar{E}}(\sigma_y - v \cdot (\sigma_z + \sigma_x)) \tag{13.4.6b}$$

$$\varepsilon_z = \frac{1}{\bar{E}}(\sigma_z - v \cdot (\sigma_x + \sigma_y)) \tag{13.4.6c}$$

The degree of multiaxiality and the influence of time and temperature are taken into account when determining the characteristic data as follows.

For a given strain, the characteristic data can be determined directly. First of all, for the deformation state, the characteristic strain rate from the quotient of principal strain and loading time is formed. Figure 13.4.7 then provides the respective Poisson's ratio.

The modulus $\bar{E} = \sigma/\varepsilon$ can be calculated from Eq. (13.4.3), whereby the coefficients D and E are determined from Figs. 13.3.2 and 13.3.3 with the aid of the characteristic strain rate. For $\bar{E}$, the sum of the forced strains is inserted in Eq. (13.4.3).

If, however, the stress state is regarded at the starting point, the calculation must be made in an iterative form, whereby a strain state is calculated from initially assumed characteristic data. From the calculated strain state, better characteristic data can now in turn be determined and used for a second iteration step.

With a method of calculation such as this, it is thus possible to determine multiaxial stress states from data determined in uniaxial tests.

In this way it is equally possible to determine the Lamé constants in order to obtain simplified equation systems for calculating a stress state from the strain state. The Lamé constants have the same dependences on the stress state.[5] It is not our intention to deal with this representation here.

It remains to be tested whether this concept can also be applied to other plastics. It has at least been found, however, that by taking into account the influences of time, temperature, and degree of multiaxiality, the material behavior can be described using the extended Hooke's law.

Besides using the concept of forced strains, it is also possible to record the degree of multiaxial stress state with the stress vector

$$\bar{\sigma} = \sqrt{\sigma_x^2 + \sigma_y^2 + \sigma_z^2} \tag{13.4.7}$$

Initial research has shown that this method covers the multiaxiality even more clearly, although it does produce additional problems in solving the equation systems for calculating the stress state. This method has, therefore, not been pursued any further for the time being.

Abbreviations

(Latin)

a_1	correlation coefficient
a_2	correlation coefficient
D	damping coefficient
E	origin modulus

$\bar{E}$	modulus
E_x	modulus in x-direction
E_y	modulus in y-direction
E_z	modulus in z-direction
HDPE	high density polyethylene
PMMA	polymethylmethacrylate
t	time
t_B	time to fracture
x	cartesian co-ordinate
y	cartesian co-ordinate
z	cartesian co-ordinate

(Greek)

ε	technical strain
ε_B	technical strain to fracture
ε_x	technical strain in x-direction
ε_y	technical strain in y-direction
ε_z	technical strain in z-direction
$\dot{\varepsilon}$	strain rate
ϑ	temperate
ν	Poisson ratio
σ	technical stress
σ_x	technical stress in x-direction
σ_y	technical stress in y-direction
σ_z	technical stress in z-direction
$\bar{\sigma}$	stress vector

References

1. A. Troost, *Einführung in die allgemeine Werkstoffkunde metallischer Werkstoffe I* Wissenschaftsverlag, Bibliographisches Institut AG, Zürich 1980.
2. R. Knausenberger, Das mechanische Verhalten isotroper und anisotroper Werkstoffe mit nichtlinearelastischen Eigenschaften, PhD thesis RWTH Aachen 1982.
3. G. Menges and E. Schmachtenberg, *Kunststoffe-German Plastics,* 1983, **9**, 543.
4. E. Schmachtenberg and T. Schmidt, *Meßtechnische Briefe,* 1983, MTB 19, **3**, 59.
5. T. Krehwinkel, unpublished work, IKV Aachen 1983.

14 Mechano-Chemical Phenomena in Polymers

Björn Terselius, Stockholm, Sweden
Ulf W. Gedde, Stockholm, Sweden
Jan-Fredrik Jansson, Stockholm, Sweden

Department of Polymer Technology, The Royal Institute of Technology
S-10044 Stockholm, Sweden

14.1 Introduction

Mechano-chemical phenomena in polymers are today well-known and well documented but poorly understood. By 1978 almost 1500 papers had been published on the subject.[1] Among the authors of more extensive reviews in this field, Ceresa,[2] Watson,[3] Baramboim,[4] Simionesco and Vasiliu-Opera,[5] Sohma and Sakaguchi,[6] Rånby and Rabek,[7] Casale and Porter,[1] Andrews and Reed,[8] and Kausch[9] deserve to be mentioned. In fact, mechano-degradation of natural rubber during mastication was detected more than 50 years ago by Staudinger.[10] That this was a radical process was first realized by Pike and Watson[11] in 1952 although it was not demonstrated until 1965 by ESR measurements carried out by Butyagin e.a.[12]

The formation of free radicals in polymers under mechanical action was first discovered in 1959 by Bresler e.a.[13] using ESR spectroscopy. Since then it has been established that mechanical treatment of polymers such as stretching, fatiguing, milling, grinding, sawing, and tearing causes chain scission and the formation of so-called mechano-radicals which lead to degradation and/or cross-linking of the polymers. Bond rupture is also often involved in environmental stress cracking, discoloration and embrittlement of polymers.[1]

When a stress is applied to a polymer material, degradation may occur either as a result of chain scission of over-stressed chains (mechano-degradation) or by activation of thermal oxidation,[14, 15] photo oxidation,[16, 17] or ozone oxidation.[18] Stress activation also plays a vital role in the chemo-relaxation of vulcanized rubber, which will be discussed in Section 14.6.

Strain-induced chain-scission and chain slippage are the dominating processes in the failure of oriented crystalline polymers. Fracture in this type of polymers is treated in Section 14.4 and by Kausch in Chapter 5.

However, failure of nonoriented uncross-linked polymers, whether they are crystalline or amorphous systems, is only to a minor extent caused by bond scission. Fracture in these polymers is instead primarily due to disentanglement of the polymer chains. This controversial subject will be dealt with in Section 14.5.

14.2 Methods of Studying Mechano-Degradation

14.2.1 Molecular Weight

Mechano-chemical reactions in polymers result in changes in the molecular topology, i.e., molecular weight distribution, degree of branching, and cross-linking. Earlier viscosity average (M_v) determinations of the molecular weight by intrinsic viscosity measurements have been largely replaced by gel permeation chromatography (GPC) techniques.

14.2.2 Electron Spin Resonance (ESR)

Free radicals are formed as a consequence of chain scission. By the ESR technique the free radicals can be counted and characterized. In 1964, Zhurkov and co-workers[19] were the first to apply ESR to the detection of macroradicals formed during tensile loading. The use of ESR to evaluate the concentration and nature of mechano-radicals has been extensively reviewed by e.g. Rånby and Rabek.[7] Assuming, first, that only two radicals are formed per bond scission and, second, that the radical decay is negligible, the kinetics of bond scission may be established.

The first assumption is generally agreed to be true by most researchers except Zhurkov and co-workers[20] and Zakrevskii and Korsukov,[21] who claim that up to a 1000 radicals (secondary bond scissions) may be formed for each primary scission. The second assumption is usually valid if the temperature is kept low enough and a normal experimental time-scale is used. Unavoidable radical decay may be corrected for if the decay kinetics are known.[22] More information about radical decay is given in Section 14.3. Trapping radicals has been used to avoid radical decay.[23] The lower concentration limit for the detection of radicals is normally 10^{12}–10^{13} radicals/g.

In Table 14.2.1 the number of radicals formed in the fracture of uniaxially stressed polymers is given. Uniaxially stressed nonoriented uncross-linked polymers produce fewer macroradicals than are detectable by ESR, with the exception of poly(phenylene oxide) in which long-lived radicals are formed.

Table 14.2.1. Number of radicals formed at fracture of uniaxially stressed polymers[a]

Polymer	Radicals/cm^3
Silk	7×10^{17}
PA-6	5×10^{17}
PE	5×10^{16}
PMMA	10^{14}

[a] Bundle of fibers in hexene at 243 K; ESR spectrum taken at liquid nitrogen temperature. Experimental data of Kausch.[9]

14.2.3 Infrared (IR) Spectroscopy

IR spectroscopy has been used to study:

— loading of chain segments before scission and
— kinetics of formation of new end groups.

In the first case, axial stress on chains has been shown by Zhurkov e.a.,[24] Statton,[25] and others to produce certain deformed and frequence-shifted absorption bands in the IR spectrum of the stressed polymer. Thus the stress-induced shift of the 975 cm^{-1} band of PP has been the subject of studies by several investigators.[26–31] The results have been compared by Kausch[9] (see Fig. 14.2.1).

In the second case, the formation of, e.g., aldehyde groups at 1735 cm^{-1} during stressing of polyolefines was studied by Zhurkov e.a.[30]

Recently, Wool[31] used this technique and identified a number of differences at the molecular level between stress relaxation and creep of highly oriented polypropylene film.

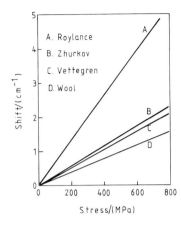

Fig. 14.2.1. Stress-induced shift of the peak of the 975 cm^{-1} absorption band of PP as found by different investigators.[9]

14.2.4 Mass Spectrometry

A time of flight mass spectrometer can readily detect the number of volatile degradation products. The technique has mainly been used by Regel e.a.[32] They observed a correlation between the amount of volatile products produced and the number of bond ruptures.

14.2.5 Small-Angle X-Ray Scattering (SAXS)

The formation of microvoids during stressing of a polymer can be detected by SAXS. Zhurkov e.a.[33] used this technique to characterize size and number of "submicrocracks" formed in uniaxially stressed samples.

More details and references on this subject can be found in reviews by Casale and Porter[1] and Kausch[9] as well as in Chapter 5 by Kausch in this volume.

14.3 Mechano-Radical Phenomenology

14.3.1 Stress-Induced Chain Scission

The application of a uniaxial stress to a macromolecular chain system will cause retractive entropy elastic forces, which are comparatively small, and energy elastic forces due to hindered rotation, bending, and stretching of bonds. The chains may release the axial stress by chain slip with respect to the surrounding matrix (enthalpy relaxation), change of chain conformation (entropy relaxation), or chain rupture through homolytic bond scission.

Chain scission may be looked upon as the result of homolytic bond cleavage of over-stressed chain segments, the potential binding energy state of which is enhanced by the action of stress.[34] The stress is thought to lower the thermal activation energy for bond dissociation and thus promotes a thermo-mechanical chain scission.[32]

14.3.2 Mechano-Radical Reactions

The mechanisms of propagation and termination of macroradicals formed by mechanical scission are essentially the same as for thermally or photo-initiated radicals.

The radical mechanism can be schematically represented as follows:

$$R\text{–}R \qquad\qquad \rightarrow R \cdot + R \cdot \qquad \text{(mechanical scission)}$$
$$R \cdot + R \cdot \qquad \rightarrow R\text{–}R \qquad \text{(recombination)}$$
$$R \cdot \quad R \cdot \qquad \rightarrow R + R \qquad \text{(disproportionation giving saturated}$$
$$\qquad\qquad\qquad\qquad\qquad\qquad\qquad\qquad + \text{ unsaturated species)}$$
$$R \cdot + R\text{–}R\text{–}R \rightarrow R + R\text{–}\dot{R}\text{–}R \text{ (transfer)}$$
$$R \cdot + A \qquad\quad \rightarrow R\text{–}A \qquad \text{(acceptor reaction)}$$

14.3.3 Transfer

Because of axial chain stresses, the new chain end radicals formed rapidly retract from each other and are thus prevented from recombination.[32] Primary radicals are then converted to secondary radicals by inter- and intramolecular hydrogen transfer (abstraction).

14.3.4 Recombination

The recombination reactions are responsible for the normal decay of radicals as observed by ESR.

Heat treatment of a fractured specimen accelerates the radical decay. This decay can be analytically described in terms of reaction rate theory, as pointed out by Naga-mura e.a.[35] so that the decay process can be described by a certain activation energy. In some cases, the decay of the radicals may be associated with thermal activation of the chain segments and segmental mobility. This has in fact been found to be the case, e.g., by Nagamura and Takayanagi[36] for the very stable primary mechano-radicals of poly(2,6-dimethyl-p-phenylene oxide), PPO, for which two regions of radical decay have been observed. The low temperature decay is assumed to be controlled by the gradual decay of peroxy radicals and is associated with the gamma relaxation.[36] The activation energy for the decay in the high temperature region has been calculated assuming a second order reaction for the process and is found to be about 18 kcal/mole.[36] This value and the temperature region of the decay are in good agreement with the corresponding parameters for the beta relaxation process of PPO, which has been assigned by Lim e.a.[37] to the combined reorientation motion of rings and connect-ing atoms of units in defect regions or in regions of poor chain packing. The decay regions are thus associated with the corresponding mechanical beta and gamma relaxa-tions.

The measurements on PPO mentioned above were made in vacuum on stretched films. However, if the heat treatment is carried out in the presence of oxygen and if the mechano-radicals are produced, e.g., by sawing the polymer at the temperature of liquid nitrogen, the decay shows anomalies.

Sohma and Sakaguchi[6] have summarized some of these results. If the relative intensi-ty of the ESR spectrum observed at 77 K is plotted versus the heat treatment tempera-ture, the curve normally shows a continuous decay of the radicals, because of low stability of the radicals which can survive only in a trapped state at low temperatures. It was however observed that the radicals produced by sawing PE and PP in liquid nitrogen in which oxygen was dissolved did not decay in this simple way. Instead the radical concentration increased during heat treatment in the temperature range 100 to 200 K as shown in Fig. 14.3.1. The radicals formed were of the peroxy type.

The important role of oxygen was confirmed by comparing the decay of the mech-ano-radicals produced in liquid nitrogen containing oxygen and radicals produced in a vacuum.

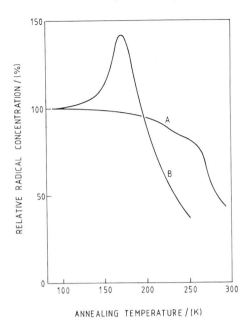

Fig. 14.3.1. Decay curve of PP radicals. The *A* and *B* curves correspond to radicals produced by gamma-irradiation and sawing, respectively.[6]

Furthermore, Sohma and Sakaguchi[6] found, by comparing the decay curves from heat treatment under different insulating conditions and by studying the effect of electron scavengers, that the excess electrical charge produced by friction during sawing also plays a vital role in the appearance of the anomalous decay. Thus, it was concluded that excess electron charge and the presence of oxygen cause an anomalous increase in the concentration of mechano-radicals during heat treatment.

14.3.5 Radical Trapping

Mechano-radicals are formed in the amorphous regions of semicrystalline polymers. The fold surfaces are effective radical trapping sites and may be used to follow the effect of quenching, annealing, and cold drawing on the micro-structure in these polymers.[38]

14.4 Fracture in Oriented Crystalline Polymers

Mechano-chemical phenomena are best studied in highly oriented crystalline polymers, primarily because the effects can be traced by the methods described in Section 5.2, since the fracture of fibrous polymers is connected with a great number of chain scissions.

Critical reviews on the subject have been given by Kausch,[9, 39] Peterlin,[40] and De Vries.[41] There has been some debate between these authors and the Leningrad group under the leadership of S.N. Zhurkov concerning the formation of mechano-radicals, especially in fibrous systems. The Leningrad group pioneered the work in this field [19−21, 24, 30, 33] and from their research some fundamental aspects of fracture in fibrous systems were established:

— A considerable amount of homolytic chain scission occurs in the stressed systems. The number of spins (radicals) is of the order of 10^{18} cm^{-3} at the moment of fracture.

— The different segments of the polymer are subjected to very different local stresses. The maximum stress on the primary bonds can be as much as 10 times higher than the average stress.

— So-called submicrocracks are formed in a stressed system. The dimension of the submicrocracks perpendicular to the axis of stress is about 200 Å and specific for the system studied. It is also independent of stress and strain. The concentration of submicrocracks increases gradually with time at constant load, reaching a limit close to failure.

Zhurkov e.a.[30] estimated the number of chain scissions on the basis of IR data. They obtained values, for the number of newly formed end groups, three orders of magnitude higher than the number of radicals observed by ESR.

This discrepancy was explained by the authors in terms of a chain radical reaction initiated by the mechanically induced chain scission. Later attempts to reproduce the results of the end group analysis of the Leningrad group have failed.[41, 42] The numbers of chain scissions obtained by IR-end group analysis on polyethylene fibers by Fanconi e.a.[42] were almost two orders of magnitude lower than those obtained by the Leningrad group. The number of bond scissions calculated from molecular weight determinations by GPC assuming a random scission model[43] are lower, by a factor of ten, than the value obtained from the IR data by Fanconi e.a.[42] Part of this discrepancy may arise from differences in the initial molecular weight of the PE samples studied, the physical state of the specimens, and the temperature of the tests.[42]

The extent of bond rupture as indicated by GPC,[43-45] assuming a random scission model, is larger by a factor of ten than the data obtained by ESR. If one assumes that the tendency for the larger molecules to rupture is greater than that of the smaller molecules, the agreement is somewhat better.[42]

Further discussion of fracture in fibrous systems requires the introduction of models to describe the fibrous structure. A great number of different models have been suggested.[46-50] For fibers of polyolefins and polyamides the microfibrillar model of Peterlin[46] shown in Fig. 14.4.1 has the best reputation.

The basic unit in this model is the microfibril, which is a few hundred Å wide and several orders of magnitude longer in the direction of the fiber axis. The microfibril consists of lamellar crystals, a few hundred Å thick or less, with amorphous interlayers between the crystal lamellae. The crystals are interconnected by a great number of almost taut interlamellar tie chains. The microfibrils are linked together by highly oriented interfibrillar tie chains forming higher order structures, the fibrils.

The submicrocracks explained by Zhurkov e.a.[33] on the basis of a radical chain reaction are of the same size as the microfibrils.[51] Peterlin[51] suggests that they originate from rupture of the microfibrillar structure at its weakest positions, which according to Peterlin are the ends of the microfibrils. At these positions very few tie chains connect the different microfibrils and large stresses act on the few tie chains present. However, Reimschuessel and Prevorsek[52] conclude from electron microscopy studies that the microfibrils form an essentially endless interwoven structure, where branching and fusion of microfibrils may be more characteristic than abrupt ends to the microfibrils. Kausch[39] agrees in principle with Peterlin that the fixed lateral dimension of the submicrocracks strongly indicates that they originate from the opening at the microfibrillar ends due to the retraction of the microfibrils.

It has been pointed out by Kausch[39] that the radical chain reaction suggested by the Leningrad group (see Fig. 14.4.2) is based on an extension of the generally

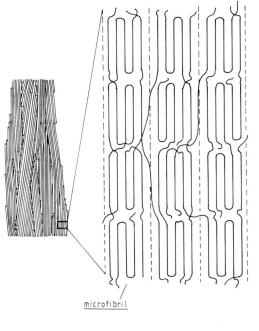

microfibril

Fig. 14.4.1. A schematical representation of the microfibrillar model of Peterlin.[46]

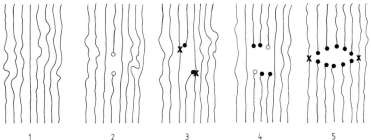

1   2   3   4   5

Fig. 14.4.2. Mechano-chemical reaction according to Zhurkov. (1) Highly stressed chain which rupture, (2) formation of chain end radicals, (3) radical reaction leading to main chain radicals, (4) scission of radicalized chain and repetition of step 3, (5) further repetition of the former steps leading to the formation of a submicrocrack; o, chain end radical; × main chain radical; ● stable end group.[33]

accepted view that a lowering of the activation energy for the thermally induced scission of a radicalized chain is valid also for a stress-induced scission of such a chain. According to Kausch[39] no termination is known that accounts for the invariant transverse dimensions of the submicrocracks. Kausch[39] also considers the influence of the microfibrils as stress concentrators to be weak and ineffective with regard to accelerating chain scission for two reasons: (1) the ends of the microfibrils are easily sheared and are thus free from large axial stresses; (2) if the submicrocracks acted as stress concentrators they would widen further laterally. This has not been observed.

In a step-strain experiment with PA-6 fiber, Kausch and Becht[53] observed that the radical concentration increased through each period of constant strain despite the stress relaxation. These data are shown in Chapter 5 (Fig. 5.2.3). The chain segments which rupture at the start of each period of constant strain cannot participate in the stress relaxation, since, if the axial segmental stress decreased as much as the macroscopic stress (a 10% reduction), there would be no chain rupture. It must also be concluded that the microfibrils do not unload through intramicrofibrillar slip, and that the lateral rigity of the crystal lamellae in the microfibrils must be large enough to allow the buildup of large stress concentrations. Furthermore, if intramicrofibrillar slip did occur, the microfibrils would disintegrate into submicrofibrils. This has not been observed. Kausch[39] suggests that the increased homogeneity of local segmental stresses gives rise to a dynamic instability at high stress levels, i.e., that a considerable proportion of the segments are loaded close to the breaking level, and that when some of them rupture the adjacent chain segments are subjected to a higher load and will fracture.

The above discussion concerns primarily PA-6 fibers. Other oriented polymers such as PE may behave in a similar manner, although there are important differences between these two polymers: PA-6 exhibits no crystalline relaxation, whereas PE has a pronounced crystalline relaxation.[54] The crystalline relaxation is associated with a certain translational mobility of the chains in the crystals. Aharoni and Sibilia[55] have studied the solid-state extrudability of a large number of polymers including PA-6 and PE. They found that polymers having a crystalline relaxation could be extruded in the solid state, as opposed to polymers lacking this ability. The rigidity of the crystals in PA-6 fibers manifested in the studies of Kausch and Becht[53] is in accordance with the findings of Aharoni and Sibilia.[55] However, SAXS data by Bonart and Hosemann[56] indicate rearrangements within the microfibrils accompanying the drawing of fibrous PE. The SAXS long period is almost constant (position of the SAXS maximum) but the sharpness of the meridional SAXS maximum rapidly decreases with increasing draw ratio. This change in the SAXS pattern may be caused by a reduced regularity of the alterations between crystalline and amorphous regions in the microfibrils and to a decreasing density difference between the two components in the microfibrils. Thus, the structure of the microfibrils of PE is not rigid and a certain amount of translational motion of chains in the crystals seems to occur.

These observations are further supported by the fact that fibrous PE can be drawn to very high draw ratios even at room temperature, provided that the molecular weight of the system is appropriate. Such a plastic deformation can only be accomplished by sliding motions of microfibrils and fibrils, which in turn require a translational mobility of the polymer chains through the crystal lamellae.

There are thus considerable experimental data supporting the idea of a translational mobility of polymer chains in the crystals of fibrous PE, which suggest that the buildup of large stresses in the tie segments can, to some extent, be reduced by the translational motion of the chains through the crystals in this polymer. It seems, in contrast to what has been observed in fibrous PA-6, that relaxation of highly stressed tie chains in fibrous PE systems may be accomplished by intra- and intermicrofibrillar slip.

14.5 Kinetic Failure Models

Eyring and co-workers,[57] considering the disappearance of secondary bonds, laid the foundation of fracture kinetics. Later Bueche[58] and Zhurkov and Tomashevskii,[59] considering instead the disappearance of primary bonds, independently arrived at the same type of expression

$$\log t_f = \log t_0 + (E_a - V\sigma)/RT \tag{14.5.1}$$

Here t_f is the time to failure, t_0 reciprocal molecular oscillation frequency, E_a activation energy for bond scission, V activation volume, σ stress, R the gas constant, and T the absolute temperature.

For different solid materials, Zhurkov was able to show that t_0 is approximately equal 10^{-12} s, i.e., is of the same order of magnitude as the thermal oscillation time for molecular bonds. For oriented semicrystalline polymers, Zhurkov obtained E_a-values of about 120 kJ/mole, which is of the same order of magnitude as the dissociation energy for a radicalized chain. These purely formal correspondances led Zhurkov to believe that chain scission is not only related to, but in fact controls, the failure of polymers by the three-stage fracture mechanism discussed in Section 14.4.

Kausch[9] has calculated E_a and t_0 values from failure plots for PVC pipes, and has found that E_a is about 3 times the dissociation energy of a radicalized chain and t_0 equals 10^{-52} s! If instead the expression in question is an ordinary Arrhenius relation describing a viscoelastic deformation process, the deviations will be natural. Since E_a is then a function of temperature, log t_0 is replaced by a temperature-dependent term responsible for extremely low values of t_0.[9]

In conclusion, there seems to be very little support for the opinion that chain scission controls failure processes of unoriented polymer materials. Instead, structural weakening of a material during loading is associated with disentanglement, chain slip, and void-opening processes.

It should, however, be emphasized that the mechano-chemical failure mechanism still receives considerable support in the Soviet Union. Recently a review was published by Smirnov and co-workers[60] which included over 100 references in favor of mechano-chemical breakdown of polymers. The increase in the lifetime of loaded polymer materials containing inhibitors and antioxidants is taken as support for the hypothesis. Since an abstract by two of the authors of this chapter was cited we will clarify this matter below.

14.5.1 Stabilization Against Mechano-Degradation of Strained Polymers

Attempts at mechano-stabilization have been made. Kuzminskii and Lyubchanskaya[61] found that phenyl-β-naphthylamine (PBNA) was consumed at a faster rate in stressed cross-linked rubber than in an unstressed one. Begunovskaya and co-workers[62] found that 2,4-diamino-difenylamine was a better stabilizer than PBNA against mechano-degradation of noncross-linked rubber milled at room temperature. However, PBNA was more effective as an antioxidant at 120° C. Potter and Scott[63] have shown that the stable galvinoxyl radical in natural rubber disappears at a faster rate when the rubber is stretched than when it is not. The authors conclude that galvinoxyl participates in a stress-activated reaction with the rubber as well as in a stress-induced chain scission. Jansson and Terselius[64] report a significant increase in the long-term strength, lifetime, and creep modulus of PVC containing 1–2% by weight of 2,6-ditert.butyl-p-cresol (DBPK). The lifetime was also increased to some extent by bisphenol and by galvinoxyl.

In a follow-up to these experiments, it was shown[65] by differential scanning calorimetry (DSC) that DBPK, despite being compounded in a roll mill at 160° C as a crystalline powder, acts as a plasticizer for PVC and is in fact just as T_g-depressing as dioctylphthalate. Although conclusive evidence is still lacking, we believe that *antiplastication* is the reason for the increased long-term modulus at least in the case of DBPK as an additive to PVC. In fact, we also suggest that the increase in the tensile strength of PVC reported by Voskresenskii and Shakirzyanova[66] for PVC containing small amounts of orthotolidine can be explained in a similar way.

14.6 Strain Dependence of Chemo-Relaxation

The pioneer work in this field was done by Tobolsky and co-workers[67] in the early forties. He observed that during chemical stress-relaxation experiments in the temperature range 100 to 150° C, cross-linked rubbers could become progressively softer or harder or, as with natural rubber, first softer then harder. It was deduced that at elevated temperatures cross-linking and chain scission occurred simultaneously, and that the rates of both cross-linking and chain scission were essentially unaffected by the elongation of the sample. It was assumed that new cross-links were formed in such a way that the network produced was in a relaxed state, the so called "two network hypothesis."[68] By performing both intermittent stress-relaxation measurements, which reflect the combined effect of chain scission and cross-linking, and continuous relaxation measurements, which to a first approximation reflect chain scission only, the two effects could be separated.

From continuous stress-relaxation measurements on sulphur-cured natural rubber in air at 100° C, Tobolsky and co-workers[67] found that the rate of chain scission was independent of elongation up to very high elongations. However, when the elongation exceeded 200% the chain scission process was accelerated. These results were confirmed much later by Murakami and Kusano[69] who found that during intermittent relaxation measurements at 88 and 100° C in air the chain scission process was independent of elongation up to 150%. At higher elongations, the process was accelerated. Bartenev and Lyalina[70] also reached this conclusion.

Further evidence for a strain-independent relaxation was given by Beatty and Juve.[71] They showed that unfilled and filled rubbers in compression did not change their relaxation rates when the deformation was increased from 20 to 60%.

Lyubchanskaya e. a.[72] reported that the relaxation rate in compression for sulphur-cured natural rubber at 90° C was independent of deformation in the range 20 to 50%, but increased somewhat when the deformation was reduced to 10%.

More recently, Tobolsky e. a.[73] reported that, at small elongations in continuous relaxation experiments in vacuum on cispolybutadiene at 230° C, the stress increased with time. In fact, at very small elongations (0.5%) the continuous relaxation curve approaches the intermittent relaxation curve, indicating an oxidative cross-linking.

When the elongation was stepwise increased up to 13%, the stress increase was gradually more delayed, a fact which was interpreted as indicating a gradual increase in chain scission at the expense of cross-linking. However, in air at the same temperature, the stress relaxation was very rapid and also independent of elongation in the range 1 to 5% elongation.

In opposition to the results of Tobolsky and co-workers,[73] Murakami and Tamura[74] found this remarkable stress increase during continuous stress-relaxation measurements at small elongations on cross-linked polyester at 190° C in air. An increase in the elongation from 1.8 to 6.3% reduced the stress increase. This was interpreted as an oxidative cross-linking decreasing with small increasing elongation.

In conclusion, at very high elongations, chain scission is accelerated by strain and, at moderate elongations, chain scission is independent of strain. At small elongations, in vacuum, chain scission is accelerated at the expense of cross-linking and in air both strain-independent chain scission and strain-activated chain scission at the expense of cross-linking have been reported.

14.7 Mechano-Degradation During Processing of Polymers

This subject has been extensively reviewed by Ceresa.[2] Baramboim,[4] and Porter e. a.[75] It has been established that the physical state of the polymer is crucial for the mechano-degradation.

In the glassy or crystalline state during comminution and vibro-milling, degradation is based on a mechanical action on the polymer material with subsequent chain scission in the peripheral regions and along the fracture surfaces of particles. In the viscoelastic (viscous) state during mastication, milling, and extrusion, degradation is based on a shearing action on the polymer material. Mechano-degradation also occurs in the gel state by swelling of the polymer in the vapor phase of the solvent. Finally, degradation occurs in solution during ultrasonic irradiation under cavitation conditions, during high-speed stirring or shaking, by forcing polymers through narrow orifices, during freezing and thawing, and during the discharge of high-voltage charges.

The mechano-degradation in the solid state proceeds by shearing action on the material. The shear-induced stresses in the polymer can only be relieved by disentanglements through relaxation or by chain scission, usually by homolytic bond cleavage.[2]

In the presence of radical acceptors such as oxygen and DPPH, the primary macroradicals are deactivated and stabilized by the acceptor.[2] In the case of oxygen, alkylperoxy radicals are formed. These are terminated by abstraction of hydrogen from adjacent molecules. In this way "permanented" molecular fragments cause degradation.

In the absence of acceptors, i.e., in an inert atmosphere, degradation is retarded due to the fact that the primary macroradicals recombine or attack adjacent main chains giving rise to chain branching and cross-linking. With synthetic dienes, gelation is a common phenomenon during mastication in nitrogen.[2] Resonance stabilization of the primary radicals, which occurs in e.g., natural rubber or polystyrene, also favors recombination.

The temperature dependence of degradation is of minor importance in the glassy or crystalline state.[75] In the viscoelastic state, on the other hand, the temperature has a twofold influence on degradation. As the temperature increases, the solubility of oxygen in polymers decreases. Hence, the deactivation of the primary radicals due to the presence of oxygen becomes less important at higher temperatures and the rate of degradation is reduced.[75]

In the presence of oxygen, the mechanism of degradation is at the same time gradually shifted toward strain-activated thermo-oxidative degradation. A temperature increase reduces the bulk viscosity of the polymer by increasing the segmental mobility.[2] Thus, the possibility of relieving internal stresses by stress relaxation is enhanced, and at a given shear rate less shear energy is imposed on the molecules.

For this reason a reduction in shearing effect also results from a decrease in the molecular weight of the polymer. Thus, the rate of degradation (molecular weight decrease) increases with original molecular weight. The rate of chain scission is, however, independent of molecular weight, as is also the lower molecular weight limit for degradation.

14.8 Mechano-Chemical Synthesis

It has also been possible to take advantage of mechano-degradation of polymers in the solid state or in solution. Thus, a mild controlled polymer degradation has proved to be useful for initiating graft and block copolymerization. Such a mechanochemical synthesis is a flexible method which can be used in processing in standard equipment with polymer blends, polymer-monomer blends, or polymer-filler systems. Excellent reviews on this topic have been presented by Ceresa[2] and by Casale and Porter.[1]

In the solid state, mechano-synthesis has been carried out by grinding or comminution, most frequently in vibro-mills. The main advantages are the almost unlimited number of combinations of different monomers and polymers and the possibility of carrying out the synthesis without a solvent. The disadvantages are the relatively high energy consumption and metal abrasion. Most of the research in this field has been conducted by Russian and East European scientists.[76-78] The composition and the properties of the products seem to be affected mainly by milling time and by the presence or absence of radical acceptors. The nature of the balls, the structure of the milling material, and the temperature have only minor effects on the product composition.

In the viscoelastic state, mastication is by far the most studied method of mechanosynthesis. As long as 50 years ago, Staudinger[10] showed that unvulcanized rubber could be plasticized by this method before compounding. Later, Watson[79] showed that mastication increased the adhesion between rubber and reinforcing fillers. Angier, Ceresa, and Watson conducted extensive research in the field of mechano-synthesis at the Natural Rubber Products Research Association about 20 years ago. This work has been summarized in a book by Ceresa.[2]

The product composition varies widely with chemical structure of the components and with temperature and efficiency of mastication. The atmosphere and the presence or absence of a radical acceptor have some influence on the composition. The critical factor for the degree of conversion is the bulk viscosity. The bulk viscosity determines the magnitude of shear stresses in the system. At the same time, the viscosity changes with the degree of conversion.

Since the discovery by Compagnon and Le Bras[80] in 1941 that a chemical reaction will occur if a mixture of a rubber and maleic anhydride is passed through a calender with a very narrow roll nip, an enormous amount of polymer modification has been carried out by means of strong mechanical action.

Thus, by graft/block copolymerization, brittle plastics have been made tougher by the incorporation of small amounts of elastomers, and rubbers have been reinforced by the addition of small amounts of glassy polymers. A great number of other combinations of properties have been achieved by mechano-chemical polymerization. A broad survey of polymers modified in this way has been presented by Lauer.[81]

According to Casale and Porter,[1] two-roll milling and coextrusion are the methods with the greatest commercial value, whereas vibro-milling is of purer scientific interest.

References

1. A. Casale and R.S. Porter, *Polymer Stress Reactions*. Academic Press, New York 1978, **1–2**.
2. R.J. Ceresa, *Block and Graft Copolymers,* p. 65, Butterworth London 1962.
3. W.F. Watson, Mechano-Chemical Reactions, in *Chemical Reactions of Polymers,* edited by E.M. Fetters, p. 1085; Wiley, New York 1964.
4. N.K. Baramboim, in *Mechanochemistry of Polymers,* edited by W.F. Watson, Maclaren, London 1964.
5. C. Simionesco and C. Vasiliu-Oprea, *Mechanochemia Compusilor Macromoleculari,* Academie Republicii Socialiste, Romania, Bucuresti 1967.
6. J. Sohma and M. Sakaguchi, *Adv. Polym. Sci.* 1976, **20**.
7. B. Rånby and J.F. Rabek, *ESR Spectroscopy in Polymer Research,* Springer, Berlin 1977.
8. E.H. Andrews and P.E. Reed, *Adv. Polym. Sci.* 1978, **27**, 3.
9. H.H. Kausch, *Polymer Fracture,* 2nd ed., Springer, Heidelberg-New York 1986.
10. H. Staudinger, *Kautschuk* 1929, **5**, 128.
11. M. Pike and W.F. Watson, *J. Polymer Sci.* 1952, **9**, 229.
12. P.Yu. Butyagin, V.F. Drozdovskii, D.R. Razgon and I.V. Kolbanov, *Fiz. Tverd. Tela* 1965, **7**, 941.
13. S.E. Bresler, S.N. Zhurkov, E.N. Kazbekov, E.M. Saminskii and F.E. Tomashevskii, *Zh. Tekh. Fiz.* 1959, **29**, 358.

14. G.L. Slonimskii, V.A. Kargin, G.N. Buiko, E.V. Retzova, and M.L. Yuis-Riera, *Starenie i Utomlenie VNITO Rezinshchikov Konf.* 1953, 100.
15. A.S. Kuzminskii, M.G. Maizels and N.N. Lezhnev, *Dokl. Akad. Nauk SSSR* 1950, 319.
16. F.S. Kaufmann, Jr., "A New Technique for Evaluating Outdoor Weathering Properties of High Density Polyethylene", in *Applied Polymer Symposium*, edited by M.R. Kamal, vol. 4, p. 131, Wiley-Interscience, New York 1967.
17. G.G. Samoilov and E.E. Tomashevskii, *Sov. Phys.-Solid State* 1968, **10**, 866.
18. E.H. Andrews, D. Barnard, M. Braden and A.N. Gent, in *The Chemistry and Physics of Rubber-like Substances*, Chapter 12 edited by L. Bateman. Maclaren, London 1963.
19. S.N. Zhurkov, A.Y. Savostin and E.E. Tomashevskii, *Dokl. Akad. Nauk SSR* 1964, **159**, 303.
20. S.N. Zhurkov, V.A. Zakrevskii, V.E. Korsukov and V.S. Kusenko, *J. Polymer Sci. A-2* 1972, **10**, 1509.
21. V.A. Zakrevskii and V.E. Korsukov, *Vysokomol. Soedin,* 1972, 955; English transl., *Polymer Sci. USSR* 1972, **14**, 1064.
22. K.L. DeVries, D.K. Roylance and M.L. Williams, *J. Polymer Sci. A-1* 1970, **8**, 237.
23. J.S. Ham, M.K. Davis and J.H. Song, *J. Polymer Sci. Phys.* 1973, **11**, 217.
24. S.N. Zhurkov, Y.I. Vettegren, V.E. Korsukov and I.I. Novak, *Proceedings Second International Conference Fracture,* Chapman A. Hall, Brighton, England 1969; p. 545.
25. W.O. Statton, Paper presented at IUPAC Symp., Rio de Janeiro, July 1974.
26. V.I. Vettegren and I.I. Novak, *J. Polymer Sci. Phys.* 1973, **11**, 2135.
27. V.I. Vettegren and I.I. Novak, *Fiz. Tverd. Tela* 1973, **15**, 1417; Engl. Transl., *Sov. Phys.-Solid State* 1973, **15/5**, 957.
28. D.K. Roylance and K.L. DeVries, *Polymer Lett.* 1971, **9**, 443.
29. S.N. Zhurkov, V.I. Vettegren, V.E. Korsukov and I.I. Novak, *Sov. Phys.-Solid State* 1969, **11**, 2133.
30. S.N. Zhurkov, V.I. Vettegren, V.E. Korsukov and I.I. Novak, *Intern. J. Fracture* 1969, **4**, 47.
31. R.P. Wool, *J. Polymer Sci.* 1975, **13**, 1795.
32. V.R. Regel, T.M. Muinov and O.F. Pozdnyakov, *Proc. Phys. Basic Yield Fracture,* p. 194, Oxford 1956.
33. S.N. Zhurkov, V.I. Vettegren, V.E. Korsukov and I.I. Novak, *Proc. 2nd Int. Congr. Fracture,* p. 47, Brighton 1969.
34. H.H. Kausch, *J. Polymer Sci. Symp.* 1971, **32**, 1.
35. T. Nagamura, N. Kusomoto and M. Takayanagi, *J. Polymer Sci. Phys.* 1973, **11**, 2357.
36. T. Nagamura and M. Takayanagi, *J. Polymer Sci. Phys.* 1975, **13**, 567.
37. T. Lim, V. Frosini, V. Zaleckas, D. Morrow and J.A. Sauer, *Polymer Eng. Sci.* 1973, **13**, 51.
38. V.R. Regel, A.V. Amelin, O.F. Pozdnyakov and T.P. Sanfirova, *IUPAC, Helsinki,* 1972, Preprint IV-29, 163.
39. H.H. Kausch, *Polymer Eng. Sci.* 1979, **19**, 140.
40. A. Peterlin, *J. Macromol. Sci. Phys.* 1981, **B 19**, 401.
41. K.L. DeVries, *J. Appl. Polymer Sci. Symp.* 1979, 439.
42. B.M. Fanconi, K.L. DeVries and R.H. Smith, *Polymer* 1982, **23**, 1027.
43. T.M. Stoechel, J. Blasius and B. Crist, *J. Polymer Sci., Phys.* 1978, **16**, 485.
44. L.N. Shen, *J. Polymer Sci. Lett.* 1977, **15**, 615.
45. R.E. Mehta, appendix to paper by K.L. DeVries and M.L. Williams, *J. Macromol. Sci.* 1973, **B 8**, 691.
46. A. Peterlin, in *Ultra-High Modulus Poymers,* edited by A. Ciferri and I.M. Ward, p. 279, *Applied Science Publishers,* London 1979.
47. D.C. Prevorsek, Y.D. Known and R.K. Sharma, *J. Mater. Sci.* 1977, **12**, 2310.
48. P.F. Dismore and W.O. Statton, *J. Polymer Sci. Symp.* 1966, **13**, 133.
49. P.J. Barkham and R.G.C. Arridge, *J. Mater. Sci.* 1976, **11**, 27.
50. E.W. Fischer and H. Goddar, *J. Polymer Sci. Symp.* 1969, **16**, 4405.

51. A. Peterlin, *J. Macromol. Sci.* 1981, **B 19**, 401.
52. A.C. Reimschuessel and D.C. Prevorsek, *J. Polymer Sci., Phys.* 1976, **14**, 485.
53. H.H. Kausch and J. Becht, *Rheol. Acta* 1970, **9**, 137.
54. N.G. McCrum, B.E. Read and G. Williams, *Anelastic and Dielectric Effects in Polymeric Solids,* Wiley, London 1967.
55. S.M. Aharoni and J.P. Sibilia, *Polymer Eng. Sci.* 1979, **19**, 450.
56. R. Bonart and R. Hosemann, *Kolloid Z. Z. Polymere* 1962, **186**, 16.
57. S. Glasstone, K.J. Laidler and H. Eyring, *The Theory of Rate Processes,* McGraw Hill, New York 1941.
58. F. Bueche, *J. Appl. Phys.* 1941, **29**, 1231.
59. S.N. Zhurkov and E.E. Tomashevskii, *Physical Basis of Yield and Fracture,* Inst. of Physics, London 1966.
60. E.V. Deyun, G.B. Manelis, E.B. Polianchik and L.P. Smirnov, *Usp. Khim.* 1980, **49**, 1574; Engl. transl., *Russ. Chem. Revs.* 1980, **49**, 759.
61. A.S. Kuzminskii, and L.I. Lyubchanskaya, *Rubber Chem. Technol.* 1956, **29**, 770.
62. L.M. Begunovskaya, V.G. Zhakova, B.K. Karmin and V.G. Epshtein, *Starenie i Utomlenie* 1953, 31.
63. W. Potter and G. Scott, *Europ. Polymer J.* 1971, **7**, 489.
64. J.-F. Jansson and B. Terselius, *J. Appl. Polymer Sci. Symp.* 1979, **35**, 455.
65. B. Terselius and J.-F. Jansson, unpublished results.
66. V.A. Voskresenskii and S.F. Shakirzyanova, *Zh. Prikl. Khim.* 1962, **35**, 1145.
67. A.V. Tobolsky, I.B. Prettyman and J.H. Dillon, *J. Appl. Phys.* 1944, **15**, 309.
68. R.D. Andrews, A.V. Tobolsky and E.E. Hansson, *J. Appl. Phys. 1946,* **17**, 352.
69. K. Murakami and T. Kusano, *Rheol. Acta* 1974, **13**, 127.
70. G.M. Bartenev and N.M. Lyalina, *Vysokomol. Soed.* 1970, **A 12**, 368.
71. J.R. Beatty and A.E. Juve, *India Rubber World* 1950, **121**, 537.
72. L.I. Lyubchanskaya, L.S. Feldshtein and A.S. Kuzminskii, *Sov. Rubber Technol.* 1962, **21**, 20.
73. A.V. Tobolsky, Y. Takahashi and S. Naganuma, *Polymer J.* 1972, **3**, 60.
74. K. Murakami and S. Tamura, *J. Polymer Sci. Lett.* 1972, **10**, 941.
75. A. Casale, R.S. Porter and J.F. Johnson, *Rubber Chem. Technol.* 1971, **44**, 534.
76. N.K. Baramboim, *Mechanochemistry of Polymers,* edited by W.F. Watson, *Maclaren,* London 1964.
77. H. Grohn and K. Bischof, *Plaste Kautschuk* 1961, **8**, 311.
78. C. Simionescu, C. Vasilu-Oprea and C. Neguleanu, *Makromol. Chem.* 1971, **148**, 155.
79. W.F. Watson, *Proc. 3rd Rubber Technol. Conf.* 1954, 553.
80. P. Compagnon and J. Le Bras. C. r. *Acad. Sci. Paris,* 1941, **212**, 616.
81. W. Lauer, *Kautschuk, Gummi und Kunststoffe* 1975, **28**, 536.

15 Mechanical Behavior in Gaseous Environments

Norman Brown

Department of Materials Science and Engineering, University of Pennsylvania
Philadelphia, PA 19104, U.S.A.

15.1 Solubility of Gases in Polymers

The fundamental reason why gases affect the mechanical behavior of polymers is that gases plasticize just as organic liquids do. Gases N_2, Ar, CO_2, and CH_3, for example, dissolve in the polymer and weaken the van der Waals bond. The magnitude of the effect depends primarily on the concentration of the gas that is dissolved in the polymer.

The concentration C_g of gases in polymers is presented by Hildebrand and Scott[1] in their book based on the theory of Flory and Huggins. However, the following simpler equation is presented since it agrees with experimental data of Van Amerongen[2] and still contains the essential features of the theoretical Flory-Huggins equation.

$$C_g = A_p P \exp[-H_v/R(1/T_B - 1/T)] \tag{15.1.1}$$

where P is the pressure in atmospheres (1 atm $= 1.013 \times 10^5$ N/m^2), T the temperature, H_v the heat of vaporization, T_B the boiling point of the gas, and A_p depends on the polymer. In the case of a crystalline polymer, A_p is proportional to the volume fraction of the amorphous region.

$$H_v = T_B S_v \tag{15.1.2}$$

where S_v is the entropy of vaporization and $S_v = 67$–71 J mol^{-1} K^{-1} for practically all gases. Combining Eqs. (15.1.1) and (15.1.2) the concentration of a gas in a polymer is well represented by the equation

$$C_g = A_p P \exp[-8.3(1 - T_B/T)] \tag{15.1.3}$$

Data from Stannett[3] for polymers like natural rubber and branched polyethylene give

$$C_g \approx 10^{-2} P \exp[8.3\, T_B/T] \tag{15.1.4}$$

in units of cm^3 of gas at STP per cm^3 of polymer and P is in atmospheres.

The effects of the dissolved gas on the mechanical behavior can be divided into two categories (1) solution throughout the bulk and (2) solution in a thin surface layer. The former produces bulk plasticization and the latter usually produces crazes especially at low temperatures.

15.2 Bulk Plasticization

In order to dissolve a gas throughout a material in a reasonable time, it is necessary for the diffusion coefficient to be sufficiently large. Thus, a temperature is required which is usually high compared to the boiling point of the gas. In accordance with Eq. (15.1.4) the pressure must be high in order to obtain a significant solubility at

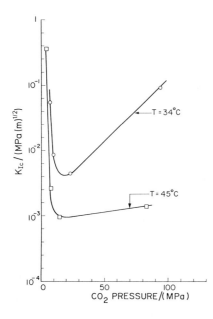

Fig. 15.2.1. Fracture toughness versus pressure of CO_2 in PS (ref. 7).

the elevated temperature. For example, in order to produce the same solubility of N_2 at 298 K as at one atmosphere and 10° C above its boiling point, a pressure of 182 atmospheres is required.

The pressure has two effects (1) it increases the solubility of the gas and (2) it exerts a hydrostatic stress on the polymer. These two effects counteract each other. The increased solubility plasticizes the material, and thereby reduces its modulus, yield stress, and T_g, whereas the hydrostatic pressure does the opposite. The solubility effect dominates at low pressure and the stress effect at high. Thus, the modulus and T_g as a function of pressure go through a minimum. The minimum value is also a function of temperature as expected from Eq. (15.1.3) since the modulus and T_g depend on the amount of dissolved gas. The yield point will follow the modulus, since in general, the yield point of polymers depends directly on the value of the elastic modulus as emphasized by Brown.[4, 5]

Wang and Kramer[6] measured the effects of high pressure CO_2 on the glass-transition temperature and mechanical properties of polystyrene. As an example[7] of the general effect, Fig. 15.2.1 shows how fracture toughness varies with CO_2 pressure at two temperatures. Young's modulus, T_g, and stress to grow a craze show the same pressure dependence as K_{Ic}. From zero pressure to the pressure where fracture toughness is a minimum, the plasticizing effect of the pressure dominates. At pressures beyond the minimum, the effect that dominates is the reduction in free volume. The effects of pressure in the absence of an environmental effect are presented in Chapter 6.

15.3 Foam

A polymer foam can be produced by dissolving gas in the polymer under a high pressure and temperature and then suddenly releasing the pressure. The dissolved gas then expands and produces foam. If the polymer is sufficiently cross-linked or held together by the random entanglement of the molecules, a continuous mass of foam will be produced; otherwise a fluff or powder will result.

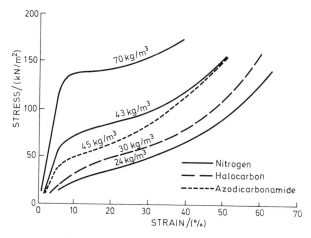

Fig. 15.3.1. Stress-strain curves of PE foamed by N_2 for various densities (ref. 8).

As described by Webster,[8] polyethylene foam is produced by exposing the cross-linked polymer to a temperature of about 150° C and at least 28 MPa of N_2. If about 3.8 Wt.% of N_2 is dissolved then the resulting foam will have a density of 32 kg/m^3. Thus, the final density depends on the amount of dissolved gas controlled by the temperature and pressure as predicted by Eq. (15.1.4). The size of the pores is about 0.2 mn. The stress-strain curves of a PE foam for various foam densities are shown in Fig. 15.3.1.

15.4   Crazing in Gaseous Environments

If a polymer such as PS crazes in an inert environment (intrinsic crazing), then gas that dissolves through the bulk of the polymer will change the nature of the crazing processes as will any other solvated molecule. This behavior has been investigated by Kambour e.a.[9] who showed that the effect of temperature on intrinsic crazing is a function of the departure from T_g and that solvated molecules which change T_g shows an analogous effect on the critical strain to initiate crazing.

Wang and Kramer[7] showed that the CO_2 under high pressure at high temperature, 307 to 318 K, modified the crazing behavior of PC relative to its intrinsic behavior and relative to its crazing behavior in hexane. The difference in behavior between the CO_2 and the liquid hexane environments stems primarily from the fact that the hexane did not diffuse as deeply into the polymer as the CO_2 and thus its plasticizing effect on craze thickening was different.

At low temperatures, every polymer that has been tested in a gas such as N_2, exhibits a change in its mechanical behavior compared to its behavior in a vacuum. This change is usually manifested by the production of crazing. Even in those cases where crazes were not observed, a change in mechanical behavior occurred in the gaseous environment as compared to a vacuum. At these low temperatures, no change in a bulk property such as the elastic modulus is observed since the diffusion coefficient is too low for the gas to penetrate very far into the polymer during the lifetime of the usual experiment. The effect of the gas is confined to a thin surface layer. Since crazes nucleate at the specimen surface of a single phase polymer, the solution of

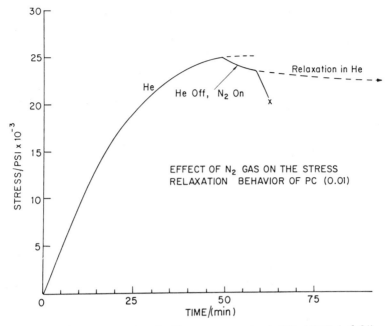

Fig. 15.4.1. Showing the time for N_2 to cause crazing in PC at 78 K (ref. 21).

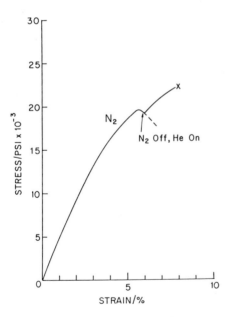

Fig. 15.4.2. Effect of switching from N_2 to HE PC $-$ 78 K, 0.01 min^{-1} (ref. 21).

gas in a thin surface layer is all that is needed to start crazing. As the craze grows the gas rapidly permeates the porous craze and needs only to plasticize a thin layer at the internal surfaces in order to promote craze growth.

At a low temperature of 77 K, where N_2 always has had an effect on all polymers that have been investigated, the diffusion coefficient is extremely low: $D = 10^{-24} - 10^{-27}$

$cm^2/s.$[3] The depth of penetration of the gas is given by $\delta \approx \sqrt{Dt}$. The effect of N_2 has been observed within a few minutes after exposure or removal of the gas at 77 K as shown in Figs. 15.4.1 and 15.4.2. Thus, for $t = 10^2$ s, $\delta \approx 10^{-12}$ cm; it appears that the gas causes crazing without penetrating the polymer to a depth greater than the diameter of the molecule. However, Olf and Peterlin[10] have pointed out that the diffusivity associated with crazing is determined by the local strain at the point where the craze is nucleated and by the strain at the internal interfaces between the craze and the bulk polymer. At these sites there are large localized strains which greatly enhance the diffusion coefficient relative to the value in the unperturbed bulk. Experimental results which indicate that the local diffusion coefficient of N_2 at a craze at 77 K can be about 10^{-9}–10^{-4} cm^2/s will be presented later. These values do not represent the permeation through the pores of the craze, which is extremely rapid, but the local diffusion coefficient at highly strained regions.

15.5 Effect of Boiling Point on Stress-Strain Curves

When evaluating the effect of a gas on the stress-strain behavior, it is useful to use the behavior in a vacuum as the reference state. For temperatures of 77 K and above it is experimentally more convenient to use a He environment rather than a vacuum. At 77 K and above, the solubility of He is negligible in accordance with Eq. (15.1.3) because T_B (He) = 4.2 K. Experiments by Brown and Parrish[11] have verified this point. H_2 ($T_B = 14$ K) has been found to produce a very small amount of crazing at 77 K in PS.[12]

A typical set of stress-strain curves in N_2 over a range of partial pressures at 77 K for polychlorotrifluoroethylene (PCTFE) is shown in Fig. 15.5.1. The effect of a partial pressures of N_2 less than about (0.02–0.05) atm at 77 K were not measurable. The behavior of Ar is similar.

The effect of temperature on the intrinsic stress-strain behavior (in He) is shown in Fig. 15.5.2a. For comparison, the stress-strain behavior in N_2 is shown in Fig. 15.5.2b. There is a temperature above which the gas ceases to have a measurable

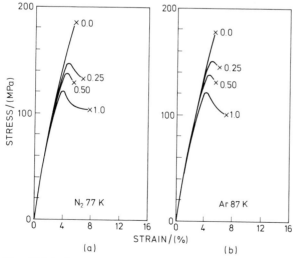

Fig. 15.5.1. Stress-strain curves at 77 for PCTFE at various partial pressures of N_2 and Ar (ref. 13).

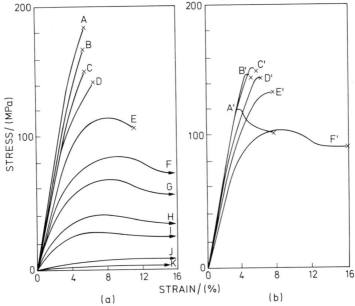

Fig. 15.5.2. Intrinsic stress-strain curves of PCTFE at various temperatures. (a) In He at 1 atmo-sphere *A*, 77; *B*, 118; *C*, 150; *D*, 169; *E*, 192; *F*, 220; and *G*, 240 K. (b) PCTFE in N_2 at 1 atm. *A'*, 77; *B'*, 99; *C'*, 138; *D'*, 160; *E'*, 179; and *F'*, 200 K (ref. 13).

effect. In the particular case of N_2 in PCTFE, no evidence of crazing is found above about 150 K at $P=1$ atmosphere. Since the crazing stems from the amount of gas that is dissolved in the polymer in accordance with Eq. (15.1.3), the critical temperature, T_{max}, above which no effect is observed increases with the pressure. Another factor that determines T_{max} is the minimum relative pressure that produces a measurable change in a mechanical property. The relative pressure (P_r), is defined as the pressure divided by the vapor pressure in equilibrium with the liquid. The minimum relative pressure, P_r (min) below which a change in the mechanical is not measurable has been determined. It depends on the sensitivity of the measuring technique and the sensitivity of the mechanical property of the particular gas-polymer combination to the concentra-tion of dissolved gas. Experimentally it has been observed for various gas-polymer combinations that, P_r (min) is 0.02–0.05. By combining Eq. (15.1.3) and the definition of P_r(min)

$$T_{max} = [1 + (1/8.3) \ln (P_r(min)]^{-1} T_B \qquad (15.5.1)$$

or, using the experimental value P_r (min) given above

$$T(max) = (1.9 - 1.6) T_B \qquad (15.5.2)$$

Equation (15.5.2) agrees with the experimental observations. For example, for PMMA, Fig. 15.5.3 shows the effects of various gases on its ultimate strength as a function of temperature. Thus, the higher the boiling point of a gas the higher the temperature range over which it is effective. At the boiling point there is no difference in behavior between immersion in the vapor or in the liquid; both have the same effect.

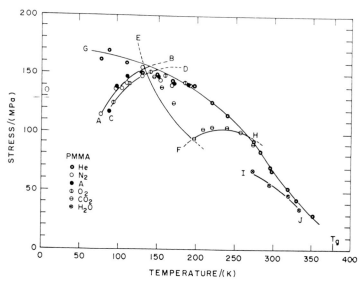

Fig. 15.5.3. Effect of temperature on tensile strength of PMMA in various environments for $P = 1$ atm. GH, He; AB, N_2; CD, O_2; FH, CO_2 gas; FE, CO_2 vapor in equilibrium with solid; and IJ, H_2O (ref. 14).

15.6 Effect of Strain Rate

According to the Eyring equation, the effect of strain rate on the *tensile strength*, σ_u, is given by

$$d\sigma_u/d\ln\dot{\varepsilon} = kT/V \tag{15.6.1}$$

where V is the activation volume. This relationship says that the sensitivity of strength on strain rate should decrease as T decreases. The yield strength becomes less sensitive to strain rate as T decreases. Also when brittle fracture occurs as it generally does at low temperatures, the strain-rate sensitivity becomes low. The reason why brittle-fracture strength is generally less sensitive to $\dot{\varepsilon}$ than shear-yield strength is based on the fact that brittle fracture is dominated by the concentrated stress that is produced at a surface defect, whereas shear yielding in polymers is relatively insensitive to points of stress concentration. The production of crazes by a gaseous environment introduces a new factor into the effect of strain rate.

Gaseous environments cause crazing by locally plasticizing the polymer at the points of stress concentration where the crazes are initiated. The degree to which the gas can diffuse into a thin surface layer and thereby plasticize it, determines the ultimate strength of the specimen. Since the strain rate controls the time during which the stress acts, it also controls the time during which diffusion can take place and thus the depth of penetration of the gas.

Figure 15.6.1 shows the stress-strain curves for PC in N_2 at 77 K at various strain rates compared to the curve in He. The curve in He changes only slightly with strain rate. Figure 15.6.2a shows σ_u versus log $\dot{\varepsilon}$ for PC in He and N_2 for tension and compression at 77 K. In comparing tension and compression in He where no crazing occurs it is noted that the compressive yield point is more sensitive to $\dot{\varepsilon}$ than for tension because PC is comparatively brittle in tension at 77 K. Figure 15.6.2b shows the behav-

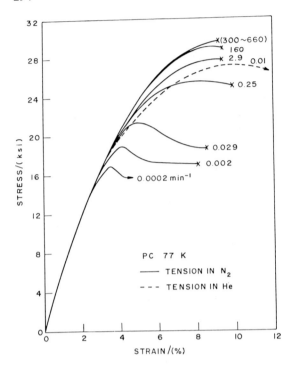

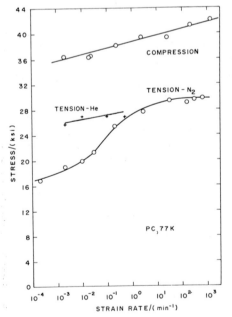

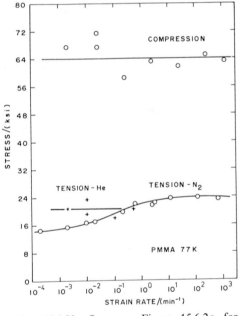

Fig. 15.6.1. Stress-strain curves in tension for PC at various strain rates (ref. 15).

Fig. 15.6.2a. Ultimate stress versus log strain rate for PC in N_2 and He. Compression strength taken at 5% offset strain (ref. 15).

Fig. 15.6.2b. Same as Figure 15.6.2a for PMMA (ref. 15).

ior of PMMA which is brittle both in tension and compression and thus $d\sigma_u/d\ln\dot{\varepsilon}$ is negligible.

The effect of N_2 on $d\sigma_u/d\ln\dot{\varepsilon}$ is very profound as shown in Fig. 15.6.2 (a and b), because the gas produces crazing. The greatest effect is at low-strain rates where the gas has time to penetrate the polymer. At very high-strain rates $d\sigma_u/d\ln\dot{\varepsilon}$ approaches the value that is observed in He. When the strain rate is sufficiently high no crazes are observed in N_2. The critical strain rate at which the crazes disappear was used by Imai and Brown[15] to calculate the effective diffusion coefficient of the gas. The effective diffusion coefficient is based on the stress-enhanced diffusion as proposed by Olf and Peterlin.[10] It was determined that the stress-enhanced-diffusion coefficient was found to be 10^{-14} to 10^{-9} cm^2/s compared to the zero stress values of (10^{-24} to 10^{-27}) cm^2/s.

It is also to be noted in Fig. 15.6.2b that σ_u at high-strain rates in PMMA in N_2 exceeds the value in He. The stress enhancement effect in N_2 will be presented in Section 15.11. It is sufficient to say at this time that the effect comes from a blunting of the points of stress concentration without producing enough crazing to cause a general reduction in σ_u.

15.7 Sensitivity of Various Polymers to N_2

It is interesting to compare the sensitivity of different polymers to the same gaseous environment. The sensitivity may be described by the change in ultimate strength relative to its intrinsic value in a vacuum as given by the following parameter: $(\sigma_{in}-\sigma_g)/\sigma_{in}$ where σ_{in} is the intrinsic value of ultimate strength and σ_g is the value in the gas. Since σ_g strongly depends on strain rate, we must compare the sensitivities at the same strain rate. Table 15.7.1 gives the sensitivity parameters of polymers in N_2 at 77 K. For all the polymers the average value is 0.29 ± 0.10 at a strain rate of 2×10^{-3} min^{-1}. It is noted that the various polyethylenes all have an average value of 0.17 ± 0.04 compared to an average value of 0.35 ± 0.05 for all the other polymers. PTFE which is also semicrystalline is more sensitive than PE to N_2. Differences in crystallinity cannot explain the lower sensitivity of PE. Both the PE and PTFE results indicate that increasing the crystallinity tends to increase the sensitivity to N_2 which also means an increase in the propensity to crazing.

The sensitivity of the polymer also depends on the gas. Work with PMMA in various gases showed that H_2S produces the greatest sensitivity. Sufficient work has not been done to generalize the relative effects of different gases on the sensitivity parameter. Probably the sensitivity of a polymer to a gas depends on the degree of plasticization per gas molecule.

15.8 Quantitative Dependence of Ultimate Strength on the Concentration of the Gas

The fractional change in ultimate strength depends on the concentration of sorbed gas and therefore on the relative pressure as has been determined for several polymers. In particular for polystyrene and polycarbonate for pressures of 0–1 atmospheres[12]

$$(\sigma_{in}-\sigma_g)/\sigma_{in}=((\sigma_{in}-\sigma_g)/\sigma_{in})\,(P_r=1)\,\{P\exp[-8.3(1-T_B/T)]\}^{1/2} \qquad (15.8.1)$$

where P is the pressure in atmospheres, and $((\sigma_{in}-\sigma_g)/\sigma_{in})\,(P_r=1)$ is the sensitivity as given in Table 15.7.1 and depends on the strain rate. If the concentration of sorbed

Table 15.7.1. Sensitivity of various polymers to N_2 at 77 K at $P=1$ atmosphere

Material	$[(\sigma_{in}-\sigma_g)/\sigma_{in}]$	Strain Rate/(min^{-1})	
PE Marlex 6006 [16]	0.19	0.002	
	0.15	0.02	
PE 3406[a][16]	0.16	0.002	
	0.16	0.02	
PE 3408[b][16]	0.22	0.002	
	0.17	0.02	
LDPE[16]	0.12	0.002	
	0.06	0.02	
PCTFE[13]	0.36	0.002	
	0.35	0.02	
PS[12]	0.43	0.001	
	0.26	0.1	
PMMA[14]	0.31	0.002	
	<0	0.5	
PC	0.30	0.002	
	0	2.0	
PET (oriented)[11]	0.37	0.0008	(stress perpendicular to direction of
	0.30	0.008	orientation)
	0.13	0.8	
Polysulphone	0.35	0.002	
PTFE[17]	0.26	0.004	49% Crystallinity
	0.06	0.4	
	0.28	0.004	59% Crystallinity
	0.14	0.4	
	0.30	0.004	66% Crystallinity
	0.07	0.4	
PP[18]	0.44	0.05	
PET[19]	0.33	0.01	

[a] Ethylene-Butene-Octene Copolymer
[b] Ethylene-Hexene-Copolymer

gas is given by Eq. (15.1.4) then the dependence of the fractional change of ultimate strength is given by

$$(\sigma_{in}-\sigma_g)/\sigma_{in}=BC_g^{1/2} \qquad (15.8.2)$$

where B depends on the sensitivity of the polymer as given in Table 15.7.1. For polystyrene, $B=0.07$. From Eq. (15.8.2) we can estimate the concentration of gas to produce a certain fractional amount of weakening of the ultimate strength. For polystyrene, if the fractional amount of weakening is 0.1 then $C_g=2$ cm^3 (STP) per cm^3 of polymer. Taking the density of PS$=1.05$ gm/cm^3, it is estimated that 0.01 molecules of N_2 per monomer of PS reduces the intrinsic ultimate strength by 10%. Equation (15.8.2) also indicates that the weakening effect per N_2 molecule decreases with increasing concentration of gas.

15.9   Stress-Strain Curves versus Temperature of Gas

Whereas the ultimate strength decreases monotonically with increasing pressure, it goes through a maximum value as a function of temperature and at constant pressure of the environmental gas. The behavior is shown in Figs. 15.5.2 and 15.5.3. The minimum value of the ultimate strength occurs at T_B where C_g is greatest. As the temperature is increased from above T_B, C_g decreases and the weakening effect of the gas decreases. At the same time, as the temperature increases the ultimate strength decreases because the intrinsic strength of a polymer always decreases with increasing temperature. Thus, the ultimate strength goes through a maximum value at a temperature above T_B.

It is also interesting to note that as the temperature is decreased below the boiling point, the ultimate strength increases. This is related to several factors (1) as the temperature is decreased the intrinsic strength increases, (2) the solubility of the liquid decreases with decreasing temperature, and (3) the diffusion coefficient decreases with decreasing temperature.

It is interesting to note that in the presence of the vapor which is in equilibrium with its solid there is an effect. Hiltner, Kastelic and Baer[19] showed the effect of solid N_2 at 50 K on PET, Olf and Peterlin[10] the effect of solid N_2 at 63 K on PP, and Imai and Brown[14] the effect of solid CO_2 on PMMA. Ultimately, there is a temperature below which the vapor of the solid has no effect because the diffusion coefficient is too low.

It is interesting to speculate on the mechanism by which solid N_2 at 50 K can produce crazing and thereby lower the ultimate strength. It is generally thought that environmental crazing is produced by the solution of the environmental agent in a thin layer of material at the surface or within the internal surface between the craze and the bulk. This absorption requires diffusion within the time of the experiment. It is interesting to speculate whether the diffusion coefficient of N_2 in a polymer at 50 K is sufficiently high for the N_2 to penetrate much beyond one molecular diameter. As pointed out previously, Olf and Peterlin[10] have suggested that the local diffusion coefficient in the highly strained region of the craze is what matters. In the absence of experimental data, it is an open question as to how deep the environmental agent must penetrate in order to nucleate a craze, especially in the case of a solid environment.

15.10 Effect of N₂ on Polyethylene

Since PE is generally studied as the representative example of a crystalline polymer, it is interesting to observe the effect of N_2 gas on various types of polyethylene. Figures 15.10.1, 15.10.2, 15.10.3 show the effect of N_2 on three general types of polyethylene: linear, long-branched linear, and short branched. As shown in Table 15.7.1, polyethylenes, as a whole, are less sensitive to the effect of N_2 in causing crazing as compared to all other polymers. Above about 143 K, the effect of N_2 disappears in accordance with Eq. (15.5.2). Between 77 and 143 K the long-branched PE is the most brittle and has the lowest ultimate strength; the linear high-density homopolymer is more ductile and has the next highest ultimate strength; and the short-branched copolymer, commercially called linear low-density PE has the highest tensile strength. It is seen that the differences in strength among the different types of PE in N_2 follow their differences in intrinsic strength. Also, each material shows a maximum in its strength at an intermediate temperature between 77 and 143 K in accordance with the general behavior of other polymers.

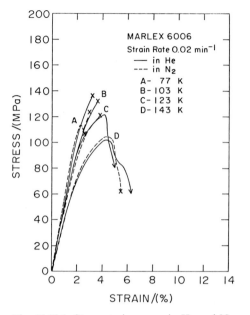

Fig. 15.10.1. Stress-strain curves in He and N_2 from 77–143 K for HDPE (linear) (ref. 16).

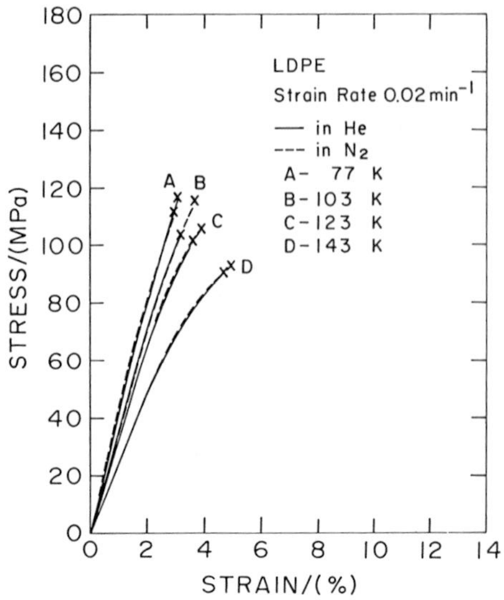

Fig. 15.10.2. Same as Fig. 15.10.1 for LDPE (branched) (ref. 16).

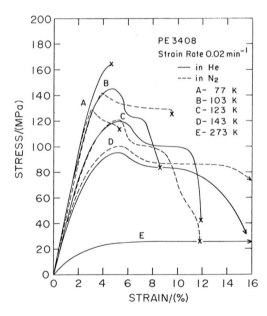

Fig. 15.10.3. Same as Fig. 15.10.1 for Ethylene-Hexene high molecular weight copolymer, Phillips M-8000 (ref. 16).

It was noted previously, as shown in Table 15.7.1, that the magnitude of the sensitivity factor $(\sigma_{in} - \sigma_g)/\sigma_g$ tends to increase with crystallinity as is the case of PTFE. It is also generally agreed that the gas dissolves in the noncrystalline regions. Thus, it seems anomalous that as the total amount of dissolved gas decreases with increasing crystallinity the sensitivity to crazing increases. This is only a superficial anomaly, because the propensity to crazing is a highly localized event. No doubt the crazes initiate in the amorphous regions that contain the gas. Probably the crystalline regions go along for the ride after the fibrillation process is started in the noncrystalline regions. Almost nothing is known about the variation in morphology of the noncrystalline regions as a function of crystallinity. The fact that the sensitivity to crazing seems to increase with crystallinity would indicate that the noncrystalline regions get weaker with increased crystallinity and/or the dissolved gas produces a greater degree of plasticization with increasing crystallinity.

15.11 Strength Enhancement by Gases

All polymers that have been tested in N_2 around 77 K, have had their mechanical behavior altered. Usually the ultimate strength decreases as described above and this decrease increases with decreasing strain rate. There are cases where the ultimate strength in a gas increased with respect to the intrinsic strength. Nylon and Polyvinylidene chloride[20] are two examples which at nominal strain rates show a higher ultimate strength in N_2 as compared to He at 77 K. Figures 15.11.1 (a and b) show the behavior of nylon and PVDC which fail in a brittle fashion without showing crazes in N_2.

This strengthing is associated with a conditioning effect which is produced by a combination of the stress and the N_2 environment. Figure 15.11.2 illustrates the conditioning effect. The PVDC is exposed to the N_2 while under a stress and then partially unloaded; it is reloaded in He where it can be seen that its fracture stress is now greater than that of specimens which were simply loaded in He. If the specimen had been exposed to the N_2 without a stress and then loaded in He, the prior exposure to N_2 would have no effect. Thus, the combination of stress plus N_2 enhanced the subsequent fracture stress in He. These experiments suggest that the enhancement in

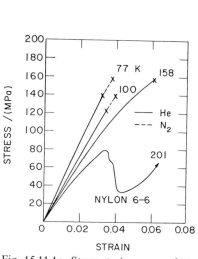

Fig. 15.11.1a. Stress-strain curves of Nylon 6-6 in N_2 (– – –) and He (———) at various temperatures (ref. 20).

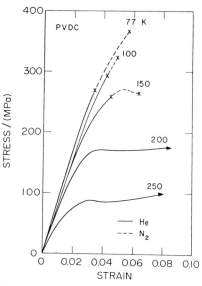

Fig. 15.11.1b. Same as Fig. 15.11.1a for Polyvinylidene Chloride (PVDC) (ref. 20).

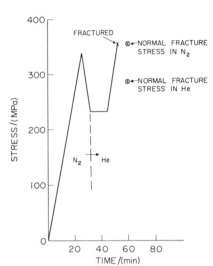

Fig. 15.11.2. The enhancement of fracture stress of PVDC at 78 K in He after being preloaded in N_2 (ref. 20).

strength by the nitrogen is produced during the loading period prior to the beginning of the fracture process. In the case of nylon and PVDC no crazing has been observed after fracture. The conditioning or strength enhancing effect must occur at the surface defects where the brittle fracture is likely to initiate.

The strength enhancement has been attributed to blunting by Parrish and Brown.[21] The fracture occurs at a point of stress concentration. The N_2, in conjunction with the stress, must blunt the points of stress concentration and thereby increase the fracture stress. The blunting process must not be so severe as to produce crazing which generally reduces the ultimate strength. It is suggested that the blunting effect occurs very close to the surface so that the gas hardly if at all penetrates the material as in the usual case of crazing. It has been pointed out that both nylon and PVCD have very low diffusion coefficients compared to other polymers, therefore, the penetration of gas is likely to be much less. If the gas could penetrate deep enough, it would plasticize the material and weaken it by producing crazes.

Strength enhancement has also been observed at high-strain rates in PMMA (Fig. 15.6.2 b). The enhancement effect supports the blunting theory since at the high-strain rates, the gas does not have time to penetrate enough to form crazes.

15.12 Effect of Gas on Craze Density and Velocity

When crazing is the primary mode of deformation then all the macroscopic mechanical properties such as the stress-strain curves,[23] creep curves,[24] and stress-relaxation curves[25] are based on two fundamental crazing parameters: (1) the craze density, and (2) the craze velocity. For example Brown e. a.[24] showed that the creep curves in gases at low temperatures are described by the equation

$$\varepsilon = A_\gamma \, C_c \, \varrho_c (l_{c0} + v_c t)^3 \tag{15.12.1}$$

where A_γ is the area/unit volume of the specimen; C_c a shape factor for the craze; ϱ_c the craze density; l_{c0} a dimension of the nucleated craze; v_c the craze velocity; and t is time. Figure 15.12.1 shows a variety of creep curves for different stresses, gas pressures, and temperatures. Figure 15.12.2 shows $\varepsilon^{1/3}$ is proportional to t in accor-

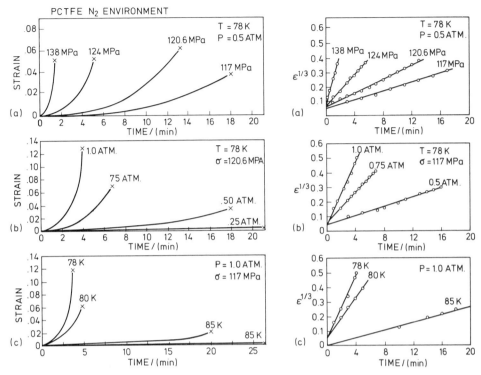

Fig. 15.12.1. Creep curves of PCTFE in N_2 (a) $T=$ 78 K; $P=0.5$ atm. (b) $T=78$ K; $\sigma=120.6$ MPa (c) $P=1.0$ atm., $\sigma=117$ MPA (ref. 24).

Fig. 15.12.2. $\varepsilon^{1/3}$ versus t for creep curves in Fig. 15.12.1 (ref. 24).

dance with Eq. (15.12.1). The slopes of the curves in Fig. 15.12.2 agree with the prediction of Eq. (15.12.1) as substantiated by separate microscopic measurements of ϱ_c and v_c.

It is also to be noticed in Fig. 15.12.1 that the creep rate decreases with decreasing pressure as is expected, but that it decreases with increasing temperature which is anomalous. The explanation of this behavior is based on the pressure and temperature dependence of the solubility as described by Eq. (15.1.1). If the creep curves in N_2 are observed from T_B to higher temperatures at a constant pressure then the creep rate goes through a minimum as shown in Fig. 15.12.3, and then increases with temperature. Below the temperature of the minimum, plasticization via gas solubility dominates, above the minimum thermally activated deformation dominates.

The craze velocity is the most important parameter that determines the creep curves, the stress relaxation,[25] and the stress-strain curves.[23] Brown and Metzger[26] have measured the affect of P, T, and σ on the craze velocity as shown in Figs. 15.12.4 and 15.12.5. It is to be noted that the slope of these curves decreases at higher velocities; this effect is attributed to a decrease in gas penetration as the gas diffuses into a moving interface.

The curves shown in Figs. 15.12.3, 15.12.4, and 15.12.5 are all based on the following equation[27]

$$\ln v_c = \ln A_0 - 1/RT[U_0 - V\sigma - A_2 P \exp\{-(H_v/R)(1/T_B - 1/T)\} \cdot \exp\{-v_c \delta_p/D_g\}] \tag{15.12.2}$$

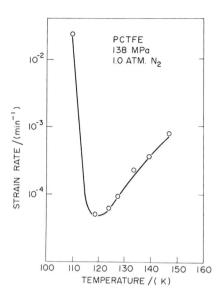

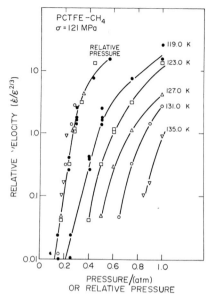

Fig. 15.12.3. Creep rate versus T in N_2 at $P=1$ atm; $\sigma=138$ MPa. Below 120 K the plasticizing effect of N_2 dominates and above 120 K thermally activated deformation dominates (ref. 26).

Fig. 15.12.4. Craze velocity of PCTFE in CH_4 versus pressure at various temperatures, $\sigma=121$ MPa. Curve on left is velocity versus relative pressure obtained by combining all the curves (ref. 26).

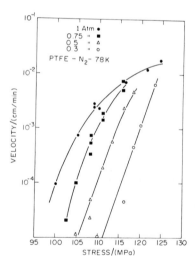

Fig. 15.12.5. Velocity of crazes versus stress for PCTE in N_2 at 78 K (●) 1 atm.; (■) 0.75 atm.; (▲) 0.5 atm.; (○) 0.3 atm. (ref. 27).

where v_c is the craze velocity; A_0 a constant; U_0 an energy barrier for thermal activation; V the activation volume; σ the stress; A_2 the constant relating concentration to relative pressure; H_v the heat of vaporization; T_B the boiling point; δ_p the depth of penetration of gas to produce the necessary plasticization for crazing; and D_g is the effective diffusion coefficient of the gas at the stress, σ. The great sensitivity of creep rate to temperature just above T_B stems from the fact that v_c varies exponentially with function of

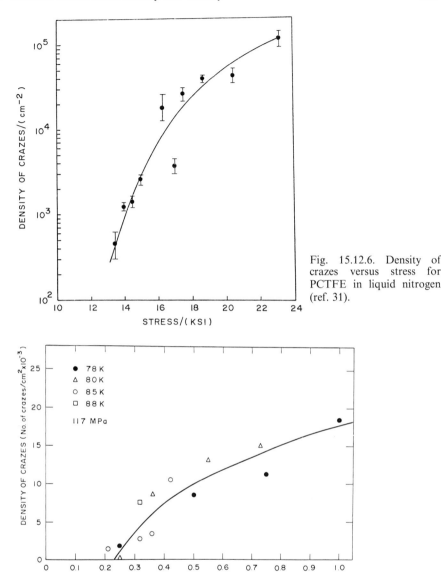

Fig. 15.12.6. Density of crazes versus stress for PCTFE in liquid nitrogen (ref. 31).

Fig. 15.12.7. Density of crazes versus relative pressure for PCTFE in N_2; $\sigma = 117$ MPa (ref. 26).

T which is itself an exponent. The last factor in Eq. (15.12.2) is associated with the reduction in concentration as the gas diffuses into a moving interface.

In a homopolymer, crazes nucleate at the surface of the specimen and particularly at points of stress concentration as shown by Sauer and Hsiao.[28] Thus, the factors that generally determine the craze density are: (1) condition of the surface as investigated by Argon e.a.[29]; (2) the level of stress; (3) time of stress application; and (4) the environment. In the case of crazing at low temperatures ~ 80 K, in gaseous environments, it was observed that the density did not depend on time for PCTFE and PMMA.

However Ziegler and Brown,[30] and Argon e.a.[29] showed the nucleation of crazes in PS in air at room temperature was time dependent. Time dependence is probably not readily observed at very low temperatures because the time constant becomes too long.

The effect of stress on craze density in PCTFE in N_2 at 77 K is shown in Fig. 15.12.6. The effect of relative pressure on craze density is shown in Fig. 15.12.7. The relative pressure is proportional to the solubility in accordance with Eq. (15.1.1). It is to be noted that there is a critical stress for a given relative pressure, and critical relative pressure for a given stress, below which no crazing is observed. These critical conditions depend on the defect which produces the greatest stress concentation at the surface of the specimen.

Acknowledgement

Support was received from the Gas Research Institute.

References

1. J.H. Hildebrand and R.L. Scott, *The Solubility of Nonelectrolytes,* **34**, Dover, New York 1964.
2. G.J. Van Amerongen, *Rubber Chem. Technol.* 1964, **37**, 1065.
3. V.T. Stannett, *Diffusion in Polymers,* edited by J. Crank and G.S. Park, Chapter 2, Academic Press, New York 1968.
4. N. Brown, *Mater. Sci. Eng.* 1971, **8**, 69.
5. N. Brown, J. *Mater. Sci.* 1983, **18**, 2241.
6. W.V. Wang and E.J. Kramer, *J. Polym. Sci. Phys.* 1980, 20, 1371.
7. W.V. Wang and E.J. Kramer, *Polymer* 1982, **23**, 1667.
8. J.G. Webster, PE Jubilee Conference, Plastics and Rubber Institute, p. C 4.9.1, London, June 1983.
9. R.P. Kambour, C.L. Gruner and E.E. Romagosa, *J. Polymer Sci.* 1973, **11**, 1879.
10. H.G. Olf and A. Peterlin, *J. Polymer Sci. A-2* 1974, **12**, 2209.
11. N. Brown and M.F. Parrish, *Recent advances in Science and Technology of Materials,* Vol. 2, edited by A. Bishay, p. 1, Plenum Press, New York 1973.
12. J.C.B. Wu and N. Brown, *J. Mater. Sci.* 1982, **17**, 1311.
13. Y. Imai and N. Brown, *Polymer* 1977, **18**, 298.
14. Y. Imai and N. Brown, *J. Mater. Sci.* 1976, **11**, 417.
15. Y. Imai and N. Brown, *J. Polymer Sci. Phys.* 1976, **14**, 723.
16. E. Kamei and N. Brown, *J. Polymer Sci. Phys.* 1984, **22**, 543.
17. S. Fischer and N. Brown, *J. Appl. Phys.* 1973, **44**, 4322.
18. H.G. Olf and A. Peterlin, *Polymer* 1970, **14**, 78.
19. A. Hiltner, J.A. Kastelic and E. Baer, *Advances in Polymer Science and Engineering,* edited by K.D. Pae, R.D. Morrow and Yu Chen, p. 335, Plenum Press, New York 1972.
20. J.B.C. Wu and N. Brown, *Mater. Sci. Eng.* 1980, **44**, 121.
21. M.F. Parrish and N. Brown, *J. Macromol. Sci. Phys.* 1973, B **8** (3–4), 655.
22. N. Brown, *J. Polym. Sci. Phys.* 1973, **11**, 2099.
23. N. Brown, *Phil. Mag.* 1975, **32**, 1041.
24. N. Brown, B.D. Metzger and Y. Imai, *J. Polymer Sci. Phys.* 1978, **16**, 1085.
25. J.B.C. Wu and N. Brown, *J. Rheol.* 1979, **23**, 2, 231.
26. N. Brown and B.D. Metzer, *J. Polymer Sci. Phys.* 1980, **18**, 1979.
27. N. Brown, *J. Macromol. Sci. Phys.* 1981, B 19(3), 387.
28. J.A. Sauer and C.C. Hsiao, *Trans. ASME* 1953, **75**, 895.
29. A.S. Argon, J.G. Hanoosh, and M.M. Salama, *Fracture* 1, ICF 4 1977, **4**, 445.
30. E.E. Ziegler and W.E. Brown, *Plast. Technol.* 1955, 1, 341, 409.
31. N. Brown and S. Fischer, *J. Polym. Sci. Phys.* 1975, **13**, 1315.

16 Environmental Stress Cracking: The Phenomenon and Its Utility

Arnold Lustiger

Polymer Science and Technology Section, Battelle Columbus Laboratories
505 King Avenue, Columbus, OH 43201, U.S.A.

16.1 Introduction

Environmental stress cracking (ESC) of polyethylene has been a failure phenomenon of great practical significance since it was first noted in the early 1950 s.[1] Much of the subsequent research on the development of new polyethylenes centered on improving the material's resistance to this mode of failure in light of a long history of environmental stress cracking incidents in cable and container applications. In this chapter, the environmental stress-cracking phenomenon will be viewed in a slightly different context: as a tool to be exploited rather than as a problem to be solved.

When polyethylene is used in a production mode for a given application, two questions are often raised. First, among the many polyethylene brands available that fall within the nominal requirements for a specific application, which grade will best resist failure in the field? Second, once a specific type of polyethylene is chosen, how can the various lots of incoming material be monitored to assure that all the material is uniform? It will be demonstrated that through appropriate utilization of ESC tests, informed judgments can be made concerning the adequacy of a polyethylene structure to meet its projected service life.

Gas piping used in natural gas distribution is a prime example of a polyethylene application in which resistance to failure, or "service" life, is a critical consideration. At present, more than 99% of all pipe being installed for natural gas distribution is made from polyethylene.[2] Because these installations are expected to last more than 50 years, polyethylene used in this application can truly be considered an *engineering thermoplastic*. Premature failure of even a small fraction of the polyethylene now in the ground for gas distribution can have extremely serious implications.

If the ESC phenomenon can indeed be used as a tool to evaluate the suitability of polyethylene for long-term service, the gas-piping application seems to be a primary area where such an approach is needed. Therefore, this chapter deals with ESC specifically as it relates to polyethylene pipe evaluation and testing. It should be noted that a similar approach has been used in the past for polyethylene container applications.[3]

16.2 Polyethylene Pipe-Field Failures

Natural gas distribution piping represents the most demanding present application for polyethylene because of the 50-year service life requirement and the severe implications of premature failure. In order to understand what governs service life and how ESC can help in its relative determination, the primary mode of failure that is encountered in the field must be described.

Figure 16.2.1 is a photograph of a plastic pipe that has failed. The crack in the pipe was caused by a rock that impinged on the outer surface. As a result of the

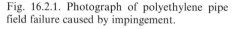

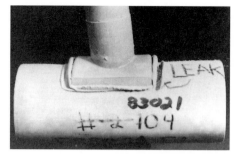

Fig. 16.2.1. Photograph of polyethylene pipe field failure caused by impingement.

Fig. 16.2.2. Photograph of polyethylene pipe field failure caused by strain in fitting.

rock impingement, tensile stresses were induced on the inside surface of the pipe, allowing initiation of the axial crack. The failure occurred one year after the pipe was installed.

Figure 16.2.2 is a photograph of another *field failure*. In this case failure initiated at the boundary between the pipe and a fitting that was melted onto the pipe, and was attributable to lateral strain of the fitting on the pipe observable as a tilt in the fitting. A cross-sectional view (Fig. 16.2.3) shows that the crack initiated at the boundary between the fitting and the *"melt bead"* formed when the fitting was fused to the pipe. This sharp boundary constituted a region of stress concentration that allowed a crack to initiate and finally propagate in the circumferential direction through the pipe wall. In this case, failure took place 10 years after installation.

Both of these failures occurred due to slow-crack growth through the pipe wall with virtually no macroscopic evidence of deformation in the crack region. This apparently brittle behavior differs from the ductile, creep-type behavior once thought to be the only failure mechanism occurring in polyethylene under long-term, low-level stresses; prior use of the term *"brittle failure"* has been generally limited to impact or fatigue type loading situations. *Brittle-type slow-crack growth* is the dominant mode of field failure for polyethylene piping systems (with the exception of external damage from excavations near an existing pipe line), and resistance to this type of failure dictates the service life of a given polyethylene material.

Given this background, the two questions raised in the introduction can be rephrased specifically in the context of polyethylene pipe:
1) How can a plastic pipe user distinguish among various polyethylene (PE) piping systems to determine the best material (i.e., the material with the highest slow-crack growth resistance)?

Fig. 16.2.3. Close-up view at point of failure cross section of pipe shown in Fig. 16.2.2.

2) Once a piping material is chosen, how can the user monitor variations among lots of this material as supplied by the extruder?

The first question is addressed in the next section.

16.3 Environmental Stress-Cracking Resistance for Material Comparison [4]

16.3.1 Experimental

The medium-density polyethylene pipes studied were 25.4 mm (1 inch) in diameter and met the PE2306 and PE3406 Plastics Pipe Institute standards designation. In identifying specific lots of pipe in this chapter, the code following the PE2306 or PE3406 designation identifies the manufacturer, the extruder, and the date of extrusion through the roman numeral, the letter, and the subsequent three- or four-digit number, respectively.

In the *constant-tensile-load test,* a ring 12.7 mm wide is cut from the pipe. The ring is axially notched on both the inside and outside walls to a depth of 1.5 mm at the region of minimum wall thickness and at 180 degrees from this point. The ring is then placed in a split-ring fixture and placed under a constant load in either the presence or absence of a 1% solution of Igepal CO-630 (GAF Corporation, New York) surfactant in water at 23° C. A specimen mounted in the constant-tensile-load test fixture is shown in Fig. 16.3.1. After failure, specimens were examined under the scanning electron microscope (SEM).

To better elucidate the brittle-failure mechanism, *environmental stress cracks* were initiated on the inside surface of a spherulitic pipe by flattening a ring cut from the pipe and then placing the ring in a fixture in a 25%. Igepal solution at 50° C. The complete procedure, known as the compressed-ring test, is described in more detail

Fig. 16.3.1. Test fixture and specimen configuration (insert) for the constant-tensile-load test.

in Section 16.5. The ring specimen, which displayed widespread microcracking on the inside surface after testing, was then etched using the permanganate technique Bassett and co-workers.[6] A portion of this specimen was examined using the SEM and then replicated for transmission electron microscope (TEM) observation. Subsequent morphological examinations of the inside pipe wall were made on a number of the pipes studied using the same etching technique. *Gel permeation chromatography* (*GPC*) and a density gradient column were used to measure molecular weight distribution and density differences between selected piping materials.

16.3.2 A Methodology for Normalizing Constant-Tensile-Load Data

A present limitation of *environmental stress-crack resistance* (*ESCR*) testing is its inability to isolate the yield-stress property as a parameter independent of the polymer's mechanical properties. Thus, for example, constant-strain tests such as the Bell Laboratories' bent-strip test and the compressed-ring test have been criticized because of stiffness variations among specimens. The variations, of course, give rise to an ambiguity when interpreting the results of these tests: Do differences between times to failure mirror a real difference in ESCR, or do these differences merely reflect the higher stress levels in the stiffer specimens?

A similar objection can also be directed against constant-tensile-load testing. Although load is more or less constant in this test, response to the load varies among materials. Therefore, specimen stiffness again enters as a complicating material property that obscures ESCR as an independent parameter. Does a material fail quickly in this test as a result of its low ESCR or because its low stiffness allows more deformation under the constant load?

A prime example of the confusion created by this situation is the case of testing high- and low-density polyethylene in the constant-strain bent-strip ESCR test: high-density polyethylene (HDPE) fails faster than low-density polyethylene (LDPE). On the other hand, the same samples exhibit opposite effects in a constant-tensile-load ESCR test (modified *Carey test*) where notched strips are subjected to a constant-tensile load, with the high-density material cracking at a later time than low-density material (see Table 16.3.1).

The reason for the difference in failure times between the two tests becomes more clear when one considers the influence of mechanical properties on a material's response to a load as shown in Fig. 16.3.2. Because of its relative stiffness, high-density polyethylene is stressed close to or beyond the yield point in a constant-strain test, and cracking takes place in the portion of the bend where the material is just below the yield strain. In contrast, low-density polyethylene does not come close to its yield point under the same test conditions. Conversely, low-density polyethylene is more susceptible to failure than the high-density material in the constant-tensile-load test for the same

Table 16.3.1. Constant Strain versus Constant Load ESCR Testing of High-Density Polyethylene (HDPE) and Low-Density Polyethylene (LDPE)

	Bent strip, F_{50} Failure time hours	Modified carey test Failure time hr	
		3.51 MPa	9.0 MPa
HDPE	<1	4.7	0.6
LDPE	20	1.9	Yield

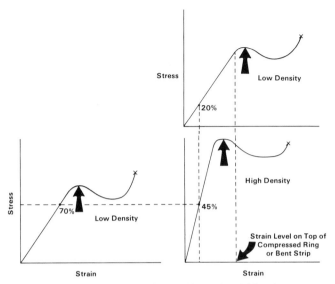

Fig. 16.3.2. ESCR testing in relationship to the yield point.

reason that the high-density material failed faster than the low-density material in the constant-strain test, i.e., the yield point was more closely reached by the less stiff, low-density material in the constant-tensile-load test. In contrast, the yield point was not even approached in the stiffer high-density material under the same loading conditions.

Thus, as the above discussion suggests, neither a constant stress nor a constant-strain test provides good criteria for discerning ESCR. The parameter deserving closer examination as the ordinate of an ESC plot in a constant-load situation is *"percentage of yield stress"* or *"reduced stress."* For a more realistic comparison of polyethylenes, this percentage should be kept constant, although the actual stress may differ widely among specimens.

16.3.3 Constant-Tensile-Load Data

Constant-tensile-load data obtained on 25.4-mm diameter polyethylene pipe in both Igepal and air are shown in Figs. 16.3.3 and 16.3.4. The ordinate of each plot is labeled *"reduced stress"* as discussed above, defined as the percentage of applied or nominal stress to yield stress.

Figure 16.3.3 graphically displays reduced stress versus failure time data for seven different lots of polyethylene piping materials in Igepal. The typically generated curve displays a shallow-sloped region followed by a more steeply sloped region. Generally, ductile-type failure showing large deformation and necking, corresponding to relatively high stress levels and short failure times, occurs in the shallow-sloped region of the curve. In the region of lower stress levels and longer failure times, a brittle-type failure occurs, characterized by little deformation at the point of failure and corresponding to the region of the curve where the slope is steeper. The point where the slope changes can be characterized as a type of *"ductile-brittle transition,"* although prior use of this term generally has been limited to impact fracture.

Comparing different medium-density polyethylene materials with differing yield points results in a wide band of scattered data in the ductile portion of the curve

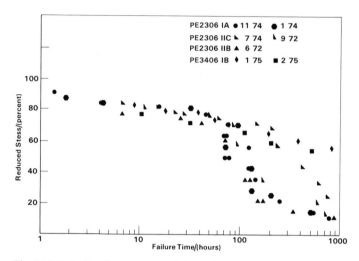

Fig. 16.3.3. Reduced stress versus failure time for seven polyethylene materials in Igepal.

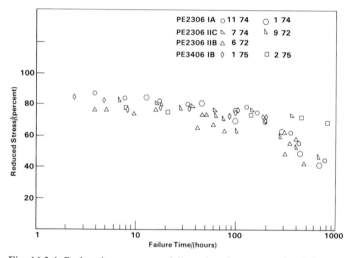

Fig. 16.3.4. Reduced stress versus failure time for seven polyethylene materials in air.

when the ordinate is labeled *"nominal stress"* rather than "reduced stress" (Figs. 16.3.5 and 16.3.6). Dividing applied stress by the yield stress for all pipes tested thus tends to normalize the data in the ductile region of the curve. Table 16.3.2, which describes the linear correlation factors in the ductile region for both sets of data, shows that this normalization is statistically significant.

Analyzing and contrasting the curves in Figs. 16.3.3 and 16.3.4 can help shed light on the utility of ESC resistance testing for evaluating polyethylene pipe.

For the materials labeled PE2306IA-1174, PE2306IA-174, PE2306IIC-972, and PE2306IIB-672, a *ductile-brittle transition* occurs at approximately 70 hours in Igepal, while the same materials in air display a transition at approximately 160 hours. For the PE2306IIC-774 material, on the other hand, the transition in Igepal occurs at about 200 hours with no apparent transition in air during the 1000 hours that the

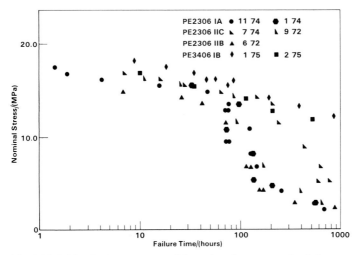

Fig. 16.3.5. Nominal stress versus failure time for seven polyethylene materials in Igepal.

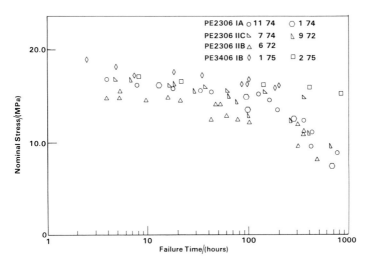

Fig. 16.3.6. Nominal stress versus failure time for seven polyethylene materials in air.

Table 16.3.2. Linear Correlation Factors for Data in Ductile Regions of Figs. 16.3.3 Through 16.3.6

	Stress versus Failure time	Reduced stress versus Failure time
Igepal (36 Data points)	−0.838	−0.940
Air (50 Data points)	−0.592	−0.790

material was tested. In light of the results for the four materials described previously (PE2306IA-1174, PE2306IA-174, PE2306IIC-972, and PE2306IIB-672), one would expect that a transition in air would in all probability have taken place in the PE2306IIC-774 material had the specimens remained on test after 1000 hours. Similarly, for the two PE3406IB materials tested, which display no transition in air or in Igepal, longer testing would probably have revealed a ductile-brittle transition.

Because there is a consistent acceleration in the onset of brittle failure in Igepal relative to the same type of failure in air, it is suggested that the relative position of the ductile-brittle transition in Igepal can be used as a method for comparing the resistance of polyethylene materials to slow-crack-brittle fracture in the field. The later this transition, the greater a given material's service life.

16.4 ESCR Phenomenon

Given this empirical correlation between time to failure in air and in Igepal, the more fundamental question as to the basic molecular mechanism of brittle failure in both air and Igepal arises. This question is addressed in this Section.

16.4.1 Microscopic Examination

The distinction between a "ductile" and "brittle" failure is not merely one of convention. The respective fracture surfaces clearly show a fibrous surface in the ductile failure (Fig. 16.4.1) as opposed to a surface which macroscopically appears smooth in the brittle failure (Fig. 16.4.2). Under the SEM, Fig. 16.4.3 shows uniform fiber pullout in the ductile-failure-fracture surface, while Fig. 16.4.4 displays the much shorter isolated fiber structure characteristic of brittle failure.

To more closely observe the initiation process for environmentally enhanced brittle failure, widespread cracks were induced on the inside surface of a spherulitic pipe using the compressed-ring procedure. This sample was etched and then examined with the SEM as shown in Fig. 16.4.5. The permanganate etching technique used clearly reveals the spherulitic rings, although it also apparently leaves fine bumpy textured artifacts on the surface. An environmental stress crack can be seen traversing the figure. Close observation of this crack indicates that its preferential path is either directly down the center of the spherulites or between them, which supports the supposition that ESC is an interlamellar failure process. The region designated A is apparently

Fig. 16.4.1. Fracture surface after ductile failure (© 1983, Butterworth Scientific).

Fig. 16.4.2. Fracture surface after brittle failure (© 1983, Butterworth Scientific).

Fig. 16.4.3. SEM micrograph of fracture surface after ductile failure (© 1983, Butterworth Scientific).

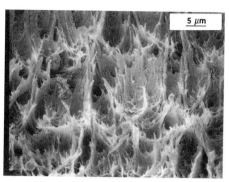

Fig. 16.4.4. SEM micrograph of fracture surface after brittle failure (© 1983, Butterworth Scientific).

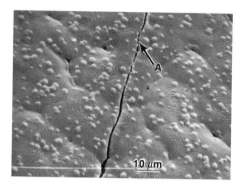

Fig. 16.4.5. SEM micrograph of ESC path in spherulitic pipe (© 1983, Butterworth Scientific).

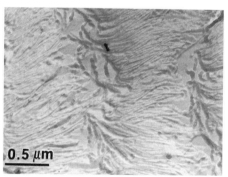

Fig. 16.4.6. TEM micrograph at low magnification of portion of spherulite in spherulitic pipe after undergoing ESC (© 1983, Butterworth Scientific).

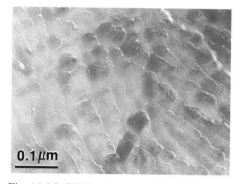

Fig. 16.4.7. TEM micrograph at high magnification of portion of spherulite in spherulitic pipe showing lamellar separation (© 1983, Butterworth Scientific).

the one exception, with cracking being observed obliquely splitting one specific spherulite. The mechanics in this latter case were evidently not favorable for interlamellar failure since the crack would have to deviate greatly to traverse either a spherulite boundary or spherulite nucleus. Similar interspherulitic and intraspherulitic environmental stress-crack initiation in films of high-density polyethylene has been previously documented.[7]

Under TEM, the spherulitic rings and the orientation of lamellae can be clearly resolved (Fig. 16.4.6). A microcrack between lamellae is apparent at higher magnification (Fig. 16.4.7). This separation process is presumably the cause of environmental brittle-fracture initiation. A similar phenomenon has been documented for intrinsic brittle fracture in air.[8]

16.4.2 A Graphic Model for Failure

All these data imply that *intrinsic brittle fracture* under long-term low-level loading conditions is related to environmental stress cracking in Igepal. In order to describe this behavior on a molecular level and help visualize which structural variables most directly influence the cracking, it would be worthwhile to review the fiber deformation process in semicrystalline polymers through a graphic presentation. This model can then be used to contrast ductile behavior with the brittle behavior generally observed at lower loading levels in polyethylene.

16.4.3 Molecular Elements

Although the exact conformation of polymer chains in lamellae is currently under study[9, 10] one can for purposes of this model make a simplifying assumption that these chains fold on each other.

In conceptualizing the failure mechanism it is important to consider the intercrystalline or amorphous polymer chains. Figure 16.4.8a is a simplified schematic showing three types of intercrystalline material 1) *Cilia*—chains suspended from the end of a crystalline chain, 2) *Loose loops*—chains that begin and end in the same lamella, and 3) *Tie molecules*—chains that begin and end in adjacent lamellae.

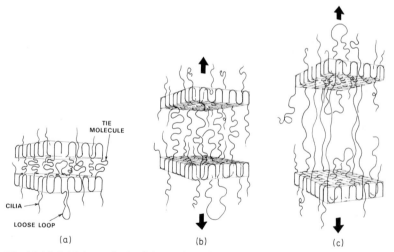

Fig. 16.4.8. Initial steps in the deformation of polyethylene (© 1983, Butterworth Scientific).

16.4.4 Ductile Deformation

If a tensile load is applied normal to the face of the lamellae, the tie molecules stretch as shown in Fig. 16.4.8b. (Note that to facilitate model visualization, the tie molecules are illustrated as continuous chains traversing adjacent lamellae. Although in much of the polymer literature tie molecules are indeed conceptualized in this way[10]

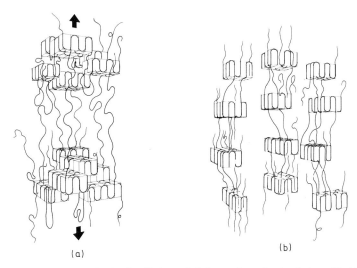

Fig. 16.4.9. Steps in the ductile (creep) deformation of polyethylene (© 1983, Butterworth Scientific).

other literature described later discusses tie molecules as entanglements of a number of separate chains.) At a certain point, however, the tie molecules can be pulled out no farther (Fig. 16.4.8c). At this time the lamellae break up into smaller units (Fig. 16.4.9a). According to this model, as advanced by Peterlin,[11] these so-called "mosaic blocks" are directly incorporated into a new fiber morphology (Fig. 16.4.9b).

For semicrystalline polymers this model of ductile fiber pullout and, by implication, of creep deformation is far from being universally accepted. For example, Corneliussen and Peterlin[12] have shown, using small angle X-ray scattering, that the fold period of chains in lamellae is usually different from the period of folded chains in the fibers. This finding suggests that the direct incorporation of mosaic blocks into fibers may be a misleading notion. Others suggest that in the ductile deformation of semicrystalline polymers folded chains actually unfold as a result of the force of the tensile stress,[13] or undergo a stress-induced phase change[14] and renucleate into the fibers shown in Fig. 16.4.9b. Whichever model is more accurate, the important point of this analysis is the role tie molecules play in ductile deformation. Because tie molecules essentially are the "cement" holding the lamellar "bricks" together, their integrity is critical for ductile-type behavior to occur.

16.4.5   Brittle Failure

Brittle-type slow crack behavior takes place over longer periods of time at lower stress levels than the ductile deformation discussed above. The first three steps in this process are similar to those illustrated in Fig. 16.4.8. However, the stress necessary to achieve large-scale fiber pullout is not attained because the material is under a lower stress level. Therefore, the loading situation can be expected to remain as shown in Fig. 16.4.8c for a relatively long time, although under long-term low-level stress, tie molecules can begin to untangle and relax. After a finite period of time, most of the tie molecules untangle, so that ultimately the load cannot be supported by the few tie molecules remaining, and as a result the material fails in a brittle manner. Thus the material passes from the state in Fig. 16.4.8 to that of Fig. 16.4.10.

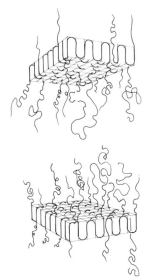

Fig. 16.4.10. Final step in the brittle failure of polyethylene (© 1983, Butterworth Scientific).

From the constant-tensile-load data above, it appears that the Igepal solution accelerates the brittle-failure process. The molecular displacement associated with brittle failure would seem to be enhanced in the presence of Igepal. Presumably, brittle failure in polyethylene is an intrinsic phenomenon, an effect that is merely accelerated by the presence of Igepal. Interlamellar failure, which is the proposed rationale for ESC in general[15,16] can be seen as a rate-dependent process. That is, given enough time at stresses below those resulting in ductile failure, tie molecule entanglements will relax, resulting in brittle failure without any environment to "lubricate" the tie molecules although the failure will be accelerated in the presence of such a "lubricant" or plasticizer. The plasticization of tie molecules has previously been ascribed specifically as the ESC mechanism.[17,18]

Thus the short nonuniform fiber formation observed for brittle failure in Fig. 16.4.4 is caused by an interference in the fiber-formation process taking place as a result of interlamellar failure.

16.4.6 Structure-Property Relationships

It would follow from this discussion that polyethylene materials containing relatively few tie molecules are more susceptible to the various brittle modes of failure. Conversely, materials with relatively high concentrations of tie molecules are more resistant to these types of failures.

However, it should be added that if the proportion of tie molecules to crystalline molecules is too high, the material will display high ductility, but also very low stiffness. Because, in the case of polyethylene pipe, the material must have a certain stiffness to resist exterior mechanical deformation, a trade-off must be established. As a result, medium-density polyethylenes are generally used in this application rather than high- or low-density polyethylenes.

Visualizing the mechanism of brittle failure in light of the above model can help identify molecular parameters of importance in optimizing piping materials for resistance to slow-crack growth. Some of these parameters are:

1) *Molecular weight.* The higher the molecular weight the longer the polymer chains, resulting in more tie molecules and ore effective tie molecule entanglements. Because polymers are polydisperse, the entire weight distribution is a critical factor.[19]

2) *Comonomer content.* Because of a small amount of comonomer such as 1-butene or 1-hexene, medium-density polyethylenes contain short branches that tend to inhibit crystallinity. Higher comonomer concentration and longer comonomer short chain branches provide better brittle-fracture resistance, because the portions of polymer chains with the longer branches (i.e., 1-hexene or longer) probably do not enter the tightly packed lamellar lattice and hence add to the intercrystalline tie molecule material.[3] Another possible effect of short-chain branching is the increased effectiveness of tie molecule entanglements because the chain, in effect, contains protrusions, thereby inhibiting the ability of the tie molecules to slip past one another.

3) *Density/degree of crystallinity.* It would be expected that the more crystalline the material, the fewer amorphous intercrystalline tie molecules that hold it together.

4) *Lamellar orientation.* If the lamellae are preferentially oriented perpendicular to the tensile-stress direction, they would be more amenable to interlamellar failure than if they are parallel to the stress. In the case of a spherulitic polyethylene, this effect would be minimized, because in spherulites the lamellae are oriented radially. (Examples of apparently nonspherulitic polyethylene are described later.)

16.4.7 Rationalizing Property Differences

The data in Fig. 16.3.3 indicated significant property differences among the seven pipes tested. Many of these difference can be rationalized based on the important structural parameters identified above. This discussion centers on rationalizing property differences in four of the seven pipes tested under constant tensile load as shown in Fig. 16.3.3. The four are:

1) PE2306IA-1174 (Ductile-brittle transition in Igepal, 70 hours)
2) PE2306IIC-972 (Ductile-brittle transition in Igepal, 70 hours)
3) PE2306IIC-774 (Ductile brittle transition in Igepal, 160 hours)
4) PE3406IB-175 (Ductile-brittle transition in Igepal, 1000 hours).

Three different piping materials are represented: PE2306IIC-972 and PE2306IIC-774 are two lots of the same piping material originating from the same extruder.

16.4.8 Density/Molecular Weight

Basic density and molecular weight data on these materials are given in Table 16.4.1. Although the density data alone would suggest that the PE3406 material would display the most brittle behavior (i.e., an early transition), the molecular weight data for this material indicate a molecular weight distribution that would retard brittle fracture. To better conceptualize the molecular weight differences, a composite plot of molecular weight distribution curves is shown in Fig. 16.4.11 for PE2306IA-1174, PE2306IIC-774, and PE3406IB-175.

These curves reveal that (1) PE2306IA-1174 has a relatively low number of low molecular weight molecules but also the lowest fraction of high molecular weight molecules; (2) PE2306IIC-774 has a higher number of short chains but a significantly higher proportion of high molecular weight molecules; and (3) although PE3406IB-175 has relatively few low molecular weight molecules, as does PE2306IA-1174, it also contains a high molecular weight fraction even larger than that for the PE2306IIC-774.

Table 16.4.1. Density and Molecular Weight Distribution Data for Four PE Piping Materials

	PE2306IA-1174	PE2306IIC-972	PE2306IIC-774	PE3406IB-275
Density/(g/cc)	0.9406	0.9406	0.9393	0.9514
Molecular weight data				
$M_w \times 10^{-4}$	10.65	11.55	12.79	27.5
$M_n \times 10^{-4}$	2.43	1.44	1.53	2.72
$M_z \times 10^{-4}$	61.02	80.67	88.44	288.4
$M_{z+1} \times 10^{-4}$	207.9	223.9	227.4	702.7
M_w/M_n	4.4	8.2	8.4	10.00

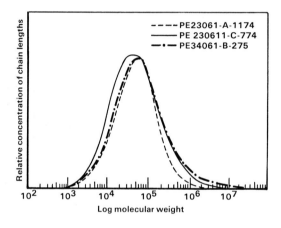

Fig. 16.4.11. Composite molecular weight distribution curve for PE2306IA-1174, PE2306IIC-774, and PE3406IB-175 (© 1983, Butterworth Scientific).

Based on the description of the graphic failure model, higher molecular weight will considerably delay brittle fracture because of the relatively large number of tie molecules that result. This explains the improved crack resistance of the PE3406 material as compared with the other materials.

16.4.9 Morphology/Comonomer Content

The property differences represented by the three PE2306II materials are not as easily explained. The situation is complicated by the significant differences in the ductile-brittle transition reflected in the two lots of the same PE2306IIC materials made by the same extruder but on different dates. Table 16.4.1 indicates no significant density or molecular weight variations to justify the difference in ductile-brittle transition between these two.

The morphology of the PE2306IA-1174 was found to be spherulitic (Fig. 16.4.12). On the other hand, both PE2306II materials where found to be apparently nonspherulitic. However, a significant difference was noted in the lamellar orientation in the nonspherulitic pipes. In the PE2306IIC-774 pipe, the lamellae are randomly oriented (Fig. 16.4.13); in contract, the PE2306IIC-972 pipe displayed regions with lamellae preferentially oriented in the extrusion (axial) direction perpendicular to the pipe surface (Fig. 16.4.14).

This preferential orientation would be expected to have a deleterious effect on brittle-fracture resistance. Because brittle-type slow-crack fracture is interlamellar in nature, the axial orientation of lamellae and the specimen orientation in the constant-tensile-

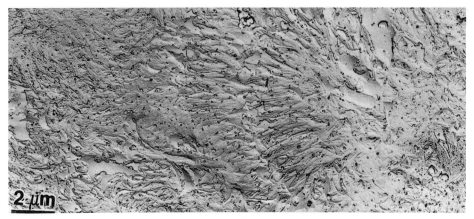

Fig. 16.4.12. TEM micrograph of spherulitic PE2306IA-1174 after permanganate etching (© 1983, Butterworth Scientific).

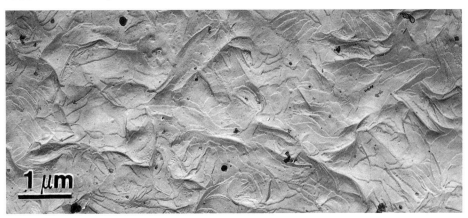

Fig. 16.4.13. TEM micrograph of apparently nonspherulitic PE2306IIC-774 showing after permanganate etching (© 1983, Butterworth Scientific).

Fig. 16.4.14. TEM micrograph of apparently nonspherulitic PE2306IIC-972 pipe showing regions of preferential orientation of lamellae (© 1983, Butterworth Scientific).

load test, which induces failure in the axial direction, would render the PE2306IIC-972 material more susceptible to brittle failure. These subtle differences in morphology may explain the perceived property differences between the two PE2306II materials, despite the fact that the resin, coming from the same manufacturer, is at least nominally the same. One would intuitively expect these differences to be related to extrusion conditions rather than basic lot-to-lot differences in polyethylene resins, although the precise extrusion conditions necessary to duplicate the oriented lamellar morphology are unknown.

However, a note of caution should be injected in reviewing these results. Although the nonspherulitic morphology has been well documented,[20] and the lack of a small-angle light scattering pattern in the PE2306II materials seems to confirm its nonspheru-litic nature, it has recently been shown that the inside surface of polyethylene pipes can often be thermally oxidized during extrusion. This oxidation can alter the morphology of this surface.[21]

In addition to significant morphological differences between the PE2306IA-1174 and the two PE2306II materials, differences in molecular weight and comonomer were also documented. PE2306IA-1174 displayed an average spherulite size of 22 microns (1 micron $= 10^{-4}$ m), in diameter, as well as relatively low weight average and z-average molecular weights. PE2306IA-1174 was found to contain the shorter 1-butene chains as opposed to the 1-hexene chains of the PE2306II materials. Apparently, the combined effect of morphology, molecular weight, and comonomer differences contributed to the relatively early ductile-brittle transition in the PE2306I materials.

16.5 ESCR for Quality Control

The second question, concerning the applicability of ESCR for quality control purposes, is addressed in this section. The primary ESCR method for this purpose in polyethylene pipe is the compressed-ring test.[22] In effect, the test approximates the stress configuration in the more common Bell Laboratories bent-strip test.

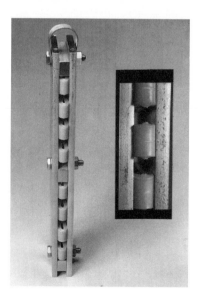

Fig. 16.5.1. Photograph of pipe specimens in the compressed-ring fixture prior to immersion in Igepal.

The *compressed-ring method* (for 1-inch pipe) consists of slicing a 12.7-mm-wide ring from the pipe and pressing a razor blade 0.64 mm deep into the outside wall at the center of the ring, parallel to the edge. The notch is 19 mm long and is consistently located in the region of minimum wall thickness.

Eight of these specimens are mounted between two compression plates with the notched area positioned parallel to the direction of compression. The specimens are compressed so that the plates are 9.14 mm apart and are then placed in a 25% solution of the stress cracking agent at 50° C. The times for crack initiation are recorded. A picture of the compressed-ring fixture with specimens prior to immersion in Igepal is shown in Fig. 16.5.1. The ring width, notch depth, and distance between compression plates varies with the pipe diameter and wall thickness.[22] Samples were removed from the solution at different times prior to complete fracture, washed, air dried, and axially cut in half so that the uncracked portion could be discarded. The cracked half was subsequently coated with gold by vacuum deposition while under the same degree of bending as in the test, and was when observed while bent in the scanning electron microscope (SEM).

16.5.1 Effect of Material Changes

As explained in the section on normalizing constant-tensile-load data, neither a constant load nor a constant strain-test provides an adequate criterion for material comparison when there are stiffness variations among the materials. Because the compressed-ring test is a constant-strain method, the test is considered invalid for determining the relative merits of different polyethylene piping materials. However, it can be very effective in detecting lot-to-lot variations for the same material. Almost any change in the pipe, whether material or process oriented, will be reflected in changes in failure time in this test. A few such typical changes and their effect on ESCR are shown in Table 16.5.1. In the compressed-ring test, if time to failure for a given lot of material is considerably shorter or longer than that for previous lots of the same material, one would suspect that there is a difference in the material that may affect service life.

Examples of ESCR failure times in this test for the same materials tested under constant-tensile load in Figs. 16.3.3 through 16.3.6 are shown in Table 16.5.2.

Table 16.5.1. Material changes and ESCR Behavior

ESCR increases with increasing:
Average molecular weight[23]
Inorganic pigment content[24]
Cooling rate after extrusion (may be transient effect)[25]
ESCR decreases with increasing:
Spherulite size[26]
Melt index[23, 27]
Organic pigment[24]
Crystallinity[23, 25]
Molecular weight distribution (breadth)[19]
ESCR varies with:
Comonomer content[3]
Various additives[28]

Table 16.5.2. Compressed Ring ESCR Data
for the same Materials Tested under Constant
Tensile Load in Figs. 16.3.3 Through 16.3.6

Materials	Failure time
	hours
PE2306I-A-1174	25.7 ± 1.8
PE2306I-A-174	44 ± 8
PE2306II-C-774	237 ± 1.0
PE2306II-C-972	27 ± 5
PE2306II-B-672	31 ± 3
PE3406I-B-175	>2000
PE3406I-B-275	>2000

16.5.2 Effect of Surface Irregularities

Potentially more significant than the change in failure time is the location of the crack origin – whether the crack is initiated at the razor notch, between the notch and the edge, or on the unnotched side of the sample. One usually finds that ESCR failure times decrease substantially when failure on the unnotched side is observed.

Using the scanning electron microscope, it has been found that imperfections on the specimen surface serve as initiation points for extra-notch cracking. Two types of surface flaws which can cause cracking outside the notch have been detected: surface voids and surface scratches.

16.5.3 Surface Voids

Although 1-inch PE2306IIC-774 and PE2306IIB-872 are the same compound made into pipe by different extruders, the C-774 pipe failed in 237 hours in the compressed-ring test, while the B-872 pipe failed in 25 hours. As is shown in Table 16.5.3, no appreciable difference in molecular weight, molecular weight distribution, or density could be found to explain this behavior.

One feature noted in the PE2306IIC-774 material was that cracks originated at the notch, whereas the PE2306IIB-872 material exhibited failures that usually started on the unnotched side of the specimen, with failures on the notched side taking place at a later time. Figure 16.5.2 shows a failure site in C-774 material. (The insert is

Table 16.5.3. Molecular Weight and Density Data in PE2306II Piping
Material

Property	Lot	
	C-774	B-872
(M_w)	127,900	128,500
(M_n)	15,300	14,500
(M_w/M_n)	8.4	4.9
(M_z)	884,400	864,300
(M_{z+1})	2,274,400	2,395,200
Density/(g/cm^3)	0.9403	0.9393

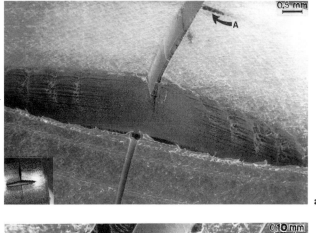

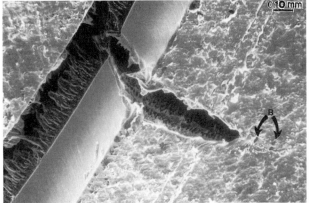

Fig. 16.5.2. SEM micrograph of notched portion of C-774 compressed-ring specimen removed from the Igepal bath during crack formation (© 1981, J. Wiley). Note: Vertical groove is the razor notch; A denotes in Fig. a small crack initiating at the notch; insert is photograph of the same specimen. In Fig. b the area A is shown at higher magnification; at B the crack propagation undercutting the specimen is shown.

a photograph of the same specimen.) The razor notch is shown in the vertical direction. The crack wall displays a wavelike pattern distributed radially around the notch, indicating crack initiation at this point. A shows a crack just starting at the notch, magnified further in Fig. 16.5.2 b.

Because the crack started at a point below the surface, the crack front is an arc that moves radially from that point. Evidence that the crack propagated from the lower notch area can be derived from the wavelike pattern on the fracture surface in Fig. 16.5.2. These waves can be interpreted as earlier crack fronts as the stress crack propagated through the pipe. Similar notch initiation and subsurface cracking have been documented previously.[29]

Figure 16.5.3 shows a B-872 failure on the notched side of the specimen. The crack is asymmetric, in contrast to the symmetry of the crack around the notch in the C-774 sample. The large crack on the right of Fig. 16.5.3 obviously grew independently of the notch. At higher magnification (Fig. 16.5.4), this large crack is seen surrounded by a number of superficial cracks, a few of which are shown at C, D, and E. These

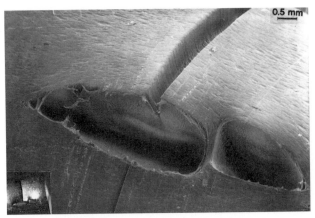

Fig. 16.5.3. SEM micrograph of notched portion of B-872 compressed-ring specimen removed from the Igepal bath during crack formation. Note: Insert is photograph of the same specimen (© 1981, J. Wiley).

Fig. 16.5.4. SEM micrograph of right portion of crack in Fig. 16.5.2 is higher magnification. Note: Superficial cracks are shown at C, D, and E (© 1981, J. Wiley).

small cracks seem to be initiated directly from the surface voids that are readily evident at this magnification and can be seen over the entire specimen surface. Presumably, the large crack in the center of the figure also originated from one of these voids and would connect with the crack to its left if the specimen had remained in Igepal. However, evidence of notch initiation can also be seen at N in Fig. 16.5.4 showing radial crack propagation from the notch root on the fracture surface.

Two major theories have been advanced to explain environmental stress cracking. Hopkins, Baker, and Howard[30] theorized that the stress-cracking agent exerts a spreading pressure, in conjunction with applied mechanical stress, and results in crack initiation. On the other hand, Isaksen, Newman, and Clark[31] ascribe the role of the stress cracking agent to interference with the uniform fiber deformation upon drawing, and it is this theory that was invoked in detailing the relationship between failure in air and Igepal described above.

Although the theory of Isaksen e.a. is now widely accepted as the most plausible ESC mechanism, present test results indicate that the influence of flaws on the polymer surface cannot be discounted. Even though, on a molecular level, interference with uniform fiber formation is the ultimate cause of failure, voids on the specimen surface greatly accelerate this effect, through stress concentrations. It was found that failures in the B-872 materials typically occurred on the unnotched side of the specimen because the notch on the other side of the specimen relieved stresses in the bending direction and retarded cracking. Thus, the biaxial stress condition effected by the notch, so important for ESC to occur in low-density and most medium-density polyethylenes,[32] actually prolonged time to failure in this case.

Surface roughness per se does seem to affect ESC, as can be discerned from the relatively rough surface of the highly ESC resistant C-774 material. Cracks on the B-872 specimen surfaces seem to originate only at voids or other irregularities that can act as stress concentrators. The evidence here may indicate that before the voids actually develop into cracks crazed material forms, because the small superficial cracks appear to be bridged by intervening fibers at C, D, and E, in Fig. 16.5.4. Crazing has been shown to be a precursor to environmental stress cracking for high-density polyethylene.[33, 34]

The voids evident in the B-872 samples appear to be oriented parallel to the specimen width, i.e., parallel to the extrusion direction. These voids may have formed as a result of a very rapid quench after the extrusion process, which would cause widespread local shrinking due to precipitous crystallization on the surface.

16.5.4 Surface Scratches

A failure mode similar to that in 2.54 cm PE2306IIB-872 has been observed in many different samples of smaller diameter (1.6 cm) tubing. These unusual failures correspond to markedly shorter ESCR times, as in the case described above. These

Table 16.5.4. Compressed-Ring ESCR Times to Failure for Tubing that Cracked Outside the Notch

Material	Failure time, hours	Average normal failure Time/hours
PE2306I-A-678	8	24
PE2306II-D-678	26	$\cong 200$
PE3406I-E-879	54	$\cong 4,000$
PE3408II-E-980	93	$\cong 8,000$

Fig. 16.5.5. SEM micrograph of PE2306IA-1174 compressed-ring specimen removed from the Igepal bath during crack formation.

failure times are shown in Table 16.5.4 along with "normal" ESCR times based on previous experience with these materials.

The PE3406IE-879 as seen in Fig. 16.5.5 displays what seem to be surface scratches which, in turn, initiate cracking. The "scratches" which connect these cracks appear to be continuous and are likely to be "die lines" as a result of extruder irregularities.

16.5.5 Outdoor Exposure

In addition to detecting extrusion-related defects, the compressed-ring method allows detection of excessive outdoor weathering in polyethylene pipe. Using scanning electron microscopy, weathering damage on the surface of polyethylene pipe can often be manifested as a series of shallow cracks in both the axial and circumferential directions, as shown in Fig. 16.5.6.

Decrease in time to failure for two materials as a function of outdoor exposure is detailed in Table 16.5.5. The data are for both notched and unnotched specimens in the compressed-ring test. It is interesting to note that use of the unnotched samples allowed a more sensitive measure of outdoor weathering degradation because, as explained above, notching the sample tends to relieve stresses on the specimen surface in the circumferential direction.

Fig. 16.5.6. SEM micrograph of PE2306IA-1174 compressed-ring specimen after 36 months of outdoor weathering, removed from the Igepal bath during crack formation.

Table 16.5.5. Effect of Outdoor Exposure on ESCR Failure Time of Polyethylene Piping Materials in the Compressed-Ring Test

Outdoor exposure time/months	PE2306I-A-374		PE2306II-C-1075	
	Notched average ESCR/hours	Unnotched average ESCR/hours	Notched average ESCR/hours	Unnotched average ESCR/hours
0	30.0±1	55.0±22	225.0±13	>8000[a]
3	27.0±3	22.0±9	204.0±17	–
6	27.0±2	5.9±0.6	209.0±16	>8000[a]
9	8.1±0.2	5.4±0.6	219.0±18	>8000[a]
12	4.4±0.2	4.2±0.8	209.0±9	–
15	3.5±0.5	2.9±0.1	183.0±30	84.0±12
18	3.9±2	3.9±1.3	76.0±2	50.0±13
24	6.2±1.0	5.6±1.1	30.2±1	33.4±7

[a] Terminated, no failures.

16.6 Fracture Surface Comparisons

As mentioned earlier, the *fracture surface* appearance of slow-crack-type failures displays an interrupted fiber morphology. The density and length of these fibers can vary depending on the applied stress and crack-growth rate. There appears to be some correlation between fracture surface appearance and failure time. A moderate ESCR PE2306IIC-774 material (ESCR = 237 hours) displays the fracture surface shown in Fig. 16.6.1. The fracture surface indicates an open fiber morphology in which the fibers are quite short, and considerable space can be seen between them. In contrast, PE3406E-176 (ESCR ≅ 4000 hours) displays a fracture surface showing fibers that are longer and denser in concentration (Fig. 16.6.2). It is interesting to note that in the latter material when the specimen is removed from the bath and observed prior to complete fracture, a large craze is apparent ahead of the crack (Fig. 16.6.3). Apparently, the length, uniformity, and high fiber density in this craze region are reflected in the fiber density and fiber length on the fracture surface itself. Numerous smaller crazelike regions are also apparent on the specimen surface.

A piece of evidence that may underscore the similarity between the ESC mechanism and the slow-crack brittle-fracture mechanism in air is a perceived qualitative similarity between the surface of the field failure shown in Fig. 16.2.1 and an ESCR fracture surface from the same section of pipe (Figs. 16.6.4 and 16.6.5). The fracture surface

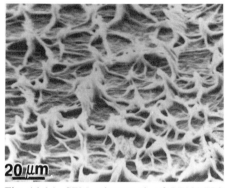

Fig. 16.6.1. SEM micrograph of PE2306IIC-774 fracture surface after failure in the compressed-ring test.

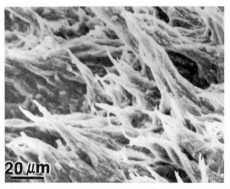

Fig. 16.6.2. SEM micrograph of PE3406IE-176 fracture surface after failure in the compressed-ring test.

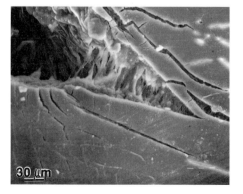

Fig. 16.6.3. SEM micrograph of PE3406IE-176 compressed-ring specimen removed from the Igepal bath during crack formation, showing large craze formation ahead of the crack.

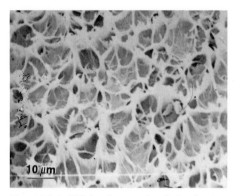

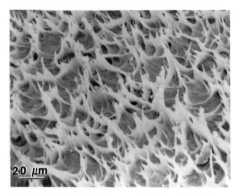

Fig. 16.6.4. SEM micrograph of the fracture surface of the field failure shown in Fig. 16.2.1.

Fig. 16.6.5. SEM micrograph of an ESCR fracture surface from the same material.

of the field failure displays respectively an open fiber formation similar to the ESCR specimen of the same material. Note that the magnifications of the two specimens vary, with the field failure photo being magnified twice as much as the ESCR specimen. This quantitative difference may be related to crack speed differences due to the two widely varying cracking conditions.

16.7 Conclusions

The environmental stress-cracking phenomenon is a useful tool for determining the relative crack resistance of various types of polyethylene, through the occurrence of a "ductile-brittle transition" during constant-tensile-load testing.

Environmental stress cracking in polyethylene is apparently an acceleration of the brittle-type slow-crack failure process that would intrinsically take place in air at a later time under similar test conditions. The relaxation of tie molecules and the resulting interlamellar failure, therefore, can take place in either air or Igepal. Use of a graphic failure model can help identify which structural variables most directly influence this property. These have been shown to include degree of crystallinity, molecular weight distribution, branch length, and lamellar orientation.

Environmental stress cracking is also valuable in a quality control function for assuring product uniformity. The compressed-ring method can be very helpful in signaling potentially undesirable material or surface characteristics, whether they are resin or process oriented.

16.8 Acknowledgement

The author thanks the Gas Reseach Institute for sponsoring this research under Contracts No. 5014-352-0152 and 5082-260-0613. I am also indebted to my colleagues Michael J. Cassady, Larry A. Smith, and Andrew Skidmore for their technical help, and to Professor Roger D. Corneliussen for technical discussion.

References

1. J.B. Howard, *Soc. Plast. Enq. J.* 1959, **6**, 397.
2. J. Watts, *Pipeline and Gas Journal* 1982, **14**, 19.
3. J.M. Hannon, *J. Appl. Polymer Sci.* 1974, **18**, 3761.
4. A. Lustiger and R.L. Markham, *Polymer* 1983, **24**, 1647.
5. ASTM Standard D3350-78, *1983 Book of ASTM Standards,* Section 8, Vol. 08.04.
6. R.H. Olley, A.M. Hodge and D.C. Bassett, *J. Polymer Sci. Phys.* 1979, **17**, 627.
7. T.W. Haas and P.H. MacRae, *Soc. Plast. Enq. J.* 1968, **24**, 27.
8. J. Petermann and J.M. Schultz, *J. Mater. Sci.* 1978, **13**, 50.
9. D.Y. Yoon and P.J. Flory, *Polymer* 1977, **18**, 509.
10. E.A. DiMarzio and C.M. Guttmann, *Polymer* 1980, **21**, 118.
11. A. Peterlin, *J. Macromol. Sci. Phys.* 1973, **B 8**, 83.
12. R. Corneliussen and A. Peterlin, *Makromol. Chem.* 1967, **105**, 193.
13. H. Gleiter, E. Horbogen and J. Petermann, in Battelle Institute Seventh Materials Science Colloquium, p. 149, Plenum, New York 1973.
14. T. Juska and I.R. Harrison, *Polymer Eng. Revs.* 1982, **2**, 13.
15. S. Bandyopadhyay and H.R. Brown, *Polymer* 1978, **19**, 589.
16. C.J. Singleton, E. Roche and P.H. Geil, *J. Appl. Polymer Sci.* 1977, **17**, 27.
17. P.D. Frayer, P.P.L. Tong and W.W. Dreher, *Polymer Eng. Sci.* 1977, **17**, 27.
18. H.R. Brown, *Polymer* 1978, **19**, 1186.
19. J.N. Herman and J.A. Biesenberger, *Polymer Eng. Sci.* 1966, **6**, 341.
20. L. Mandelkern, M. Glotin and R.A. Benson, *Macromolecules* 1981, **14**, 34.
21. B. Terselius, U.W. Gedde and J.F. Jansson, *Polymer Eng. Sci.* 1982, **22**, 424.
22. ASTM Proposed Practice, Section 8, Volume 08.04, 1983 Book of ASTM *Standards.*
23. P. Hittmair and R.Ullman, *J. Appl. Polymer Sci.* 1962, **6**, 1.
24. A. Lustiger, "The Stress Cracking Behavior and Crack Morphology of Pigmented Polyethylene", B.S. Thesis, Drexel University, 1976.
25. J.B. Howard and W.M. Martin, *Soc. Plast. Eng. J.* 1960, **16**, 68.
26. M.E.R. Shanahan, C. Chen-Fargheon and J. Schultz, *Makromol. Chem.* 1980, **181**, 1121.
27. J.B. Howard and H.M. Gilroy, *Soc. Plast. Eng. J.* 1968, **24**, 68.
28. L. Spendadel, *J. Appl. Polymer Sci.* 1972, **16**, 2375.
29. A. Lustiger, R.D. Corneliussen and M.R. Kantz, *Mater. Sci. Eng.* 1978, **33**, 117.
30. I.L. Hopkins, W.D. Baker and J.B. Howard, *J. Appl. Phys.* 1950, **21**, 207.
31. R.A. Isaksen, S. Newman and R.J. Clark, *J. Appl. Polymer Sci.* 1963, **7**, 515.
32. J.B. Howard, in *Crystalline Olefin Polymers, Part II,* edited by R.A.V. Raff and K.W. Doak, p. 47, Wiley-Interscience, New York 1964.
33. A. Lustiger and R.D. Corneliussen, *J. Polymer Sci. B* 1979, **17**, 269.
34. S. Bandyopadhyay and H.R. Brown, *Polymer Eng. Sci.* 1980, **20**, 720.
35. *Plastic Pipe Line* 1982, **2** (4) 2.

17 Environmental Stress Failure: An Irreversible Thermodynamics Approach

Hiroshi Okamoto
Yoshihito Ohde

Department of Engineering Sciences, Nagoya Institute of Technology
Gokisomachi, Shouwaku, Nagoya, 466, Japan

17.1 Introduction

Many polymeric materials, loaded mechanically and immersed in certain kinds of liquids, undergo failures by *crazing*[1] and/or cracking. The loads required are much less than those required of failures in air. The failure promoting liquids are nonsolvents and chemically inert for polymers. Failures like these are called *environmental stress crazing* (ESCR), environmental stress *cracking* (ESC), and environmental stress *failure* (ESF) which includes both. Fig. 17.1.1 shows ESC occurring in low-density polyethylene (LDPE) specimens loaded and immersed in *n*-propanol.

Many studies have been made for both scientific and engineering interests. They have covered 1) the causes of ESF, 2) the initiation and the growth kinetics of ESF, 3) the microscopic structures of polymers in crazes and around cracks, 4) the stress fields in and around crazes and around cracks, and 5) the establishment of test methods to evaluate ESCR and ESC endurance of polymeric materials.

The studies prior to 1973 were reviewed by Kambour[2] and those prior to 1979 by Kramer.[3] According to these authors, although much knowledge about the phenomena had been accumulated, many problems due to their complexities still remain unsolved.

This chapter discusses the scientific understanding of ESF. It does not give another review of the field but presents a new idea which is simple and orthodox, yet is capable of accounting for the complicated nature of ESF. The idea starts with the obvious fact that failure or fracture processes of materials are not reversible but are irreversible in the sense of thermodynamics.

It is generally accepted that the most important role of failure promoting liquids is in the selective plasticization of the polymer matrix ahead of craze or crack where

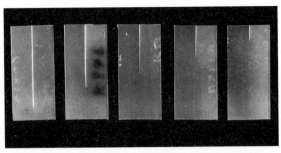

Fig. 17.1.1. Environmental stress cracking occurred in LDPE specimens bent and immersed in n-propanol. Crack grows linearly without branching. (See Fig. 17.2.1)

a polyaxial stress field is developed by stress concentration. Failure initiates or grows as a result of *plasticization-induced weakening* of the polymer matrix in the region.[4]

Let us consider the region where this mechanism operates. The time evolution of the thermodynamic state of the region is, according to the thermodynaic rule, governed by the thermodynamic potential of the relevant system. The failure processes are interpreted as the state shift in the potential diagram.

The thermodynamic potential, as a function of liquid concentration, is constructed by the Flory-Huggins solution theory. The changes of the potential curves by the changes of relevant parameters, such as the polymer ~ liquid interaction, the intensity of the stress field, and the molecular sizes of the polymer and liquid are consistent with the complicated nature of the ESC kinetics observed. In addition, they help us to understand several of the problems left unexplained in the reviews referred to above.

17.2 Characteristic Behaviors of ESC Growth Kinetics

Before discussing our thermodynamic picture, we describe the characteristic behaviors of ESC growth kinetics in this section. Most of these behaviors are taken from the ESC data obtained in our laboratory.[5-7] Those behaviors taken from findings reported in the literature are also included.

We used a version of ESC experiments under constant deformation in our methodology. The growth rate of a growing crack in a thin LDPE plate bent and immersed in a failure promoting liquid was studied. The arrangement of the experiments is depicted in Fig. 17.2.1. A crack started from the tip of the notch N and grew linearly along the ridge of the bent specimen without branching or being arrested (Fig. 17.1.1). In some LDPE and liquid pairs, the growth rates were constant as the crack traversed through the specimens. In other pairs, they increased or decreased as the crack traversed. Fortunately, the changes were smaller than those caused by temperature changes.

All of the growth rate versus reciprocal temperature relations that were found were classed into two groups—type 1 and type 2. Figure 17.2.2 shows the type 1 behaviors obtained by LDPEs of three different molecular weights immersed in NS215. The LDPEs are identified in Table 17.2.1 and the NS215 in Table 17.2.2. Figure 17.2.3 shows the type 2 behaviors obtained by the LDPEs immersed in propanol.

The curve obtained by PE-B in Fig. 17.2.2 represents type 1 behavior consisting of three temperature ranges. In the curve obtained by PE-A, the high temperature range of type 1 is outside of the experimental range. In the curve obtained by PE-C, the medium temperature range is not well developed.

In the low temperature range of type 1, there is a sharp increase in the growth rate caused by a slight elevation of temperature. The activation enthalpies estimated

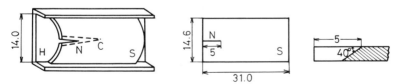

Fig. 17.2.1 Experimental arrangement. A notched specimen (S) of a specified size $(31.0 \times 14.6 \times D) \times 10^{-3}$ m was bent and set in a holder (H) with a constant width of 14.0×10^{-3} m. After immersion in a failure promoting liquid, a linear crack (C) grew at the notch. Its length was measured. The length of the cut-in notch was 5×10^{-3} m and its cross-sectional view is shown in the right figure. Reproduced after reference 7 by permission of the publishers; © , Butterworth & Co (Publishers) Ltd.

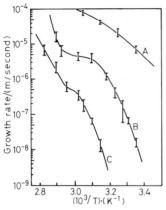

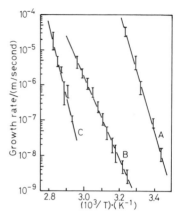

Fig. 17.2.2. Growth rates of the ESCs in PE-A, B, and C immersed in NS215 versus reciprocal temperature. The specimen thickness was 0.55×10^{-3} m. Reproduced from reference 6 by permission of the publishers; ©, Butterworth & Co (Publishers) Ltd.

Fig. 17.2.3. Growth rates of the ESCs in PE-A, B, and C immersed in n-propanol versus reciprocal temperature. The specimen thickness was 0.55×10^{-3} m. Reproduced from reference 6 by permission of the publishers; ©, Butterworth & Co (Publishers) Ltd.

Table 17.2.1. Specification of Polyethylene

Polyethylene[a]	Melt index	Density	Yield strength	Molecular weight[b]
	$(10^{-5}\ kg\,s^{-1})$	$(10^3\ kg\,m^{-3})$	$(MN\,m^{-2})$	
A	3.33	0.918	8.8	32 000
B	0.67	0.920	10.8	45 000
C	0.17	0.920	10.8	55 000

[a] Manufactured by Mitshubishi Petrochemical Co.

[b] Molecular weights were estimated through viscosity measurements of p-xylene solutions at 354 K by using the relation,[17] $\eta(d\lg^{-1}) = 0.105 \times 10^{-2}\ M^{0.63}$.

Reproduced from reference 6 by permission of the publishers; ©, Butterworth & Co (Publishers) Ltd.

Table 17.2.2. Identification of Surfactant

	Commercial name	Rational formula
	NS206[a]	$C_9H_{19}C_6H_4O(CH_2CH_2O)_6H$
	NS210[a]	$C_9H_{19}C_6H_4O(CH_2CH_2O)_{10}H$
	NS215[a]	$C_9H_{19}C_6H_4O(CH_2CH_2O)_{15}H$

[a] These materials are surfactants manufactured by Nippon Oil & Fats Co. NS210 is nearly the same material known as Igepal CO630 or Antarox CO630. Reproduced from reference 6 by permission of the publishers; ©, Butterworth & Co (publishers) Ltd.

by the usual way amount to 90 kcal/mol, proximate to that of PE *chemical bond scission*. It is, however, difficult to understand how the bond scission mechanism could operate under these chemically mild conditions. It may be said that there is a threshold temperature for ESC growth or an ESC initiation temperature. The polymer molecular weight dependence of the growth rates in this range is very strong. Figure 17.2.2 indicates that the growth rates vary with more than 20 powers of polymer molecular weight. It differs significantly from the usual molecular weight dependence of polymer rheological properties. The situations mean that the threshold temperature increases with the increase in polymer molecular weight.

In the medium temperature range of type 1, the growth rate increase with the elevation of temperature decreases to a moderate level. The values of the activation enthalpies are around 40 kcal/mol (1 cal = 4.184 J). The polymer molecular weight dependence of the growth rates also decreases.

In the high temperature range of type 1, we found another abrupt increase in the growth rate. The apparent activation enthalpies were also about 90 kcal/mol.

The growth rate versus reciprocal temperature relations of type 2 are linear showing a single growth mechanism (Fig. 17.2.3). The activation enthalpies are unreasonably high. The growth rates vary with many powers of polymer molecular weight. The findings again mean that there is a threshold temperature for ESC growth changing with polymer molecular weight. It is similar to type 1 behavior in the low temperature range.

The failure promoting liquids are more or less responsible for which of the two − type 1 or type 2 − is to be exhibited. However, type 2 ESC often occurred in a higher molecular weight polymer, even if type 1 ESC occurred in a lower molecular weight polymer immersed in the same liquid. Compare the PE-C curve with the PE-B curve in Fig. 17.2.4. Another feature that should be noticed is that type 2 ESC occurred under a mild deformation condition in a polymer and liquid pair, even if type 1 ESC occurred under a greater deformation in the same pair. Figure 17.2.5 shows the effects of the bending conditions on the growth rate behaviors of the ESC in the LDPE and polydimethylsiloxane (PDMS) pairs.

Table 17.2.3 summarizes the types of the growth rates versus reciprocal temperature relations obtained in our laboratory.

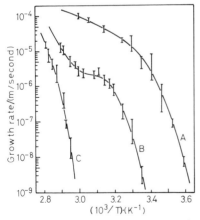

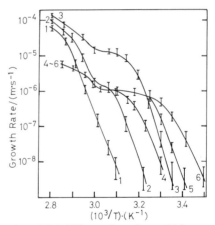

Fig. 17.2.4. Growth rates of the ESCs in PE-A, B, and C immersed in NS210 versus reciprocal temperature. The specimen thickness was 0.55×10^{-3} m. Reproduced from reference 6 by permission of the publishers; ©, Butterworth & Co (Publishers) Ltd.

Fig. 17.2.5. Effects of specimen thicknesses on growth rate versus reciprocal temperature relations of PE-B ESC induced by 30 cSt PDMS (curves 1~3) and by 10^4 cSt PDMS (curves 4~6). The specimen thicknesses are for the curve 1, 0.43×10^{-3} m; 2, 0.55; 3, 0.90; 4, 0.43; 5, 0.55; and 6, 0.90. Reproduced from reference 7 by permission of the publishers; ©, Butterworth & Co (Publishers) Ltd.

Table 17.2.3. The types of the ESC growth rate versus reciprocal temperature relations in LDPEs

	PE-A	PE-B	PE-B	PE-B	PE-C
Specimen thickness $(10^{-3}$ m)	0.55	0.43	0.55	0.90	0.55
PDMS 2 cSt	–	–	2	–	–
20 cSt	–	2	2	1	–
30 cSt	–	2	1	1	2
50 cSt	1	–	1	–	2
10^2 cSt	–	–	–	–	1
10^3 cSt	1	1	1	1	1
10^4 cSt	1	1	1	1	1
NS206	–	–	2	–	–
NS210	1	–	1	–	2
NS215	1	–	1	–	1
n-propanol	2	–	2	–	2

Just as measurable growth does not occur below the threshold temperature (under a prescribed load or deformation), it has been known that there is a threshold load below which crack does not grow (under a prescribed temperature). Often in fracture mechanics, stress intensity factor, K_I, is used as a unique parameter to designate the loading condition. For a crack of length h in a plate under tension, K_I is given by

$$K_I = \sigma_\infty \sqrt{\pi h},\tag{17.2.1}$$

where σ_∞ is the stress far from the crack. Kramer referred in his review to the works by Marshall e.a.[8] and Mai[9] as examples showing the threshold stress intensity factor for ESC growth. The former was concerned with the relations, K_I versus crack growth rate, for PEs immersed in methanol and in ethanol. The latter was concerned with the same relations for a glassy polymethylmethacrylate (PMMA) in air, in CCl_4, and in ethanol. The ESC growth rates decreased significantly below the respective threshold K_I values for the pairs examined.

The strong polymer molecular weight dependence of *time to failure* in ESC tests is well-known. Time to failure is the time from onset to failure of specimens in ESC tests under a prescribed deformation or load. A PE with number average molecular weight, M_n, 11.1×10^3 and weight average molecular weight, M_w, 47.0×10^3 exhibited an ESC time to failure, t_f, of 2 hours; a PE with $M_n = 15.5 \times 10^3$ and $M_w = 157 \times 10^3$ exhibited a t_f of 180 hours; whereas a PE with $M_n = 700 \times 10^3$ and $M_w = 6.4 \times 10^6$ exhibited a $t_f > 4000$ hours.[3]

Rudd[2] imposed a fixed strain on various polystyrenes (PSs) and immersed them in *n*-butanol. Time to failure of PSs in this experiment increased with the eleventh power of PS molecular weight. Kambour said in his review[2] that the chain entanglement concept cannot be unaltered by the breakdown mechanism of the crazes.

Kramer and Kambour, together with other investigators, recognized the importance of the strong polymer molecular weight dependence of time to failure. Time to failure is, by its definition, a rough measure of a combined rate of the ESC initiation and growth. Its molecular weight dependence is a resemblance of our finding in the growth rate behaviors in the low temperature range of type 1 and in type 2.

Several biaxial loading experiments revealed that under the negative mean stress, crazing is generally suppressed.[3] On the other hand, the experiments by Matsushige e.a.[10] showed that crazing can occur in air and in liquids even under hydrostatic pressure. Kramer seemed to feel a contradiction between the two experimental facts.[3] A reasonable picture accommodating the two is required.

A desirable model of ESC growth should account simultaneously for the various problems encountered in *ESC (and ESCR) kinetics* described heretofore. They are:

(1) What is the mechanism of the threshold temperature for ESC growth and the strong molecular weight dependence of the growth rates near that temperature?

(2) Why is the strong molecular weight dependence remedied in the medium temperature range?

(3) Why does the abrupt growth rate increase occur in the high temperature range in type 1?

(4) Above all, what factors determine the type of growth rate versus reciprocal temperature behavior of a given polymer and liquid pair?

(5) What is the mechanism of the threshold load or the threshold stress intensity factor for ESC growth?

(6) How can the hydrostatic pressure contribution be interpreted in a reasonable way?

The task seems to be difficult, but elementary thermodynamics and a very simple model provide a unified picture in approaching the problems.

17.3 Sketch of ESF Processes by Thermodynamics

Before discussing the construction of the relevant thermodynamic potential, we sketch the processes involved in the *ESF initiation and growth.*

Let us consider a notched specimen subjected to a deformation or load and focus our attention on the small region, S, just ahead of the notch. In the region S, a dilative polyaxial stress field due to stress concentration is developed. The failure promoting liquid, though it is a nonsolvent for an unstressed polymer, can swell the polymer matrix in S by virtue of the dilative stress field. In the following discussions, except in cases otherwise stated, the changes in the specimens induced by the liquids are assumed to be confined solely to that region.

The thermodynamic potential of a system, a specimen and a liquid, at a fixed temperature T is depicted schematically by the curves with the stress as a parameter in Fig. 17.3.1. The ordinate is the thermodynamic potential, G, and the abscissa is the concentration of the absorbed liquid in S.

The state of the system is defined by its temperature, stress, and the liquid concentration. When the state of the system is thermodynamically unstable, its thermodynamic potential is not minimum. It will shift along a path where its thermodynamic potential decreases, and stabilize at the minimum if it exists.

An explicit form of the potential will be given in the next section; it confirms that each curve has a minimum at a concentration, ϕ_M, provided the stress is below a limit, and that ϕ_M increases with the increase in the stress.

Curve (1) corresponds to a milder deformation (load) condition, curve (2) to a medium deformation (load), and curve (3) to a greater deformation (load). In this stage, it is assumed that the greater a deformation (load) is, the higher the stress in the region S becomes.

The ESC initiation and growth processes are considered in this chapter as an isothermal state shift in the diagram. At the start of the experiment, the state of the system in the diagram is represented by the origin. By the application of the deformation (load), the state shifts to A where OA corresponds to the elastic energy stored in

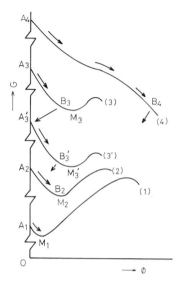

Fig. 17.3.1. Thermodynamic potential diagram showing the processes of environmental stress failure. The ordinate is the thermodynamic potential of a system composed of a specimen and its environmental liquid. The abscissa is the concentration of the liquid absorbed in the region ahead of a crack or a flaw. For labeling of points indicated in the figure, see the text.

the specimen. The state then shifts to the minimum M by way of the diffusional flow of the liquid molecules into S. Under the milder deformation (load), the state of the system stabilizes at the minimum of the curve (1) M_1 and no crack will initiate. To the naked eye, the swelled state will be seen as a craze provided that it accompanies orientation of polymer matrix.

In the greater deformation (load), the state is at A_3 of curve (3) $(OA_1 < OA_3)$ just after deformation (loading). The state then shifts to the minimum M_3. In this case, the concentration at M_3 is so high that the region S is weakened by the liquid absorption, and breaks at a point B_3 somewhere between A_3 and M_3. After the breaking, the state is at A'_3. The distance $A_3 A'_3$ is the energy released by the breaking of the region S. The breaking yields another stress concentrated region, the processes will be repeated, and the crack will grow. The relevant thermodynamic potential is depicted by the curve (3').

There must be a situation depicted by curve (2), between (1) and (3). In curve (2), the concentration at the minimum potential coincides with the concentration at the breaking point, B. With a load less than that corresonding to curve (2), the system will be stabilized at the minimum, and the crack will not initiate or grow. Under a load higher than that, the crack will progress. Therefore, the load is at the level specified by the threshold stress intensity factor for ESC growth.

The precise nature of the breakdown point B is unknown, but the liquid concentration at B, ϕ_B, must be an increasing function of polymer molecular weight. The polymer molecular weight dependence of the threshold stress intensity factor and the accompanied extraordinary molecular weight dependence of the growth rates are qualitatively explained by the nature. (See below.)

With a further increase in the deformation (load), the G curve becomes monotonous decreasing as shown by curve (4). The path of the state shift leading to failure will abruptly change at that moment.

Another possible answer is that the stress field, which is in the yielding state, is not so much sensitive to the applied deformation (load) but that ϕ_B is more affected by it. The value of ϕ_B may be decreased by the increase in the deformation (load). The relations, K_I versus growth rate, will envisage the threshold K_I as well. We may not discriminate between the two possible answers at a glance.

As described in the next Section, the shapes of the G curves are not changed as much by the molecular size of the polymer as they are by the other parameter values. The changes are responsible for the variations in ESC kinetics.

In general, the state shift A to B or A to M proceeds by way of the diffusional flow of liquid molecules described earlier, and the shift A to A' proceeds by way of the viscoelastic flow of the mixture of polymer and liquid.

Though the potentials in Fig. 17.3.1 are depicted as *isostress curves*, the supposition that the stress is maintained constant during the state shift carried out by the liquid flow is perhaps incorrect. The transformation of the polymer matrix to the craze structure will occur in the state shift. The inclusion of these is a formidable task and is abandoned at present.

Here, we consider the case where the ESC growth rates are determined by the state shift A to B rather than by that of B to A'. Let $f(i, t)$ be the probability that there are i liquid molecules in the region S at a time t. Then,

$$\partial f(i, t)/\partial t = \{f(i+1, t) \Lambda_{i+1, i} - f(i, t) \Lambda_{i, i+1}\}$$
$$- \{f(i, t) \Lambda_{i, i-1} - f(i-1, t) \Lambda_{i-1, i}\} \tag{17.3.1}$$

where, $\Lambda_{i, i\pm 1}$ means the probability that a liquid molecule migrates in ($+$sign) or out from ($-$sign) the region S with i liquid molecules in a unit time interval. According to the Eyring's rate theory, $\Lambda_{i, i\pm 1}$ may be written by

$$\Lambda_{i, i\pm 1} = \Lambda_0 \exp\{-(\Delta H + (1/2) \Delta G^{\pm}(i))/kT)\} \tag{17.3.2}$$

where

$$\Delta G^+(i) = G(i+1) - G(i)$$
$$\Delta G^-(i) = G(i-1) - G(i)$$

Λ_0 is a constant, weakly dependent on temperature, and ΔH is the activation enthalpy to be surmounted in the migration.

The mean time of state shift A to $B, \langle t \rangle$, is given through solving Eq. (17.3.1) with respect to $f(i, t)$ under the initial condition

$$f(i, 0) = \delta(i)$$

and under the boundary conditions that $f(i, t)$ is not defined for negative i and that at the point B

$$f(i_B, t) = 0$$

where $\delta(i)$ is the delta-function, and i_B is the number of liquid molecules in S at the point B. A convenient time t', $t' = \Lambda_0 \exp\{-\Delta H/kT\} t$, is introduced. The mean time of state shift A to B is thus

$$\langle t_0 \rangle = \langle t_0 \rangle \Lambda_0^{-1} \exp\{\Delta H/kT\} \tag{17.3.3}$$

where $\langle t_0 \rangle$ is given by

$$\langle t_0 \rangle = \exp\{\Delta G^+(i_B - 1)/2kT\} \int_0^\infty t' f(i_B - 1, t') dt' \tag{17.3.4}$$

In the present case, ESC growth rate, g_r, is assumed as being inversely proportional to $\langle t \rangle$

$$g_r \propto \langle t_0 \rangle^{-1} \Lambda_0 \exp\{-\Delta H/kT\} \tag{17.3.5}$$

The dependence of $\langle t_0 \rangle$ on temperature is often so large, especially near the threshold temperature, that the slope of the Arrhenius plot of growth rate does not always give the activation enthalpy ΔH.

The readers should refer to the Appendix in which numerical calculations of $1/\langle t_0 \rangle$ are given for those systems simulating the experiments.

17.4 Formulation of Thermodynamic Potential

To obtain the explicit form of the thermodynamic potential, the model shown in Fig. 17.4.1 is considered. Domain I corresponds to the region S just ahead of a notch (or a flaw or a crack) and domain II to its environmental liquid. The former consists of n_r polymer molecules of r elements and $n_{s,1}$ liquid molecules of s elements. It is subjected to a uniform polyaxial stress induced by the combined effect of the stress concentration and the environmental hydrostatic pressure, the mean principal stress of which is p_1. The latter consists of $n_{s,2}$ liquid molecules, and is subjected to a mean principal stress p_2 due to the hydrostatic pressure. Domains I and II are in contact, and the liquid molecules can migrate between the two domains. The sum of the number of the liquid molecules and the sum of the volumes of the two domains are fixed. The system is maintained at a temperature T. A fundamental assumption is that the thermodynamic equilibrium is preserved within each domain but not between them. The thermodynamic potential of the system, G, is defined as the sum of the Gibbs function of each domain.

$$G = n_r \mu_r + n_{s,1} \mu_{s,1} - p_1 V_1 + n_{s,2} \mu_{s,2} - p_2 V_2 \tag{17.4.1}$$

where μ_r and $\mu_{s,1}$ are the Helmholz chemical potentials of a polymer molecule and of a liquid molecule in I respectively, $\mu_{s,2}$ is that of a liquid molecule in II, and V_1 and V_2 are the volumes of I and II respectively.

The chemical potentials are conveniently approximated by the *Flory-Huggins lattice theory*.[11] Designating the thermodynamic potential in its reduced form $G/(kTr\,n_r)$ as $\tilde{G}$, after subtracting a constant term we obtain

$$\tilde{G} = (1/r)\ln(1-\phi) + (1/s)(\phi \ln \phi)/(1-\phi) + \chi \phi - \Delta p \, \phi/(1-\phi) \tag{17.4.2}$$

Where, ϕ is $s n_{s,1}/(r n_r + s n_{s,1})$, the volume fraction of the liquid molecules in I, χ is the interaction parameter defined for a pair, a polymer molecule element and a liquid molecule element, and Δp is $(p_1 - p_2)/kT$ (the element volume being 1). The stress term in the thermodynamic potential is given by the form of a difference between the stresses in the two domains and not by the stress in domain I. (This is the answer for the problem (6) in Section 17.2.)

Although extensive work has been done on the interaction parameter, its original form is given by

$$\chi = V_m (\delta_r - \delta_s)^2 / RT, \tag{17.4.3}$$

where δ_r and δ_s are the solubility parameters of the polymer and of the liquid, and V_m is the molar volume of the element.

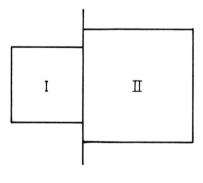

Fig. 17.4.1. Model representing the region in which ESC occurs. Domain I corresponds to the region just ahead of a crack or a flaw in a loaded polymer specimen and domain II corresponds to its environmental liquid. Reproduced from reference 6 by permission of the publishers; ©, Butterworth & Co (Publishers) Ltd.

We regard the thermodynamic potential of the whole system, a specimen and a liquid, as given by the sum of the Gibbs function, G described above and the elastic energy stored in the specimen body. Thus the curves in Fig. 17.3.1 are given by the sums.

17.5 Changes of the Shapes of the Thermodynamic Potential Curves

First, the changes of the $\tilde{G}$ versus ϕ curves produced by the changes of Δp values are described by the numerical calculations of the formula (17.4.2). Our thermodynamic picture sketched in Section 17.3 will be reconsidered.

Figure 17.5.1 shows the $\tilde{G}$ versus ϕ curves of the system with $r=1000$, $s=8$, $\chi=0.35$, and a variety of Δp values. The system roughly simulates a pair of LDPE and NS210 (For NS210, see Table 17.2.2). If r is beyond 100, the curves of different r values are not separate in the figure. Thus, to assign an arbitrary value 1000 to r for any LDPE used is not a problem. The value of $s=8$, was estimated from a convention that a propanol molecule occupies one site in the Flory-Huggins lattice. (This convention which was first used in calculating the $\tilde{G}$ curves of a LDPE and propanol pair is followed throughout.) The χ value of the pair at room temperature was estimated by Eq. (17.4.3) in which the solubility parameters were evaluated by the method of Small and Hoy.[12]

When Δp is negative or zero, ϕ_M is close to zero, i.e., the pair is almost incompatible. For a nonnegative Δp, $\tilde{G}$ exhibits a minimum and a maximum. With the increase in Δp, ϕ_M increases. These are parallel to the curves in Fig. 17.3.1. When Δp is equal to a value Δp_t, or the applied load attains a value W_t, the threshold load, ϕ_M is equal to ϕ_B and ESC will start to grow.

When Δp exceeds Δp_c, $\tilde{G}$ becomes monotonous decreasing. We call Δp_c the critical mean stress.

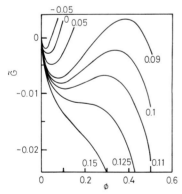

Fig. 17.5.1. Reduced thermodynamic potential $\tilde{G}$ versus liquid concentration ϕ calculated from the formula (17.4.2). The parameters are $r=1000$, $s=8$, and $\chi=0.35$. The values of Δp are indicated in the figure. Reproduced from reference 6 by permission of the publishers; ©, Butterworth & Co (Publishers) Ltd.

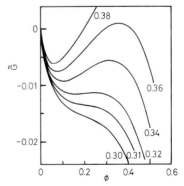

Fig. 17.5.2. Reduced thermodynamic potential $\tilde{G}$ versus liquid concentration ϕ calculated from the formula (17.4.2). The parameters are $r=1000$, $s=8$, and $\Delta p=0.1$. The values of χ are indicated in the figure. Reproduced from reference 6 by permission of the publishers; ©, Butterworth & Co (Publishers) Ltd.

Second, the changes of the $\tilde{G}$ curves by the changes of temperature are described. The growth rate versus reciprocal temperature relations, type 1 and type 2, will be discussed.

Figure 17.5.2 shows the $\tilde{G}$ versus ϕ curves of the same system, as treated in Fig. 17.5.1, calculated for a fixed $\Delta p (=0.1)$ and a variety of χ values. The system simulates a pair of LDPE and NS210 at several different temperatures. The conversion of χ values to temperatures can be made by Eq. (17.4.3). If we assume that the polymer in S is in the state of yielding in air, and that the environmental hydrostatic pressure contributes as an additive term in the mean principal stress, we may approximate Δp as $(\sigma_y / 3RT) V$ where σ_y is the yield stress of the polymer, and V the molar volume of propanol. It follows that $\Delta p \sim 0.1$ for the LDPEs used. The changes in Δp values induced by the temperature changes are not included.

At a lower temperature, in Fig. 17.5.2, where χ is larger, ϕ_M is close to zero and the pair is almost incompatible. With a decrease in χ, ϕ_M increases. When $\chi = \chi_t$, ϕ_M is equal to ϕ_B, ESC will start to grow at the corresponding temperature, the threshold temperature for ESC growth. Different ϕ_B values corresponding to different polymer molecular weights yield both different threshold temperatures and strong molecular weight dependence of the growth rate around those temperatures. The polymer molecular weight dependence of the thermodynamic potential does not concern them if $r > 100$. The value of $\langle t_0 \rangle$ in Eq. (17.3.5) is very much temperature dependent in this range, and it makes the apparent activation enthalpy unusually high. The numerical calculations following the schema in Section 17.3 support the interpretation (see Appendix).

With the further decrease in the χ value or the elevation of T, ϕ_M increases beyond ϕ_B. ESC will grow steadily following the rules of the diffusion of liquid molecules and of the flow of a polymer and liquid mixture. These just correspond to the moderate polymer molecular weight dependence of the growth rate in the medium temperature range. The value of $\langle t_0 \rangle$ is not very much dependent on temperature and the moderate apparent activation enthalpy in this temperature range results (see Appendix).

To the extent that the potential has a minimum, ϕ_B should be less than ϕ_M for crack growth because the state shift beyond the potential minimum is almost inhibited. For χ values smaller than χ_c, or at temperatures higher than the critical temperature T_c, the potential minimum vanishes and the $\tilde{G}$ versus ϕ curves are monotonous decreasing. The state can shift to an indefinitely dilute state in principle. The restriction for ϕ_B vanishes. Another easier path to the breakdown $A_4 \rightarrow B_4 \rightarrow A_4'$ will set in abruptly (Fig. 17.3.1). An abrupt increase in the ESC growth rate will occur. This is the onset of the high temperature range in type 1 behavior (Fig. 17.2.2).

Further, if the polymer molecular weight is high enough, ϕ_B will always exceed ϕ_M. ESC will not start to grow until the critical temperature is reached. The low and medium temperature range behaviors are missing in that case. The growth rate curve by PE-C, the highest molecular weight polymer examined in NS210, behaves just in that manner (Fig. 17.2.4).

In short, the expected changes of the path of the state shift leading to failure by the changes of the shapes of the $\tilde{G}$ curves with temperature are quite consistent with type 1 growth rate behavior.

Figure 17.5.3 shows the $\tilde{G}$ versus ϕ curves for the system $r = 1000$, $s = 1$, $\Delta p = 0.1$, and a variety of χ values. The parameter values are so chosen that the system simulates the LDPE and propanol pairs in a temperature range covering the experimental one. In contrast to those for the system simulating the LDPE and NS210 pairs, every curve has a minimum. The χ value at the critical condition of the pair is 1.1 which is outside of the experimental range. The growth rate versus temperature relations should reflect the single mechanism. This is just consistent with the experimental findings in the LDPE and propanol pairs (Fig. 17.2.3).

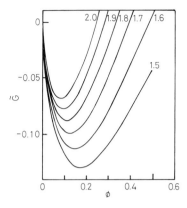

Fig. 17.5.3. Reduced thermodynamic potential $\tilde{G}$ versus liquid concentration ϕ calculated from the formula (17.4.2). The parameters are $r = 1000$, $s = 1$, and $\Delta p = 0.1$. The values of χ are indicated in the figure. Reproduced from reference 6 by permission of the publishers; ©, Butterworth & Co (Publishers) Ltd.

The polymer molecular weight dependence of the threshold temperature for ESC growth and the resulting very significant molecular weight dependence of the growth rates are interpreted in the same way as in the low temperature behavior in type 1.

We already have the qualitative answers for problems (1), (2), (3), (5), and (6) in Section 17.2. The remaining problem (4) is also obvious. Those pairs having T_c in the experimental range will exhibit type 1 behavior if $\phi_B < \phi_M$ below T_c. They will exhibit type 2 behavior if $\phi_B > \phi_M$. Those pairs not having T_c in the experimental range will also exhibit type 2 behavior. The LDPEs and propanol pairs in Table 17.2.3 correspond to the last.

17.6 Estimation of Parameter Values

The interpretation in the preceding section states that ESC initiation and growth kinetics can be depicted by the state shift in the relevant thermodynamic potential diagram. The shapes of the $\tilde{G}$ curves and the values of the breakdown concentrations play the primary roles in the ESC kinetics.

However, the estimations of the values of the parameters involved in the $\tilde{G}$ formula (17.4.2) and of the breakdown concentration are not an easy task. The crude estimations made for LDPE and NS210 pairs and for the LDPE and propanol pairs yield successful results, but they are fortunate.

The polymer and liquid pairs with which we should be concerned are incompatible in the stress free state. They are the ones which have been outside of the studies of polymer solutions. Furthermore, it is not totally assured that the way to estimate the χ values by way of Eq. (17.4.3) and the Small and Hoy's method can be applied to otherwise incompatible pairs. To estimate p_1 is also an unsettled problem in fracture mechanics at present.

What we can do in these circumstances is to establish parameter values in such a way that the theory accommodates itself to the ESC kinetics observed and to check whether the values are reasonable ones.

The ESC growth rate versus reciprocal temperature relations for the LDPE and PDMS pairs were studied in our laboratory. The viscosities of the PDMSs used ranged from 2 cSt to 10^4 cSt (see Table 17.2.3), and their molecular weights are from 384 to 5.7×10^4 according to the conversion formulae by Warrick e.a.[18] Their growth rate versus reciprocal temperature relations were classed as type 1 and type 2 depending on the molecular weights of the LDPEs, the molecular weights of the PDMSs, and the thickness of the specimens.

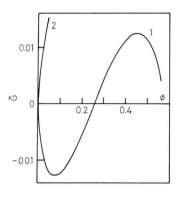

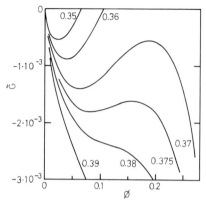

Fig. 17.6.1. Reduced thermodynamic potential $\tilde{G}$ versus liquid concentration ϕ calculated from the formula (17.4.2). The parameters are $r = 1000$, $\Delta p = 0.1$, $\chi = 0.5$; $s = 5$ (curve 1) and $s = 50$ (curve 2) respectively. Reproduced from reference 7 by permission of the publishers; ©, Butterworth & Co (Publishers) Ltd.

Fig. 17.6.2. Reduced thermodynamic potential $\tilde{G}$ versus liquid concentration ϕ calculated from the formula (17.4.2) and concentration-dependent interaction parameter $\chi = 0.4 + 0.5\phi$, for the pair of $r = 1000$ and $s = 50$. The values of Δp are indicated in the figure. Reproduced from reference 7 by permission of the publishers; ©, Butterworth & Co (Publishers) Ltd.

The direct application of Eq. (17.4.3), together with the method of Small and Hoy, gave χ approximately 0.5. The ϕ_Ms of the $\tilde{G}$ curves with $\chi = 0.5$ and $s = 50$ are very close to zero. We cannot expect from our theory that ESC might be induced by larger viscosity PDMSs contrary to the experiments. See Fig. 17.6.1 where the value of s, 50, roughly corresponds to the 50 cSt PDMS.

If we put $\chi = 0.4 + 0.5\phi$ and $\Delta p = 0.37$ for PDMSs with viscosities higher than 30 cSt, we get the $\tilde{G}$ curves capable of accounting for the ESC kinetics observed (Fig. 17.6.2). The concentration dependence seems artificial. But it is a mere modification of the findings of Sugamiya e.a.,[13] i.e., $\chi = 0.4 + 0.2\phi$, for the PDMS ~ n-alkane system and of Flory and Shih,[14] i.e., $\chi = 0.5 + 0.3\phi$, for the PDMS ~ benzene system.

By the introduction of these, we can understand the polymer molecular weight dependence of the growth rate behaviors of the ESC in LDPE and PDMS pairs in the same way as described heretofore.

The ESC threshold temperature rose with the decrease in the specimen thickness, whereas the critical temperatures were nearly unaffected by the thickness (Fig. 17.2.5). A consistent speculative conclusion based on this is that the Δp is not affected by the specimen thickness whereas the breakdown concentrations are increased by thinner specimens. The former is natural because the crack tip region is in a yielded state. The latter, too, is reasonable because, in the bending deformation, the elastic energy density stored in specimen body is decreased by a thinner specimen.

The measured values of the loads to hold specimens in PDMSs at the specified deformation in the ESC experiments were decreased by low viscosity PDMSs. It indicates a nonnegligible swelling in a specimen body by the low viscosity PDMSs. Since the swelling will result in two simultaneous effects — lowering Δp and increasing ϕ_B — the comparison of the growth rate behaviors obtained by PDMSs of different viscosities involves difficulties.

17.7 Discussion and Summary

The processes in environmental stress failures in polymers are analyzed in terms of irreversible thermodynamics. The systems concerned are composed of the polymer at a crack (or a flaw) tip region and its environmental liquid. Its thermodynamic potential is constructed as a function of the concentration of the liquid absorbed in the region and the dilative stress due to stress concentration. A state of the system is represented by a point in the thermodynamic potential diagram. The path of the state shift leading to failure in the diagram is governed by the shape of the thermodynamic potential curve, the liquid concentration giving the potential a minimum, the critical concentration, and the breakdown concentration introduced here.

The formulation of the thermodynamic potential is made by a very simple model to overcome the otherwise inaccessible complexities involved in the actual processes. Inclusion of polymer crystallinity effect is conceivable only through the term Δp. Difficulties still remain in assigning definite values to the relevant parameters which are, at present, hard to estimate closely. The adjustment of the parameter values in plausible ranges is inevitable.

In spite of the ambiguities introduced, the success in understanding the variations of the ESC kinetics found in our laboratory and reported in the literature is remarkable. It tells us that the essentials of the ESC kinetics are certainly included in the proposition.

In general, the description in this Chapter concentrates on the thermodynamic force leading to failure and the accompanied diffusional flow of liquids. The flow of the polymer and liquid mixtures by the force, shifting the state $B_3 \rightarrow A'_3$, has not been our main concern. The flow of the mixtures, however, must be closely related to craze structures, craze breakdown, and fracture surface morphology. The authors acknowledge the importance of the nonlinear behavior involved in the flow as Argon and Salama emphasized before.[15]

Appendix

Figure 17.A.1 shows examples of the numerical calculations made for the relation $1/\langle t_0 \rangle$, proportional to the growth rate, versus reciprocal temperature with ϕ_B as a parameter. The relation resembles the observed growth rate versus reciprocal temperature relations. The solid curves were calculated for the systems simulating pairs of LDPE and NS210, and the dotted curve was for a pair of LDPE and propanol. Their $\tilde{G}$ curves are depicted in Figs. 17.5.2 and 17.5.3.

Each of the calculated $1/\langle t_0 \rangle$ curves reflects well the existence of the threshold temperature for ESC growth. Because the effect of polymer molecular weight enters through ϕ_B, the strong molecular weight dependence of the growth rate near the threshold temperature is envisaged by comparing the curves of different ϕ_B s with each other.

The increase in the growth rate or in $1/\langle t_0 \rangle$ by the decrease in $1/T$ decreases in the medium temperature. The growth rate differences by the values of ϕ_B or by polymer molecular weights are diminished in that range. These are consistent with the type 1 behavior in the medium temperature range. Here, the difference in the values of the factor $\Lambda_0 \exp\{-\Delta H/kT\}$ of the three is discarded because our interest is only in the shapes of the curves.

The calculated growth rate increase with the temperature elevation or with the decrease in χ, may appear too small when compared with the experiments. The numerical calculations were performed for small systems having a size of $rn_r = 2200$ which is too small compared with real ones. Increasing rn_r will make the calculated ones more like the experimental ones.

The numerical calculations of $1/\langle t_0 \rangle$ versus Δp relations with a constant χ yielded a similar curve showing the existence of threshold stress for ESC growth.[16]

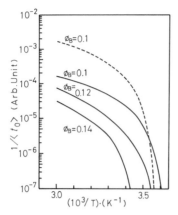

Fig. 17.A.1 Reciprocal of mean time of state shift A to B $1/\langle t_0 \rangle$ versus reciprocal temperature $1/T$. Calculations were performed for the two systems, the thermodynamic potentials of which are depicted in Figs. 17.5.2 and 17.5.3. The solid curves were calculated for the former and the dotted curve for the latter. The parameters are $r = 100$, $n_r = 22$, and the values of ϕ_B are indicated in the figure. Reproduced from reference 6 by permission of the publishers; ©, Butterworth & Co (Publishers) Ltd.

References and Notes

1. Crazes are narrow plastically deformed regions perpendicular to the applied tension. They consist of thin bands of porous oriented matter still capable of bearing stress. If the stress is maintained, cracks often form preferentially in the weak crazed material. A.N. Gent, *J. Mater. Sci.* 1970, **5**, 925.
2. R.P. Kambour, *J. Polymer Sci. Macromol. Rev.* 1973, **7**, 1.
3. E.J. Kramer, *Developments in Polymer Fracture*, edited by E.H. Andrews, Chapter 3, Appl. Sci. Publishers, London 1979.
4. Another conceivable role of the liquids is the surface energy reduction in fracture. For example, see E.H. Andrews and L. Bevan, *Polymer* 1972, **13**, 337. If the effect is nonnegligible, the potential curve (3′) in Fig. 17.3.1 should shift vertically by an appropriate amount. Our experiments have not indicated the need to include the effect.
5. Y. Ohde and H. Okamoto, *J. Mater. Sci.* 1980, **15**, 1539.
6. H. Okamoto and Y. Ohde *Polymer* 1982, **23**, 1204.
7. Y. Ohde and H. Okamoto, *Polymer* 1983, **24**, 723.
8. G.P. Marshall, L.E. Calver and J.G. Williams, *Plast. Polymers* 1969, **37**, 95.
9. Y.W. Mai, *J. Mater. Sci.* 1975, **10**, 943.
10. K. Matsushige, S.V. Radcliffe and E. Baer, *J. Mater. Sci.* 1973, **10**, 833; K. Matsushige, E. Baer and S.V. Radcliffe, *J. Macromol. Sci. Phys.* 1975, **B 11**, 565.
11. P.J. Flory, *Principles of Polymer Chemistry*, Cornell Univ. Press, Ithaca, New York, 1971. Use of the new Flory theory involving free volume was not tried because our present interest is limited to the qualitative nature of the thermodynamic potential.
12. H.Burrel, *Polymer Handbook*, edited by J. Brandrup and E.H. Immergut, IV, Wiley-Interscience, New York 1975.
13. K. Sugamiya, N. Kuwahara and M. Kaneko, *Macromolecules* 1974, **7**, 66.
14. P.J. Flory and H. Shih, *Macromolecules* 1972, **5**, 761.
15. A.S. Argon and M. Salama, *Mater. Sci. Eng.* 1976, **23**, 219.
16. H. Okamoto and Y. Ohde, *Polymer* 1980, **21**, 859.
17. Q.A. Trementozzi, *J. Polymer Sci.* 1957, **23**, 887.
18. E.L. Warrick, W.A. Piccoli and F.O. Stark, *J. Amer. Chem. Soc.* 155, **77**, 5017.

18 Fatigue Failure

Abdelsamie Moet

Department of Macromolecular Science, Case Western Reserve University
Cleveland, OH 44106, U.S.A.

18.1 Introduction

It is generally recognized that cyclic load applicaton to a material causes its prema-ture failure at stresses well below its tensile strength. This phenomenon is known as "fatigue failure". The fact that fatigue loading is frequently encountered in many engineering applications, coupled with an ever increasing use of plastics as load bearing components, spurred active research in the field of fatigue behavior of polymers and their composites. Following the methodologies well established for metals, fatigue be-havior of plastics is addressed through three complementary approaches: (1) fatigue life time prediction using Wöhler's technique[1] commonly accepted as the S-N (stress range versus number of cycles to final failure) curve, (2) mechanistic studies, and (3) fa-tigue crack propagation analysis.

Although the study of fatigue behavior of plastics was enriched with the wealth of information on their behavior under other loading conditions such as creep, stress relaxation and continuous deformation, fatigue introduces a number of additional fac-tors, such as loading frequency, upper and lower loading limits and wave form. These features appear to produce failure mechanisms otherwise not encountered. Additionally, the dissipative nature of polymers results in high mechanical hysteresis. Because of their low thermal conductivity, a large portion of the mechanical work done is converted into heat; a problem which complicates the analysis of fatigue data particularly at high loading frequency.

Recent results on fatigue behavior of polymers have been summerized in a series of publications.[2-5] In this chapter, more recent advances are reviewed together with the relevant background from previous results. Chapter 2 by Kenner on failure mecha-nisms presents a useful introduction to the phenomenon.

18.2 Fatigue Life Time Prediction

The traditional approach to fatigue life time prediction involves the use of the "endurance limit" concept developed from the S-N curve (Whöler diagram). A speci-men is subjected to constant amplitude cycling until failure at $N(\sigma)$ cycles. The applied stress σ is usually expressed as the stress amplitude; $\sigma_a = (\sigma_{max} - \sigma_{min})$, or the mean stress $\sigma_m = (\sigma_{max} + \sigma_{min})/2$. A plot of σ against N is known as the S-N curve (Fig. 18.2.1).[6] The letter S is almost invariably used in this context to denote stress (σ). Some materials are believed to possess a stress limit below which failure does not occur after a large number of cycles, usually in the order of 10^7–10^8 cycles. This stress is usually defined as the "endurance limit" and is hoped to be a material property.

Inspite of its empirical nature, the S-N approach is a commonly accepted design criterion for fatigue resistance in engineering plastics. For example, ASTM method D671 for plastics specifies repeated flexural stress (fatigue) as a standard test.

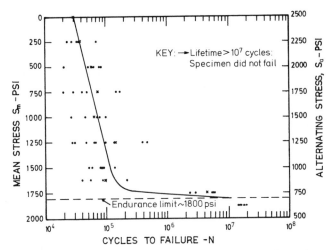

Fig. 18.2.1. Mean stress S_m and alternating stress S_a vs. log N for polystyrene specimens subject to a fixed maximum stress $S_{max} = 2500$ psi. Reproduced with permission.[6]

Experiments to construct an S-N curve are time consuming. Several samples are required to account for the statistical nature of the data obtained. It is common that fatigue life-time data from well controlled samples spread over few orders of magnitudes. This, in fact, reflects the complex nature of fracture processes involved. In this regard, it is useful to consider the unique experiment conducted by Clyton-Cave and co-workers[7a] in the mid fifties and compiled recently by McClintock and Argon[7b] (Fig. 18.2.2). Four hundred identical specimens of EN-24 steel were tested near their endurance limit. Although the life time measured as the number of cycles to fail, mostly clusters within one decade, the scatter spreads over three decades. On the other hand, the error in the fatigue stress limit falls within a reasonable range of less than $\pm 5\%$.

Heterogeneities inherent in the microstructure of most materials result in a random field of defects whose geometry, size and orientation are also random. Such random field of defects influenced by the imposed stress gives rise to a complex process of growth and interaction of defects which ultimately lead to the initiation of macroscopic cracks. A crack propagates first in a stable manner to a stage at which it undergoes a transition to unstable (uncontrolled) propagation. A structure's life time is accordingly composed of two stages, namely, crack initiation and crack propagation. Depending on the severity of defects, crack initiation may comprise 20 to 80% of the total life time. Hence, sound life time prediction ought to rely on knowledge of the law of

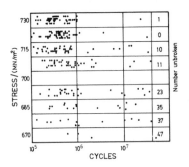

Fig. 18.2.2. S-N behavior of four hundred specimens of EN-24 steel tested near the endurance limit. Reproduced with permission.[7b]

crack initiation and that of slow crack propagation. This notion perhaps prompted research on fatigue mechanisms, cumulative damage, and fatigue crack propagation. Presently, fatigue damage on the microscale leading to crack initiation, and crack propagation in polymers can be measured quantitatively. Knowledge of submicroscopic events such as diffusion of chain molecules, disentanglement, fibrillation, and chain scission which constitute the underlying phenomena of damage formation remains qualitative in nature yet significantly important. Chapter 5 by Kausch reviews the current status of the molecular origin of fracture.

18.3 Mechanisms of Fatigue Fracture

Depending on the stress amplitude and the frequency of load application, fatigue failure of some polymers has been observed to occur via one of two general mechanisms. The first involves thermal softening (or yielding) which precedes crack propagation leading to ultimate failure. In certain materials, this mechanism dominates at large stress amplitudes within a particular range of frequency of load application.[8-10] At a lower stress amplitude, on the other hand, a conventional form of fatigue crack propagation (FCP) mechanism is generally observed. Low frequency is also found to cause fatigue fracture by conventional crack propagation at high stress amplitude. The interrelation of the two mechanisms, stress amplitude, and frequency for polyoxymethylene (acetal) is shown in Fig. 18.3.1.

The high damping and low thermal conductivity of polymers cause a strong dependence of temperature rise on the rate of load application (frequency) and on the deformation level (stress or strain amplitude). From a thermodynamic point of view,[11] part of the mechanical work done during cyclic loading is spent on irreversible *molecular processes* (see again Chapter 5 by Kausch) *leading to* microscopic deformations such as crazes, shear bands, voids, etc. The other part evolves as heat. Both processes are

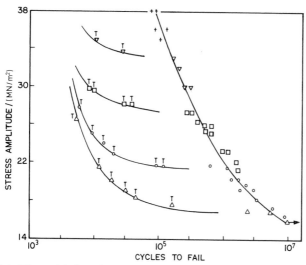

Fig. 18.3.1. Thermal fatigue failure (*T*) and conventional FCP fracture observed during reversed load cycling of acetal. +, 0.167; ∇, 0.5; □, 1.67; ○, 5.0; △, 10 Hz.[9] Reproduced from R.J. Crawford and P.P. Benham, *Polymer*, 1975, **16**, 908 by permission of the publishers, Butterworth & Co. (Publishers) Ltd© .

obviously interdependent and relate to the specific nature of the relaxation time spectrum of the macromolecules considered. The total work done is measurable from the hysteresis loop encountered in fatigue testing. Attempts were thus made to characterize fatigue damage from the evolution of the stress-strain relationship reflected in the hysteresis loop.[8, 12, 13] This type of mechanistic analysis is particularly useful for uncracked specimens with a dominant single fatigue damage mechanism.

FCP mechanism frequently involves craze formation which precedes crack initiation and propagation. On the other hand, the micromechanism underlying yielding (softening or thermal failure) remains unclear. Recent fatigue experiments on polycarbonate indicate that cyclic softening is caused by profuse crazing prior to fracture.[12] Microscopic examination of the same polymer[14, 15] shows that fatigue crazes are terminated by, and interact through, pairs of shear bands. A crack ultimately initiates and propagates within one of the crazes. Related studies on PS and PMMA, show that long fatigue life at low stress amplitude is associated with more profuse crazing along the gauge length.[16]

Current evidence suggests that large deformation (softening) can precede crack initiation under certain loading conditions. In this case, subsequent crack propagation may occur in a localized region of the transformed (softened) material. Although the nature of such deformation depends on the molecular structure and its interaction with the stress field, specimen separation remains to occur via fatigue crack propagation. A similar situation is encountered in the creep fracture of polyethylene where brittle fracture is observed at low load and ductile fracture is observed at high load.[17] The latter is none but a "brittle" crack propagation through a large yielded zone.

18.4   Fatigue Crack Initiation

The initiation of macroscopic cracks in the order of 10^{-3} m under fatigue loading is studied through two complementary approaches. A fracture mechanics approach characterizes fatigue crack initiation (FCI) by a threshold value of the stress intensity factor K_{th} or its range ΔK_{th}. This type of study involves the use of fracture mechanics concepts, the basics of which are discussed in Chapter 7. Generally, fatigue load is applied to a notched specimen and the first measurable crack (notch) extension Δh is recorded. Fatigue threshold signifies that not every precrack (notch) will extend, so that a critical condition must be met for $\Delta h / \Delta N$ to exist. Accordingly, the threshold value ΔK_{th} is interpreted as that minimum of ΔK which is required to make the precrack grow at all. The hypothesis[18] is that the crack growth is linearly related to the crack opening displacement (COD). If COD during loading exceeds a threshold value, a permanent step Δh of the crack is assumed to remain open upon unloading. In other words, fatigue threshold describes the "first" crack jump. The corresponding ΔK, i.e., ΔK_{th} is thought of as a material property characterizing the resistance to crack initiation. Alternatively, the related energy release rate $\Delta G_{th} \left(= \dfrac{\Delta K^2}{E} \right)$ has also been considered.[19]

On the other hand, mechanistic investigations of initially uncracked and initially cracked polymer specimens emphasize the role of crazing in FCI in glassy and semicrystalline polymers. For example, in a 6 mm thick extruded unnotched, polycarbonate specimen exposed to high strain fatigue, it is noted that the formation of microcrazes terminated by shear bands precedes crack initiation[14] (Fig. 18.4.1 a). The crazing density appears to reach a critical level at which the main fatigue crack initiates within one of the crazes. Once initiated, subcritical crack popagation occurs through a craze surrounded by a pair of shear bands (Fig. 18.4.1 b) forming what is known as an epsilon

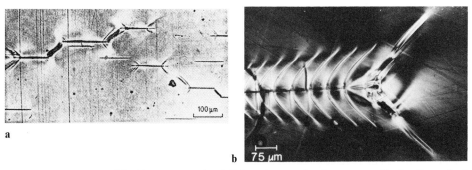

Fig. 18.4.1. (a) Normal incidence transmitted light micrograph of crazes on the tensile surface of a polycarbonate specimen. Crazes are terminated by pairs of shearbands. Reproduced with permission.[14]

(b) Crack propagation through a craze surrounded by a pair of shear bands (an Epsilon crack) in polycarbonate. Reproduced with permission.[15]

crack.[15] Similar FCI behavior is observed in tension-compression fatigue of unnotched polycarbonate sheet.[12] In HIPS (High impact polystyrene) and ABS (Acrylonitrite butadiene styrene), analysis of hysteresis loops infers that FCI occurs due to crazing and shear banding, respectively.[13] The magnitude of crazing developed prior to crack initiation depends on the stress level and on test frequency. From optical interference measurements recently reviewed by Döll,[20] it is inferred that crack initiates and propagates in glassy polymers under certain conditions through a single craze. Optical micrography on the other hand, shows that a few[2] or a myriad[21] of crazes precede FCI. What appears to be common in all these observations is that a critical level of damage ought to be reached to cause crack initiation. This critical level of damage seems to correspond to the sudden crack jump characterized by ΔK_{th}. Efforts are underway to develop techniques for quantitative damage analysis.[21] A thermodynamic approach by Chudnovsky[22] treats the phenomenon as local instability and proposes a framework to establish the law of crack initiation.

18.5 Fatigue Crack Propagation

Advances in fracture mechanics in the recent past inspired tremendous interest in fatigue crack propagation (FCP) which evolved as an independent discipline. Attempts to formulate the law of subcritical (slow, stable, or quasistatic) crack propagation under intermittent load application play a central role in this effort. Inspite of the mechanistic differences in FCP of metals and polymers, the formal approach remains the same as it is founded on the ideas of fracture mechanics. Excellent reviews of the laws of FCP include those published recently by Erdogen,[23] Radon,[24] and more recently by Tam and Martin.[25]

FCP experiment usually involves measurements of the average incremental crack length (Δh) from a sharp notch of a known depth (h) in a specimen of a defined geometry. The average crack speed is given by ($\Delta h/\Delta N$), where ΔN is the number of cycles corresponding to a crack extension, Δh. Commonly used geometries include single edge notched (SEN) and compact tension (CT) specimens (see Chapter 8, Fig. 8.6.7). A double cantilever geometry is better suited for the studies of FCP in adhesive bond lines. Although a variety of loading cycles may be applied, it is common to study FCP under tension loading programs of different wave forms, e.g., sinusoidal, triangular, or rectangular. The majority of FCP experiments, however, are conducted under tensile sinusoidal loads. The frequency of load application, the load amplitude, and the stress level determined by its maximum or mean values represent the basic loading variables.[2] The load amplitude is usually expressed as the load ratio which is the ratio of minimum stress to its maximum, i.e., $R = \sigma_{min}/\sigma_{max}$.

18.5.1 Fracture Mechanics Approach

From the discussion of Chapters 2 and 7, we note that the stress intensity factor K expresses the stress field associated with a sharp crack in an elastic continuum. Thus, for a crack propagating by opening mode (mode I) in a SEN specimen, the stress intensity factor K_I is given by

$$K_I = \sigma \sqrt{h} \cdot f(h/W) \tag{18.5.1}$$

where W is the width of the specimen and σ is the stress applied remotely, i.e., at the grips. The function $f(h/W)$ is a geometric correction factor whose solution is obtained from the boundary value problem (Chapter 7). Solutions for various geometries can be found in stress analysis hand books.[26] In fatigue, a maximum and a minimum of the stress intensity factor corresponds to the stress limits, i.e., σ_{max} and σ_{min}. Thus, a stress intensity factor range ($\Delta K = K_{max} - K_{min}$) is usually considered.

The ideal sharp planner crack which assumingly separates two adjacent rows of atoms ought to be compared with a real crack tip geometry (Fig. 18.5.1). Clearly, the difference is fantastic. It is therefore instructive to consider the quantities calculated from linear fracture mechanics in view of such differences. Parameters such as K or ΔK are useful as correlative tools particularly since they possess an invarient nature. The energy release rate $G = K^2/E$, although obtained from similar calculations, is more appropriate to correlate with the rate of crack propagation from geometric and thermodynamic view points.

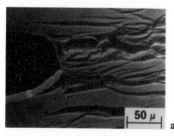 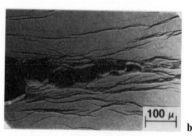

Fig. 18.5.1. (a) SEM view of the front of a fatigue crack in compact tension polypropylene specimen. The crack front is headed by "root crazes".
(b) The same as in (A) after cyclic load application showing crack advance into several crazes. Reproduced with permission.[21]

18.5.2 The Paris-Erdogan Equation

Of the numerous relations proposed to correlate the rate of FCP (dh/dN) with loading and specimen parameters, the equation proposed by Paris and Erdogan[27] gained the widest acceptance. It states

$$\frac{dh}{dN} = A(\Delta K)^m \tag{18.5.2}$$

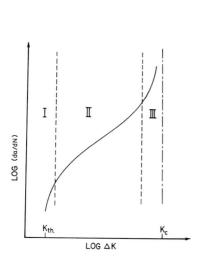

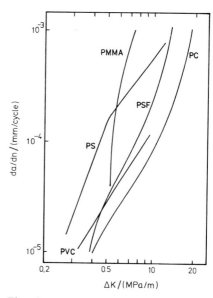

Fig. 18.5.2. A schematic of S-shaped FCP curve indicating its three characteristic regions, I, II, & III.

Fig. 18.5.3. FCP behavior of various polymers. Reproduced with permission.[29]

This equation suggests that the rate of FCP is a logarithmically linear function of ΔK. In fact, typical FCP behavior as illustrated in Fig. 18.5.2 falls into three distinct regions. Region I starts with a threshold value of the stress intensity factor range ΔK_{th}, below which propagation of the crack is not *observed*. The value of K_{th} has been attributed to the attainment of a sufficient level of activity in the notch tip region to cause its propagation.[28] The initial slope of region I is usually very steep. As the crack becomes longer, i.e., ΔK becomes larger, reduced crack acceleration occurs leading to region II. The FCP curve is effectively linear in region II in the majority of cases. The rate of FCP approaches its asymptotic value at $K = K_c$ where a transition from stable to avalanche-like crack propagation occurs.

The commonly observed linearity of FCP rate within region II promoted the general acceptance of Equation 18.5.2 to describe the phenomenon. A lack of linearity in some polymers is immediately obvious when the test is conducted over a wide range of ΔK. Nevertheless, Paris plots remain to be used to evaluate the relative resistance of materials to FCP (Fig. 18.5.3).[29] This is achieved by examining the rate of FCP at a particular value of ΔK; the higher the dh/dN the lower the FCP resistance. Alternatively, the higher the ΔK for a particular dh/dN, the more resistant the material is supposed to be.

Careful examination of the results in Figure 18.5.3 indicates that region II is not necessarily observed within the same ΔK span (e.g., PS and PMMA), hence the comparison could be misleading since curve crossover is observed. Had the entire FCP been recorded, a more certain assessment of the resistance to FCP would have been concluded. In this respect, we examine the FCP behavior of the two PVC composites[30] shown in Fig. 18.5.4. Comparison of the two curves addresses the resistance to FCP in terms of two questions: How long does it last, and how strong is it? The large reduced crack acceleration observed in the case of 10% glass fiber (low gradient of region II) results in a higher fracture toughness as measured from the respective critical energy release rate J_c. Life time of FCP, on the other hand, is evaluated from the

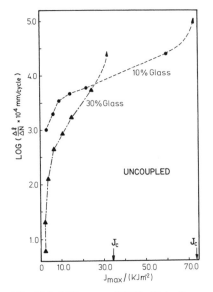

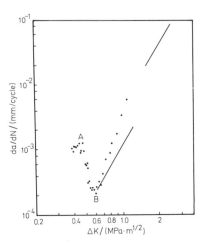

Fig. 18.5.4. The rate of FCP of injection molded glass reinforced PVC composites containing 10% and 30% glass as a function of the energy release rate J. Arrows indicate the critical energy release rate J_c for each.

Fig. 18.5.5. FCP rates at 10 Hz as a function of ΔK in low density polyethylene (LDPE). Note the decrease in dh/dN with increasing ΔK. Solid lines represent FCP data of LDPE reported earlier.[34] Reproduced from P.E. Bretz, R.W. Hertzberg and J.A. Manson, *Polymer*, 1981, **22**, 575 by permission of the publishers, Butterworth & Co. (Publishers) Ltd©.

speed at which reduced crack acceleration occurs. Thus, the 30% glass fiber composite lasts longer under the same fatigue conditions, although it displays lower J_c.

The importance of more complete characterization of FCP is further dramatized by the recently reported fatigue crack deceleration.[21, 31-34] This behavior is shown in Figure 18.5.5 for LDPE.[34] Solid lines represent FCP previously reported. The data points representing the rate of FCP in the same material examined over a wide range of ΔK qualitatively deviate from our conviction based on Paris equation and related power laws.[23-25] A decrease in (dh/dN) is observed with increasing ΔK.

18.6   The Crack Layer Theory

Microscopic examination of FCP in polymers, and other materials as well, clearly demonstrates that the crack is accompanied by a zone of damage. Figures 18.6.1–3, for example, show optical micrographs of fatigue cracks in a SEN specimen of polystyrene,[35] a CT specimen of polypropylene[21] and a SEN of injection molded glass reinforced PVC composite.[30] Kinetic measurements indicate that damage progressively accumulates within the zone ahead of the crack tip to a certain level at which the crack and the associated damage advance. This process is repeated until crack propagation becomes unstable causing ultimate failure. The crack layer theory treats the crack and the associated damage as a single entity, i.e., a crack layer (CL). A schematic of the CL is illustrated in Fig. 18.6.4. The CL, lead by its active zone, may translate, expand, rotate or change shape. The law of fatigue crack layer propagation is established by identifying the thermodynamic forces (causes) reciprocal to the rates of active zone movements. This is achieved from entropy and energy balance considerations.[22] Thus,

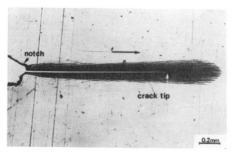

Fig. 18.6.1. An optical micrograph showing a fatigue crack in SEN polystyrene specimen surrounded and preceded by extensive crazing.

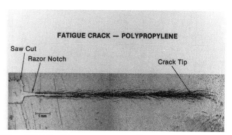

Fig. 18.6.2. Crazes surrounding and preceding fatigue crack in CT polypropylene specimen.

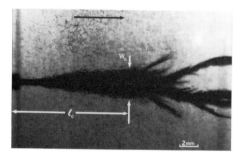

Fig. 18.6.3. Two halves of a fatigue fractured SEN specimen of 10% glass reinforced PVC; l_c and $\mathcal{W}_c$ indicate the critical length and width of the crack layer, respectively.

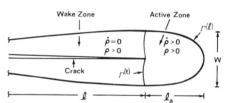

Fig. 18.6.4. A schematic illustrating a crack layer.

the rate of fatigue crack layer translation (dl/dN), which corresponds to conventional FCP, (dh/dN), is derived as[11]

$$\frac{dl}{dN} = \frac{\beta_1 \, W_i}{\gamma^* \, R_1 - J_1} \tag{18.6.1}$$

The numerator expresses the portion $\beta_1 \, W_i$ of the irreversible work W_i spent on molecular processes leading to damage formation and growth related to active zone translation. The rest of W_i is spent on other movements of the active zone and converted into heat. Experimentally, W_i is directly measured from the load-displacement hysteresis recorded during the fatigue experiment and analysis of active zone evolution.

The denominator expresses the energy barrier for CL translation, i.e., the difference between the energy *required* to accumulate the damage $(\gamma^* \, R_1)$ necessary for active zone translation, and the energy *available* (energy release rate) J_1 in the system. γ^* is the *specific enthalpy* of damage, that is, the difference between the enthalpy densities of damaged and initial states of the material (multiplied by the thickness of an element of damage, e.g., a craze). Thus γ^* has a unit of J/m². The *translational resistance moment* R_1 accounts for the amount of damage increment associated with a CL translational advance, and is given by

$$R_1 = \int_{\Gamma^{(t)}} \varrho n_1 \, d\Gamma \tag{18.6.2}$$

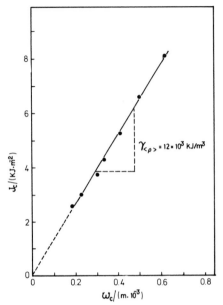

Fig. 18.6.5. Relationship of the critical energy release rate J_c to the critical crack layer width W_c for polystyrene plotted in terms of Eqs. (18.6.3) and (18.6.4).

where $\Gamma^{(t)}$ is the trailing edge of the active zone (Fig. 18.6.4), ϱ is the damage density expressed in a dimensionless form, e.g. (m³ of crazed matter)/(m³ of initial material), and n_1 is the projection of the unit normal vector on the tangent to the crack trajectory at the crack tip. Evaluation of R_1 is generally obtained from microscopic damage analysis of thin sections of the CL.[21, 35] The damage density may be averaged over the trailing edge as $\langle\varrho\rangle$, thus the magnitude of R_1 may be obtained as

$$R_1 = \langle\varrho\rangle W \tag{18.6.3}$$

where W is the CL width measured at the crack tip.

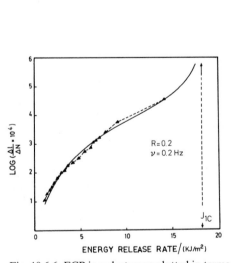

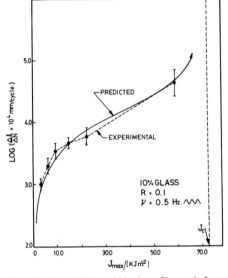

Fig. 18.6.6. FCP in polystyrene plotted in terms of Eq. (18.7).

Fig. 18.6.7. FCP in 10% glass fiber reinforced PVC plotted in terms of Eq. (18.6.5).

The specific enthalpy of damage is evaluated from global instability conditions. Sufficient and necessary conditions for instability, for a SEN specimen under constant amplitude fatigue loading, are met when[35]

$$J_1 = \gamma^* \, R_1 \tag{18.6.4}$$

In view of Eq. (18.6.3), the above condition corresponds to avalanche-like CL propagation, i.e., $dl/dN \to \infty$, Thus, a plot of J_{1c} *versus* R_{1c} (the subscript c indicates critical conditions) is a straight line from the slope of which γ^* may be obtained. Such a plot for PS[36] is illustrated in Figure 18.6.5.

Equations (18.6.1), (18.6.3) and (18.6.4) yield the law of fatigue crack layer propagation (FCLP) in the form

$$\frac{dl}{dN} = \frac{\beta_1 \, W_i}{J_{1c} \dfrac{\mathscr{W}}{\mathscr{W}_c} - J_1} \tag{18.6.5}$$

The law expressed above does not require microscopic damage analysis (R_1 or $\langle \varrho \rangle$) to be applied. This formalism assumes self similar damage distribution within the CL during its evolution. Figures 18.6.6 and 18.6.7 display applicability of Eq. (18.6.5) to FCLP in PS[11] and in a 10% glass fiber reinforced PVC composite.[30]

It is instructive at this point to discuss the implications of Eq. (18.6.1) in terms of Griffith theory. An ideal elastic system does not possess any dissipative character. Thus, according to the energy balance proposed by Griffith (Chapter 7) the crack is in equilibrium when $J = 2e_s$ (Eq. (7.4.8), where $\sigma^2 \, \pi \, h/E = J$), and $\dfrac{\partial J}{\partial h} > 0$. These represent the necessary and sufficient conditions of equilibrium. Of course, $J = 2e_s$, and $\dfrac{\partial J}{\partial h} > 0$ are the necessary and sufficient conditions of instability (Griffith criterion). When instability conditions are satisfied, the crack propagates at undetermined speed. In the geometry considered by Griffith the sufficient condition of instability is met prior to the necessary condition. Thus, $J = 2e_s$ becomes necessary and sufficient. However, to describe the rate of subcritical crack propagation the underlying dissipative processes must be considered. These are expressed in Equation (18.6.1) as $\beta_1 \, W_i$.

18.7 Closing Remarks

In this chapter, some newly addressed features of the phenomenon of fatigue failure of polymers are presented. In addition to its technological importance, fatigue failure presents a fundamental problem for science and engineering. Understanding of fatigue mechanisms (damage) and development of constitutive equations for damage evolution leading to crack initiation and growth as a function loading history is a task which can be handled through interdisciplinary research. This is true for fatigue of engineering materials in general. The rate at which polymers and their composites are being used in advancing vehicles of modern civilization ought to sharpen our interest in fatigue fracture of polymers.

References

1. A. Wöhler, *English Abstract in Engineering* 1871, **11**, 199.
2. R.W. Hertzberg and J.A. Manson, *Fatigue in Engineering Plastics*, Academic Press, New York 1980.

3. J.A. Sauer and G.C. Richardson, *Internat. J. Fracture* 1980, **16**, 499.
4. J.C. Radon, *Internat. J. Fracture* 1980, **16**, 533.
5. *Fatigue in Polymers,* International Conference, The Plastic and Rubber Institute, London, June 29 & 30, 1983.
6. J.A. Sauer, A.M. McMaster and D.R. Morrow, *J. Macromol. Sci.-Phys.* 1976, **B 12**, 535.
7a. J. Clyton-Cave, R.J. Taylor and E. Ineson, *J. Iron Steel Inst.* 1955, **180**, 161.
7b. F.A.M. McClintock and A.S. Argon (eds.), *Mechanical Behavior of Materials,* Addison-Wesley Publishing Company, London 1966.
8. S. Rabinowitz and P. Beardmore, *J. Mater. Sci.* 1974, **9**, 81.
9. R.J. Crawford and P.P. Benham, *Polymer* 1975, **16**, 908.
10. J.F. Mandell, K.L. Smith and D.D. Huang, *Polymer Eng. Sci.* 1981, **21**, 1173.
11. A. Chudnovsky and A. Moet, *J. Mater. Sci.* 1985, **20**, 630.
12. M.E. Mackay, T.G. Teng and J.M. Schultz, *J. Mater. Sci.* 1979, **14**, 221.
13. C.B. Bucknall and W.W. Stevens, *J. Mater. Sci.* 1980, **15**, 2950.
14. N.J. Mills and N. Walker, *J. Mater. Sci.* 1980, **15**, 1832.
15. M.T. Takemori and R.P. Kambour, *J. Mater. Sci.* 1981, **16**, 1108.
16. J.A. Sauer, private communications.
17. H.H. Kausch, *Polymer Fracture,* Springer-Verlag, Berlin 1978.
18. K. Hellan, *Introduction to Fracture Mechanics,* McGraw-Hill, New York 1984.
19. S. Mostovoy and E.J. Ripling, *Polymer Sci. Tech.* 1975, **9 B**, 513.
20. W. Döll, *Adv. Polymer Sci.* 1983, **52/53**, 105.
21. A. Chudnovsky, A. Moet, N.J. Bankart and M.T. Takemori, *J. Appl. Phys.* 1983, **54**, 5562.
22. A. Chudnovsky, NASA Contractor Report # 174634, March 1984.
23. F. Erdogan, *"Crack Propagation Theories",* Chapter 5 in H. Liebowitz (ed.) *"Fracture",* vol. II, Academic Press, New York 1968.
24. J.C. Radon, *Internat. J. Fracture* 1980, **16**, 533.
25. E.P. Tam and G.C. Martin, *J. Macromol. Sci.-Phys.* 1985, **B 23**, 415.
26. H. Tada, P.C. Paris and G. Irwin, "The Stress Analysis of Cracks Handbook", Del Research Corp., Hellertown, Pennsylvania 1973.
27. P. Paris and F. Erdogan, *Trans. ASME* December 1963, p. 528.
28. J.F. Knott, *Fundamentals of Fracture Mechanics,* Butterworth, London 1973.
29. M.D. Skibo, R.W. Hertzberg, J.A. Manson and S.L. Kim, *J. Mater. Sci.* 1977, **12**, 531.
30. P.X. Nguyen and A. Moet, *J. Vinyl Technology,* 1985, **7**, 160.
31. Y.W. Mai, *J. Mater. Sci.* 1974, **9**, 1896.
32. J.W. Teh, J.R. White and E.H. Andrews, *Polymer* 1979, **20**, 755.
33. A. Sandt and E. Hornbogen, *J. Mater. Sci.* 1981, **16**, 2915.
34. P.E. Bretz, R.W. Hertzberg and J.A. Manson, *Polymer* 1981, **22**, 575.
35. J. Botsis, Ph.D. Thesis, Case Western Reserve University, 1984.
36. N. Haddaoui, A. Chudnovsky and A. Moet, *Amer. Chem. Soc. Polym. Mater. Sci. Eng.* Prepr. 1983, **49**, 117.

19 Friction and Wear

Roelof P. Steijn

Engineering Technology Laboratory
E.I. du Pont de Nemours & Co. Inc., Wilmington, DE 19898, U.S.A.

19.1 Introduction

Only a quarter of a century ago, plastics were called "the wonder engineering materials of the modern age."[1] Indeed, the last 25 years have verified this pronouncement. Because of their unique physical and mechanical properties, they have won a foremost place in modern engineering and technology. In many branches of mechanical and aeronautical engineering, in biomechanics and the life sciences, in electronics and agronomy, plastics are now being used as engineering materials and as such enjoy an outstanding reputation. They are accepted without much thought — as common household items be it as flatware, luggage, TV consoles, or toothbrushes. It is difficult to think of a world without plastics, in particular when one ponders the fact that in 1982 total U.S. plastics sales came close to 16 million metric tons.[2]

In the world of *tribology,** plastics have been revolutionary. Plastic bearings and bushings, cams and gears, seals and slides, shanks and shafts, and numerous other parts vital in rolling and sliding are now standard in automobiles, washing machines, aeroplanes, typewriters, etc. The relative ease with which plastics can be fabricated has contributed to this popularity in no small measure. But the real reason is found in the inherent or built-in lubricity that most polymers display, as compared with metals.

Suppose we were to place a five-pound block of steel, copper, or zinc on a table and push it. It would require a certain force to initiate and maintain sliding. Now if we do the same with nylon, acetal resin, or high-density polyethylene, we find that it takes significantly less force to initiate and maintain sliding than with the steel, copper, and zinc. This simple observation which can be easily demonstrated in the laboratory, workshop, or classroom is known to many. Add to this property the inherent noise-damping capacity, the relatively low density or light weight, and the often pleasing appearance and color, and one has pinpointed the major characteristics responsible for the dramatic rise in use and applicability of plastics in tribology.

On a fundamental level, what do we know now about the friction and wear of plastics that may be helpful to the potential user and the practicing engineer? To answer this question one must reflect first on the theories of friction and wear that existed at the time plastics made their entry into the engineering field. Actually, these theories, especially those on wear, are still in a stage of modification and refinement. Thus to include plastics added another complexity, particularly when it soon became evident that the friction and wear behavior of plastics differed substantially from that of metals. What we know now has been the result of painstaking work by numerous research workers all over the world. Nevertheless, as is so often noticeable in the historical development of science and technology, some workers, groups of workers, and associated schools of thought have especially distinguished themselves.

---

* Tribology is defined as "the science and technology of interacting surfaces in relative motion and of the practices related thereto."[3]

The pioneering work by Professor Tabor and his many co-workers at the University of Cambridge, England, deserves special mention. Following their classical work on the friction of metals, Tabor and his school turned to polymers in the early fifties. Over the next 25 years, they investigated and interpreted many complicated aspects of *polymer friction*. Their work has led to a unique and broad understanding of what happens to plastics in sliding and rolling contact and has served as a springboard to others. Many new and important contributions have been made since then by Dr. Lancaster in England, Professor Tanaka and his students in Japan, Professor Eiss e.a. at Virginia Polytechnic Institute, and Professor Bahadur, e.a. at Iowa State in the USA. New and significant work has emerged from the laboratories of Professor Briscoe at the Imperial College, London; Professor Dowson at the University of Leeds; and Professor Czichos in Berlin.

Fundamental studies of the friction and wear of polymers are now being pursued at a growing rate. Some of these investigations are beginning to pay off. The relative ease of sliding the blocks of nylon, acetal resin, and HDPE, as mentioned earlier, has resulted in the use of sliding bearings of nylon in small appliances, acetal gears in meters and instrumentation, and acetabular cups of HDPE for artificial hip joints. Compared with the theory of polymer friction, corresponding wear theories are far less developed. But here, too, in view of the enormous complexity, splendid progress has been made.

Let us now consider the technical aspects of this fascinating field, describe in some detail the response of polymers in friction and wear systems, and review the underlying fundamentals on which that response is based. This indeed is the mission of this chapter: to review and illustrate for the uninitiated reader the *friction and wear behavior of polymers*. In doing so, we credit the comprehensive reviews on the subject published elsewhere[4-8] which have helped to shape the one written below. In reviews of this kind, repetition of certain facts and figures is unavoidable and seems pardonable.

19.2 Polymer Friction

When two polymer surfaces slide together, the ensuing friction and wear are the direct result of surface contact and surface interactions. No matter how smooth the surfaces are, contact will be made over an area much smaller than the nominal area because asperities touch and deform under the load either elastically or both elastically and plastically. Thus a contact area large enough to carry the load is established, see Fig. 19.2.1. This actual area of contact is the critical quantity that both surfaces share. It is also the area which, according to the adhesion theory of friction, "cold-welds" and, upon sliding, will have to be sheared. The force necessary to shear the welded or adhering junctions is the adhesive component of friction. Junction material

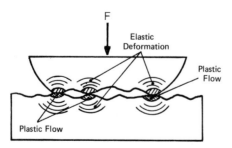

Fig. 19.2.1. Nature of contact between two metal surfaces, after Bowden and Tabor, Ref. 3.

that becomes detached from both sides in this shearing process across the actual contact area constitutes what is called *adhesive wear*.

19.3 Load Dependency of the Adhesive Component of Polymer Friction

The actual area of contact is determined by elastic and plastic deformation mechanics. For fully plastic deformation, the true area of contact A_c may be written as

$$A_c = L/p_m \tag{19.3.1}$$

where L is the load and p_m the yield pressure of the softer of the two materials. This is the familiar concept of Bowden and Tabor[9, 10] popularized in Fig. 19.2.1. If we further assume that the force of friction, F, must shear the adhesive junctions and therefore

$$F = A_c s \tag{19.3.2}$$

in which s is the specific shear strength, we can further write for the adhesive force of friction

$$F = \frac{s}{p_m} L \tag{19.3.3}$$

However, if elastic deformation persists and no plastic deformation occurs, the true area of contact as a function of load depends on the surface topography. Archard[11, 12] using hard-sphere models showed that $A_c = k L^n$ in which $n = 2/3$ for single-asperity contact; for multiple asperity contact points of increasing complexity, n varies from 0.66 to 1.00. The more complex and realistic the model, the higher the n value, see Fig. 19.3.1. Again, using Eq. (19.3.2), we have

$$F = KL^n \tag{19.3.4}$$

We then may write for the coefficient of friction μ, which is by definition the ratio of the force of friction F to the normal load L:

$$\mu = \frac{F}{L} = KL^{n-1} \tag{19.3.5}$$

Experimental work on the friction of polymers has shown that, indeed, the above equations are obeyed. Much of the earlier verification comes from workers in the textile industry concerned with fiber friction.[13–15] A noteworthy example is the friction versus load curve of PTFE assembled by Allan[16] from a large collection of literature

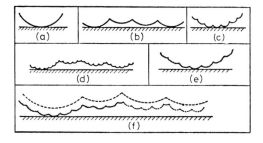

Fig. 19.3.1. Models of surfaces containing asperities of different size; a) $A_c \propto W^{2/3}$, b) $A_c \propto L^{4/5}$, c) $A_c \propto W^{8/9}$, d) $A_c \propto L^{14/15}$, e) $A_c \propto W^{26/27}$, f) $A_c \propto L^{44/45}$. (Archard; Ref. 12.)

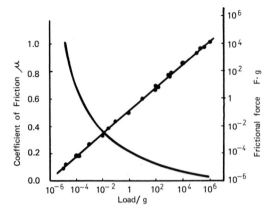

Fig. 19.3.2. Variation of frictional force and coefficient of friction with load for PTFE. (Reprinted by permission of the American Society of Lubrication Engineers. Allan, Ref. 16.)

Table 19.3.1. Experimental values of n for several plastics

Plastic	n
Cellulose acetate	0.96
Nylon (undrawn)	0.90
Nylon (drawn)	0.80
Viscose rayon	0.91
Polyvinylidene chloride	0.87
Polyethylene	0.70
Polytetrafluoroethylene	0.86

data, see Fig. 19.3.2. Allan clearly showed that the friction of this polymer as a function of load over nine decades is accurately described by $F = KL^n$ where $n = 0.86$. Other experimentally determined values of n for a number of plastics are listed in Table 19.3.1.

19.4   Elastic versus Plastic Deformation

The subject of elastic versus plastic deformation in the contact of real surfaces has been modeled more rigorously for metals by Greenwood and Williamson[17] and Whitehouse and Archard.[18] Their work led to the definition of a parameter Ψ, termed the plasticity index

$$\Psi_{GW} = \frac{E'}{H'} \left(\frac{\sigma'}{R} \right)^{1/2} \tag{19.4.1}$$

and

$$\Psi_{WA} = \frac{E'}{H'} \left(\frac{\sigma}{\beta'} \right) \tag{19.4.2}$$

In these equations, E' is the reduced modulus (from $1/E' = (1 - v_1^2)/E_1 + (1 - v_2^2)/E_2$ in which E and v are the elastic moduli and Poisson's ratios); H' is the hardness or yield pressure; and σ', R and β' represent topographical parameters measurable by computer-assisted surface profilometry. Recently, Bushan[19] modified the GW analysis for a polymer in contact with a hard material and defines a "polymer" *plasticity*

Table 19.4.1. Reduced modulus-to-hardness ratio for various materials (after Lancaster, Ref. 7)

Materials	E'/H'
Cemented carbides	2.1– 2.9
Ceramics	1.9– 3.9
Thermosetting polymers	7.3–14.3
Thermoplastic polymers	9.7–39
Steels	19.4–83
Soft metals	145–830

index Ψ_p as

$$\Psi_p = \frac{E'}{\sigma_y}\left(\frac{\sigma_p}{\beta_p}\right)^{1/2} \tag{19.4.3}$$

in which σ_y is the yield strength of the softer material and σ_p and β_p, again, surface topographical parameters. The meaning and practical significance of these plasticity indices is that, for metals, asperity deformation upon contact is plastic when $\Psi > 1$ and primarily elastic when $\Psi < 0.6$.[20] For polymers we have plastic deformation upon contact when $\Psi_p > 2.6$ and primarily elastic deformation when $\Psi_p < 1.8$.[19]

The plasticity index is made up of a material property part E'/H' or E'/σ_y and a surface topographical part $(\sigma'/R)^{1/2}$, σ/β', or $(\sigma_p/\beta'_p)^{1/2}$. Thus, for the same topography, whether the asperity contact is elastic or plastic, depends on the ratio E'/H' or E'/σ_y. Table 19.4.1 lists the E'/H' ratio for various material categories. If we estimate that engineering surfaces, depending on how they are prepared and finished, have values for $(\sigma'/R)^{1/2}$ ranging from 0.005 to 0.400, we see that plastics, especially the thermosets, are more apt to behave elastically in surface contact than soft metals, or even steels. It then appears logical that, in the general friction formula $F = KL^n$ (Eq. (19.3.4)), for metals $n = 1$ but for plastics $n < 1$, as shown in Table 19.3.1. Consequently, the coefficient of friction for polymers decreases with increasing load.

The implications are considerable. For the first time we have a class of materials that has consistently broken with the classical laws of friction. These laws are well known:

1) Friction is proportional to normal load.
2) Friction is independent of the apparent area of contact.
3) Friction is independent of sliding speed.

The laws of friction were developed over a time span of three centuries – from observations by Da Vinci (1452–1519) to empirical verification by Amonton (1663–1705) and Coulomb (1736–1806). It is amazing that these laws have been successful for so long. Indeed, for most materials of the prepolymer period, these laws were in agreement with experiment and, for most engineering purposes of that time, adequate.

19.5 Ploughing or Deformation Component of Friction: Its Relation to Rolling Friction

In addition to purely *adhesive effects* at the interface, sliding involves a certain amount of deformation from the interpenetration of opposing asperities or, rather, ploughing of asperities through opposite surfaces. Therefore, the total force of friction,

F, is the sum of the adhesive or shearing force of friction and the deformation or ploughing force of friction, often expressed as

$$F_{total} = F_A + F_D \qquad\qquad (19.5.1)$$

The viewpoint of both shearing and ploughing is explicit in the theory of metallic friction developed by Bowden and Tabor.[10] However, in metallic friction the adhesion term is considered the dominant factor so that the friction coefficient becomes approximately equal to the ratio of shear strength to yield pressure. For many metals and alloys, this ratio does not vary by a large factor and lies between 0.6 and 1.2. Plastics, however, are viscoelastic. The *ploughing* or deformation component therefore is dependent on time (sliding speed) and temperature and could significantly modify the overall friction behavior.

To gain insight into this behavior, we can learn from earlier studies on *rolling friction*. The deformation of a flat surface by a hard asperity ploughing through within the elastic range is the same as that produced by a hard sphere rolling across that surface. Tabor[21] has shown that the rolling friction of hard spheres on rubber is determined primarily by elastic hysteresis losses in the rubber associated with cyclic stresses below the roller. Flom and Bueche,[22-24] taking a different approach, proposed that rolling friction is directly related to dynamic mechanical losses in the polymer. The strong dependence on rolling speed and temperature, then, can be appreciated, and friction peaks can be related to known molecular transitions. To strengthen this point, Flom[24] has shown that a clear one-to-one correspondence can be found between rolling friction versus temperature and the loss factor or tan δ versus temperature for various thermoplastics such as nylon, PE, PVC, and PTFE.

In *sliding friction,* in which ploughing remains within the elastic range, the same considerations are valid. Bueche and Flom[25] have shown this for several plastics by carrying out well-lubricated sliding experiments at different temperatures as a function of sliding speed. By lubricating they effectively minimized the adhesive or shear component so that the measured friction was predominantly ploughing. Their results for steel against PMMA are shown in Fig. 19.5.1. They are matched in the same graph (broken lines) by tan δ versus frequency data obtained from dielectric loss measure-

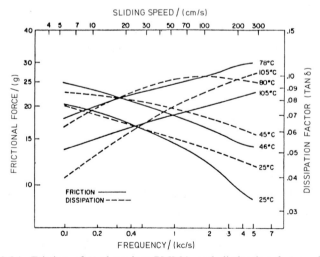

Fig. 19.5.1. Friction of steel against PMMA and dissipation factor of PMMA. (Bueche and Flom, Ref. 25.)

ments. The excellent one-to-one correspondence of the two sets of data is remarkable and indeed implies that the ploughing friction of polymers is governed by the dynamic mechanical response of the viscoelastic material.

Of course, when ploughing progresses into the plastic range, these considerations no longer apply. In that case the friction force must also contain a component for plastic grooving of the surface for one or multiple passes in a manner similar to plastic grooving by rolling as described by Eldredge and Tabor[26] for metals.

19.6 Effect of Temperature and Sliding Speed

The effect of temperature and sliding speed on the friction of plastics has been studied in detail. As a consequence, it appears that many plastics behave in friction as pictured schematically in Fig. 19.6.1. The chief characteristic is that both friction-temperature and friction-sliding speed curves exhibit a maximum that shifts to the right when either temperature or speed is raised. Experimental evidence for the existence of such a hump or friction peak can be found throughout the literature; for instance, see the work on PAN, nylon 6-6, CA, and PET by Fort,[27] on PVC by James,[28] on PE and PP by McLaren and Tabor,[29] and on nylon 6 by Watanabe, Karasawa and Matsubara.[30] Some of these results are reproduced in Figs. 19.6.2–19.6.3. Examples of the shift of the friction peak with increasing temperature or speed are shown, respectively, in Fig. 19.6.4 for P-α-C, PCTFE, and POM[31] and in Fig. 19.6.5 for acrylonitrile butadiene rubber.[31]

An adequate and all-encompassing theory of this generally observed behavior is still missing, however. Instead, we have satisfactory explanations only for restricted

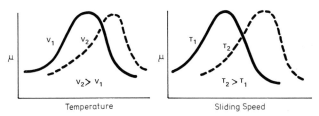

Fig. 19.6.1. Typical effect of temperature and sliding speed on the friction of polymers.

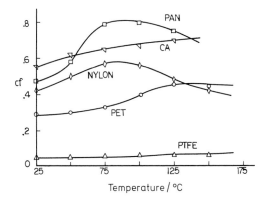

Fig. 19.6.2. Effect of temperature on the friction of several polymers. (Reprinted with permission from Fort, Ref. 27. Copyright 1962 American Chemical Society.)

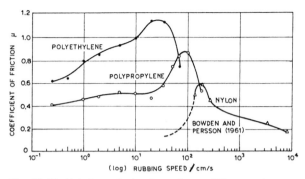

Fig. 19.6.3. Friction velocity dependence of linear polymers at 22° C. (McLaren and Tabor, Ref. 29.)

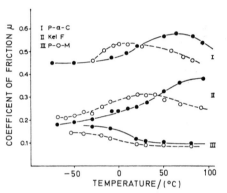

Fig. 19.6.4. Sliding friction of three polymers as a function of temperature at two different speeds. ● 0.3 cm/s. ○ 0.0003 cm/s. (Ludema and Tabor, Ref. 31.)

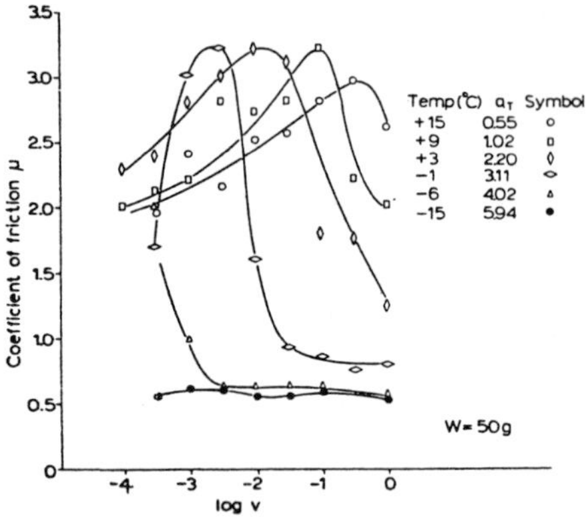

Fig. 19.6.5. Friction of glass against acrylonitrile butadiene rubber. (Ludema and Tabor, Ref. 31).

scopes of the problem or for limited polymers or groups of polymers. Grosch[32] has shown that in rubbers the time-temperature dependence of friction is viscoelastic. The shift of friction curves such as those shown in Fig. 19.6.5 can then be quantitatively accounted for by the WLF transformation (Williams-Landel-Ferry) relative to a specific reference temperature. Ludema and Tabor[31] have postulated that the fundamental equation in the adhesion theory of friction, $F = A_c s$, may be similarly affected by speed and temperature, A_c being dependent on modulus and s, the interfacial shear strength, on strain rate. Ludema and Tabor showed that their argument worked reasonably well in explaining the friction of rubbers; and, in fact, they derived an $F = A_c s$ versus sliding speed curve roughly congruent with their friction peak data on acrylonitrile butadiene rubber. However, they were unable to deal with polymer friction because of the lack of viscoelastic data commensurate with the extremely high strain rates that operate in the thin polymeric friction zone. Thus the existence of friction peaks in plastics still poses many unanswered questions and will undoubtedly require more experimentation.

19.7 Frictional Heat

One of the often underrated difficulties is the effect of frictional heat on friction. Compared with metals, frictional heat is not easily carried away. On the average, the thermal conductivity of plastics is 0.05% that of pure copper. Over the years, a number of mathematical treatises have been published on the temperature at sliding interfaces. Archard,[33] answering the need for rather simple guidelines for the experimentalist, has presented a shortened version to calculate maximum flash temperatures at the contact area. When he applied this to PMMA rubbing against PMMA in a crossed-cylinder apparatus, he calculated temperature rises of about 84° C above the ambient under conditions of load and speed that varied from about 20 grams and 356 cm/s to about 3.5 kg at only 11.4 cm/s. Tanaka and Uchiyama[34] also considered the role of softening and incipient melting during sliding in the friction and wear of thermoplastics. They deduced, from changes in the originally spherulite structure at the interface, when softening or melting had occurred and were able to determine the actual depth over which melting had taken place in PP, HDPE, nylon 6, and POM sliding against steel and glass. Their results are shown in Fig. 19.7.1. It is generally believed that thermal softening and the associated increase in the real area of contact cause the friction to rise at first. When sliding continues and melting begins to occur, the friction will go down precipitously because of lubrication by a molten film of polymer. This phenomenon was clearly demonstrated by Vinogradov e.a.[35] in dry sliding tests of various metals and alloys against PP at different loads. Some of their results are reproduced in Fig. 19.7.2. Seen in this light, some of the friction peaks attributed earlier to viscoelastic responses may actually be thermally induced friction transients.

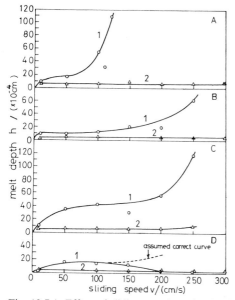

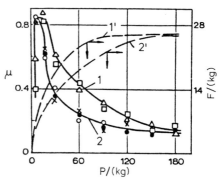

Fig. 19.7.1. Effect of sliding speed on depth of melting. 1. Glass; 2 Steel; A. PP; B. HDPE; C. Nylon 6; D. PAC. (Tanaka and Uchiyama, Ref. 34. Reprinted by permission of Plenum Publishing Corp., New York, NY.)

Fig. 19.7.2. Effect of load on metal-to-PP friction. 1. Cu; Al. 2. Iron and Steel. (Vinogradov e.a., Ref. 35.)

Films

the *mating surfaces* of polymers sliding at temperatures above the
revealed that the act of sliding produces thin polymer films stretched
er the interface. Bowers, Clinton and Zisman,[36] carrying out sliding
PTFE, were the first to draw attention to the formation of such
they have been examined and discussed at length by many others.[37-40]
Rather dramatic optical and electron micrographs have been published to show the
formation of such films and to describe their role in the friction process. Figure 19.8.1
is such a micrograph by the author. The photo shows how adhesive junctions of PTFE
formed during sliding are stretched, drawn, and oriented into thin films and bands
across the interface.[40]

In PTFE we find a prime example of this behavior. Only two other plastics, HDPE
and UHMWPE, have been shown to form transfer films with the same ease and low
expenditure of energy.[39,41] Friction in these plastics seems truly interfacial. Pooley
and Tabor[39] have reasoned that the smooth molecular profile of these two polymers
constitutes a major contributing factor in their friction behavior. In most other plastics,
such as LDPE, nylon 6-6, POM, and PET, the transfer film is far less continuous
and often lumpy and fragmentary.[34,42,43] Their formation and behavior seem also
to depend on external conditions of temperature and sliding speed.[44]

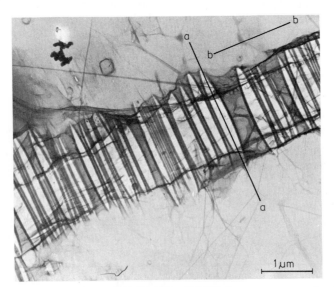

Fig. 19.8.1. Electron micrograph of bands of PTFE drawn across an abrasion groove in PTFE.
a) Direction of sliding; b) Direction of abrasion. (Steijn, Ref. 40.)

19.9 Pressure Effect on Shear Strength

In actuality, the problem of asperities in sliding contact is more complex than the
above simple description allows. In the first place, there is an interdependence between
s and p which has been discounted so far. Yet, in some of the early work on friction
with nylon fibers,[45,46] a slight increase of shear strength with load was inferred. Later
work on thin polymer films sheared between hard substrates[47,48] indicated that shear

stress increases linearly with pressure according to

$$s = s_0 + \alpha p \tag{19.9.1}$$

where s is the shear strength, s_0 a constant, p the pressure, and α the pressure coefficient. The coefficient of friction μ can be written as

$$\mu = \frac{s}{p} = \frac{s_0}{p} + \alpha \tag{19.9.2}$$

It is worthwhile to note that shear strength determined from friction measurements on thin polymer film is 5–10 times smaller than when determined on bulk polymers under hydrostatic pressure. It is believed the high strain rates encountered in interfacial shear, estimated to be as high as 10^5–10^6/s, are the underlying cause.

Equation (19.9.2) above implies much more than can be explained here. The reader should consult the excellent articles, by Bowers,[47] Briscoe and Tabor,[48,49] and Towle.[50]

19.10 Polymer Friction Related to Critical Surface Tension and Work of Adhesion

The question may well be asked if, in an *adhesion theory* of friction, friction should not reflect the differences in surface energy between the surfaces and the materials involved. Rabinowicz,[51] who has taken a broad surface-energy approach to friction and wear phenomena, has reasoned that the work of adhesion must be taken into account when surfaces and asperities penetrate one another. Accordingly, he derived the following expression for the coefficient of friction between materials a and b

$$\mu = \frac{s}{p}\left(1 + \frac{2\,W_{ab}\cot\Theta}{p\,r}\right) \tag{19.10.1}$$

In this equation, s and p are the shear strength and yield pressure, W_{ab} the surface energy of adhesion ($W_{ab} = \gamma_a + \gamma_b - \gamma_{ab}$), Θ the average surface roughness angle, and r the average junction radius. Rabinowicz showed data on unlubricated metallic friction in general support of this equation. To date, the equation has not been tested sufficiently for polymeric materials.

More recently, Spurr[52] also dealt with the friction of polymers. This investigator, who much earlier had augmented the adhesion theory of metals with a special surface term,[53] now applied the same approach to polymers. His basic friction equation reads

$$\mu = \frac{\sigma}{p_0}\cot\Psi \tag{19.10.2}$$

where σ is the yield stress, p_0 the hardness of the polymer, and $\cot\Psi$ a surface term determined by surface forces and surface conditions. From his work with PMMA, nylon, PVC, PE, and PTFE he concluded that the above equation applies to polymers as well. But, in addition, Spurr showed that the $\cot\Psi$ term for polymers was proportional to adhesion as measured by a peel test. Thus, Spurr related the friction of polymers quantitatively to adhesion by showing that $\cos\Psi$ is directly proportional to the critical surface tension γ_c. It should be pointed out that the early work by Zisman e.a.[54,55] on adhesion, chemical constitution, and static friction of halogenated polyethylenes also suggested a ranking order between coefficient of friction and critical surface tension. Table 19.10.1 from Ref. 54 shows this order. Evidence for a relation between friction of polymers and adhesion comes from Czichos and Feinle.[56] These

Table 19.10.1. Effect of constitution on friction and wettability of halogenated polyethylenes (Zisman, Ref. 54)

Polymer	Structural formula	Static coefficient of friction	Critical surface tension (dynes/cm)
Polyvinylidene chloride	$\begin{bmatrix} & H & Cl \\ & \mid & \mid \\ -&C-&C \\ & \mid & \mid \\ & H & Cl \end{bmatrix}_n$	0.90	40
Polyvinyl chloride	$\begin{bmatrix} & H & Cl \\ & \mid & \mid \\ -&C-&C \\ & \mid & \mid \\ & H & H \end{bmatrix}_n$	0.50	39
Polyethylene	$\begin{bmatrix} & H & H \\ & \mid & \mid \\ -&C-&C \\ & \mid & \mid \\ & H & H \end{bmatrix}_n$	0.33	31
Polyvinyl fluoride	$\begin{bmatrix} & H & F \\ & \mid & \mid \\ -&C-&C \\ & \mid & \mid \\ & H & H \end{bmatrix}_n$	0.30	28
Polyvinylidene fluoride	$\begin{bmatrix} & H & F \\ & \mid & \mid \\ -&C-&C \\ & \mid & \mid \\ & H & F \end{bmatrix}_n$	0.30	25
Polytrifluoroethylene	$\begin{bmatrix} & F & F \\ & \mid & \mid \\ -&C-&C \\ & \mid & \mid \\ & H & F \end{bmatrix}_n$	0.30	22
Polytetrafluoroethylene	$\begin{bmatrix} & F & F \\ & \mid & \mid \\ -&C-&C \\ & \mid & \mid \\ & F & F \end{bmatrix}_n$	0.04	18

investigators carried out sliding experiments of PA, POM, PP, and PTFE against SAN, PS, and PMMA and found a fairly linear relationship between the frictional work and published values for the work of adhesion. An extensive investigation of the relation between polymer friction and the work of adhesion was recently reported by Erhard.[57] After obtaining data points in experimental work with more than 60 polymer combinations, this investigator, by curve fitting, found an exponential relationship between the work of adhesion and the coefficient of friction.

19.11 Rubber Friction

While rubbers and elastomers fall outside the real scope of this chapter, their frictional behavior is unique, and brief mention seems in order.

According to Moore,[58] the interaction between an elastomer and a hard, rough surface is marked by a *"draping"* of the elastomer about the rigid asperities of the counterface. See Fig. 19.11.1. The fundamental equation for the force of friction expressed earlier in Eq. (19.5.1) remains valid. However, the deformation or ploughing component of friction F_D now becomes a hysteresis term F_H owing to the delayed recovery of the bulk asperity indent into the elastomer. The adhesion or shearing term F_A is purely *surface adhesion*. Thus, for elastomers, Eq. (19.5.1) is rewritten as

$$F = F_H + F_A \tag{19.11.1}$$

The hysteresis as well as the adhesion components follow viscoelastic behavior and are related to the loss modulus over real modulus ratio E''/E' or dissipation factor $\tan \delta$.

Moore writes for F_A

$$F_A = K_2 \frac{E'}{p^r} \tan \delta \tag{19.11.2}$$

and for F_H

$$F_H = K_3 \left(\frac{p}{E'}\right)^n \tan \delta \tag{19.11.3}$$

where $n \geq 1$ or, in total,

$$F = F_A + F_H = K_2 \left[\frac{E'}{p^r} + K_4 \left(\frac{p}{E'}\right)^n\right] \tan \delta \tag{19.11.4}$$

in which $K_3 = K_2 K_4$, p the nominal pressure, r an exponent, and K_2 and K_3 constants.

The strong dependence in an elastomer of viscoelastic losses on temperature and frequency is reflected in the effect of temperature and sliding speed on friction. This aspect was thoroughly explored in a classical investigation by Grosch.[32] Grosch slid various types of rubber against hard countersurfaces over a wide range of sliding velocities and temperatures. He then applied the WLF transformation to shift his isothermal friction versus speed curves into one single master curve referenced against

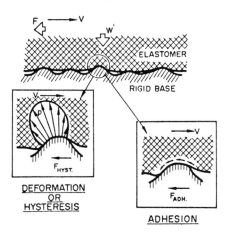

Fig. 19.11.1. Principal components of elastomeric friction according to D.F. Moore, Ref. 58.

the glass transition temperature. By using smooth and rough surfaces and cleverly separating adhesion from the process, Grosch was able to identify adhesion with spaced molecular movements of roughly 10 nm and hysteresis friction with the macro roughness of the track.

One of the most interesting friction phemonena in elastomers is the *wave propagation* discovered by Schallamach[59] during the sliding of soft rubbers. It was found that folds in the rubber surface, probably caused by localized buckling, would run across the contact area in the direction in which the rubber moves relative to the counterface. The motion is somewhat like the movement of a caterpillar and can be visualized as a macro dislocation. The striking feature is that there is no real sliding. Instead, the rubber is lifted locally from the counterface, moved over a finite distance, and put down again. Thus no bonds are broken in the direction of sliding but instead they break normally to the areas of contact.

The above views show that the friction of rubber is multifaceted. As new results have emerged from the laboratories, revisions of earlier theories have become necessary. A case in point is Schallamach's outstanding work[59-62] on dynamic rubber friction, which he viewed as a thermally activated rate process of non-Newtonian flow according to

$$V = V_0 \exp - \frac{U - \gamma F}{kT} \tag{19.11.5}$$

where V is the sliding speed, F the tangential or sliding force, U an activation energy, and γ and V_0 constants.

The monotonic increase in friction with increased sliding speeds predicted by Eq. (19.11.5) is at odds with experimental evidence of a pronounced maximum in the friction versus speed curve. Schallamach reconciled this anomaly by assuming that the sliding of rubber is more than a simple adhesion phenomenon and consists of separate thermally activated processes for bond formation and bond breaking. The number of bonds then is not constant, and is dependent on temperature and speed.[62] This reasoning led to a rather complicated expression for the friction force but one, nevertheless, that reproduces the friction maxima dealt with so adequately by Grosch. Aside from these molecular theories, different macroscopic theories on rubber friction have also been postulated, notably by Savkoor[63] and, as referred to earlier, by Ludema and Tabor.[31] For more detailed information on these studies and on the subject of rubber friction in general, the reader is referred to the specific references, review articles, and general textbooks.[58, 64]

19.12 Earlier Viewpoints

The notion that adhesion plays an important role in friction in general was not new. Both Holm[65] and later Ernst and Merchant[66] entertained this concept, but it was Bowden and Tabor who put the idea into a logical theory. To date, it has been the most satisfying explanation by far and is generally well accepted.

This is not to say that other viewpoints never arose. Coulomb himself offered a qualitative explanation that was based on the mechanical work needed for one surface to climb over the hills of the other surface. In that case, $\mu = F/L = \tan \Theta$, where Θ is the average slope to be scaled. This indeed satisfies the empirical laws of friction. Some experts, notably Bikerman,[67, 68] defended this theory, known as the Coulomb theory of friction, ad infinitum. However, its chief shortcoming, namely that it basically represents a nondissipative process, had rendered it generally unacceptable. Clear experimental evidence by radioactive methods and special microscopic techniques that strong, local adhesion occurs during both the sliding of metals[4] and the sliding of plastics[69-71] has further tipped the scales in favor of the adhesion theory of friction.

19.13 Polymer Wear

The practical importance of wear does not need elucidation. Anyone who drives a car, hikes up a mountain path, uses a typewriter, runs a washing machine, is all too familiar with the mechanical breakdown and deterioration of equipment in use, and with the cost to repair or to replace it.

Wear, as the definition denotes, is the progressive loss of material from the surface of a body in rubbing contact with another one. Burwell,[72] examining the wear of metals and alloys, was the first to systematically categorize various forms of wear, in order of decreasing importance: 1 Adhesive or galling wear, 2 Abrasive and cutting wear, 3 Corrosive wear, 4 Surface fatigue, and 5 Others and minor types.

For plastics, the same types of wear prevail although, because of progress made over the years in diagnosing and defining underlying mechanisms, the list has expanded and the ranking order may be somewhat different. Because of the lower melting and softening point and the much poorer heat conductivity, an important case can be made for "thermal wear" in plastics. And because of the considerably lower modulus, elastic deformation of the contacting surface asperities rather than plastic deformation may set the tone of the wear process.

And so we see that the response of polymers to a specific wear situation is heavily colored by physical, mechanical, and chemical properties. Hardness, compressive strength, modulus, melting point, thermal conductivity, all suggest something about the probable behavior of a polymer in a given tribological situation. Hardness and compressive strength indicate load-bearing capacity, penetration and abrasion; tensile modulus accounts for deformability and cold-flow properties; melting point shows temperature limitations. Thus, a look at the basic engineering properties of a plastic is a good beginning in assessing its probable wear behavior. Many of these properties are listed in manufacturers' data sheets and other authoritative sources.[73] Table 19.13.1, based on data from the *Modern Plastics Encyclopedia,* lists for various classes of polymers, the physical and mechanical properties of tribological significance.

Recently a more elegant and fundamental subdivision of polymer wear processes has been suggested by Briscoe,[6] who sees two general classes of wear: *cohesive and interfacial.* This subdivision seems truer in light of everything we have learned since Burwell but also reflects the keen insight and learned approach of the British School of thought into fundamental tribology.

Cohesive wear processes, according to Briscoe, involve those mechanisms that occur in relatively large volumes adjacent to the interface. Abrasive wear and fatigue wear controlled by the cohesive strength properties of the polymers belong in this category. On the other hand, interfacial wear involves friction and wear in much thinner regions. Almost by definition, it includes the classical adhesive wear mechanism right at the sliding interface.

This chapter will deal mainly with adhesive wear, abrasive wear, and fatigue. These three types have emerged over the years as the most significant ones in polymeric materials. They rank importantly in Burwell's original classification but, more than that, they are the essence of our current understanding of the wear of polymers. As we shall see, they may overlap at times because, as Briscoe rightly remarked, "Wear processes are not monomechanistic."[6] Keeping this in mind will undoubtedly help us' to see polymer wear, as surveyed below, in proper perspective.

Table 19.13.1. Physical and mechanical properties of some polymers Credit Line: "Copyright,

	Nylon 6		Nylon 6,6	
	Unmodified	30–35% Glass-Filled	Unmodified	35% Glass-Filled
Melting/°C				
T_m (crystalline)	225	225	265	265
T_g (amorphous)	...	...	...	...
Density/(gm/cm³)	1.12–1.14	1.35–1.42	1.13–1.15	1.38
Tensile modulus/10^5 psi	3.8 –1.0	14.5 –8.0	...	...
Tensile strength/psi	11800–10000	25000–13000	12000–11000	28000–22000
Elongation/%	100–300	3	60–300	3.5
Compressive strength/psi	13000	19000	15000	29400
Impact strength/(ft lb/in.)	1.0–3.0	3.0	1.0–2.1	2.2–2.6
Hardness, Rockwell	R119	M101–78	R120, M83	M100
Heat deflection temperature/°F (at 264 psi fiber stress)	155	410	167	485
Thermal conductivity/5.8 (10^{-4} cal/s/cm³/1° C/cm)	5.8	5.8	5.8	5.1

	Polypropylene Unmodified	Polyimide	
		Unmodified	PTFE-Filled
Melting point/°C			
T_m (crystalline)	176	...	...
T_g (amorphous)	...	235	...
Density/(gm/cm³)	0.902–0.910	1.43	1.42
Tensile modulus/10^5 psi	1.6–2.25	1.88	...
Tensile strength/psi	4300–5500	17000	5000
Elongation/%	200–700	10	<1
Compressive strength/psi	5500–8000	29000	20000
Impact strength/(ft lb/in.)	0.5–2.2	0.7	0.25
Hardness, Rockwell	R80–110	E99	M115
Heat deflection temperature/°F (at 264 psi fiber stress)	125–140	270–280	>550
Thermal conductivity/ (10^{-4} cal/s/cm³/1° C/cm)	2.8	2.3–2.6	5.2

19.14 Adhesive Wear

Burwell, in his original classification of wear mechanisms has termed *adhesive wear* the most fundamental of all types of wear; it occurs whenever two solid surfaces are in sliding contact, whether lubricated or not, and prevails when all other types of wear are eliminated. This quote from Burwell[72] denotes the difficulty of studying adhesive wear, since abrasion, fatigue, delamination, and so forth, are rarely totally eliminated. Early experimental work on the wear of metals by Archard and Hirst,[74] carried

ASTM, 1916 Race Street, Philadelphia, PA, 19103. Reprinted with permission." (Steijn, Ref. 76)

Acetal		ABS 20–40% Glass-Filled	Fluoroplastic		Polycarbonate	
Unmodified	20% Glass-Filled		PTFE	FEP	Unfilled	15–40% Glass-Filled
181	181	...	327	275	...	...
...	...	110–125	...	...	150	150
1.42	1.56	1.23–1.36	2.14–2.20	2.12–2.17	1.20	1.24–1.52
5.2	10.0	5.9 –10.3	0.58	0.50	3.0–3.5	5.0–17.0
10000	8 500–11 000	8 500–19 000	2000–5000	2700–3100	8000–9500	12 000–25 000
25–75	2–7	2.5–3.0	200–400	250–330	100–130	0.9–5.0
18 000/10% defl.	18 000/10% defl.	12 000–22 000	1700	2200	12 500	13 000–21 000
1.4	0.8	1.0–2.4	3.0	2.5–2.7	12.0–18.0	1.2–4.0
M94, R120	M75–M90	M65–M100	D50–55 (Shore)	D60–65 (Shore)	M70–78	M88–95
255	315	210–240	250 (at 66 psi)	158 (at 66 psi)	265–285	284–300
5.5	...	...	6.0	6.0	4.6	4.9–5.2

Polyethylene		Polyester (Aromatic)	Phenol-formaldehyde
LDPE	HDPE		
110–130	120–140	232–267	Thermoset
0.910–0.925	0.941–0.965	1.31–1.38	1.69–1.95
0.14 –0.38	0.6–1.8	2.8	19.0–33.0
600–2300	3100–5500	8200	5000–18 000
90–800	20–1300	50–300	0.2
...	2700–3600	8600–14 500	16 000–70 000
No break	0.5–20.0	0.8–1.0	0.3–18.0
R10	R15	M68–98	E54–E101
90–105	110–130	122–185	300–600
8.0	11.0–12.4	4.2–6.9	8.2–14.5

Conversion Factors: 1 psi = 6.9 KPa, and 1 ft lb/in. = 53.38 J/m.

out on a pin-ring machine, included the thermoplastics PMMA, PTFE, PE, and several thermosets. They found that, for constant load and sliding speed, wear was independent of the apparent area of contact and proportional to sliding distance, but the effect of load was not clear. Eventually, for polymers sliding against polymers and other hard counterfaces, the simple wear equation for metals[75] was also adopted:

$$W_v = k\, L\, s / 3\, p_m \tag{19.14.1}$$

in which W_v is the total wear volume, L the load, s the sliding distance, p_m the indentation hardness, and k the so-called wear coefficient. Short-range data typical of this behavior are shown in Fig. 19.14.1 for nylon.[76] On a fundamental basis we may question the validity of Eq. (19.14.1) for polymers. Earlier we have seen that, for polymers, the real area of contact A_c and the load L are exponentially related as $A_c = k\,L^n$ with $n < 1$ unless surface contact is fully plastic. Thus a linear relationship of wear with load is difficult to rationalize. Rhee[77] suggested a more general wear equation

$$W_v = \alpha\, P^a\, v^b\, t^c \qquad\qquad (19.14.2)$$

in which W = wear, P = load, v = sliding speed, t = sliding time, and α is a proportionality constant; a, b, and c are constants, with a and b differing substantially from unity while c is close to 1.

The engineering analog of Eq. (19.14.1) is an expression for the depth of wear h after Lewis[78]

$$h = K(Pv)\,t \qquad\qquad (19.14.3)$$

In this equation, P is the apparent pressure (from $P = L/A$), v the sliding speed, t the time of sliding, and K a proportionality constant called the wear factor having dimensions (in.)^3 (min)/ft lb hr. This equation contains the product of load and speed Pv, which is a useful design parameter in bearing technology.

The wear Eqs. (19.14.1) and (19.14.3) are restricted in scope and therefore of limited utility in industrial practice, as many engineers have found out in utter frustration. They are valid only for so-called "simple" sliding systems or within the "useful" or "limiting pV" range. An excellent description of the usefulness of the wear coefficient in general has been published by Rabinowicz.[79]

Temperature rises by frictional heat prompted by increases in load and, especially, sliding speed are considered the most noteworthy causes of unpredictable changes in the wear rate. Lancaster[7] measured the change in steady-state wear rate of PE, PP, PMMA, acetal homopolymer, and nylon 6-6 sliding against smooth steel by heating up the steel counterface. His results are shown in Fig. 19.14.2. This graph shows that unacceptably high wear rates can result from relatively small temperature changes.

It has been recognized that, in many instances of adhesive wear, a film of polymer is formed upon the smooth counterface. Its genesis is adhesion to the counterface, deformation, and breakage of the junction in the polymer itself. Earlier we briefly mentioned their role in polymer friction. Now we draw attention to the fact that the transferred layers break up upon continued sliding, become detached, and thereby

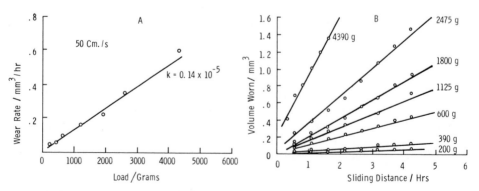

Fig. 19.14.1. Wear of nylon on steel. A) Wear rate versus load; B) Wear versus distance. (Copyright ASTM, 1916 Race Street, Philadelphia, PA 19103. Reprinted with permission. Steijn, Ref. 76.)

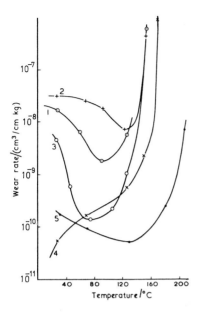

Fig. 19.14.2. Variation of steady-state wear rate on steel with temperature. 1. PE; 2. PMMA; 3. PP; 4. PAC; 5. Nylon 6.6. (Lancaster, Ref. 7. Reprinted by permission of North Holland Publishing Company, Amsterdam.)

generate loose wear debris while, at the same time, new transfer takes place. Thus the original polymer surface wears away by a transfer mechanism perhaps somewhat similar to the classical wear and transfer process described by Kerridge[80] for brass sliding against smooth steel unless back-transfer as suggested by Briscoe[6] becomes operative on a sizable scale and induces zero wear.

Polymeric transfer films have received considerable attention ever since their discovery in 1953 by Bowers, Clinton, and Zisman[36, 81] during friction experiments with PTFE and PE. PTFE in particular has been studied in great detail. Makinson and Tabor[38] found that, at high sliding speed or low temperature, PTFE transfers in the form of lumps and slabs at least 0.1 μm thick, but at low speed and moderate temperatures the transferred PTFE consists of thin, highly oriented films and fibers. Steijn,[40] sliding hardened steel sliders on flat, abraded PTFE, observed also the unique capacity of PTFE to be drawn and stretched into thin, highly oriented films and fibers several hundreds Angstrom thick. Pooley and Tabor[39] subsequently found that unbranched HDPE also behaves in much the same way as PTFE and felt that the ease of forming thin transfer films on the counterface had much to do with the molecular profile which, for these polymers, is smooth on an atomic scale. Tanaka and Miyata[41] established that UHMWPE also behaved like PTFE and HDPE.

In contrast to the very thin, highly oriented transfer films of PTFE, HDPE, and UHMWPE, much lumpier transfer has been reported by Rhee and Ludema[42] for nylon 6-6, POM, and PTFE-filled POM. Transfer from polymer to another polymeric counterface has been studied by Jain and Bahadur.[43] For PE sliding against PTFE, PVC, PP, and PMMA and for PET against PE and PVC, they found that transfer took place under all conditions of sliding but was dependent on load, sliding speed, and time.

Thus, there is no question as to the reality of transfer. What is not yet clear is the exact mechanism by which adhesive or transfer wear ultimately produces loose wear debris, or in particular, what happens after the transfer film is formed against the counterface. Several investigators have looked into this problem. Kar and Bahadur[44] studied the micromechanism of wear and concluded that, for HDPE and POM,

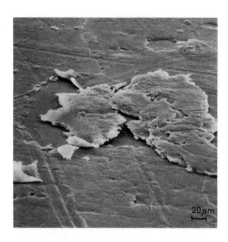

Fig. 19.14.3. Wear surface of FEP thrust washer (scale mark indicates 20 μm). (Copyright ASTM, 1916 Race Street, Philadelphia, PA 19103. Reprinted with permission. Steijn, Ref. 76.)

breakdown of the transferred film occurred by thermal softening or even melting but for PTFE by continued shear.

Tanaka[82] also studied the connection between transfer and wear. He carried out wear experiments on PTFE, HDPE, LDPE, and nylon 6 pins sliding against steel. He concluded that the polymer transfer layer attains a finite equilibrium thickness that is controlled by the balance between removal at the leading edge of the pin and replenishment of the film by transfer. This mechanism appears very much like the transfer and wear process described by Kerridge[80] in his classical work with brass sliding against steel. If so, the steady-state wear rate of the polymer against the polymer-covered counterface would be controlled by the rate of film removal which, in turn, depends on its mechanical strength and degree of anchoring onto the substrate. Briscoe[6] suggested that modification of these aspects of the transfer film is one of the paramount roles of fillers dispersed throughout a polymer matrix. Indeed, experimental evidence can be seen in the improved durability of transfer films of CuO- and Pb_3O_4-filled HDPE.[83] When no smooth transfer film is formed and polymer aggregates, lumps or flakes appear on the sliding surfaces, the wear mechanism is obviously different in scale, although not necessarily in principle. In that case polymer particles are loosened from the polymer surface and pulled along only to be shortly released by the mating counterface. An example is shown in Fig. 19.14.3 for FEP sliding against a smooth, hardened steel ring.

It is interesting at this point to question how smooth a counterface should be to guarantee adhesive wear and not abrasive wear. The detailed work by Eiss e.a.[84-87] on polymer wear, both adhesive and abrasive, indicates the extreme complexity of this problem. Their work confirmed that, to define the real surface-contact interaction, the mechanical-strength properties of the polymer must be taken into account. External sliding conditions such as load and rubbing speed must also be considered. Finally, surface topography is important, not only in terms of roughness but also because it is characterized by profile height distribution with skewness and kurtosis, asperity curvature, and a penetration parameter based on the bearing area curve and polymer yield strength.

19.15 Abrasive Wear

To most people, abrasive wear hardly needs definition. As implied by the word *"abrading,"* from the Latin "abradere" — to scrape off, abrasive wear, or wear by abrading, means gouging. As defined by the Research Group on Wear of Engineering

Materials of the OECD (Organization for Economic Cooperation and Development), abrasive wear is "wear by displacement of material caused by hard particles and protuberances." Thus abrasive wear occurs when a hard, rough surface rubs against a softer surface and gouges out by cutting, grooving, and ploughing slivers or fragments from the softer surface. A hardened and rough surface of steel, therefore, abrades a surface of polyethylene, polycarbonate, and other plastics. This type of abrasion is termed *2-body abrasion.* In *3-body abrasion,* abrasive grains introduced between two sliding surfaces do the cutting and gouging on one or both surfaces.

The simplest expression for abrasive wear is derived by considering how an abrasive grain, in the form of an inverted cone, indents a softer surface and ploughs out a chip.[5] The *wear rate,* expressed as removed volume V per unit sliding distance l can then be written as

$$\frac{dV}{dl} = K \frac{L \tan \Theta}{\pi \, p_m} \tag{19.15.1}$$

in which L is the load, p_m the indentation hardness of the softer surface, Θ the slope of the cone, and K an accommodation factor to imply that not all material in the groove is swept out or removed as loose debris. Rabinowicz pointed out that the form of this equation for abrasive wear is similar to that for simple adhesive wear, and we may broaden Eq. (19.15.1) into the general form

$$W_v = k_{abr} \, Ll/p_m \tag{19.15.2}$$

This equation implies that abrasive wear is directly proportional to load, inversely proportional to hardness, and the severity of abrasion as a function of the shape and geometry of the abrasive grains is reflected in the magnitude of the nondimensional constant k_{abr}. Moore[88] arrived by abstract reasoning at an analogous expression for wear W_v per unit load and per unit sliding distance

$$W_v = K_1 \, K_2 \, K_3 \, \sigma \, H'^{-1} \tag{19.15.3}$$

in which σ is the load per unit area, H' the indentation hardness, K_1 the probability of debris generation, K_2 the mean proportion of the groove volume removed, and K_3 a shape factor depending on the geometry of the abrasive particles.

In Eq. (19.15.2), K_1, K_2, and K_3 appear capped together as k_{abr}. A more detailed analysis of the $K_1 K_2 K_3$-term and in particular of the effect of the abrasive-particle geometry has been rendered recently by Moore and Swanson.[89] This work was done in reference to rigid-plastic materials with elastic limits of strain 1–10% of those of polymers. Verification of the simple abrasive wear equation for plastics in general, however, can be found throughout the literature. Deviations become apparent as soon as the experimental conditions stray from the simple assumptions originally made. As an example, the linearity of wear with time or sliding distance is clearly shown by the single-pass data of Roberts and Chang[90] on nylon 6-6 and polycarbonate abraded against SiC abrasive paper, as illustrated in Fig. 19.15.1. However, when multiple-pass data were collected and plotted, the wear was found to fall off rapidly with time, see Fig. 19.15.2. The single-pass wear rates were measured to be 20–40 times greater than the equilibrium multiple-pass wear rates. In multiple passes the abrasive paper gradually loses its cutting power and is loaded with plastic wear debris.

The plots of Fig. 19.15.1 and 19.15.2 also show the effect of abrasive-particle size. In general, the larger the mean abrasive particle size, the higher the wear. This effect, reflected in the slopes of the curves in Figs. 19.15.1 and 19.15.2, of course influences the magnitude of k_{abr}.

The wear-load relationship for abrasion of plastics has been examined in the work of Ratner and Farberova.[91] Their results of wear rate against pressure are shown

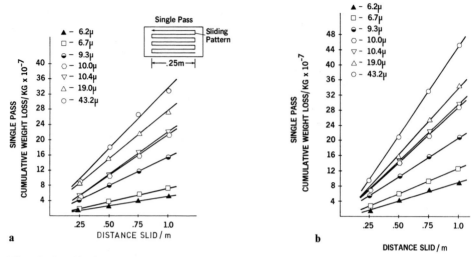

Fig. 19.15.1. Single-pass cumulative mass loss versus distance for two polymers sliding against SiC particles of various sizes: 6.2 μm; 6.7 μm; 9.3 μm; 10.0 μm; 10.4 μm; 19.0 μm; 43.2 μm. a) Nylon 6-6, b) Polycarbonate. (Roberts and Chang, Ref. 90.)

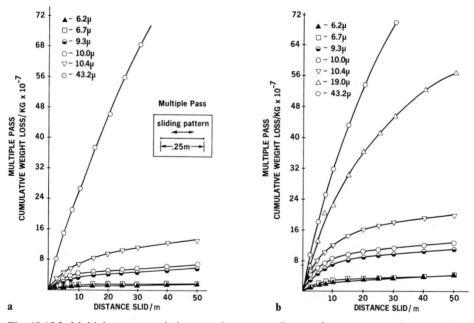

Fig. 19.15.2. Multiple-pass cumulative mass loss versus distance for same two polymers against SiC particles. Symbols as in Fig. 19.5.1, a and b. a) Nylon 6-6, b) Polycarbonate. (Roberts and Chang, Ref. 90.)

in Fig. 19.15.3 for a number of plastics abraded against emery paper. In these tests pressure is analogous to load. It can be seen that, indeed, wear is directly proportional to load. However, these investigators also used metal gauze as a counterface in abrasive wear tests. With metal gauze, they invariably found that abrasive wear is proportional to load raised to the n-th power. For most plastics n was found to vary from 1.1 to as high as 3.5. The explanation for this behavior by Ratner and Farberova is that abrasion against metal gauze is not gouging or cutting but instead the well-rounded protrusions of the gauze induce predominantly elastic encounters. Ratner and Lure[92] have taken this further and proposed that these repetitive encounters on gauze induce fatigue. They have also proposed that fatigue processes may be interpreted as being thermally activated, and therefore this type of wear could be represented by a rate equation of the form

$$\text{Wear Rate} = K \exp \left[-(E - kL)/RT \right] \qquad (19.15.4)$$

in which E is an activation energy, L the load, k a constant, and K a proportionality constant. Abrasion tests with HDPE, PVC, and PMMA against metal gauze have yielded activation energies for these plastics in reasonable agreement with values determined by other means. It may be argued that the experiments against metal gauze are not abrasion experiments, but basically fatigue wear experiments. Perhaps this question is one of definition. Ratner contends that abrasion encompasses all events leading to removal of the surface layer.[93]

In further work on the abrasion of plastics, Ratner e.a.[93] introduced another interesting concept. They reasoned that the separation of wear particles from the bulk requires an amount of mechanical work to be supplied by the force of friction equal to the area under the stress-strain curve. This amount of work is approximately the product of tensile stress at break s and elongation at break ε, or $s\varepsilon$. On that basis, abrasive wear over a contact area determined by the indentation hardness H' would be directly proportional to the frictional force divided by the product of H' and $s\varepsilon$, or the specific wear rate W_L would be

$$W_L \propto \frac{\mu}{H s \varepsilon} \qquad (19.15.5)$$

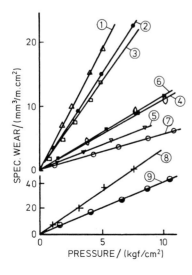

Fig. 19.15.3. Wear of various plastics as a function of pressure during abrasion on emery paper. 1) "Antifrik", 2) High-pressure PE, 3) Glass-fiber laminate, 4) Nylon 6, 5) Polyamide AK-7, 6) Polyamide AK-7 plus 5% talc, 7) Babbitt (metal), 8) PTFE, 9) PVC-based flooring. (Ratner and Farberova, Ref. 93. Reprinted by permission of Applied Science Publishers, Ltd.)

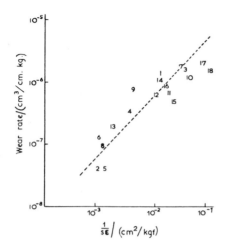

Fig. 19.15.4. Correlation between wear rates during single traversals over steel of 1.2 μm R_a and $1/S$. 1) PMMA, 2) LDPE, 3) PS, 4) Acetal copolymer, 5) Nylon 6-6, 6) PP, 7) Epoxy, 8) PTFE, 9) PMMA-acrylonitrile copolymer, 10) Polyester, 11) PTCFE, 12) PC, 13) Nylon 11, 14) ABS copolymer, 15) PPO, 16) Polysulfone, 17) PVC, 18) PVDC. (Lancaster, Ref. 94.)

Lancaster[94] measured s and ε by tensile testing a large variety of polymers. When these data are plotted against abrasive wear measured on a rough steel surface (approximately 1.25 μm AA), the graph shown in Fig. 19.15.4 is obtained. The points follow a line at slope 1 — indeed in good agreement with Eq. (19.15.5).

It should be clear from the foregoing that in abrasive wear the mode of deformation, i.e., elastic or plastic depending on the topography of the countersurface, has a significant bearing on the failure process. The plasticity index mentioned earlier is an important criterion. Lancaster, in his comprehensive article on abrasive wear of polymers,[94] is emphatic on this point. The cutting wear of rigid polymers by abrasives or hard, rough countersurfaces, is clearly plastic by nature. A wear formula such as Eq. (19.15.1) may well describe the process.

For smoother countersurfaces or, as the Russians claim,[92, 93] for rounded metal gauze, the deformation will shift to the elastic mode. For this condition, of course, Eq. (19.15.1), which focuses on the plastic parameter H' or p_m, is invalid. The ultimate wear failure here may well be fatigue. In this respect, the work by Eiss and Bayraktaroglu[84] on the wear of LDPE deserves particular attention. They showed that, indeed, sliding between LDPE and steel surfaces as rough as 1.16 μm R_a, generated large wear particles by abrasive wear. Against smooth steel surfaces of 0.065 μm surface finish, the wear of LDPE was nonabrasive and took place by the formation and subsequent breakup of thin transfer films on the steel surface.

19.16 Wear by Surface Fatigue

Wear in plastics by surface fatigue is recognized as a real source of failure. In the previous section on abrasive wear, we have referred to this point of view. We recall the claims by Ratner e.a.[92, 93] that plastics sliding over the rounded asperities of metal gauze wear by fatigue. Indeed, the Russian school working on abrasion has lent a powerful incentive to this thought. A theoretical basis for fatigue in highly elastic materials like rubbers was worked out by Kraghelsky and Nepomnyashchi.[95] Ratner has gone so far as to contend that the wear of plastics against smooth steel is, in principle, the same as against metal gauze.[96] With neither material is microcutting involved.

Despite the emphasis on surface fatigue by these Russian tribologists, relatively little experimental support for this viewpoint has appeared in Western publications. Only recently have those involved in long-time wear testing of polymers for artificial human joints come upon instances of failure interpreted as fatigue from microscopic examinations. Most of this work was conducted at the University of Leeds by an interdisciplinary research team.[97-100] Their tests were run on a 3-pin-on-disc machine over sliding distances as long as 1500 km. The pin specimens were of surgical grade UHMWPE; the disc was stainless steel lapped to a finish of approximately 0.015 μm R_a. A typical wear curve is shown in Fig. 19.16.1.

In essence, the curve consists of two sections, A and B. The investigators concluded that A represents adhesive wear, and a thin transfer film of UHMWPE was found on the counterface. At B, a new type of wear becomes operative and adds to the adhesive wear. Scanning electron microscopy, transmission electron microscopy, and optical microscopy revealed small cracks perpendicular to the wear direction and other surface defects characteristic of fatigue damage.

Figure 19.16.2 shows a microstructure in UHMWPE characterized by shallow pits extended in the direction of sliding from which thin flakes seem to have been lost. This feature, noted only in the B-part of the wear curve, has been recognized as a surface fatigue failure.[98] A plot of the nominal stress at the onset of curve B versus the log sliding distance was found to have a shape that is typical for S-N curves. The fine powdery debris that emerged only during the B-stage was attributed to fatigue wear.

Admittedly, evidence of fatigue wear during sliding of polymers on hard, smooth counterfaces has been slow in coming. So far, it seems limited to test pins of UHMWPE, although Walker and Erkman[101] found fatigue cracking in used prosthetic hip joints made of UHMWPE. However, there is little doubt that repetitive elastic encounters between pin and counterface, no matter how smooth, weaken the surface strata and, upon continued sliding, will cause material breakdown.

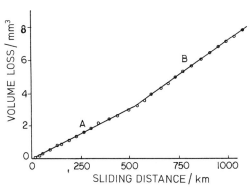

Fig. 19.16.1. Typical wear vs. distance curve for UHMWPE sliding against surgical grade stainless steel under dry conditions. (Atkinson, Brown and Dowson, Ref. 98.)

Fig. 19.16.2. Surface fatigue in UHMWPE. (Courtesy of Atkinson, Brown and Dowson, Ref. 98.)

19.17 Friction and Wear of PTFE

Because of its low friction and exemplary sliding characteristics, PTFE has been the subject of numerous investigations since its discovery. Reports and reviews on its frictional behavior have appeared regularly in the technical and scientific literature.[5, 7, 102] More recently its wear properties have also received considerable attention.[103] With this much already in print, we will limit ourselves here to the essential and unique aspects of PTFE's friction and wear behavior.

As pointed out before,[5] PTFE possesses, aside from its unusual frictional characteristics, other unique properties. It has a high crystalline melting point of 327° C, a molecular weight as high as 10 million if not 100 million, and therefore an unusually high melt viscosity – some 10^{11} poises at 380° C. The chemical inertness of PTFE is unparalleled among the polymers, its dielectric and electrical properties are superb. The cohesive energy density of 30 cal/cm^3 and the critical surfce tension of 18.5 dynes/cm are some of the lowest among the polymers. The crystal structure of PTFE, determined by Bunn and Howells,[104] is purely linear and consists of a zigzag backbone of $- CF_2CF_2 -$ with a 180° twist every 13 CF$_2$ units, see Fig. 19.17.1. The lateral packing of these rodlike molecules is hexagonal with $a_0 = 5.62$ Å. Pooley and Tabor[39] have drawn attention to the smooth molecular profile of this structure – also quite unique among the polymers. Hanford and Joyce,[105] the first ones to seriously contemplate PTFE's low friction, have suggested a minimal interchain bonding between the rods due to the tight and massive shielding power by the large and closely fitting fluorine ions. These features are depicted in Fig. 19.17.1. The microstructure of PTFE is another unusual feature among polymers. Bunn, Cobbold and Palmer[106] have shown that it is not spherulitic, but rather is made up of finely striated bands. Speerschneider and Li[107, 108] have attempted to correlate the band structure with its mechanical behavior, including friction.

An accurate picture of the measured coefficient of friction of PTFE sliding against itself was presented previously (Fig. 19.3.2). It shows that the coefficient of friction is lowest when the load is high. Under these conditions a thin, highly oriented film of PTFE is formed between the sliding surfaces either as a true transfer film[36, 38]

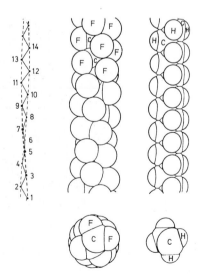

Fig. 19.17.1. Molecular structure of PTFE. Left, twisted zigzag chain with 13 carbon atoms per 180° burst and repeat distance 16.9 Å (1 Å = 1 × 10^{-10} m). Center, side and end views of a molecule of PTFE. Right, ribbon-like PE molecule for comparison. (Bunn and Howells, Reprinted by permission from *Nature,* 1954, **174**, 549, copyright © MacMillan Journals, Ltd.

or, on the PTFE side, in the form of drawn films, webs, and fibers.[40] Examples of these are shown in Figs. 19.17.2 and 19.17.3. The friction is now truly interfacial; there is no bulk shear and the expenditure of energy is low. The films are quite thin; early estimates were in the order of 200–500 Å (20–50 nm), but later work by Pooley and Tabor[39] suggests even thinner films—perhaps as thin as 25 Å (2.5 nm).

Apparently, these films are readily generated. Steijn[40] showed that they are fully formed after only one single pass of a steel slider on flat PTFE. If a second pass is made at 90° to the first pass, new films are laid down atop and at right angles to the first ones. When he evaporated a heavy metal like Cr, Pt, or Au on the flat PTFE and then slid a clean steel ball across, he observed the films to be clear and free of metal. This condition indicated that the extracted film was not the surface itself but new and freshly drawn from the adhesion sites. Electron diffraction at very-low-beam intensities showed the films to be crystalline, with long molecular chains oriented in the sliding direction like a fiber pattern. When two single-pass sliding tracks cross perpendicularly, films extracted from the intersection indicate their layered structure by showing both orientations at 90° from one another. In a cleverly designed pin-on-disc experiment in which the preferred orientation of a PTFE transfer film was intentionally altered by adding a rotation to the pin, Briscoe and Stolarski[109] showed that the erstwhile low friction rises sharply as soon as the oriented film is disturbed.

Thus it appears that the thin PTFE film in its highly oriented state is the *sine qua non*—the essential condition for very low friction. In a sense the surfaces slide over a roadbed of thin PTFE films continuously drawn from the substrate. In this context we recall the investigation by Briscoe and Tabor[48] who measured the shear strength of thin films of PMMA, LDPE, HDPE, and PTFE between glass as a function of pressure. Some of their data are reproduced in Fig. 19.17.4. From these curves it can be seen that PTFE films rubbed onto glass have the lowest shear strength. The low friction of interfacial PTFE films generated by sliding the polymer over a hard substrate or even against itself then seems a natural consequence. When no such smooth and oriented films are established, even the friction of PTFE can be relatively high. Makinson and Tabor[38] have observed that, at higher sliding speeds and/or lower temperatures PTFE transfers more in the form of lumps and fragments than as continuous and highly oriented films.

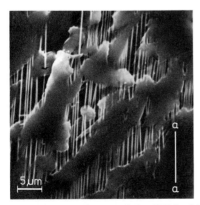

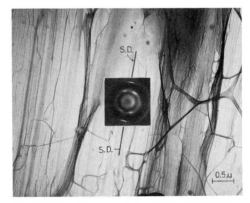

Fig. 19.17.2. Micrograph by SEM of PTFE fibers drawn across PTFE surface by steel slider.

Fig. 19.17.3. Micrograph by TEM of PTFE film on sliding surface of PTFE. Extraction replica. Center insert: electron diffraction pattern. (Steijn, Ref. 40.)

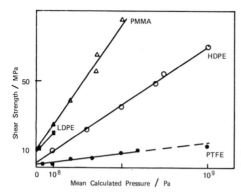

Fig. 19.17.4. Plot showing linearity of shear strength with contact pressure. (Briscoe and Tabor, Ref. 48.)

There are still many aspects of the friction of PTFE that deserve further study. Steijn[110] has shown that the kinetic friction of PTFE is sensitive to the history of the sliding surface and that time and relaxation effects play a role in sliding behavior. For instance when, in a 3-pin-on-disc apparatus, the friction of PTFE on PTFE is continuously measured while sliding velocity is gradually raised and then lowered, a friction loop results as depicted in Fig. 19.17.5. Obviously, the sliding surface "remembers," and the shear strength of the interfacial film is conditioned by the preceding events. Numerous other "irregularities," time effects, and friction transients have been reported, many of which attest to a viscoelastic contribution in the sliding behavior of PTFE.[111]

Taking advantage of the positive velocity coefficient of PTFE friction, Steijn also measured sliding speed as a function of traction at various temperatures by pulling a sled with PTFE runners over a flat PTFE slab.[110] When he plotted log sliding speed against the force of traction, as well as log speed versus the reciprocal of the absolute temperature, he got straight lines. Thus his results could be described in the form of a rate equation

$$v = v_0 e^{-\frac{u_0 - \gamma f}{kT}} \tag{19.17.1}$$

in which v is the sliding velocity, f the force of friction per unit effective area, T the absolute temperature, k the Boltzmann factor, u_0 the activation energy, γ the activation volume, and v_0 is a constant. Kinetically, therefore, the sliding of PTFE on PTFE, or rather the shearing of the interfacial films, could be viewed as a non-Newtonian viscous flow process such as described in the theory of rate processes[112] and somewhat

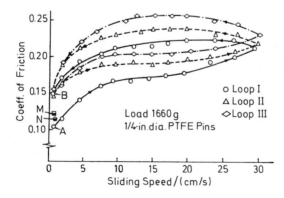

Fig. 19.17.5. Consecutive friction loops in friction versus sliding speed experiments of PTFE on PTFE. (Reprinted by permission of the American Society of Lubrication Engineers. Steijn, Ref. 110.)

similar to the early views of Schallamach[60] on the friction of rubber. One could perhaps visualize this process as slippage between adjacent molecular rods. By substituting the experimental results in Eq. (19.17.1), activation energies of approximately 7 kcal/mol were calculated and activation volumes were about 3800 cubic angstroms $(3.8 \cdot 10^{-21}$ cm$^3)$. These values are in the range for breaking van der Waals bonds which, by itself, is not unreasonable. The physical meaning of an activation volume of 3800 cubic angstroms $(3.8 \cdot 10^{-21}$ cm$^3)$ is, however, open to much speculation.

The wear of PTFE cannot be viewed separately from the low-friction properties of PTFE and its ability to shear off thin films and slivers with the greatest of ease. Tanaka, Uchiyama, and Toyooka[103] recognized that the wear rate of PTFE is about 100 times as fast as the wear of other crystalline polymers. They concluded that the microscopic band structure provides a unique mechanism for generating such films. Their model for this mechanism is pictured in Fig. 19.17.6. This Japanese research team also calculated an activation energy of about 7 kcal/mol, which they assigned to the slippage between crystalline slices within the bands. They concluded that band width, rather than crystallinity, affects wear rate.

Hu and Eiss,[113] on the other hand, concluded that crystallinity and molecular weight (MW) are the key elements in the wear process of PTFE. They found that, for a fixed crystallinity, wear decreases with increasing MW, and for a fixed MW, wear decreases with higher crystallinity. Using a pin-on-disc wear apparatus, Uchiyama[114] investigated in detail the actual formation of PTFE wear particles. He traced most of the debris as coming from the transferred and conglomerated particles and transfer film on the counterface, and not directly from the PTFE pin. Arkles and Schireson[115] showed that the MW of PTFE debris is typically reduced 100-fold in the wear process, an observation confirmed by Hu and Eiss.[113] As Arkles suggested, it is highly probable that chain scission occurs on a gross scale during wear. In view of the long length of the PTFE molecule, about 0.026 mm for an MW of 10,000,000, this hypothesis appears reasonable.

In summary, despite minor uncertainties, the mechanism by which PTFE wears so easily is fairly well understood. Like its low friction, the high wear is facilitated by its special quality to generate and form thin, highly oriented films, fibers, and fragments between sliding surfaces, either against the counterface or on the PTFE surface itself. These films and fragments eventually break up and are pushed out as loose debris. The ease of formation of such films is caused, in turn, by the unique molecular structure of PTFE itself.

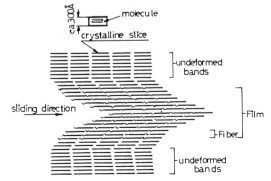

Fig. 19.17.6. Proposed mechanism of the formation of PTFE films due to the destruction of the banded structure. (Tanaka e.a., Ref. 103.)

19.18 Filled Polymers and Composites

Plastics selected for applications in tribology are often modified to give them greater mechanical and thermal strength, higher heat conductivity, and better wear and friction properties. For these reasons, fillers and reinforcing fibers are used to make composite structures. The improvement in hardness, compressive strength, impact resistance, tensile strength, moduli, and so forth, thereby providing enhanced load capacity, abrasion resistance, stiffness, and the like, is often dramatic. This improvement can be seen in Table 19.13.1, which compares modified and unmodified polymers. Further addition of solid lubricants or even impregnation with oils vastly improves the frictional behavior and performance as a "dry" bearing material.

It is beyond the scope of this chapter to describe the technology and engineering of industrial composites and filled or reinforced plastics developed over the years for tribological end uses. References 116–127 contain a number of excellent reviews on the subject and should be consulted for detailed information.

Table 19.18.1 gives examples of the composites and combinations of material that have been developed and are being used, especially in dry bearing applications. It can be seen that PTFE is widely used as a matrix material with additives such as glass, lead, and bronze, to promote wear and heat flow; and as a dry-lubricant filler material in PAC, PI, and the nylons. One may say that the addition of suitable fillers and strengtheners improves the wear resistance of virgin PTFE by several orders of magnitude. The bar diagram of Fig. 19.18.1 by Evans and Lancaster[127] illustrates this point. Detailed information on the wear, friction, and strength characteristics of fluorocarbon composites can be found in the excellent paper by Arkles, Gerakaris, and Goodhue.[124]

Table 19.18.1. Engineering plastic composites in tribology

PTFE/glass	*NYLON*/glass
/mineral (mica, etc.)	/glass, PTFE
/glass, mineral	/glass, MoS$_2'$
/graphite	/PTFE, graphite, PI
/graphite, carbon fiber	/FEP
/lead, bronze	/MoS$_2$, minerals
/glass, CdO, Ag	/oil- and silicone-impregnated
/PPS	
/PI	*THERMOSET*
	phenolic/bronze powder
POLYACETAL	phenolic/PTFE, bronze
/PTFE	phenolic/bronze
/glass, PTFE	
/bronze, PTFE	
/carbon fiber	*MISC.*
/graphite	PFA/carbon fibers
/oil – impregnated	PVDF/carbon fiber
	PEEK/PTFE, carbon fiber
POLYIMIDE	PPS/PTFE, glass, carbon fiber
/PTFE	PES/PTFE, glass, carbon fiber
/PTFE, graphite	PEI/PTFE, glass, carbon fiber
/MoS$_2$	
/MoS$_2$, WS$_2$, WSe$_2$, TaSe$_2$	
/Cu powder	

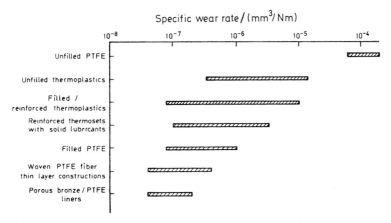

Fig. 19.18.1. Order of magnitude values for the specific wear rates of various groups of polymers and composites as dry bearing materials. (Evans and Lancaster, Ref. 127. Reprinted with permission of Academic Press, Inc., New York, N.Y.)

On the negative side, certain fillers and additives, glass fibers especially, bring an abrasive element to the composite that can easily scratch and damage the mating surface. Lancaster[7, 94] has measured the abrasiveness of a number of polymers with and without fillers against bronze and shown that the effect of glass, asbestos, mica, and minerals can be detrimental indeed.

Numerous other aspects of fillers and composites — too many to describe here — enter into the ultimate performance of the composite. Filler size, shape, aspect ratio, orientation, hardness, optimum concentration, chemical reactivity — these are some of the particulars that have been studied by various investigators and must be considered to ensure maximum performance and service life in the machine component.

19.19 Other Wear Processes

Aside from adhesive, abrasive, and fatigue wear, polymers in sliding or rolling contact are, of course, subject to various other types of wear and degradation. Obviously, thermal wear takes place when insufficient care is taken during sliding, and friction heat accompanied by high flash temperatures brings the plastic to the point of accelerated surface degradation. As has been pointed out earlier, in crystalline thermoplastics this degradation can take the form of softening and incipient surface melting.[34, 44] In thermosets, it may lead to thermal decomposition.

Environmental influences also affect the friction and wear behavior of plastics. Reichenbach[128] observed that the relative humidity (R.H.) of the air affects the friction of PTFE against steel. At zero humidity, both friction and wear were found to be very high. Introduction of air of 50% R.H. caused a rapid drop in friction and formation of a transfer film. Steijn[111] also noted that PTFE sliding on PTFE in nitrogen exhibited poor sliding behavior that was quickly remedied by flushing the test chamber with moist air. A satisfactory explanation for these effects of humidity is still lacking. As for chemical wear, Richardson[129] commented on several cases of potential polymer degradation by rubbing contact with metals, and Richardson and Pascoe[130] claim that the wear of PVC against mild steel is associated with a mechano-chemical comminution process at the sliding surface generating low-MW debris.

More recently fretting wear, or fretting corrosion, has been recognized as a form of wear to which plastics are mildly susceptible. Stott, Bethune and Higham[131-133] found from fretting tests against steel that PC, PVC, PSP, nylon 6-6, and PE suffered significant surface damage by plastic flow in the contact zone, with extrusion of polymer fibers. Several polymers like PC, PVC, and PSP also showed cracks while PTCFE and PMMA did not suffer any apparent damage. Transferred iron oxide, mostly α-Fe_2O_3, was present on all polymer scars. Fretting damage to the steel was substantial; it was most severe against nylon 6-6, followed by PC, PMMA, PVC, PSP, PCTFE, PVDF, and PE with no damage to the steel by PTFE. Stott, Bethune, and Higham suggested that damage to the steel depends on adhesion and surface energetics. In support of this idea, they showed that wear of the steel decreased with the critical surface tension of the polymers involved.

Still other basic wear processes to which polymers may be subjected are solid particulate erosion, cavitation, impact wear, and delamination. Although *erosion* has been extensively studied for metals and ceramics, interest in the erosion of plastics and rubbers has only begun, including some work to measure relative erosion rates.[134-136] Cavitation, delamination, and impact wear of polymers are not being described here.

From a global point of view, the friction and wear of plastics, or rather the tribology of polymeric materials, offers a wide-open field for further study. We have seen that many wear mechanisms of great complexity are operative. As we learn more, new wear phenomena are being discovered. In an age in which industrial technology grows ever more wear-intensive, temperature seems, perhaps, the major and most serious limitation to the use of plastics in tribology. Hence the emphasis on high-temperature plastics like fluoro- and perfluorocarbons, polyphenylene sulfides, and polyimides in filled and composite form. These compounds are being investigated in research and development centers throughout the world. And it is here where the next advance in polymer tribology must be made.

Polymer Acronyms used in Chapter 19

CA	cellulose acetate	PFA	perfluoroalkoxy-modified tetrafluoroethylene
FEP	fluoroethylene polymer		
HDPE	high-density polyethylene	PI	polyimide
LDPE	low-density polyethylene	PMMA	poly(methyl methacrylate)
P-α-C	polycyclohexyl-α-chloro-acrylate	POM	polyoxymethylene
		PP	polypropylene
PA	polyamide	PPS	polyphenylene sulfide
PAC	polyacetal	PS	polystyrene
PAN	polyacrylonitrile	PSP	polysulfone
PC	polycarbonate	PTFE	polytetrafluoroethylene
PCTFE	polychlorotrifluoroethylene	PVC	poly(vinyl chloride)
PE	polyethylene	PVDE	poly(vinylidene chloride)
PEEK	polyetheretherketone	PVDF	poly(vinylidene fluoride)
PEI	polyetherimide	SAN	styrene acrylonitrile
PES	polyethersulfone	UHMWPE	ultrahigh-molecular-weight polyethylene
PET	poly(ethylene terephthalate)		

References

1. G.F. Kinney, *Engineering Properties & Applications of Plastics,* Wiley, New York 1957.
2. "Materials 82-83," *Modern Plastics,* January 1983, 51.
3. F.P. Bowden and D. Tabor, *Friction — An Introduction to Tribology,* Anchor Press, Doubleday, Garden City, New York 1973.

4. L.-H. Lee, Ed., *Advances in Polymer Friction and Wear, Polymer Science & Technology, 5A and 5B,* Plenum Press, New York 1974.
5. R.P. Steijn, "Friction and Wear of Plastics," *ASM Metals Engineering Quarterly* 1967, **7**, 9.
6. B. Briscoe, *Tribol. Internat.* Aug. 1981, **13**, 231.
7. J.K. Lancaster, *Polymer Science,* Chapter 14, **2**, edited by A.D. Jenkins, p. 958, North-Holland, Amsterdam 1972.
8. B.J. Briscoe, *Proc. 7th Leeds-Lyons Symposium on Tribology; University of Leeds, 1980;* Mechanical Engineering Press, Burg St. Edmunds 1980.
9. F.P. Bowden and D. Tabor, *The Friction & Lubrication of Solids − Part I,* The Clarendon Press, Oxford 1954.
10. F.P. Bowden and D. Tabor, *The Friction & Lubrication of Solids − Part II,* The Clarendon Press, Oxford 1954.
11. J.F. Archard, *Proc. Royal Soc. A.* 1957, **A 243**, 190.
12. J.F. Archard, *J. Appl. Phys.* 1961, **32**, 1420.
13. A.S. Lodge and H.G. Howell, *Proc. Phys. Soc. B* 1954, **67**, 89.
14. M.W. Pascoe and D. Tabor, *Proc. Royal Soc. A* 1956, **235**, 210.
15. S.C. Cohen and D. Tabor, *Proc. Royal Soc. A* 1966, **291**, 186.
16. A.J.G. Allan, *Lubr. Eng.* May 1958, **14**, 211.
17. J.A. Greenwood and J.B.P. Williamson, *Proc. Royal Soc. A* 1966, **295**, 300.
18. D.J. Whitehouse and J.F. Archard, *Proc. Royal Soc. A* 1970, **316**, 97.
19. B. Bushan, *Trans. ASME,* 1984, **106**, 26.
20. J.F. Archard, *Tribol. Internat.* Oct. 1974, **7**, 213.
21. D. Tabor, *Phil. Mag.* 1952, **43**, 1055.
22. D.G. Flom and A.M. Bueche, *J. Appl. Phys.* 1959, **30**, 1725.
23. D.G. Flom, *J. Appl. Phys.* 1960, **31**, 306.
24. D.G. Flom, *J. Appl. Phys.* 1961, **32**, 1426.
25. A.M. Bueche and D.G. Flom, *Wear* 1959, **2**, 168.
26. K.R. Eldredge and D. Tabor, *Proc. Royal Soc. A* 1955, **229**, 181.
27. T. Fort, Jr., *J. Phys. Chem.* 1962, **66**, 1136.
28. D.I. James, *Wear* 1959, **2**, 183.
29. K.G. McLaren and D. Tabor, Institution of Mechanical Engineers, Paper No. 18, *Lubrication and Wear Convention,* Bournemouth, May 1963.
30. M. Watanabe, M. Karasawa and K. Matsubara, *Wear* 1968, **12**, 185.
31. K.C. Ludema and D. Tabor, *Wear* 1966, **9**, 329.
32. K.A. Grosch, *Proc. Royal Soc. A* 1963, **274**, 21.
33. J.R. Archard, *Wear* 1958/59, **2**, 438.
34. K. Tanaka and Y. Uchiyama, in *Advances in Polymer Friction and Wear,* edited by L.H. Lee, Part 5 B, p. 499, Plenum Press, New York 1974.
35. G.V. Vinogradov, V.A. Mustafaev and Yu. Ya. Podolsky, *Wear* 1965, **5**, 358.
36. R.C. Bowers, W.C. Clinton and W.A. Zisman, *Modern Plastics* 1954, **31**, 131.
37. R.C. Bowers and W.A. Zisman, NRL Report 5945, U.S. Naval Research Lab, Washington, DC, 1963.
38. K.R. Makinson and D. Tabor, *Proc. Royal Soc. A* 1964, **281**, 49.
39. C.M. Pooley and D. Tabor, *Proc. Royal Soc. A* 1972, **329**, 251.
40. R.P. Steijn, *Wear* 1968, **12**, 193.
41. K. Tanaka and T. Miyata, *Wear* 1977, **41**, 383.
42. S.H. Rhee and K.C. Ludema, *Wear* 1978, **46**, 231.
43. V.K. Jain and S. Bahadur, *Wear* 1978, **46**, 177.
44. M.K. Kar and S. Bahadur, *Wear* 1978, **46**, 189.
45. H.G. Howell and J. Mazur, *J. Text. Inst.* 1953, **44**, T 59–69.
46. N. Adams, *J. Appl. Polymer Sci.* 1963, **7**, 2075.
47. R.C. Bowers, *J. Appl. Phys.* 1971, **42**, 4961.
48. B.J. Briscoe and D. Tabor, *Wear* 1975, **34**, 29.
49. B.J. Briscoe and D. Tabor, *Brit. Polymer,* March 1978, **10**, 74.

50. L.C. Towle, *Advances in Polymer Friction and Wear*, **5 A**, edited by L.-H. Lee, p. 179, Plenum Press, New York 1974.
51. E. Rabinowicz, *Friction and Wear of Materials*, Wiley, New York-London-Sydney 1965.
52. R.T. Spurr, *Wear* 1982, **79**, 301.
53. R.T. Spurr, Paper II in *Proceedings 7th Leeds-Lyon's Symposium on Tribology*, University of Leeds, 1980. Mechanical Engineering Press, Bury St. Edmunds 1980.
54. W.A. Zisman, *Ind. Eng. Chem.* Oct 1963, **55**, 18.
55. R.C. Bowers, W.C. Clinton and W.A. Zisman, *J. Appl. Phys.* 1953, **24**, 1066.
56. H. Czichos and P. Feinle, *Tribologisches Verhalten von thermoplastischen, gefüllten und glasfaserverstärkten Kunststoffen*, Forschungsbericht 83, Bundesanstalt für Materialprüfung; Berlin 1982.
57. G. Erhard, *Wear* 1983, **84**, 167.
58. D.F. Moore, *The Friction and Lubrication of Elastomers*, Pergamon Press, New York 1972.
59. A. Schallamach, *Wear* 1971, **17**, 301.
60. A. Schallamach, *Proc. Phys. Soc. B* 1953, **66**, 386.
61. A. Schallamach, *Wear* 1957/58, **1**, 384.
62. A. Schallamach, *Wear* 1963, **6**, 375.
63. R. Savkoor, *Wear* 1965, **8**, 222.
64. A.D. Roberts, *Tribol. Internat.* Apr. 1976, **7**, 75.
65. R. Holm, *Electrical Contacts*, Uppsala 1946. Almquist & Wiksells.
66. H. Ernst and M.E. Merchant, *Special Summer Conf. Friction & Surface Finish*, Tech. Press, MIT, 1940.
67. J. Bikerman, in *Advances in Polymer Friction and Wear*, edited by L.-H. Lee, p. 149, *Polymer Science & Technology*, **5 A**, Plenum Press, New York 1974.
68. J. Bikerman, *Wear* 1976, **39**, 1.
69. K.V. Shooter and D. Tabor, *Proc. Phys. Soc. B* 1952, **65**, 661.
70. R.F. King and D. Tabor, *Proc. Phys. Soc. B* 1953, **66**, 728.
71. E. Rabinowicz and K.V. Shooter, *Proc. Phys. Soc. B* 1952, **65**, 671.
72. J.T. Burwell, Jr., *Wear* 1957/58, **1**, 119.
73. *Modern Plastics Encyclopedia*, McGraw-Hill, New York.
74. J.F. Archard and W. Hirst, *Proc. Roy. Soc. A* 1956, **236**, 397.
75. J.R. Archard, *J. Appl. Phys.* 1953, **24**, 981.
76. R.P. Steijn, in *Wear Tests for Plastics: Selection and Use*, ASTM SPT 701, edited by R.G. Bayer, ASTM, 1979.
77. S.K. Rhee, *Wear* 1970, **16**, 431.
78. R.B. Lewis, *Mech. Eng.* 1964, **86**, 32.
79. E. Rabinowicz, *JOLT*, Apr. 1981, **103**, 188.
80. M. Kerridge, *Proc. Phys. Soc. B* 1955, **68**, 400.
81. R.C. Bowers, W.C. Clinton and W.A. Zisman, *Lubr. Eng.* Aug. 1953, **9**, 204.
82. K. Tanaka, *Wear* 1982, **75**, 183.
83. B.J. Briscoe, A.K. Pogosian and D. Tabor, *Wear* 1974, **27**, 19.
84. N.S. Eiss and M.M. Bayraktaroglu, *ASLE Trans.* 1980, **23**, No. 3, 269.
85. J.H. Warren and N.S. Eiss, *Trans. ASME* January 1978, **100**, 92.
86. N.S. Eiss and K.A. Smyth, *Trans. ASME* April 1981, **103**, 266.
87. N.S. Eiss and S.C. Milloy, in *Wear of Materials 1983*, edited by K. Ludema, p. 650, ASME, 1983.
88. M.A. Moore, in *Fundamentals of Friction and Wear of Materials*, edited by D.A. Rigney, p. 73, ASM, 1981.
89. M.A. Moore and P.D. Swanson, in *Wear of Materials 1983*, edited by K. Ludema, p. 1, ASME, New York 1983.
90. J.C. Roberts and H.W. Chang, *Wear* 1982, **79**, 363.
91. S.B. Ratner and I.I. Farberova, *Sov. Plast.*, Sept. 1960, **9**, 51 (Eng. Transl.); also in *Abrasion of Rubber*, edited by D.I. James, p. 297, MacLaren, London 1967.

92. S.B. Ratner and E.G. Lure, in *Abrasion of Rubber,* edited by D.I. James, p. 155, Palmerton Publishing Co., New York 1967.
93. S.B. Ratner, I.I. Farberova, O.-V. Radyukevich and E.G. Lure. *Sov. Plast,* July 1964, **7**, 37; also in *Abrasion of Rubber,* edited by D.I. James, p. 145, MacLaren, London 1967.
94. J.K. Lancaster, *Wear* 1969, **14**, 223.
95. I.V. Kraghelsky and E.F. Nepomnyashchi, *Wear* 1965, **8**, 303.
96. S.B. Ratner, in *Abrasion of Rubber,* edited by D.I. James, p. 23, Palmerton Publishing Co., New York 1967.
97. D. Dowson, J.R. Atkinson and K. Brown, in *Advances in Polymer Wear,* **5 B**, edited by L.-H. Lee, Plenum Press, New York-London 1974.
98. J.R. Atkinson, K.J. Brown and D. Dowson, *Trans. ASME* Apr. 1978, **100**, 208.
99. K.J. Brown, J.R. Atkinson and D. Dowson, *Trans. ASME* January 1982, **104**, 17.
100. K.J. Brown, J.R. Atkinson, D. Dowson and V. Wright, *Wear* 1976, **40**, 255.
101. P.S. Walker and M.J. Erkman, *Advances in Polymer Friction and Wear,* **5 B**, edited by L.-H. Lee, *553*, Plenum Press, New York-London 1974.
102. B.J. Briscoe, C.M. Pooley and D. Tabor, in *Advances in Polymer Friction and Wear,* **5 A**, edited by L.-H. Lee, p. 191, Plenum Press, New York-London 1974.
103. K. Tanaka, Y. Uchiyama and S. Toyooka, *Wear* 1973, **23**, 153.
104. C.W. Bunn and E.R. Howells, *Nature* 1954, **174**, 529.
105. W.E. Hanford and R.M. Joyce, *J. Amer. Chem. Soc.* 1946, **68**, 2082.
106. C.W. Bunn, A.J. Cobbold and R.P. Palmer, *J. Polymer Sci.* 1946, **28**, 365.
107. C.J. Speerschneider and C.H. Li, *J. Appl. Phys.* 1962, **33**, 1871.
108. C.J. Speerschneider and C.H. Li, *J. Appl. Phys.* 1963, **34**, 3004.
109. B.J. Briscoe and T.A. Stolarski, *Trans. ASME,* 1981, **103**, 503.
110. R.P. Steijn, *ASLE Trans.* 1968, **11**, 235.
111. R.P. Steijn, *ASLE Trans.* 1966, **9**, 149.
112. S. Glasstone, K.J. Laidler and H. Eyring, *The Theory of Rate Processes,* MacGraw-Hill, New York 1941.
113. T.-Y. Hu and N.S. Eiss, in *Wear of Materials 1983,* pp. 636–641, ASME, New York 1983.
114. Y. Uchiyama, *Wear* 1981–82, **74**, 247.
115. B.C. Arkles and M.J. Schireson, *Wear* 1976, **39**, 177.
116. P.H. Pinchbeck, *Wear* 1962, **5**, 85.
117. J.K. Lancaster, *Tribol. Internat.* 1972, **5**, 249.
118. J.C. Anderson, *Tribol. Internat.* 1982, **15**, No. 5, 255.
119. D.C. Evans and G.S. Senior, *Tribol. Internat.* 1982, **15**, No. 5, 243.
120. M.N. Gardos, *Tribol. Internat.* 1982, **15**, No. 5, 273.
121. B. Bhushan and D.F. Wilcock, *Wear* 1982, **75**, 41.
122. J.K. Lancaster, *Tribol. Internat.* 1979, **12**, 65.
123. K. Tanaka, *Trans. ASME,* 1977, **99**, No. 4, 408.
124. B. Arkles, S. Gerakaris and R. Goodhue, in *Advances in Polymer Wear,* **5 B**, edited by L.-H. Lee, p. 663, Plenum Press, New York-London 1974.
125. B. Arkles, J. Theberge and M. Schireson, *Lubr. Eng.* 1977, **33**, 33.
126. M.P. Wolverton, J.E. Theberge and K.L. McCadden, *Mach. Des.* 1983, **55**, No. 3, 111.
127. D.G. Evans and J.K. Lancaster, in *Material Science & Technology, Wear,* edited by D. Scott, Academic Press, New York 1979.
128. G.S. Reichenbach, *Lubr. Eng.* 1964, **20**, 409.
129. M.O.W. Richardson, in *Advances in Polymer Friction and Wear,* edited by L.-H. Lee, Plenum Press, New York-London 1974, 785.
130. M.O.W. Richardson and M.W. Pascoe, in *Advances in Polymer Friction and Wear,* edited by L.-H. Lee, Plenum Press, New York-London 1974, 585.
131. F.H. Stott, B. Bethune and P.A. Higham, *Tribol. Internat.* 1977, **10**, No. 4, 211.
132. P.A. Higham, B. Bethune and F.H. Stott, *Wear* 1978, **46**, 335.

133. P.A. Higham, F.H. Stott and B. Bethune, *Wear* 1978, **47**, 71.
134. A.I. Marei and P.V. Izvozchikov, in *Abrasion of Rubber,* edited by D.I. James, Mac-Laren & Sons, London 1967, 274.
135. G.P. Tilley, in *Treatise on Materials Science and Technology, Wear,* **13**, edited by Douglas Scott, Academic Press, New York-San Francisco-London 1979, 287.
136. F.G. Hammitt, E.E. Timm, J.B. Hwang and Y.C. Huang, in *ASTM STP 567,* edited by A. Thiruvengadam, ASTM, Philadelphia 1974, 197.

20 Dielectrical and Dynamic Mechanical Properties of Rubbers

Sture Persson
Skega AB, S-93402 Ersmark, Sweden

Bengt Stenberg
Department of Polymer Technology, The Royal Institute of Technology
S-10044 Stockholm, Sweden

The *dielectrical properties* of polymers have been the subject of considerable research since the 1920s. As a result, a number of standard works have been published over the years. Their main direction has been toward the development of dielectrical theories within this subject area. Of these, the original monographs of Debye[1] and Fröhlich[2] which still give a valuable introduction to the theories of dielectricity deserve special mention. The measurement of dipole moment has been treated extensively by Smith.[3] A later work, the summary of data on polymers by McCrum, Read and Williams[4] must be mentioned, and A.R. Blythe[5] gives a condensed summary of the electrical properties of polymers.

The measurement of dielectrical properties using bridge, resonance, and wave guide methods is treated by von Hippel.[6] Boyd[7] gives a valuable summary which covers both dielectric theories and measuring methods.

Practical information concerning the preparation of specimens and the arrangement of electrodes appears in national and international standard books such as the *Annual Book of ASTM Standards,* section "Electrical Insulating Material D150" *British Standard Instruction,* "Method of Testing Plastics, Part 2 Electrical Properties" (B.S. 2782: 1970); and *IEC Publication 250,* "Recommended Methods for the Determination of the Permittivity and Dielectric Dissipation Factor of Electrical Insulating Materials at Power, Audio and Radio Frequencies including Metre Wavelengths." Hedvig[8] and McCrum e.a.[4] are the only works in which the influence of *cross-linking reactions* on the dielectrical properties of polymer materials is given any attention.

20.1 Vulcametry

In 1953, A.G. Bayer developed a method for the continuous measurement of cross-linking reactions in rubber compounds. This method, which was subsequently patented,[9] formed the beginning of modern *vulcametry.*[10] A characteristic of the method is that a specimen of unvulcanized rubber is, during the vulcanizing process, subjected to a continuous or periodic movement or force (tension, compression, shear, or torsion). The force or movement response is measured at the same time. The force/movement is usually transferred by a rotor or by a linearly moving paddle. The method soon became popular, and as a result several successors to the original instrument were developed. The Wallace-Shawbury curometer,[11] the Cepar apparatus,[12] the Viscurometer,[13] the Zwick-Schwing elastometer,[14] and the Vuremo[15] are some of the successors. The most well-known instrument is, however, the *Monsanto rheometer.*[16]

After the Second World War, the production of synthetic rubber expanded rapidly. The purpose of vulcametry was originally to function as a control method for synthetic rubber, but it was soon shown to be a useful method for studying the reaction kinetics

Table 20.1.1. 90% cure times in seconds

Temperature/(°C)	Isothermal	Curometer	Rheometer
120	3120	4800	7800
140	870	1100	1800
160	280	320	460
180	72	105	195
200	17	47	97

After reference 21

of the vulcanizing process. Many papers have been published on this matter; among them are the papers by Scheele,[17] Kaiser e.a.,[18] Härtel,[19] and Shershnev.[20]

There has been some criticism of traditional vulcametry. Norman[21] has shown that

A) the heating of the specimen is fairly slow, and even after temperature equilibrium has been reached temperature gradients still exist through the specimen.

B) Undesired slippage can occur between the specimen cavity and the rotor or paddle.

C) During the test some rubber materials have a tendency to become porous.

Table 20.1.1 shows that the Monsanto rheometer gives a much longer cure time than the Curometer, which in turn gives a longer cure time than the isothermal cure. The great differences at low temperature are unexpected and are most likely due to a considerable difference between the true average temperature of the specimen and the measured temperature even after a long period of time. Great variations at high temperatures are not, however, unexpected.

Another disadvantage associated with traditonal vulcametry is the difficulty of interpreting the rheometer data when they are used to determine cure times for voluminous rubber products such as rubber dampers, off-the-road tires, and mill linings.

There is, therefore, a great need for a method which makes it possible to measure the cross-linking reaction during the actual vulcanizing of the products.

20.2 Dielectric Vulcametry – General Outline

The dielectric measuring methods by which it is possible to study the manner in which the *cross-linking reaction* influence the dielectrical properties in different rubber materials are rarely dealt with in the literature.

Research groups in the USA,[22] Czechoslovakia,[23] Great Britain,[24] Romania,[8] and Japan[25] have published a number of papers concerned with the subject. Studies dealing with the changes in the dielectrical properties of rubber during vulcanization (in situ observations) have been published by Hedvig,[8] Yalof and Zika,[26] Yalof and Brisbin,[27] Rothenpieler,[28] Siemon e.a.,[29] and Senturia e.a.[30] During the last few years some research has been done on how the cross-linking reaction influences the dielectrical properties of polyurethanes. Thus, Shteinberg e.a.[31] and Baturin e.a.[32] have been studying polyether-urethanes cross-linked with 2.4-toluene diisocyanate (TDI) and trimethylolpropane (TMP). Safonov e.a.[33] have studied polyether-urethanes cross-linked with both TDI and hexamethylene diisocyanate (HMDI). Gowri-Krishna e.a.[34] has studied isocyanate-terminated polyurethanes which were cross-linked with 4,4'-methylene-bis-ortochloroaniline (MOCA) and has found that both the permittivity ε' and the dielectric loss factor ε'' decrease when the cross-link density increases. Lawandy e.a.[35] have found the same dielectric behavior with thermoplastic polyurethanes, i.e., polyurethanes containing only physical cross-links.

In spite of the fact that the earliest work was done as early as the 1920s, the dielectric methods, as yet, attained no real importance within vulcametry. The most obvious reason is that the way in which the curing reaction influences the dielectrical properties of rubber is not yet sufficiently understood to be utilized in practical vulcametry.

20.3 Dielectrical Properties of Sulphur-Cured Natural Rubber

At the end of the 1920s and in the beginning of the 1930s. Scott e. a.[22] and Kitchin[36] carried out pioneer work by studying how the permittivity, power factor and conductivity were influenced by the amount of bound sulphur in natural rubber at different frequencies and temperatures. The measurements were carried out at five different frequencies between 60 and 300 kHz within the temperature range of -75 to $+235°$ C.

Natural rubber (cis 1,4-polyisoprene) is in principle a nonpolar polymer[37, 38] but, despite the fact that cis 1,4-polyisoprene contains no polar groups, the natural rubber gives dielectric losses. The reason for this is that natural rubber always contains a sufficient amount of, e.g., *carbonyl groups* ($>C=O$), hydroperoxides ($-OOH$), and so forth created by oxidation[8, 37] for dielectric losses to appear.

Norman[39] shows that the dielecric losses increase with increasing oxidation of natural rubber. In an otherwise nonpolar polymer, carbonyl groups act as probes by which the movement of the molecules and hence the molecular relaxation can be studied.

Even the purest and most carefully prepared nonpolar polymer (such as e. g. polyethylene) contains enough polar groups to show measurable dielectric losses.

In order to study the influence of cross-links on the dielectric relaxation spectrum, it is necessary to survey the dielectrical properties of the uncured polymers at the relevant temperatures and frequencies.[8]

The original method of vulcanizing rubber is to mix rubber and sulphur, after which the mixture is heated to 120–180° C. This results in the rubber molecules being cross-linked with mono-, di-, and polysulpidic sulphur bridges. Besides cross-links, heterocyclic groups are created in the polymer chains[38–41] and carbonyl groups, in addition to those which can already be found in the uncured rubber, are formed (see Fig. 20.3.1).

At the low cross-link densities which characterize rubber, the average distance between cross-links is greater than the average length of a chain segment where mobility increases considerably when the temperature exceeds the glass-transition temperature,

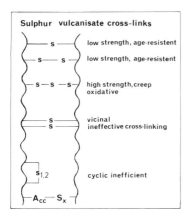

Fig. 20.3.1. Schematic diagram of structure of rubber vulcanized with sulfur and accelerator.

T_g (α'-relaxation). A chain segment of average length includes 50–100 C—C links.[42] At a *sulphur content* of 10–15%, the distance between two cross-links is estimated to be 50–100 monomer units. This indicates that a lightly cross-linked rubber with less than 10% sulphur should show only minor changes in both the glass temperature T_g and the permittivity. However, experiments have shown[8] that both the glass temperature T_g and the permittivity are already influenced fairly strongly at sulphur contents noticeably below 10%. This shows that the formation of the previosly mentioned heterocyclic groups probably results in a limitation of the molecular segmental mobility, with an increase in the glass temperature, T_g as a consequence. The size and number of heterocyclical groups have a direct connection with the permittivity and dielectric loss characteristics of the material. At sulphur contents below 10%, the shape of the dielectric spectrum for *sulphur-vulcanized rubber* is mainly determined by the heterocyclical groups. At sulphur contents exceeding 10% the influence of the cross-links starts to increase.

20.4 Dielectric Properties of Sulphur/accelerator-Cured Natural Rubber

Let us review some earlier work on the subject. Scott e.a.[22] have carried out studies at only one or a few primarily low frequencies. Schallamach[24] and Kitchin[36] show the necessity of using a sufficiently wide frequency range when recording dielectric absorption spectra. Schallamach[24] has examined how tan δ varies as a function of bound sulphur content in natural rubber compounds vulcanized with
a) sulphur only, b) sulphur + tetramethylthiuramdisulphide (TMTD), c) sulphur + zinc diethyldithiocarbamate (ZDC), and d) tetramethylthiuramdisulphide (TMTD) only. The measurements were performed at room temperature and covered the frequency range of 200 Hz to 40 MHz.

The loss tangent (tan δ) increases in sulphur-vulcanized rubber at the same time as its maximum f_{max} is moved toward lower frequencies when the content of bound sulphur increases in accordance with Fig. 20.4.1. It is mainly the heterocyclically bound sulphur which is responsible for these effects.

Sulphur/TMTD-vulcanized natural rubbers with different quantities of bound sulphur show great similarities to the sulphur-vulcanized natural rubber in the sense that the loss tangent (tan δ) increases at the same time as tan δ_{max} moves toward lower frequencies.

With the same content of bound sulphur as in previous examples and using zinc diethyldithiocarbamate (ZDC) as an accelerator, somewhat higher f_{max} and slightly

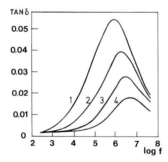

Fig. 20.4.1. Dependence of loss tangent on frequency for natural rubbers cured with ZDC as accelerator. Combined sulfur: 1, 4.7%; 2, 3.5%; 3, 2.17%; 4, 1.17%. After reference 24.

lower tan δ_{max} values are obtained. This shows that the ZDC-accelerated cross-linking reaction gives less heterocyclical groups in the molecule chains.

TMTD-cured natural rubber gives very low tan δ-values at high frequencies. The TMTD sulphur atoms are probably mainly linked to the rubber molecules in the form of intermolecular monosulphide and disulphide cross-links.

The results shown can be summarized as indicating that *intermolecularly linked* sulphur (cross-links) contributes very little to the dielectric relaxation as a sulphur atom or atoms which form cross-links that are anchored at both ends in the molecular chains. There are, therefore, strong reasons to believe that the dielectric losses are mainly caused by sulphur which is linked to the molecule in other ways than in the form of cross-links. This should mean that the dielectric losses for a given content of bound sulphur should decrease with increasing modulus of elasticity. This has also been shown to be the case, which illustrates that in sulphur-cured natural rubber ZDC favors the formation of cross-links instead of the (ineffective) heterocyclic combinations associated with TMTD.

20.5 Dielectrical Properties of Peroxide-Cured Natural Rubber

In addition to sulphur and sulphur donors, *peroxides* can be used to cross-link rubber. The cross-linking reaction takes place by way of a radical mechanism[43] and gives C—C cross-links.[44] The most important difference between peroxide and sulphur vulcanizing is that the peroxides do not form polar groups either in the molecular chain or in the cross-links. Consequently the weak shift of the relaxation maximum towards higher temperatures in a peroxide-cured rubber is exclusively an effect of the influence of the cross-links in preventing the motion of the polymer chains.

Hedvig[8] also gives examples of peroxide-initiated cross-linking reactions where the transition temperature shows a strong shift toward higher temperatures as a result of curing.

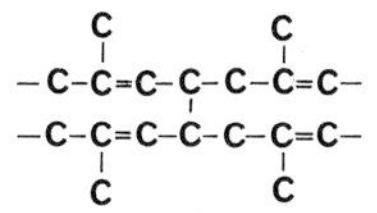

$$
\begin{array}{cc}
\text{C} & \text{C} \\
| & | \\
-\text{C}-\text{C}=\text{C}-\text{C}-\text{C}-\text{C}=\text{C}- & \\
-\text{C}-\text{C}=\text{C}-\text{C}-\text{C}-\text{C}=\text{C}- & \\
| & | \\
\text{C} & \text{C}
\end{array}
$$

Fig. 20.5.1. Carbon-carbon cross-links in peroxide vulcanized rubber.

20.6 Fillers

Fillers, organic as well as *inorganic,* have a wide usage in rubber technology. The fillers influence the physical and chemical properties of the rubber. The electrical characteristics of rubber are also influenced.[45] The mixing of *carbon black* in rubber gives rise to complex dielectrical characteristics due, among other things, to the conductivity of carbon black. Addition of carbon black and other fillers means that the rubber material becomes dielectrically heterogeneous which leads to the development of charges in the border lines between the rubber and the filler. This phenomenon is known as the *Maxwell-Wagner-Sillars* (MWS) *effect.*[8]

When the dielectric spectrum is considered as a function of the frequency, the MWS polarization usually appears as a background associated with an increase in the absolute value of the permittivity. If the temperature is used as a variable, the MWS-effect is evident as a powerful increase in the relaxation intensity when the frequency is reduced.

20.7 Carbon Black

Rubber is very seldom used without fillers, and carbon black is without comparison the most important filler in rubber. Not only the mechanical characteristics of rubber but also the dielectrical properties are influenced. High contents of special types of carbon black with very high specific surface gives low resistivity values. The lowest resistivity values which can be obtained are about 1Ω cm. Norman[45] and James[46] have published broad surveys of the area.

Usually the carbon black particles are separated by insulating polymer layers which means that the Maxwell-Wagner-Sillars polarization becomes very strong. In a dielectric spectrum this appears as a powerful maximum at low frequencies. With carbon black contents of 30–40% permittivity values of approximately 100 can be obtained. Apart from the frequency- and temperature-dependent MWS-effect, Lukomskaya and Dogadkin[47] have found that in carbon black-filled vulcanizates there is a temperature and frequency-independent polarization which is dependent on the quantity of carbon black and on the form of the carbon black particles. The permittivity at high frequencies in carbon black-filled rubber material can be expressed

$$\varepsilon_h = \varepsilon_h(U)[(1 + 3\phi_s C_b)] \tag{20.7.1}$$

where $\varepsilon_h(U)$ is the high frequency permittivity of rubber material not filled with carbon black, C_b the content by weight of carbon black, ϕ_s the shape factor, 1 is for completely spherical particles, and >1 for carbon black particles with nonspherical shapes.

Bueche[48] has studied the permittivity and resistivity as a function of temperature and frequency in a three-phase system consisting of styrene-butadiene rubber (SBR), HAF-carbon black, and wax (tetracosane $C_{24}H_{50}$).

Kumar[49] has investigated how the real and imaginary parts (ε' and ε'') of permittivity are influenced by the quantity of carbon black in synthetic rubber at various temperatures.

20.8 Inorganic Fillers

Hanna e.a.[50] have investigated how the real and imaginary parts (ε' and ε'') of permittivity are influenced by different quantities of SiO_2 in TMTD-cured SBR rubber in the frequency range from 60 Hz to 10^8 Hz at room temperature.

Hilal e.a.[51] have investigated how SiO_2 and $CaCO_3$ and mixtures of these influence ε' and ε'' in natural rubber at room temperature in the frequency range 10^2 Hz to 10^{10} Hz.

Hanna e.a.[52] have also investigated how different quantities of SiO_2 in nitrile rubber influence ε' and ε'' in the frequency range of 500 Hz to 10^{10} Hz at room temperature.

20.9 Dielectric Vulcametry

Work of Bakule e.a.,[23, 37, 42, 54 – 56] Honskus e.a.,[38] and Nedbal e.a.[51] deserve special notice as they are the only modern works, apart from those carried out by Adachi e.a.,[25] specifically directed to the study of how the cross-link density and structure influence the dielectrical properties of various rubber materials.

These studies form the basis of what may be called *dielectric vulcametry*. A relatively detailed account of the results obtained is therefore called for. Bakule, Honskus, and Nedbal and their collaborators have studied the dielectrical properties of natural rubber, polyisoprene, and polybutadiene cross-linked with sulphur or dicumylperoxide or mixtures of these. Cross-link densities between 1×10^{-5} and 3×10^{-3} mol/cm^3 were determined by equilibrium swelling in benzene according to methods suggested by Meissner, e.a.[43]

Cross-linking of rubber with sulphur is very inefficient, that is, a very small part of the added sulphur forms cross-links. The chemistry of vulcanization and network structure is covered well by Eirich[58] and Brydson.[44] About one sulphur atom in 100 forms an intermolecular cross-link. The addition of accelerators, however, increases the efficiency somewhat. The rest of the linked sulphur forms heterocyclic poly-, di- and monosulphidic groups which are randomly distributed along the molecular chain.[40,41] The heterocyclic sulphur groups have relatively high dipole moments per atom of linked sulphur.[38] Due to the low cross-link efficiency of sulphur, sulphur-cured rubbers have low cross-link densities but at the same time a large quantity of polar groups.

The cross-linking of rubber with dicumylperoxide has a high efficiency close to 1. At very low dicumylperoxide concentrations, the efficiency is somewhat reduced.[4] The efficiency of cross-linking is reduced when sulphur is added to a rubber dicumylperoxide mixture as sulphur takes part in the radical reaction.[37]

By cross-linking rubber with sulphur or peroxide separately or in combination with each other, the number of polar groups per unit volume and the cross-link density can be varied more or less independently of each other within relatively wide boundaries. Dielectric measuring methods are especially suited to the study of how the relaxation of the polar groups is influenced by the number of polar groups and by the cross-link density.

20.10 Dipole Moment

If the imaginary part (ε'') of the permittivity is plotted as a function of the frequency ($\log f$) at various temperatures, curves of the type shown in Fig. 20.10.1 are obtained.[42]

The shape of the curves and their position on the frequencey axis depend, apart from the temperature, on the structure of the material.

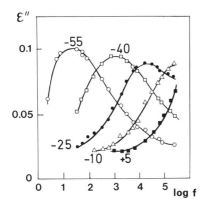

Fig. 20.10.1. Frequency dependence of ε'' in the main dispersion zone (zone I) measured at the temperatures (°C) indicated. After reference 42.

The complex permittivity ε^* of rubber can be fairly well described using the Cole-Cole distribution function.

$$\varepsilon^* - \varepsilon_h = \varepsilon_l - \varepsilon_h / 1 + (i \omega t^*)^{B_1} \tag{20.10.1}$$

where ε_l and ε_h are the permittivities measured at very low and very high frequencies respectively, i is the square root of -1, and ω is the angular frequency; t^* is the relaxation time and B_1 is a characteristic parameter for the material.

The following parameters are of special interest:

A) the relaxation strength $(\varepsilon_l - \varepsilon_h)$ which is proportional to the area under the absorption curve,

B) the relaxation time (t^*) which constitutes a measure of the position of the absorption curve on the frequency axis, and

C) the parameter B_1 which constitutes a measure of the breadth of the absorption curve.

With the help of Onsager's equation, the product $\mu_0^2 N$ can be calculated from the relaxation strength $(\varepsilon_h - \varepsilon_l)$

$$\mu_0^2 N_{Pl} = 9 kT (2\varepsilon_l - \varepsilon_h)(\varepsilon_l - \varepsilon_h)/4 \pi \varepsilon_l (\varepsilon_h + 2)^2 \tag{20.10.2}$$

where k is the Boltzmann constant, T the absolute temperature, N_{Pl} the number of polar groups per volume unit, and μ_0 is the dipole moment for a polar group.

The dipole moment per monomer unit of polyisoprene or polybutadiene is very low.[37] When sulphur is linked to the rubber molecules, polar groups are formed whose dipole moment is approximately 10 times greater than that of the corresponding monomer units.[23, 37] How the product $\mu_0^2 N_{Pl}$ is influenced by the quantity of sulphur or by a combination of a constant quantity of dicumylperoxide and a varying quantity of sulphur is shown in Fig. 20.10.2. The correlation between dipole moment $(\mu_0^2 N_{Pl})$ and sulphur quantity or sulphur quantity plus a constant quantity of dicumylperoxide is linear. The value of the product $\mu_0^2 N_{Pl}$ is higher for the test specimen which is cross-linked with a combination of sulphur and peroxide than for the test specimen which is only cross-linked with sulphur. The difference between the values obtained for $\mu_0^2 N_{Pl}$ is independent of the number (N_{Pl}) of polar sulphide groups. This can be interpreted as indicating that reaction products (probably oxygen) from the peroxide reaction give rise to new polar groups.

The dipole moment is independent of cross-link density in spite of the fact that it varies by two orders of magnitude. The highest cross-link density, 1.2×10^{-3} mol/cm^3 for the test specimen shown in Fig. 20.10.2, has been obtained for the specimen which was cross-linked with dicumylperoxide alone. The test specimen which was cross-linked

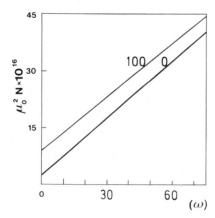

Fig. 20.10.2. Dependence of $\mu_0^2 N$ on the content of combined sulfur. The content is expressed as mg of sulfur per 1 g of rubber. Parameters on the curves denote dicumylperoxide concentrations in mg of dicumylperoxide per 1 g of rubber. After reference 23.

with 100 mg dicumylperoxide and 50 mg sulphur per 1 g rubber has a cross-link density of about 1×10^{-4} mol/cm^3, whereas the highest cross-link density of 5.2×10^{-5} mol/cm^3 for the specimens which were cross-linked with sulphur alone was obtained with 60 mg sulphur per 1 g rubber. The cross-link density of the other specimens cross-linked with sulphur was in all cases lower than 5.2×10^{-5} mol/cm^3.

The difference in cross-link density in the investigated specimen is so great that the number of monomer units between two adjacent cross-links varies from ten to several thousand. It can, therefore, be established that even if there are not more than ten monomer units between two adjacent cross-links, these cross-links are not capable of preventing, to any measurable degree, the movement of the polar groups.

20.11 Relaxation Time

The dependency of the relaxation maximum on temperature can be described by the standard coefficients C_1 and C_2 of the WLF-equation (the WLF-equation is also treated in Chapters 4 and 10)

$$\log f_{\max}(T)/\log f_{\max}(T_s) = (C_1(T-T_s)/C_2 + T - T_s) \tag{20.11.1}$$

where $f_{\max}(T)$ is the frequency of absorption maximum at an arbitrarily chosen temperature T and $f_{\max}(T_s)$ is the frequency of absorption maximum at a standard reference temperature T_s. One has to note the limitations in the use of the WLF-Equation discussed in Section 10.6.

$$C_1 = 8.86 \quad \text{and} \quad C_2 = 101.6 \quad \text{for} \quad T_0 = T_g + 45°\text{C}$$

The difference between the standard reference temperature, T_s, and the glass-transition temperature, T_g, is about 45 K. The influence of temperature on $f_{\max}$ for some of the materials investigated is shown in Fig. 20.11.1. The curves in the figure satisfy the Eq. (20.11.1). The reference temperature is indicated by open circles in the figure. Curve 1 shows the test specimen with 100 mg dicumylperoxide per 1 g rubber, i.e., 3.3×10^{-4} mol peroxide per 1 g rubber. Curve 2 shows the test specimen with the same quantity of sulphur, i.e., 3.3×10^{-4} mole sulphur per 1 g rubber. Although the differences in cross-link density between these specimen is very large — 1.2×10^{-3} mol/cm^3 for specimen 1 (curve 1) and less than 10^{-5} mol/cm^3 for specimen 2 (curve 2) — the shift between the curves is very small.

The shift between curves 2 and 4 is also very small. The cross-link density for specimen 3 (curve 3) cross-linked with 60 mg sulphur plus 50 mg dicumylperoxide per

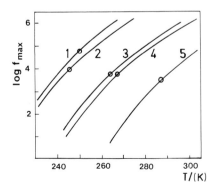

Fig. 20.11.1. Temperature dependence of log $f_{\max}$ for (1) natural rubber with 100 mg of dicumylperoxide per 1 g of rubber, (2) natural rubber with 12 mg of sulfur per 1 g of rubber, (3) natural rubber with 50 mg of dicumylperoxide and 60 mg of sulfur per 1 g of rubber, (4) natural rubber with 60 mg of sulfur per 1 g of rubber, and (5) natural rubber with 100 mg of sulfur per 1 g of rubber. After reference 23.

1 g rubber is 2.5×10^{-4} mol/cm^3 whereas the cross-link density for specimen 4 cross-linked with only 60 mg sulphur per 1 g rubber is 0.52×10^{-4} mol/cm^3.

The small shift to higher frequencies which was obtained for the peroxide-cured specimen probably depends on the plasticizing effect of the reaction product from the cure rather than on differences in cross-link density.

Specimen 5 (curve 5) has been cross-linked with 100 mg sulphur per 1 g rubber, which corresponds to a cross-link density of 1.1×10^{-5} mol/cm^3. If the shift between the curves for different cross-link densities is compared with the shift between the curves for different sulphur contents, it is evident that the position of the relaxation zones is almost entirely independent of cross-link density and is dominated mainly by the quantity of sulphur in the mixture. This corresponds well will with the results obtained by Heinze e.a.[62] especially the fact that *intramolecular* bound sulphur and *intermolecular* sulphur links have the same influence on the shift of the glass-transition temperature, T_g.

20.12 Width of Absorption Curve

The width of the absorption curve (see Fig. 20.10.1) can be written

$$\Delta_h = 1/2 (\log f_2 - \log f_1) \qquad (20.12.1)$$

In Eq. (20.12.1), $f_2 > f_1$ are the frequencies at which the absorption curve has the values $h \varepsilon''_{max}$ $(h < 1)$, where ε''_{max} is the maximum value of the absorption curve.

The width of the absorption curve increases with increasing quantity of sulphur when the rubber is cross-linked with sulphur alone. A slight deviation from this rule arises only at the very lowest sulphur quantities.

The absolute values for the parameter Δ_h are different for the specimens manufactured from natural rubber, polyisoprene, and polybutadiene. These differences are in accord with the differences in cross-link densities normally found for specimens manufactured from these three rubber types. The cross-link density of a cis-polyisoprene specimen is approximately twice as high as that of natural rubber specimens whereas the polybutadiene specimen has about ten times as high a cross-link density as the natural rubber specimen.

The influence of the quantity of sulphur on the Δ_h-parameter for specimens which contain not only sulphur but also dicumylperoxide is of an entirely different character than for the specimens which contain only sulphur, since the width of the absorption curve decreases when the quantity of sulphur is increased. The Δ_h-parameter ceases to decrease when the quantity of dicumylperoxide in the specimen is much lower than the content of sulphur. The cross-link density is influenced to only a limited degree by the presence of peroxide and the width of the absorption curves is approximately the same as for a rubber-sulphur system.

The strong influence of cross-link density on the width of the absorption curve is shown in Fig. 20.12.1.

The width increases linearly with increasing content of dicumylperoxide for specimens with a constant content of sulphur.

The width of the absorption curve $(\Delta_{0.7})$ is shown as a function of the logarithm of cross-link density (p_{cl}) in Fig. 20.12.2.

The $\Delta_{0.7}$-value increases with increasing cross-link density for all investigated combinations of rubber (natural rubber, cis-polyisoprene, polybutadiene), and curing system (sulphur, sulphur/peroxide, peroxide).

The relationship between the width of the absorption curve $(\Delta_{0.7})$ and the cross-link density $(p_{cl}$ mol/cm$^3)$ can be approximately described by the following linear empirical

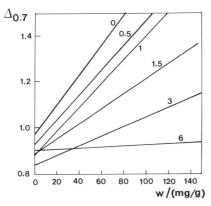

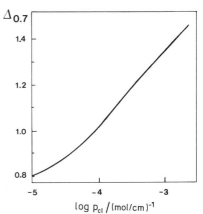

Fig. 20.12.1. Dependence of the absorption curve width ($\Delta_{0.7}$) on dicumylperoxide concentration w. The concentration w is expressed in mg of dicumylperoxide per 1 g of rubber. Parameters on the curves denote sulfur concentrations in mg of sulfur per 1 g of rubber. After reference 23.

Fig. 20.12.2. Dependence of the absorption curve width ($\Delta_{0.7}$) on logarithm of cross-link density (log p_{cl}) of natural rubber vulcanized with sulfur, dicumylperoxide, and mixtures of sulfur and dicumylperoxide. After reference 23.

equation

$$\Delta_{0.7} = 0.311 \log p_{cl} + 2.26 \tag{20.12.2}$$

in the cross-link range from 1×10^{-5} to 1×10^{-3} mol/cm^3.

It is apparent in Fig. 20.12.2 that there is no marked difference between the different combinations of rubber and curing system although they vary strongly with regard to the number of polar groups per unit volume.

It can, therefore, be established that the cross-link density is the factor which mainly determines the width of the absorption curve. The widening of the dielectric absorption curve with increased cross-link density agrees with Mason's[60] theory of free volume. Using his interpretation, the following can be established: Mason's dispersion function of the free volume increases with increasing cross-link density and as a result of this the dielectric absorption curve is widened.

20.13 Types of Cross-Links

Honskus and Bakule[38] have suggested that it is possible with the aid of dielectric measurements to distinguish between sulphur which is bound to the molecule in the form of cross-links and sulphur which is bound to the molecule in other ways, e.g., heterocyclic groups in accelerated or nonaccelerated natural rubber compounds.

It has been found that tan δ_{max} gradually increases with curing time and that the maximum value is reached after a relatively long curing time.

The dielectric loss tangent is dependent not only on the number of polar groups but also on the dipole moments of these groups. This means that the dielectric loss tangent is dependent on the chemical composition of the groups that are linked to the molecule in other ways than as cross-links.

The dipole moments of disulphide and polysulphide groups are not twice as large as, nor any direct multiple of, the dipole moment of a monosulphide. The dipole moment per sulphur atom is considerably lower in polysulphides due to the fact that the sulphur atoms interfere with each other.[42]

These results agree well with earlier knowledge of accelerated sulphur vulcanization, which is characterized by the fact that the sulphur is rapidly linked to the molecule in the form of polysulphides. These polysulphides are then broken down to lower polysulphides, disulphides, and monosulphides during the continued curing reaction. This decomposition occurs in the cross-links as well as in sulphur groups linked in other ways to the rubber molecule.

The decomposition of polysulphidic cross-links causes the cross-link density to decrease while the decomposition of polysulphides linked to the molecules in other ways than cross-links increases the dielectric losses.

20.14 High Temperature Relaxation Zones

Bakule, Nedbal and Havranek[42] have studied how the complex permittivity (ε^*) of natural rubber cross-linked with sulphur or with a combination of sulphur and dicumylperoxide varies with the frequency and temperature. Three absorption zones were detected within the examined frequency range from 1.5 to 3.0×10^5 Hz and 3.0 to 1.0×10^{-4} Hz. Data for the main transition zone (I) have been obtained using the time-temperature superposition principle from measurements made at lower temperatures.[61]

Figure 20.14.1 shows $\log \varepsilon''$ as a function of $\log f$. Dispersion zone (I) corresponds to the main transition that is normally associated with the glass transition, T_g. This dispersion zone reflects the fact that molecule segments with about 150 carbon atoms start to move. ε'' is shown as a function of $\log f$ at temperatures between -55 and $+5°$ C in Fig. 20.10.1. The f_{max} values of the dispersion curves increase about one frequency decade per $10°$ C increase in temperature. The height of the dispersion curves increases linearly with the amount of bound sulphur in the rubber. From this the conclusion can be drawn that the dispersion zone (I) is caused by movements in the molecule segments that contain polar sulphur groups.

The dipole moment per sulphur atom was estimated from the highest values of the dispersion curves using Onsager's Equation.[4] The value obtained is independent of the amount of bound sulphur as well as the cross-link density.

Besides dispersion zone (I) another dispersion zone (II) was detected in the same frequency range, but at a higher temperature (-10 to $+95°$ C). The values of the absorption peaks in zone (II) are approximately as large for specimens cross-linked with sulphur as for those cross-linked with a mixture of sulphur and dicumylperoxide. The peak values that are dependent only on cross-link density decrease with increasing cross-link density.

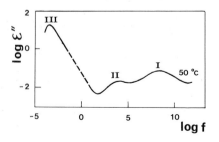

Fig. 20.14.1. Frequency dependence of the imaginary part of the dielectric constant $\log \varepsilon''$ at $50°$ C for one of measured samples. The three dispersion zones are designated I, II, and III. After reference 42.

Dispersion zone (III) appears at very low frequencies. The peak values of the dispersion curve increase with the amount of sulphur in the same way as for dispersion zone (I), but the peak values also increase with the amount of dicumylperoxide. This means that in dispersion zone (III) the peak values of the curve increase as a function of the cross-link density. In contrast to dispersion zone (I), the peak values of the dispersion curves in zone (III) are very sensitive to temperature. The peak values decrease rapidly with increasing temperature. The existence of dispersion zone (III) with its very long relaxation times and strong dispersion peaks can be explained by two molecular mechanisms. One explanation of the observed phenomenon lies in a coordinated orientation polarization of large molecular segments containing more than 100 polar groups.

The other explanation lies in the volume polarization that is caused by ions that cannot be discharged through the electrodes. From the measurement carried out it is impossible to decide definitely which of these mechanisms is responsible for dispersion zone (III).

Adachi e.a.[25] have studied the influence of the cross-link density on the α'-relaxation in peroxide-cured nitrile rubber with 29% acrylonitrile content in the frequency range 3.125 Hz to 1.6348 MHz. Results obtained agree, on the whole, with the results that Bakule and Havranek[23] have obtained for natural rubber, polyisoprene, and polybutadiene. Contrary to Bakule and Havranek,[23] Adachi e.a.[25] obtain a lower relaxation strength $(\varepsilon_l - \varepsilon_h)$ at a higher cross-link density. This circumstance might be explained by the fact that the relaxation of polar heterocyclic groups that are formed during the sulphur vulcanization of polydienes can have completely camouflaged the reduction in the relaxation strength that the cross-links give. Results obtained show that when the cross-link density increases

A) the relaxation strength decreases $(\varepsilon_l - \varepsilon_h)$,

B) the half-peak width Δ_h of the absorption curve increases, and

C) the absorption maximum is shifted toward higher temperatures.

Certain general conclusions can be drawn from the studies carried out. As a result of every factor, e.g., cross-linking, that reduces the mobility of the molecule segments.

A) The loss maximum (ε''_{max}) at constant temperature moves toward lower frequencies or that (t^*_{max}) moves toward higher temperatures,

B) a widening of the relaxation curve (Δ_h) occurs, and

C) the in relaxation strength $(\varepsilon_l - \varepsilon_h)$ decreases.

These effects can be provided by a decrease in free volume[60] or by an increase in the interaction between partially oriented chain segments,[64] so called intermolecular effects, and/or changes in the stereo chemical conformations toward more unstable or unfavored positions, so-called intramolecular effects.

Mason[60] observed that in dicumylperoxide-vulcanized natural rubber the specific volume decreases, the glass-transition temperature, T_g, increases, and the transition range widens when the cross-link density increases. Mason suggests that these changes are a result of restrictions in the micro-Brownian movements in the molecular network depending on contraction as well as wider distribution of the available free volume.

The results of the studies of Adachi have been summarized in Fig. 20.14.2 where T_{max}, Δh, and $T\Delta\varepsilon = T(\varepsilon_l - \varepsilon_h)$ are plotted as functions of the effective cross-link density (p_{cl}).

Both the T_{max} and the half-peak width of the dispersion curves increase linearly with the cross-link density. The product, $T\Delta\varepsilon$, that decreases approximately linearly with the cross-link density reflects a decrease in the number of dipoles that take part in the α'-relaxation. The decrease in the product, $T\Delta\varepsilon$, is most probably a result of a decrease in the free volume rather than restrictions in large scale movements in the molecular chains (intermolecular movements).

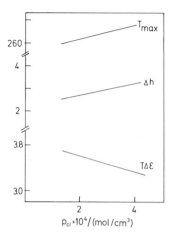

Fig. 20.14.2. Dependence of the position of the loss maximum T_{max} (1.6 kHz), the half-width Δh of the absorption curve, and the strength of relaxation $T \Delta \varepsilon$ on cross-link density.

Cross-links do not influence the mobility of the molecular segments except for those segments which are very close to the cross-links. This decrease in mobility is due to the local decrease in the free volume close to the cross-links, which leads to a general decrease in the free volume and an increase of its heterogeneous distribution.

20.15 Influence of Pressure on the Dielectric Properties of Rubber Materials

Williams e.a.[65] have studied how isostatic pressure affects the α-relaxation in three noncross-linked nitrile rubbers with different acrylo-nitrile contents. Hydrostatic pressure causes complex phenomena in the relaxation spectra for nitrile rubbers with low acrylo-nitrile content so that two relaxation processes appear and ε''_{max} varies with pressure. In the case of nitrile rubber with 40% acrylo-nitrile content only one α'-relaxation appears.

ε''_{max} values and the half-peak width of the dispersion curve are independent of pressure. This is characteristic of a long range of homopolymers.[4]

It is thought to be common for all analyzed materials that f_{max} moves toward lower frequencies at increased pressure if the temperature is constant.

Questad e.a.[66] have studied how far cross-linked polyurethane rubber is affected by the pressure within the pressure range from 0.1 to 650 MPa.

In addition to chemical cross-links, chain entanglements in the molecules also contribute to the measured cross-link density. In rubbers filled with carbon black, links between carbon black and rubber will also be apprehended as cross-links. Thus, it follows that an absolute estimate of the cross-link density in rubbers filled with carbon black is very difficult to make. An empirical correction factor for carbon black content and type has been suggested by Kraus.[67]

20.16 Dielectric Properties: Summary

Dielectric properties, such as permittivity (ε') and loss tangent (tan δ) have been reported for a large number of polymers at different frequencies and temperatures. How the dielectrical properties of rubbers are influenced by cross-linking has been on the other hand, very sparsely researched. Peroxide and/or compounds of some polydienes (NR, IR, BR, and NBR) have been studied. It has been found that the width of the dielectrical relaxation curve correlates fairly well with the cross-link density, independently of the composition of the curing system.

How the cross-link density affects the dielectric properties of technical rubber compounds does not appear to have been studied so far.

20.17 Vibrorelaxation of Rubbers: Introduction

The name rubber or the more general and more modern term elastomer is used for materials having rubberlike properties, i.e. a high degree of reversible deformability under the action of comparatively small stresses. Elastomers recover almost completely and instantaneously on release of the deforming stresses. This property of high elasticity is not a characteristic of a particular chemical substance but derives from a particular kind of molecular structure. The requirements for a material to be an elastomer are as follows:

A) The molecules must be very long and able to rotate freely about the bonds joining the neighboring molecular units.

B) The molecules must be joined at a number of sites to form a three-dimensional network either by chemical bonding or mechanical entanglement (cross-linked network).

C) Intermolecular attractions (secondary and van der Waals forces) must be small.

The theory of rubber elasticity is based on these molecular assumptions and will not be discussed here. The reader is referred to many excellent surveys.[68-71]

The method of stress relaxation is based on set measurements which are usually defined as the amount by which a standard test piece fails to return to its original shape or thickness after being subjected to a standard load or deflection for a fixed period of time at constant temperature. The set (C') is expressed as a percentage of the original deflection and is calculated according to[72]

$$C' = 100*(l_0 - l_f)/(l_0 - l_{max}) \tag{20.17.1}$$

where C' is the set expressed as percentage of the original deflection, l_0 the original dimension of specimen, l_f the final length of specimen, and l_{max} is the length of specimen at maximum deflection.

At elevated temperatures, set us caused by chemical relaxation processes. For practical reasons, measurements of set are usually carried out with compressed rubber specimens. Compression set is a rapid, simple, and cheap method but the results are rather difficult to translate to other relaxation processes. Southern[73] has mentioned an example where the content and type of carbon black filler do not affect compression set very much but do affect stress relaxation and creep behavior. Compression set is nevertheless a method which is widely used[74] in the technical assessment of rubber materials, and efforts have been made to determine how chemical structure and set properties are related.[75, 76]

20.18 Static Relaxation

The measurement of stress relaxation in elastomers as a function of time has been shown by Tobolsky[77] to be a useful tool in the investigation of relaxation. Tobolsky's well-known theories in this field have been used by many workers for investigations of

A) The relative extent of scission and cross-linking reactions by intermittent and continuous stress-relaxation measurements. It was observed early that during chemical stress-relaxation experiments in the temperature range from 100 to 150° C some rubber vulcanizates became softer while others became harder. Natural rubber first softened and then hardened. It was deduced that at these elevated temperatures both cross-linking and scission reactions were occurring simultaneously. It was also found that the rates of these processes were essentially unaffected by stress or elongation of the specimen. Data from continuous and intermittent stress-relaxation measurements are interpreted as assuming the validity of the two-network theory which was proposed by Tobolsky.[77] It is assumed in this theory that, when a rubber specimen is maintained in a deformed state at constant length, scission reactions cause a decay of stress whereas cross-linking reactions do not affect the stress. This is equivalent to saying that the new cross-links form in the network in a relaxed condition. In a continuous stress-relaxation experiment, the specimen is maintained at a fixed deformation and the stress is measured as a function of time. This method is assumed to measure the effect of scission alone. In an intermittent stress-relaxation experiment the specimen is maintained in an unstretched condition. At widely spaced intervals, the rubber is rapidly deformed and the stress is rapidly measured. The specimen is then immediately returned to its unstretched length. The intermittent stress measurements reflect the combined effects of scission and cross-linking. The difference between intermittent and continuous relaxation values thus shows the extent of cross-linking. Nearly all the scission and cross-linking have been found to occur while the network is in the undeformed state.

B) Whether the scission occurs mainly as random cleavage at the links of the network chains or as cleavage at the cross-link sites. The p_{cl}-(initial cross-link density) dependence of $F(t)/F(0)$, where $F(t)$ and $F(0)$ are the initial force and the force at time t respectively), indicates that scissions occur preferentially at random on main chains, in accordance with Tobolsky's theory. On the other hand, p_{cl}-independence is taken as an indication of cross-link scission.

C) The different chemical reactions which cause the relaxation. In networks with, for example, polysulfidic cross-links, exchange reactions occur at elevated temperatures during the early stages of the relaxation. A valuable tool in these investigations has been the so-called Procedure X presented by Tobolsky and Murakami.[78] This method has also been applied by Russian workers to analyze relaxation curves of different rubbers with respect to physical and chemical mechanisms.[79] Murakami and Ono[80] and Murakami[81] have reported excellent and extensive work on the analysis of the chemical stress relaxation of rubbers. Stress-relaxation experiments are usually carried out by the simple elongation of specimens sufficiently thin so that the thermo-oxidative degradation at high temperatures will not be dependent on the rate of diffusion of oxygen from the air. Compression-relaxation investigations in which thick-walled specimens are used have led to different opinions regarding the relaxation mechanisms.[82–86] Technically, used rubbers are thick-walled and loaded in combinations of compression and shear, but few studies have been concerned with these types of loadings. In most types of civil engineering applications, however, the rubber components are designed to operate under compression thus avoiding exposure of the main part of the rubber bulk to oxygen.

20.19 Dynamic Stress Relaxation (Vibrorelaxation): Details

In many applications, rubbers are subjected to cyclic deformations. Murakami and Kusano have reported that an oscillating deformation affects the relaxation properties of elongated thin-walled specimens at different temperatures and amplitudes.[87] They proposed that the influence of these parameters on the relaxation could be summarized in an exponential function. The influence on the stress relaxation of thick-walled rubber specimens of a dynamic deformation superimposed upon a static compression has been reported by Bartenev and Schelkownikowa.[88] The influence of filler, temperature, amplitude, and frequency were reported. Breakage of bonds between carbon black and the rubber molecules was reported to be responsible for the relaxation effects. Stenberg and co-workers[89] have reported preliminary results on how the compression relaxation of TMTD-vulcanized unfilled natural rubber cylinders is influenced by superimposing a small sinusoidal deformation on a static compression (see Fig. 20.19.1) and further results are presented here. Rubber cylinders with height-diameter $= 20$ mm and with the composition shown in Table 20.19.1 were used. The vulcanization time was 17 minutes at 160° C.

The cross-linking system is one of the most important factors which affect the properties of a rubber material and this is described in Fig. 20.3.1. System A is stable toward thermo-oxidative degradation but system B has a lower thermo-oxidative stability at high temperature. The TMTD sulfurless vulcanizate involves the formation of a zinc dialkyldi-thiocarbamate which is not only a strong antioxidant but also very involatile.[90] Relaxation data are usually shown as logarithmic relaxation plotted against linear time. The basis for this type of representation is that a straight line is obtained in a time-region where one relaxation time prevails. Using Procedure X, the relaxation curve can be subdivided into different relaxation times and intensities. In these measurements 10 minutes is chosen as normalization time since after 10 minutes the physical relaxation is negligible or very low and the continued relaxation is mainly chemical. Figure 20.19.2 shows the maximum and minimum values of stress as functions of linear time, from which R_M and R_m were calculated according to Eqs. (20.19.2) and (20.19.3)

$$R_M(t) = 100 \times \sigma_{max}(t)/\sigma_{max}(t_N) \qquad\qquad (20.19.2)$$

$$R_m(t) = 100 \times \sigma_{min}(t)/\sigma_{min}(t_N) \qquad\qquad (20.19.3)$$

where $\sigma_{max}(t)$ and $\sigma_{min}(t)$ are respectively the maximum and minimum values of the stress at time t, and $\sigma_{max}(t_N)$ and $\sigma_{min}(t_N)$ are the stresses at time $t_N = 10$ minutes.

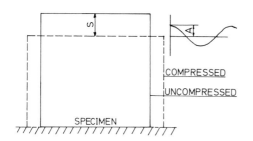

Fig. 20.19.1. Maximum and minimum values of stress. A is the dynamic strain amplitude superimposed on a static compression, S, of 20%.

Table 20.19.1. Materials used

	A	B	C
Natural rubber	100	100	100
Carbon black	0	0	40
ZnO	5	5	5
S	0	3	0
CBS	0	0.6	0
TMTD	4	0	4
Stearic acid	1	1	1

Figure 20.19.2 shows that the initial stress values are independent of the measuring temperature and that the relaxation rate is higher at shorter than at longer times.

In Figs. 20.19.3 and 20.19.4 logarithmic relaxation is plotted against linear time for materials A and B respectively at different amplitudes at 85° C and 2 Hz. The influence of amplitude is strong and it can generally be said that amplitude- and temperature-dependence are of the same magnitude, an observation which is in accordance with the findings of Murakami[87] and Bartenev.[88] A comparison between Figs. 20.19.3 (material A) and 20.19.4 (material B) reveals the influence of the vulcanization system. The amplitude-dependence is stronger for material B than for material A. The maximum values are relatively more insensitive to amplitude than the minimum values. It must be observed that the minimum value determines the sealing effect of an O-ring or a gasket and that this value decreases much more sharply at a higher amplitude. Examples of the dependence on temperature of the vibrorelaxation are shown in Fig. 20.19.5. At 50° C a small difference is obtained between the maximum and minimum values and this difference increases with increasing temperature. A reasonable explanation is that the increasing difference between the maximum and minimum values has a chemical origin. Figure 20.19.6 indicates that rearrangement reactions are taking place in material B and that these reactions start at a temperature as low as 50° C since the difference between the maximum and minimum values is larger in Fig. 20.19.6 than in Fig. 20.19.5. At the highest temperature in Figs. 20.19.5 and 20.19.6, the maximum values level off. This is caused by the formation of a hard layer of oxidized

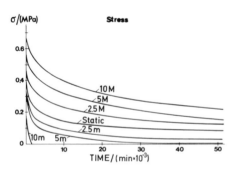

Fig. 20.19.2. Stress plotted against linear time. Number denotes amplitude (%), *M* is the maximum, and *m* the minimum value of stress. Static is the static relaxation at 20% compression. Material B at 85° C, 2 Hz.

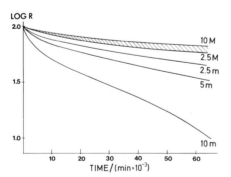

Fig. 20.19.3. Logarithmic relaxation for material A at different amplitudes at 85° C and 2 Hz plotted against linear time. Symbols as in Fig. 20.19.2. i.e., 10 m means minimum value at 10% amplitude.

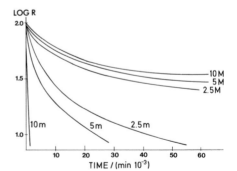

Fig. 20.19.4. Logarithmic relaxation for material B at different amplitudes at 85° C and 2 Hz plotted against linear time. Symbols as in Fig. 20.19.2.

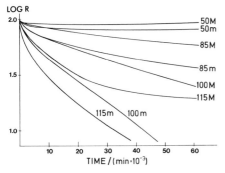

Fig. 20.19.5. Logarithmic relaxation for material A at different temperatures at 5% amplitude and 2 Hz. Symbols: number means temperature in °C, M is the maximum value, and m is the minimum value.

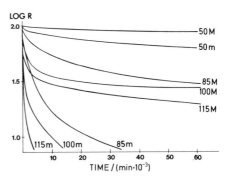

Fig. 20.19.6. Logarithmic relaxation for material B at different temperatures at 5% amplitude and 2 Hz. Symbols as in Fig. 20.19.5.

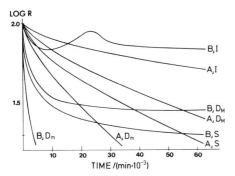

Fig. 20.19.7. Logarithmic relaxation plotted against linear time at 100° C and 2 Hz. The first letter represents material, I_R is the intermittent relaxation, D_M the maximum value of dynamic relaxation, D_M the minimum value of dynamic relaxation, and S is the static relaxation at 20% compression.

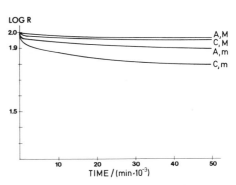

Fig. 20.19.8. Logarithmic relaxation for materials A and C plotted against linear time. The first letter represents material, M is the maximum value, and m is the minimum value.

rubber which is impermeable to oxygen[85] and increasingly with increased aging time prevents oxygen in the surrounding air from diffusing into the bulk. This proposed mechanism has been further investigated in [86]. Investigations of frequency-dependence have shown only a weak dependence on frequency between 0.001 to 36 Hz. A slight increase in relaxation rate was noted with increasing frequency in accordance with the Murakami observations of [87] and Bartenev.[88] Figure 20.19.7 shows relaxation data at 100° C for materials A and B representing:
— dynamic mechanical relaxation
— static relaxation
— intermittent relaxation.

The maximum in the intermittent relaxation curve for material B is caused by the formation of an oxide skin which partly cracks later on. As can be seen in Fig. 20.19.7, no simple correlation can be found between the dynamic relaxation and the intermittent and static relaxations. The large difference between maximum and minimum values is caused by set effects due to scission and cross-linking reactions. Hitherto, measurements on unfilled natural rubbers have been discussed. As can be seen in Fig. 20.19.8 the addition of carbon black complicates the behavior. Carbon black causes the difference between maximum and minimum value at low temperature, e. g., 40° C to increase compared with the value for the unfilled material. Bartenev[88] has reported similar results. The measurements discussed in this Chapter have been performed in air.

In collaboration with Swedish industry, the vibrorelaxation effect will be investigated for other rubber materials used in O-rings and gaskets in contact with oil and water.

In order to hinder the vibrorelaxation effect, the set properties of the rubber material must be improved and here work described in [91–96] can serve as a guide. But it must be remembered that improved set properties at high temperatures often give impaired properties at temperatures below 0° C.

20.20 Mechanical Properties: Summary

Natural rubber cylinders (height-diameter = 20 mm) were compressed to 20%. Sinusoidal strains of 2.5, 5, and 10% respectively, were superimposed on the static compression and the decrease of the maximum and minimum values of the stress recorded. A difference between the curves representing the maximum and minimum values was obtained. This difference increased with increasing dynamic amplitude. The maximum values showed no dependence on the amplitude and the curves asymptotically reached the same relaxation value at long times. The difference between the maximum and minimum values increased with increasing temperature.

Acknowledgment

We gratefully acknowledge the support of the National Swedish Board for Technical Development. We also wish to thank Professor Bengt Rånby of this Departement for continuous support and interest in this work.

References

1. P. Debye, *Polar Molecules,* Chemical Catalog Co., reprinted by Dover, New York 1929.
2. H. Fröhlich, *Theory of Dielectrics,* Oxford University Press, 1949.
3. J.W. Smith, *Electric Dipole Moments,* Butterworths, London 1955.
4. N.G. McCrum, B.E. Read and G. Williams, *Anelastic and Dielectric Effects in Polymeric Solids,* Wiley, New York 1967.
5. A.R. Blythe, *Electrical Properties of Polymers,* Cambridge University Press 1979.
6. A.R. von Hippel, Ed., *Dielectric Materials and Applications,* Wiley, New York 1965.
7. R.H. Boyd, Meth. Experim. Phys. 1980, **116C**, 397.
8. P. Hedvig, *Dielectric Spectroscopy of Polymers,* Adam Hilger, London 1977.
9. Amer. Pat. 3039297, March 19th, 1958.
10. J. Peter and W. Heidemann, *Kautschuk Gummi* 1957, **7**, 168.
11. A.R. More, S.H. Morrell and A.R. Payne, *Rubber J. Internat. Plast.* 1959, **136**, 858.
12. W.E. Claxton, F.S. Conant and J.W. Liska, *Rubber World* 1961, **144**, 71.
13. A.E. Juve, P.W. Karper, L.O. Schroyer and A.G. Veith, *Rubber World* 1963, **149**, 43.
14. International Synthetic Rubber News 1968, **6**, Oct. 16.
15. Z. Bartha, P. Ször and S. Ambrus, *Plaste Kautschuk* 1964, **11**, 670.
16. G.E. Decker, R. Wise and D. Guerry, *Rubber Chem. Technol.* 1963, **36**, 451.
17. W. Scheele, *Kautsch. Gummi Kunstst.* 1965, **18**, 138.
18. G. Kaiser, K. Hummel and W. Scheele, *Kautsch. Gummi Kunstst.* 1966, **19**, 347.

19. V. Härtel, *Symp. Proc. Scandinavian Rubber Conf.* 1979, **54**.
20. V.A. Shershnev, *Rubber Chem. Technol.* 1982, **55**, 537.
21. R.H. Norman, *Polymer Testing* 1980, **1**, 247.
22. A.H. Scott, A.T. McPherson and H.L. Curtis, Nat. *Bureau* Stand. *J. Res.* 1933, **11**, 173.
23. R. Bakule and A. Havranek, *J. Polym. Sci. Symp.* 1975, **53**, 347.
24. A. Schallamach, Inst. Rubber Ind. Trans. 1951, **27**, No. 1, 40.
25. H. Adachi, K. Adachi and T. Kotaka, *Polymer J.* 1980, **12**, No. 5, 329.
26. S. Yalof and K. Zika, *Rubber Age* 1976 June, p. 31–35; July, p. 43–49.
27. S. Yalof and D. Brisbin, *Amer. Lab.* 1973, Jan. 65.
28. A. Rothenpieler, Symp. Proc. Scandinavian Rubber Conf. 1975, **47**.
29. J.T. Siemon, Z.N. Sanjana and J.F. Meier, *Elastomerics* 1982, **114**, March 25.
30. S.D. Senturia, N.F. Sheppard, Jr., H.L. Lee and D.R. Day, Adhesion Society Abstracts Fifth Annual Meeting, paper: 7a, 1982.
31. V.G. Shteinberg, Y.A. Ol'khov, A.G. Melentev and S.M. Baturin, *Polymer Sci. USSR* 1980, **22**, 269.
32. S.M. Baturin, G.B. Manelis, A.G. Melentev, E.M. Nadgornyi, Y.A. Ol'khov and V.G. Shteinberg, *Polymer Sci. USSR* 1976, **18**, 2808.
33. G.P. Safonov, Y.A. Ol'khov and S.G. Entelis, *Polymer Sci. USSR* 1975, **17**, 395.
34. J. Gowri-Krishna, O.S. Josyulu, J. Sobhanadri and R. Subrahmaniam, *J. Phys. D.* 1982, **15**, 2315.
35. S.N. Lawandy and K.N. Abd-el-Nour, *Europ. Rubber J.* 1982, **164**, March, 53.
36. D.W. Kitchin, *Ind. Eng. Chem.* 1932, **24**, 549.
37. R. Bakule, J. Honskus, J. Nedbal and P. Zinburg, *Coll. Czechoslov. Chem. Commun.* 1973, **38**, 408.
38. J. Honskus and R. Bakule, Internat. Rubb. Symp. Gottwaldow Sept. 1–5, 1975.
39. R.H. Norman, *Proc. Inst. Electr. Eng.* 1953, 100, **2A**, 41.
40. E.H. Farmer and F.W. Shipley, *J. Chem. Soc.* 1947, **2**, 1519.
41. G.F. Bloomfield, *J. Chem. Soc.* 1947, **2**, 1547.
42. R. Bakule, J. Nedbal and A. Havranek, *J. Macromol. Sci. Phys.* 1971, **5**, 233.
43. B. Meissner, I. Klier and S. Kucharik, *J. Polymer Sci. C.* 1967, **16**, 793.
44. J.A. Brydson, *Rubber Chemistry*, Chapter 8, Applied Science Publ., London 1978.
45. R.H. Norman, *Conductive Rubber*, MacLaren & Sons, London 1959.
46. D.I. James, Ed., "Recent developments in Conductive Rubbers", *Proc. of RAPRA Seminar*, Oct 28, 1976.
47. A.I. Lukomskaya and B.A. Dogadkin, *Kolloid-Z.* 1960, **22**, 576.
48. F. Bueche, *J. Polymer Sci.* 1973, **11**, 1319.
49. A. Kumar, *IEEE Trans. Electr. Insul.* 1979, **EI-14**, No 3, 175.
50. F.F. Hanna, A.A. Yehia and A. Abou-Bakr, *Brit. Polymer J.* 1973, **5**, 83.
51. M. Hilal, A. Abou-Bakr and S. Mahmoud, *Plaste Kautschuk* 1970, **17**, 580.
52. F.F. Hanna, A.A. Yehia and A. Abou-Bakr, *Plaste Kautschuk* 1973, **20**, 435.
53. F.F. Hanna and A. Abou-Bakr, *Brit. Polymer J.* 1973, **5**, 49.
54. R. Bakule and B. Stoll, *Colloid Polymer Sci.* 1977, **255**/No 12, 1176.
55. R. Bakule, K. Hajek and J. Nedbal, *Polymer Bull.* 1981, **4**, 399.
56. R. Bakule, *Internat. Rubb. Symp. Gottwaldow,* Sept. 1–5, 1975.
57. J. Nedbal, R. Bakule and V. Müllerova, *Collect. Czechoslov. Chem. Commun.* 1970, **35**, 2861.
58. F.R. Eirich, *Science and Technology of Rubber,* Academic Press, New York 1978.
59. K.S. Cole and R.H. Cole, *J. Chem. Phys.* 1941, **9**, 341.
60. P. Mason, *Polymer* 1964, **5**, 625.
61. J.D. Ferry, *Viscoelastic Properties of Polymers,* John Wiley, New York 1970.
62. H.D. Heinze, K. Schmieder, G. Schnell and K.A. Wolf, *Rubber Chem. Technol.* 1962, **35**, 776.
63. J.H. Kallweit, *Kolloid-Z.* 1963, **188**, 97.
64. A. Peterlin and J. Elwell, *J. Mater. Sci.* 1967, **2**, 1.

65. G. Williams, D.C. Watts and J-P. Nottin, *Faraday Trans.* 1972, **68**, 16.
66. D.L. Questad, K.D. Pae, B.A. Newman and J.I. Scheinbeim, *J. Appl. Phys.* 1980, **51**, 5100.
67. G. Kraus, *J. Appl. Polymer Sci.* 1963, **7**, 861.
68. L.R.G. Treloar, *The Physics of Rubber Elasticity,* Clarendon Press, Oxford 1975.
69. J.E. Mark, *J. Chem. Educ.* 1981, **58**, 898.
70. T.L. Smith, "Molecular Aspects of Rubber Elasticity," in *Treatise on Materials Science and Technology,* Vol. 10, *Properties of Solid Polymeric Materials,* edited by J.M. Schultz, Academic Press, New York 1977.
71. P.J. Flory, "The Molecular Theory of Rubber Elasticity," in *Contemporry Topics in Polymer Science,* Vol. 2, edited by E.M. Pearce and J.R. Schaefgen, Plenum Press, New York 1977.
72. ASTM, D-395-78, Standard Test Methods for Rubber Property-Compression Set, in *Annual Book of ASTM Standards,* Part 347, Philadelphia 1981.
73. E. Southern, in *Elastomers: Criteria for Engineering Design,* Chapter 16, edited by C. Hepburn and J.W. Reynolds, Applied Science Publ. London 1979.
74. D.K. Thomas, *Plast. Rubber Internat.* **8**, No. 2, 53.
75. D.M. Chang, *Rubber Chem. Technol.* 1981, **54**, 170.
76. H.J. Jahn and H.H. Bertram, *Rubber Chem. Technol.* 1973, 305.
77. A.V. Tobolsky, *Properties and Structure of Polymers,* Wiley, New York 1960.
78. A.V. Tobolsky and K. Murakami, *J. Polymer Sci.* 1959, **40**, 443.
79. G.M. Bartenev and Y.V. Zelenev, in *Relaxation Phenomena in Polymers.* Chapter 4, John Wiley, New York 1974.
80. K. Murakami and K. Ono, Polymer Science Library 1. *Chemorheology of Polymers,* Elsevier, Amsterdam 1979.
81. K. Murakami, "Mechanical Degradation" in *Aspects of degradation and stabilisation of polymers,* Chapter 7, edited by H.H.G. Jellinek, Elsevier, Amsterdam 1978.
82. C.J. Derham, *Rubber India* 1974, **26**, 16.
83. G.T. Knight and H.S. Lim, International Rubber Conference, Kuala Lumpur 1975.
84. P.B. Lindley and S.C. Teo. *Plastics and Rubber: Materials and Applications* 1977, **2**, 83.
85. B. Stenberg, Y. Shur and J.F. Jansson, *J. Appl. Polymer Sci.* 1979, **35**, 511.
86. B. Stenberg and O. Dickman, *J. Appl. Polymer Sci.* 1983, **28**, 2133.
87. K. Murakami and K. Kusano, *Rheol. Acta* 1974, **13**, 127.
88. G.M. Bartenev and L.A. Schelkownikowa, *Plaste Kautschuk* 1973, **20**, 343.
89. B. Stenberg, T. Björkman and O. Dickman, *Polymer Testing* 1981, **2**, 287.
90. J.E. Stuckey, "High Temperature Stability of Rubber Vulcanisates," in *Developments in Polymer Stabilisation −1,* Chapter 3, edited by G. Scott, Applied Science Publ. London 1979.
91. K. Muniandy and S.T.M. Sand, *J. Rubber Res. Inst. Malays.* 1979, **27**, 46.
92. C.K. Das, *Rubber India* 1978, **30**, No. 11, 13.
93. D.J. Eliot and O. Atmawinata, *Rubber India* 1977, **29**, No. 11, 25.
94. A.L. Moran and D.B. Pattison, *Rubber Age* 1971, **103**, No. 7, 37.
95. F.P. Baldwin, *Rubber Chem. Technol.* 1940, **43**, 1040.
96. E.W. Bergstrom, U.S. Govt. Res. Develop. Rept. 1966, **41 (2)**, 50.

21 Knit-Lines in Injection Molding and Mechanical Behavior

Ralf Michael Criens
Hüter-Georg Moslé

Department of Materials Engineering, University of Duisburg
Lotharstraße 1–21, D-4100 Duisburg 1, West Germany

21.1 Introduction

Knit-lines are areas in injection molded parts made of thermoplastics in which, during manufacturing, separate polymer melt flows arise, meet, and weld more or less into one another. This process of flowing and melting together causes a disturbed morphology and thereby a nonhomogeneity of the structure and a change in the mechanical properties of the part.

The knit-lines appear not only in injection-molded parts but also in those obtained by extrusion processing, for example in tubes, as well as during the manufacturing of hollow pipes by blow molding.[1-3] In the following we discuss principally the effects of knit-lines on injection-molded parts.

A word on terminology is in order. Knit-lines are sometimes called weld-lines. However, the word "welding" is used by some authors as synonymous with crack healing. To avoid possible confusion, we use exclusively the term knit-lines.

21.2 Formation of Knit-Lines in Injection Molding

In manufacturing injection molded parts with a more complex geometry, the formation of several separate melt streams is often inevitable. The formation is due to the following:

— *double- or multi-grating* of a part (Fig. 21.2.1),
— pins in the mold for the production of holes or inserts in the mold (Fig. 21.2.2),
— variation in the wall-thicknesses within a part (Fig. 21.2.3), and
— *jetting*.

Knit-lines, especially those due to the first two reasons, cannot be avoided and the location of the knit-lines in the part can be modified only within limits.

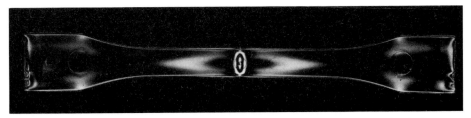

Fig. 21.2.1. Double-gating standard tensile bar; polarization photograph, crossed polarizers.

Fig. 21.2.2. Film-gated plate with holes; polarization photograph, crossed polarizers.

Fig. 21.2.3. Angle body with basis; thickness of the basis: 3, 2, and 1 mm (from left to right), polarization photograph, crossed polarizers.

21.2.1 Thickness Differences

In the case of large differences in the thicknesses within a part, the designer of the tool can influence the emplacement of the knit-lines. Figure 21.2.3 shows the modification of the position of the knit-line by choosing different thicknesses of the basis area of an angle body with basis. Even if thicknesses is defined in advance, that is to say thickness differences are prescribed according to function, some options are open. The tool designer can decide, by chosing the position of the gate, that knit-lines will be located in areas of the part which are subject to the lowest stress.

21.2.2 Jetting

According to Sarholz[7] and Malguarnera[9] jetting is the injection of the polymer melt by a free plastic jet into the mold at the beginning of the mold filling stage. This mold filling results from pin-point gate molding by using high injection speed and melt with low viscosity. During this jetting and refilling of the melt, knit-lines can arise, as shown by Malguarnera.[9] These effects can be avoided by the right tool design and processing parameters (lower injection speed, lower processing temperatures).

As described above, the knit-lines produced by wall thickness differences and by jetting can be limited by the designer or by the manufacturer. Therefore, in the following these types of knit-line formation will not be discussed.

21.2.3 Influences of Types of Knit-Line Formation

Differences in flow conditions during the converging of the melt fronts have specific consequences for the knit-line properties. In the case of multigating, primarily independent ground currents meet each other; however, in the case of flow around barriers a single ground current is split and rejoined following a relatively short distance of flow without direct impingement. Boundy and Boyer[12] found in their analysis of polystyrene (PS) that the two different mechanisms of producing knit-lines have different influences on the mechanical strength of the components. They distinguish between knit-lines produced by impingement (*"cold welds"*) and knit-lines obtained by flowing around barriers inside the mold (*"hot welds"*). They stated that PS samples containing "hot welds" had about 75% of the strength of samples containing no knit-lines; whereas samples with "cold welds" only had a 50% strength.

For this reason knit-lines are weak regions in injection-molded thermoplastic parts. Using tensile standard bars with and without knit-lines, Ehms and Bussian[13] demonstrated that the degree by which the strength is reduced depends on the processing conditions and on the type of thermoplastic used.

21.3   Damages Produced by Knit-Lines

In injection molding processing the thermoplastic melt is projected into the mold as a *"source flow."* On its way it cools down, the viscosity increases, and two or more melt flows when converge they mix together and weld. The intensity of the mixing and welding is dependent on the viscosity and the temperature of the melt as well as on the pressure in the mold and the temperature of the tool.

Hagerman[14] and Hobbs[15] have shown that when two separate melt fronts meet, the mold filling by a "source flow" leads to *V-notches* in the knit-line regions. The V-notches are formed by the compression of the air in the mold (Fig. 21.3.1). The air is pushed by the melt fronts from the center towards the walls of the mold. In this area the top film of the melt flow has already cooled down and a circular V-notch is formed. The size of this notch depends on the quantity of air compressed.

As Hagerman,[14] Hobbs,[15] as well as ourselves[10] have discussed, these surface defects indicate that knit-lines can be recognized as V-notches with microscopes,[14, 15] scanning electron microscopes, and feelers.[19]

These V-Notches have been found in doublegated specimens[14, 19] and also in tile-shape samples with holes.[15, 19]

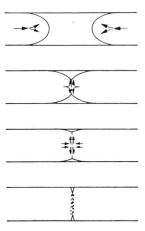

Fig. 21.3.1. Volcano flow (adapted from reference 15).

If these components are mechanically loaded, the notch causes a stress concentration at its bottom, thus producing cracks and subsequently widening and leading to a fracture of the part. This breaking mechanism depends on the speed of deformation of the component and the break resisting properties of the thermoplastic used. We must take into account that the emergence and growth of such a crack takes place in an area of morphological disturbance. Values obtained for the mechanical properties of nondisturbed moldings to *not* necessarily apply to conditions to which knit-lines are exposed.

21.4   Morphologic Defects Inside Knit-Lines

The morphology found in knit-line zones of thermoplastics differs largely from that found in nonknitted areas. Depending on the processing conditions and the properties of the thermoplastic used, the melt flows weld more or less intensively in the knit-line area. The morphology of the remaining zones of the component is determined by freezing a continuous streaming melt flow. Figure 21.3.1 shows how knit-lines adopt orientations vertically drawing upon the direction of injection as described for ABS samples by Hagerman[14] and for polypropylene (PP) samples by Hobbs.[15]

The orientation of knit-lines is produced by a so-called *"volcano"* or *"fountain" effect* which occurs when the mold is filled by "source flow." The flowline consists of a tough frozen top film cracking like a "volcano" at the flow center, pushing back border areas to mold rims and forming a fresh top film. Two of these flow fronts moving toward each other, meet at their center and form a nucleus area where the *"plastic heart"* of the two ground currents merge. Outer areas of the flow front pushed toward the mold rims are pressed against each other and merge into the external areas oriented sideways and form the V-notches. The expansion of the outer and the nucleus areas depends on the viscosity of the melt along both flowlines, the amount of the energy available for joining the nucleus areas, and the thickness of the cooled top film.

In Fig. 21.4.1 one can see, by shrunken double-gated standard bars, the reduction of the length of the samples caused by the orientation of the nonknitted regions. The reduction of the width in the knit-line area in the middle of the specimens is caused by the cross-orientation at the knit-line region.

Size and welding of the knit-lines can, therefore, be influenced by the way the injection molding takes place, such as:

— pressure inside the mold
— injection speed
— melt temperature
— tool temperature

and construction of the tool as it affects length and cross section of the flowlines and air extraction (V-notch!).

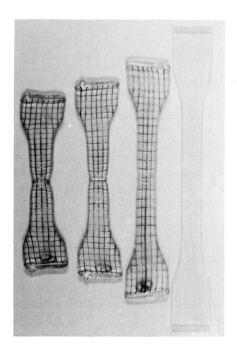

Fig. 21.4.1. Shrunken double-gated standard bars of PS produced at melt temperatures: 210, 230, and 250° C (from left to right); tool temperature, 40° C

In *amorphous thermoplastics,* it is mainly the degree and the direction of the orientation of the macromolecules that will influence the structure (morphology) of the molded parts and the mechanical properties of the component.

In *semicrystalline thermoplastics,* orientation influences are thought to be superimposed by the crystalline superstructures regarding size, number, and distribution.[15, 16, 19] The cooling conditions, the time for which the component stays in the mold at a higher temperature as well as the geometry of the tool and of the part itself are far more important than they are in amorphous thermoplastics.[17]

This correlation between mold design, processing conditions, and geometry of the component will have to be observed when knit-lines are examined.

21.5 Mechanical Properties of Knit-Lines

Taking processing conditions into account, we will outline the influences of knit-lines on the mechanical properties of injection-molded thermoplastics based on published test results.[4, 7 – 19] The samples to be examined will be those designed for the production assuring the absence of knit-lines as well as samples involving defined types of knit-lines, namely produced by *multi-gating* or by flow around barriers.

Knit-lines generated by impingement were examined by using standard tensile bars (Fig. 21.5.1) produced in a mold fitted with one or two gates.

Knit-lines produced by flowing around obstacles were generated when a plate containing two round holes was injection molded by a film gate (Fig. 21.5.2). Of the two knit-lines produced, one was located between the two holes, the other was located between the hole and the end of the plate farthest away from the gate.

In order to have comparable nonknitted samples, plates containing no holes were molded under equivalent processing conditions. Holes of the same diameter and location were than drilled carefully. The measurements of the plate were 60 by 30 by 2 mm; the hole diameter was 10 mm.[17, 18]

The test materials used were the amorphous thermoplastics PS, SAN, ABS, and PC and semicrystalline POM.

Fig. 21.5.1. Standard tensile bar (double-gated); knit-lines formed by impingement.

Fig. 21.5.2. Plate with holes; knit-lines formed by flowing around pins.

21.5.1 Processing Conditions

Results of experiments reported in [5] and [6] indicate that it is mainly the temperature of the thermoplastic melt which has a major influence on the mechanical properties of a component. In the investigation described below, melt and tool temperature were altered according to the respective areas of the processing temperatures.

The maximum value reached by the internal pressure in the mold was maintained at a constant level in order to avoid side effects arising from interference between injection pressure and processing temperatures. Injection intervals were set to assure sufficient cooling time for the samples and to prohibit any remaining internal pressure

Table 21.5.1. Test Materials and Processing Parameters (1 bar $= 10^5$ Pa)

Material	Melt temperature/°C			Tool temp/°C		Pressure in the mold/bar	
	1.	2.	3.	1.	2.	Bar	Plates
PS PS 168 N, BASF	210	230	250	40	60	300	700
SAN Luran 378 P, BASF	220	235	250	40	60	300	700
ABS Terluran 877 T, BASF	200	230	260	60	90	300	700
PC Macrolon 3200, Bayer	300	320	340	90	120	300	500
POM Delrin 500, DuPont	190	210	230	60	90	300	700

at the de-molding. Tools were equipped with devices ensuring permanent control and adjustment of melt and tool temperature and of the internal pressure in the mold at different points. Processing parameters for individual test materials are listed in Table 21.5.1.

The mechanical behavior of samples containing knit-lines was evaluated through various tensile tests. The execution of the tests depended on the shape of the sample objects.

In addition to tensile tests, one type of samples was submitted to impact-bending tests in order to obtain further information about the influence of the V-notches located at the surface of the knit-line areas.

21.5.2 Experimental Methods

In order to examine knit-lines produced by multi-gating, tensile standard bars subjected to uniaxial tensile stress were used. Tests were carried out at a constant cross head speed of 5 mm/min. (Speeds differing from this will be indicated below.) Parallel tests of single- and doublegated samples were conducted in order to quantify the influence of the knit-lines.

The tensile test samples were put under stress until the first criterion of technical failure emerged. This was either a yield point, or else the point of fracture if a sample broke without reaching a yield point.

In addition to these tests involving a slow deformation rate, we determined the behavior under impact bending loading (collision speed of about 4 m/s). These tests were carried out by using an instrumented pendulum (working capacity: 4 J) which permitted direct evaluation of the bending force. All tests were carried out at room temperature.

The second type of sample, the perforated plate, is not a standardized test object but a construction component working as a stabilizing or joining element. As stated above, these plates contain two knit-lines produced by flow around barriers. Therefore, different methods of evaluating the strength values of this sample are possible.

The first method to test the complete plate in a way related to the technical application of the component is the mandrel tensile test.

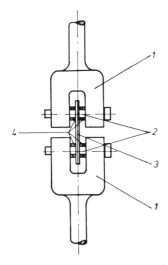

Fig. 21.5.3. Test rig for the mandrel tensile test, 1) tensile holder, 2) core blocks, 3) plate with holes, and 4) distance plate.

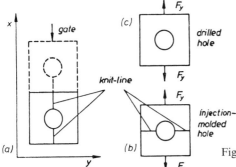

Fig. 21.5.4. Tensile test on divided half plates.

A test rig (Fig. 21.5.3) with specially fitted core blocks in both holes ensured that the force field was homogeneous and unable to bend the sample during the test. In order to separate the influence of the knit-lines, specimens with both molded holes and drilled holes were tested in the same way.

The second tensile test was designed specifically to evaluate the strength of the knit-lines in comparison to the strength of the same regions in specimens without knit-lines. This tensile test was made on the half of the plate which was farthest from the gate. As can be seen in Fig. 21.5.4a, the gate opposite the half plate is a square specimen with a hole in its center. Tensile test on half plates perpendicular to the molding direction define the strength of the knit-lines in case of molded holes (Fig. 21.5.4b), while tests conducted on specimens with drilled holes (Fig. 21.5.4c) measure the strength of the same region without knit-lines.

21.5.3 Experimental Results

We discuss now the results of the different tests for all types of specimens used. Mathematically, the influence of the knit-lines can be expressed by a so-called *knit-line factor* a_{kl}, which is defined[17] as

$$a_{kl} = \frac{\text{test result of specimens with knit-lines}}{\text{test result of specimens without knit-lines}}$$

The advantage of using the factor so defined consists in enabling an easy comparison of the influences of knit-lines dependent on the different test materials, types of specimens, and on the processing parameters possible. Therefore, the results will be discussed in terms of variations of these knit-line factors. In addition, a few representative stress-strain curves will be presented for the standard test bars.

21.5.4 Standard Tensile Bars

In both cases, for the injection molded tensile bars with and without knit-lines the stress-strain behavior is most dependent on the type of thermoplastic used, followed by the influences of the processing parameters (melt and tool temperature). The effects of the knit-lines on the mechanical behavior depend on the processing temperatures; a basic classification of the test materials into ductile and brittle ones is possible.

SAN and PMMA samples with low macromolecular orientation failed by fracture without any yielding. PS samples and highly oriented PMMA samples fractured slightly after the yield point. The possibility of these thermoplastics to reduce by yielding local stress concentrations caused by microcracks is limited.

As an example, the stress-strain curves of PS standard bars, with and without knit-lines, produced at three different melt temperatures are shown in Fig. 21.5.5. Similar diagrams were found for SAN and PMMA.

In the nonknitted samples made from these thermoplastics, a crazing could be seen before yielding or fracture. From the diagram a clear difference of the influence of the different melt temperatures can be seen for samples with and without knit-lines.

The higher macromolecular orientation caused by lower melt temperatures leads to higher strength in the samples without knit-lines for decreasing melt temperatures.

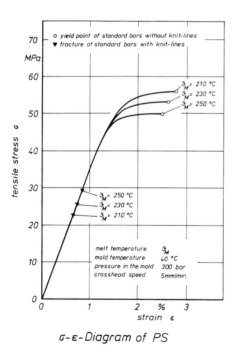

σ-ε-Diagram of PS

Fig. 21.5.5. Stress-strain diagram of PS samples with and without knit-lines (1 bar = 10^5 N m^{-2}).

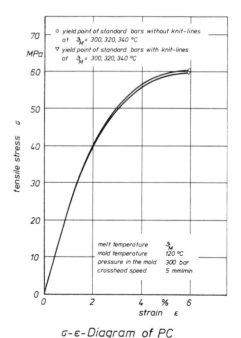

σ-ε-Diagram of PC

Fig. 21.5.6. Stress-strain diagram of PC samples with and without knit-lines (1 bar = 10^5 N m^{-2}).

Starting with the same initial modulus, the change to a strong viscoelastic behavior can be observed at higher values of stress and strain.

The stress-strain curves of the specimens with knit-lines and of the nonknitted specimens start with equal initial modulus. They follow a common curve up to the point of fracture without leaving the nearly elastic behavior. While the different melt temperatures do not influence the trend of the stress-strain curves, increasing melt temperature leads to fractures at higher values of stress and strain.

This behavior can be caused by better welding and merging of the knit-line regions, and also by producing a smaller V-notch by a melt front with a lower viscosity at higher temperatures. An increase of the tool temperature has a similar effect.

Unlike the specimens with knit-lines, no crazing could be seen before the fracture of the unknitted samples.

As an example of an amorphous thermoplastic with ductile strength behavior, the stress-strain diagram of PC samples can be seen in Fig. 21.5.6. No influence of the different melt and tool temperatures could be found for PC. Even the occurrence of knit-lines in the doublegated specimens caused only a negligible reduction of the yield stress and strain.

The possibility of this material reducing the stress concentration at the bottom of the V-notch by yielding, leads to a fracture of the knitted samples only after high elongation with necking. Even at higher or lower deformation speeds, there is no influence of the knit-lines on the yield point of PC. This has been shown in tensile tests conducted at cross head speeds of 0.5 and 50 mm/min; see Figs. 21.5.7 and 21.5.8.

In Fig. 21.5.9 a classification of the effects of the knit-lines on the stress and strain of the points of failure of standard tensile bars is shown by the variation of the knit-line factors as a function of the melt temperatures.

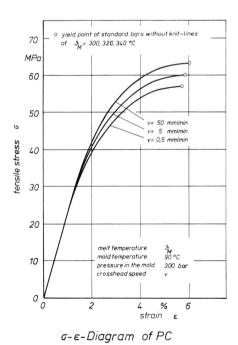

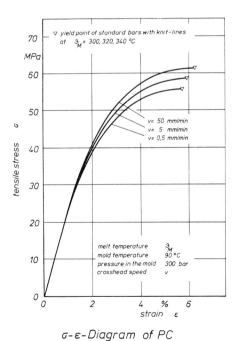

Fig. 21.5.7. Stress-strain diagram of PC samples without knit-lines ($1 \text{ bar} = 10^5 \text{ N m}^{-2}$).

Fig. 21.5.8. Stress-strain diagram of PC samples with knit-lines ($1 \text{ bar} = 10^5 \text{ N m}^{-2}$).

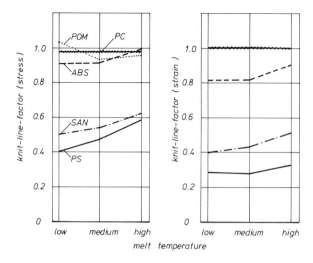

Fig. 21.5.9. Influence of knit-lines on the tensile stress and strain of double-gated standard tensile bars, knit-line factor versus melt temperature.

The curves of POM differ from the general trend. This is caused by the more complex correlation between the cooling characteristic during injection molding and the mechanical properties of a part made from a semicrystalline thermoplastic such as POM. There will be a different quality and quantity of the crystalline regions depending on the processing temperatures, the cooling speed, the cooling time, and the geometry of the tool and of the part itself. However, more investigations will be necessary here.

Clearly, it can be seen by comparing the knit-line factors of the different thermoplastics investigated that the tensile properties are reduced more by the occurrence of knit-lines than by the absence of yielding.

The knit-line factors of thermoplastics showing a ductile stress-strain behavior (PC, POM, ABS) carry values higher than 0.9 in the case of failure stress and higher than 0.8 in the case of failure strain. (For POM it was impossible to define an unequivocal yield strain.)

In contrast to this behavior, the unyielding polymers PS and SAN show knit-line factors which are lower than 0.6 for stress and lower than 0.5 for strain. The values of the failure strain are decreased more by the knit-lines than the values of stress are.

For all thermoplastics investigated, the knit-line factors increase with increasing melt temperature. The exception is PC: here no influence of the melt temperature variation could be found.

The results of the impact-bending test (Fig. 21.5.10) indicate tendencies similar to those of the tensile tests in respect to the influences of the processing temperatures (melt and tool temperature). Generally, the knit-line factors show lower values than those calculated from the results of the tensile tests.

In unyielding thermoplastics (PS and SAN) the impact strength of the samples with knit-lines is reduced to about 20% of those without knit-lines. Resulting from high deformation speed during impact bending, the stress concentration at the bottom of the V-notch cannot be wholly reduced. Even by the results of these tests it can be clearly determined that the more unyielding the polymer, the higher the influence of the knit-lines on the mechanical properties.

The reduction of strength caused by knit-lines generally decreases with increased processing temperatures. This is produced by better welding and merging of knit-lines,

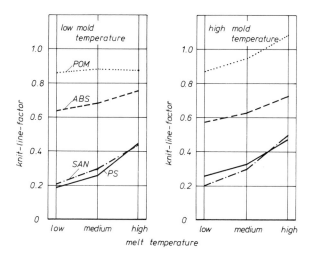

Fig. 21.5.10. Influence of knit-lines on the impact bending force, knit-line factur versus melt temperature.

as described above. The results of the tensile tests show that there are two *oppositing* tendencies for specimens with and without knit-lines working with the processing temperatures. On one hand, there is the increasing strength of the knitted samples with increased temperatures. On the other hand, the degree of the macromolecular orientation decreases with the higher temperatures of the melt and/or of the tool, which causes lower strength in the direction of the orientation. In these tests the direction of the orientation is also the loaded one.

These opposite trends of increasing strength of knitted samples and decreasing strength of nonknitted ones, both with higher processing temperatures, lead to the exceptional increase of the knit-line factor with respect to higher temperatures. We have to bear in mind this synthesis of temperature effects on orientation and on knit-line strength.

In amorphous thermoplastics a variation of the melt or tool temperature has the same effects, whereas in semicrystalline polymers there is a strong influence of the cooling conditions on the growth and distribution of the crystalline regions. At the low tool temperature, no affects of the melt temperature variation could be found for POM. Even at the higher tool temperature the knit-line factor increases with increasing melt temperature. Samples made from PC − with or without knit-lines − do not break during impact bending tests.

21.5.5 Perforated Plate

In such a specimen, knit-lines are produced by a melt flow which splits when passing a pin in the mold and then rejoins the melt stream.

The mechanical behavior of these specimens has been tested by the *tensile mandrel test,* which is functionally practical, and by the direct measurement of the knit-line strength by the tensile test on half plates. To make a comparison possible between the strength of plates with and without knit-lines, a series of specimens with molded and with drilled holes were tested by the same methods.

The results of the tensile mandrel tests are depicted as the variation of the knit-lines factor and as a function of the melt temperature in Fig. 21.5.11, left.

Similar to the results determined for knit-lines produced by doublegating, the example in this test shows that a stronger influence of knit-lines can be observed for the unyielding test materials.

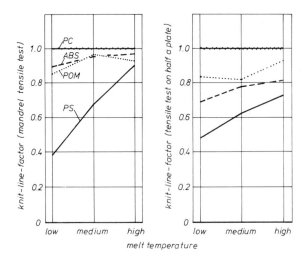

Fig. 21.5.11. Influence of knit-line on the strength of perforated plates, knit-line factor versus melt temperature; left: mandrel tensile test, right: tensile test on half plates.

In PS the knit-lines reduce the strength of the plates to about 40% of that of the plates without knit-lines. The influence of the knit-lines decreases with higher melt temperature. Plates with knit-lines manufactured at the highest melt temperature investigated have about 90% of the strength of the ones without knit-lines.

In ductile ABS, smaller strength-reductions can be observed as well as lower effects of the melt temperature variation.

For semicrystalline POM only a negligible influence and for the ductile amorphous PC no influence of the knit-lines on the mechanical properties could be found for any of the test parameters.

Tensile Test on Half Plates

The knit-line factors resulting from this test (Fig. 21.5.11, right) have lower values than those established by the tensile mandrel test, even if the tendency in respect to the melt temperature variation is the same. The influence of the knit-lines on the strength of the parts decreases with increasing melt temperature.

An extraordinary role is played by PC. Here the specimens reach the same yield point whether they have knit-lines or not. Also, the variation of the melt temperature does not affect the mechanical behavior.

A comparison made between diagrams in Fig. 21.5.11 indicates the strong influence of the kind of loading in respect to the effects of knit-lines on the final strength of an injection molded part.

The force flow during the tensile mandrel tests is produced by using homogeneous core blocks in both holes. The direction of the main stresses is 45° to the loading direction. The knit-line farthest away from the gate is loaded by the pressure of the core block, which causes pressure stress at the border of the hole and tensile stress at the gate opposite to the end of the plate. This complex kind of loading (in relation to a uniaxial tensile one) causes a crack at the bottom of the knit-line's V-notch at the end of the plate. Due to the crack growing, this knit-line failed first (Fig. 21.5.12, left).

On the other hand, in plates with drilled holes and therefore without knit-lines, a crack initiation could be found at the border of a hole in the direction of the main stresses (Fig. 21.5.12, right).

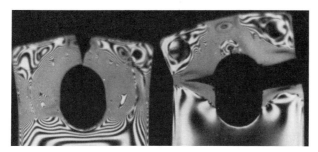

Fig. 21.5.12. Perforated plates loaded by mandrel tensile test; left: plate with molded holes, right: plate with drilled holes.

From a comparison of the measurd failure tensile forces in this test one can find knit-line factors between 0.9 and 1.0 for plates made from POM, ABS, and PC, and between 0.4 and 0.9 for PS plates. The values depend on the melt temperatures (Fig. 21.5.11, left).

Testing the strength at uniaxial tensile loading, the comparison made between plates with and without knit-lines leads to lower knit-line factors; see Fig. 21.5.11, right.

One can see from these results that the influence of knit-lines on the mechanical behavior of a part varies with the kind of loading (direction and type of loading as well as forces of conduction) and the geometry of the element. If one deals with direct loading of knit-line region, a failure intension can occur at a value of about 50% of the strength corresponding to specimens without knit-lines.

On the other hand, failures may occur in the areas of a part subjected to a complex stress distribution where the knit-lines are not tensile-stressed by the main stress direction; then their influence on the final strength of the part may be negligible.

21.6 Summary and Conclusions

Knit-lines in an injection molded part are often cannot be avoided because of either design needs or economics. The two most important mechanisms of knit-line formation are:

— *Double- or multi-gating* a part; two or more separate melt flows arise and stream together.
— *One melt flow* is divided by an insert and rejoins after having passed the barrier.

However, in any case the melting, merging, and welding of separated melt flows lead to areas in the injection-molded parts of thermoplastics which are distinguished by considerable inhomogeneities with respect to the morphology and the mechanical properties.

A V-notch in the surface of a part of the knit-line, and therefore a mechanical defect which causes a stress concentration, is caused by the compression of air in the mold by the melt flow streaming together.

The merging and welding of the melt streams lead to a macromolecular *"knit-line structure,"* which differs considerably from the morphology of the residual part. The knit-line itself is not a two-dimensional area but a three-dimensional region having a finate extensibility. The most important characteristics of the macromolecular structure constitute the orientations of the macromolecules in the knit-line region. An orientation can be found which is rectangular to the direction of the melt flows at the outer areas of the part. The morphology of the nucleus area is determined by the thoroughly merged melt streams.

These internal defects, as well as those on the surface, influence the mechanical behavior of a part with knit-lines.

The investigation of these factors and their effects upon mechanical strength was done by uniaxial tensile tests on samples with knit-lines (produced by multigating or by splitting one melt stream by an insert) compared to samples of the same geometry without knit-lines. All these samples were injection-molded under the same processing parameters.

For unyielding thermoplastics (PS, SAN) it was found that the tensile strength of parts with knit-lines is reduced to about 50% of the strength of ones without knit-lines, when the knit-lines are produced by doublegating. Knit-lines caused by flowing around an insert reduce the tensile strength of knitted samples to about 75% of the non-knitted ones.

The influence of the knit-lines on the tensile properties of samples made from ductile amorphous (ABS, PC) or semicrystalline (POM) thermoplastics is minor because of the ability of these materials to reduce stress concentrations by yielding. Negligible effects of the knit-lines have been found in the yield stress and strain of PC samples.

A correlation between these influences of knit-lines differing from thermoplastic to thermoplastic and the fracturing behavior of the material used (unyielding, ductility, crack growing behavior) is possible. Thus, there is a correlation between the ability of the polymer to reduce stress concentrations and the influences of knit-lines on mechanical failure in the parts made from a given polymer.

For amorphous, unyielding thermoplastics there is a strong influence of the orientation of macromolecules on the strength. The processing parameters which determine the degree of orientation also affect the dimensions of the knit-line region as well as the shape of the V-notch and the degree of the cross-orientation in the knit-line area. By testing samples manufactured at different melt and tool temperatures while the other processing parameters were kept constant, it has been shown that the influence of knit-lines on the mechanical strength is reduced with increased processing temperatures the more the thermoplastic used is unyielding.

This theory is supported by the results of additional tests conducted at high deformation speeds (impact-bending tests). In all thermoplastics investigated a large reduction of the impact-bending force resulting from the occurrence of knit-lines in the part could be found. By contrast, the respective reduction of the tensile load measured at lower deformation speeds (cross head speeds: 0.5, 5, and 50 mm/min) was relatively smaller.

Further, there is, of course, a substantial influence from the viscosity of the polymer melt flows streaming together on the welding and merging of the knit-lines and on the formation of the V-notches.

The reduction of the strength caused by the V-notches in the knit-line region cannot be excluded absolutely by processing and constructive precautions. It is, however, possible to decrease these effects by providing venting at the knit-line or by gating so as to minimize the flow length of the melt flows streaming together.

When manufacturing a part with knit-lines, the parameters of the injection molding process have to be selected in a way which guarantees viscosity of the joining melt flows as low as possible, so as to have sufficient energy for a thorough merging. Both effects can be obtained by increasing processing temperatures, injection speed or injection pressure.[9]

By designing the molding and the tool, the influences of knit-lines on the mechanical properties can be reduced by minimizing the flow length and increasing the part thickness (coordination between flow length and wall thicknesses).

Mechanically loaded parts with knit-lines should be produced by using a ductile thermoplastic.

The above recommendations are based on results of other investigators as well as on our own studies, the latter both published and unpublished. Up to now, examinations of the long-term behavior of knitted parts injection molded from thermoplastics (stress relaxation and creep behavior) do not exist. Also, very little is known about the effects of crystalline structures and of semicrystalline structures of thermoplastic parts with knit-lines on mechanical properties. Finally, more information is needed on the effects of fillers on the knit-line properties. These are subjects of today's and of future investigations of the authors.

References

1. R.A. Worth, *Polymer Eng. Sci.* 1980, **20**, 551.
2. B. Procter, *Mod. Plastics* 1973, 319.
3. C.J. Chung, *Trans. N.Y. Acad. Sci.* 1974, 319.
4. H.G. Moslé, O.S. Brüller and H. Dick, *Proc. 38th ANTEC, SPE*, p. 290, 1980.
5. H.G. Moslé, O.S. Brüller and H. Dick, *Proc. 39th ANTEC, SPE*, p. 721, 1981.
6. H.G. Moslé and H. Dick, *Erarbeitung von Kennwerten für das kunststoffgerechte Konstruieren*, Opladen 1980.
7. R. Sarholz, *Spritzgießen*, München-Wien, p. 123, 1979.
8. S.C. Malguarnera and A. Manisali, *Polymer Eng. Sci.* 1981, **21**, 587.
9. S.C. Malguarnera, *Polym.-Plast. Technol. Eng.* 1982, **18**, 1.
10. S.C. Malguarnera and A. Manisali, *Proc. of the 38th ANTEC, SPE*, p. 124, 1980.
11. S.C. Malguarnera and A. Manisali, *Proc. of the 39th ANTEC, SPE*, p. 27, 1981.
12. R.H. Boundy and R.F. Boyer (eds.), *Styrene: Its Polymers, Copolymers and Derivaties*, p. 490, New York 1952.
13. E. Ehms and M. Bussian, *Plaste Kautschuk* 1972, **19**, 214.
14. E.M. Hagerman, *Plast. Eng.* 1973, **29**, 67.
15. S.Y. Hobbs, *Polymer Eng. Sci.* 1974, **14**, 621.
16. R.C. Thamm, *Rubber Chem. Technol.* 1977, **50**, 24.
17. R.M. Criens and H.G. Moslé, *Proc. of the 40th ANTEC, SPE*, p. 22, 1982.
18. H.G. Moslé and R.M. Criens, *Kunststoffe* 1982, **72**, 222.
19. R.M. Criens, *Mater. Chem. Phys.* 1986, **14**, 69.

22 Failure Mechanisms in Polymeric Fibers

Joseph Zimmerman

P.O. Box 4042, Greenville, DE, 19807, U.S.A.

22.1 Introduction

In the many end uses for *fibrous materials,* it is expected that they will withstand for some respectable time period the mechanical and environmental stresses to which they are subjected. As our understanding of the structural and chemical factors which govern the pertinent properties of fibers has improved, we have seen remarkable growth in our ability to tailor polymer compositions and processing technology to achieve new levels of properties. In this chapter we will review a number of different aspects of the durability of polymeric fibers. These include *tensile and compressive failure, photo and thermal degradation,* damage from high energy radiation, and *hydrolytic degradation.*

22.2 Units

It is intended that units to be used in this chapter will conform to the SI system. For units of stress, strength, and modulus, it is the preference of the author that, for the most part, specific quantities are used (i.e., reflecting unit weight rather than unit cross-sectional areas). Following the recommendation of the American National Metric Council, we will use decinewtons per tex (dN/tex) for specific stresses where tex is mass (in grams) of 1000 meters of fiber. To convert from this unit to gigapascals (GPA), which is based on cross-sectional area, one multiplies by $0.1 \varrho_f$ (where ϱ_f is fiber density). Alternatively, to convert to the older grams per denier (g/d) unit, one simply divides by 0.8826 or multiplies by 1.133.

22.3 Tensile Failure

In considering the tensile strength of a polymeric fiber, one must remember that the reported values are not unique for the material being described. Thus, the strength depends on such factors as molecular weight, degree of orientation, defect level, uniformity, temperature of testing, rate of testing, relative humidity (for some materials), sample length, and the manner in which the filaments making up the fibrous assembly are arranged.

In the preparation of fibers from flexible chain polymers such as nylon 66, nylon 6, and polyethylene, the extruded fibers are oriented in a drawing process, and the tensile strength increases directly with the draw ratio.[1] The tensile strength of nylon 6 or 66 filaments has tended to correlate with *amorphous orientation*[2, 3] rather than *crystallite orientation*. These studies have indicated, by extrapolation, strengths of 18–21 dN/tex at perfect amorphous orientation. Similar studies have shown much lower "ultimate" strengths of 9–10 dN/tex for polyethylene terephthalate[2,4] and 12–13 dN/tex for poly-propylene.[5] With gel-spun linear polyethylene, draw ratios of 32 (in one study) have given measured tensile strengths of about 30 dN/tex,[6] and these workers have estimated *"ultimate"* strength for this polymer of 74–85 dN/tex based on extrapolation of a strength versus modulus relationship to the theoretical modulus of 2500–3000 dN/tex.[7] Fibrillar crystal growing of ultra-high molecular weight polyethylene from dilute flowing solutions has given tensile strengths as high as 47 dN/tex with high draw ratios (15 or higher).[8,9]

It is generally known that *number average molecular weight* has an important effect on the tensile strength which can be attained with a given composition. The end groups are considered to act as defects. However, beyond a given molecular weight, there is some evidence that molecular weight, per se, no longer has a major effect on tensile strength measured along the direction of orientation.[10,11] On the other hand, increased molecular weight apparently results in an ability to draw to higher draw ratio, and this results effectively in a significant dependence of attainable strength on molecular weight. Thus, for linear polyethylene, at a given width of the molecular weight distribution, tensile strength increased as the 0.4 power of molecular weight over the range of weight average molecular weight (M_w) from 54,000 to 4,000,000.[12] Reduction in polydispersity by a factor of 8 increased the tensile strength by about 85% at a given M_w.

For many fibers such as nylon or rayon, increasing draw ratio and tensile strength (σ_f) results in a trade-off with elongation to break (ε) with $\sigma_f \, \varepsilon^q = $ constant for many cases with q ranging from 0.5 to 0.7.[3,13] However, for fibers based on stiff, *p*-linked high melting amide-forming intermediates, such as Kevlar aramid, which are prepared by spinning of liquid crystalline solutions,[14] high orientation, tensile strength and modulus are obtained in the as-spun fibers with strength more than twice that of the previously available materials (e.g., 18–25 dN/tex versus 8–9 for nylon or polyester).[15] These relationships are shown in Fig. 22.3.1 for nylon 66 having M_n about 20,000 and for Kevlar aramid with M_n in the same range.

It is frequently observed, with polymeric fibers, that in the absence of degradation or crystal transitions or percent crystallinity changes, tensile strength decreases about linearly with increasing temperature. It is also known that when stresses are applied which are less than the nominal breaking stresses, the fibers will eventually break

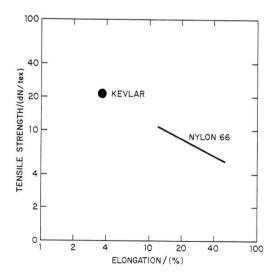

Fig. 22.3.1. Tensile strength versus elongation for nylon 66; comparison with typical KEVLAR aramid.

with a linear relationship with a negative slope between the logarithm of time-to-break and stress. There is, in general, as strong relationship between the extrapolated zero strength temperature and the extrapolated log-time-to-break at zero stress.

These observations are consistent with an activated failure process whose rate increases exponentially with stress. There are two major views of this process. The first considers it as a slippage of polymer chains through crystalline regions until enough flaws accumulate in a given cross section to result in a catastrophic failure.[1,15-17] The second views the main process as being the cleavage of intramolecular bonds.[18-20] While it is clear that bonds are broken during tensile deformation of fibers, and the literature in this field is extensive,[21-36] it is still the opinion of the author, for reasons which will be described below, that the principal mechanism governing the failure process of many polymeric fibers involves breaking of intermolecular bonds or creep of stressed chains through crystallites. The interesting subject of reactions produced by stress will not be discussed in detail in this chapter but are extensively covered in Chapter 5.

The absolute rate theory for the kinetics of the failure process is based on the application to failure of the theory[15,16] of flow phenomena described by Tobolsky and Eyring.[37] For failure under a fixed applied load the dependence of time to break t_b on stress σ is given by

$$\ln t_b = \ln(h/kT) + \ln j + \Delta G^*/RT - \sigma A_c d_1/2k Tw_l \qquad (22.3.1)$$

where h is Planck's constant (6.626×10^{-34} J·s), k is Boltzmann's constant (1.38×10^{-23} J·K^{-1}), T is in K, j is the number of jumps in the direction of the applied stress before catastrophic failure occurs, ΔG^* is defined as the "Gibbs function of activation" (frequently referred to in the literature on this subject as the free energy of activation or ΔF^*) for the failure process (in J/mol). R is the gas constant (8.314), A_c is the cross-sectional area per polymer chain (in cm), d_1 is the *"jump distance"* (in cm) or about twice the distance involved in movement to the top of the barrier, and w_l is the effective fraction of chains bearing the load in the real and imperfectly oriented fiber which does not have a uniform distribution of lengths of tie chains between crystallites and/or has some chains which re-enter the crystallites instead of continuing on to the next one. In this formulation, stress is in units of dynes/cm^2.

It should be noted that the general form of the lifetime equation is very similar for the treatment based on bond rupture and that the term $\ln(kT/h)$ is about -29.5 (or -12.8 in base 10) at 25° C. The effective form of Eq. (22.3.1) in base 10 is

$$\log t_b = a - b\sigma \qquad (22.3.2)$$

where a is the extrapolated zero stress intercept and is clearly related to ΔG^*. The slope b can be seen to be inversely related to the fraction of chains bearing the load. Since the tensile strength at 1 s breaking time σ_1 is a/b, Eq. (22.3.2) can be rewritten as follows:

$$\log t_b = a(1 - \sigma/\sigma_1) \qquad (22.3.3)$$

The extrapolated zero strength temperature T_0 measured at 1 s testing time is related to ΔG^* by

$$\Delta G^*/(2,303 R T_0) = \log(k T_0/h) - \log j \qquad (22.3.4)$$

However, ΔG^* estimated from the extrapolated zero stress intercept (a) at ambient temperature and from extrapolated zero strength temperature can differ by the quantity $\Delta S^*(T_0 - T)$ where ΔS^* is the entropy of activation for the failure process.

It is instructive to compare the generally agreed upon values of ΔG^* for tensile failure with the approximate extrapolated zero strength temperatures and the approxi-

Table 22.3.1. Gibbs functions of Activation for Failure versus Bond Energies

Fiber	a (avg)	ΔG^{*} [a] (kJ/mol)	T_0 (°C)	ΔG^{*} [b] (kJ/mol)	Bond energy (kJ/mol)
Nylon	15	160–190	380	155	293
Kevlar aramid	28	205–240	640	230	335
Polyethylene[1, 38]	5.5	120–126	150	100	348

[a] Or bond energy, U_0, in the bond breaking theory of Zhurkov.
[b] Estimated from T_0 with assumption of $J=10$ for nylon and polyethylene and $J=1$ for Kevlar.

mate bond energies for the weakest intramolecular bonds for nylon, Kevlar aramid, and polyethylene. These are summarized in Table 22.3.1.

Clearly, there is no correlation between ΔG^{*} for failure and known bond energies. The literature dealing with these questions often quotes lower values for the bond energies based on degradation experiments which can involve such reactions as oxidation, β-elimination, and cyclization rather than homolytic chain scission. The correlation seems to be better with the properties of the crystal, involving intermolecular forces. Thus a random copolymer of 66 and 6 nylons (80/20) has a lower zero stress intercept[1] (and ΔG^{*}) in spite of no reduction in bond energies. This general conclusion is supported by other studies and discussions of *time-dependent fracture*.[39–42]

Departures from a linear temperature dependence can occur at low temperatures or from the logarithmic time dependence at high testing rates because of the onset of brittle fracture, possibly exacerbated by flaws. However, in cyclic tensile fatigue experiments at modest rates of cycling (e.g., ≤ 30 Hz), one observes similar time dependence as in static loading except that one considers the maximum stress to be applied for something less than one-half of a cycle. Thus in a log-cycles-to-failure versus stress plot for Kevlar aramid single filaments,[15, 43] the extrapolated zero stress intercept for log-cycles-to-failure was in the range of 32–34. However, when the minimum load of the cycle was decreased to zero,[43] there was a 1–2 decade reduction in cycles to failure versus testing with a significant minimum load. The mechanism of this effect has not been resolved, but the possibility of compressive effects resulting from orientation, modulus, and stress gradients across the filament should be considered. This reduction in fatigue life, however, corresponds to a reduction in working strength of only about 7% or less.

The absolute rate theory predicts that little loss in strength will occur in static or cyclic fatigue until the fiber is close to its destined breaking time. The equation which describes this is

$$\sigma_{f,t} - \sigma_{f,0} = [\log(1 - t/t_b)]/b \tag{22.3.5}$$

where $\sigma_{f,0}$ and $\sigma_{f,t}$ are the initial breaking stress and breaking stress after an effective fatigue time, t.

The predictions of the theory for $\sigma_{f,0} = 6.6$ dN/tex and $b = 2.27$ are shown in Table 22.3.2.

On cyclic testing on an Instron tensile tester[44] of nylon or polyester fibers at 25° C to 80% of the average time to break the observed strength loss was about 6%, in reasonably good agreement with the predictions. In similar testing at 120° C, the strength losses ranged from 0 to 4%. This type of behavior has also been noted by other workers.[10]

The mechanisms of tensile failure and fatigue have been studied in great detail with the aid of scanning electron microscopy by J.W.S. Hearle and co-workers,[45–48]

Table 22.3.2. Predicted Strength Loss in Creep Failure

T/t_b	$\dfrac{\sigma_{f,0} - \sigma_{f,t}}{\text{(dN/tex)}}$	% Strength loss
0.2	0.044	0.7
0.5	0.133	2.0
0.8	0.31	4.7
0.9	0.44	6.7
0.99	0.88	13.3
0.999	1.32	20.0

and different types of fracture morphology have been described and classified which depend on fiber type and loading history. For example, the importance of axial splitting for high modulus aramid fibers has been shown, especially in fatigue testing, reflecting relatively low transverse and shear strengths.

There are some indications, for polyethylene, that the creep failure behavior may be improved at very high molecular weights in that the flow of the viscous element responsible for *"irreversible growth"* is reduced.[49] If the dependence of yield stress on strain rate can be used as an indication, the log time, zero stress intercept can possibly be as high as 9.0.[50] The estimated time-to-break relationships versus reduced stress σ/σ_1 for Kevlar aramid, nylon, and polyethylene are shown in Fig. 22.3.2. The upper line for polyethylene is the optimistic estimate.

In general, one can see that the usual practice of judging tensile strength based on 0.1–10 s testing time and at 25° C does not necessarily provide an accurate assessment of resistance to tensile failure if the loads are to be borne for some significant length

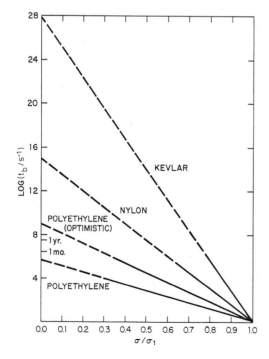

Fig. 22.3.2. Effect of fixed load on average failure time for several fiber types.

of time. As an example, if we compare polyethylene, nylon, and Kevlar aramid fiber at a steady load which is sustained for 1 month ($\log t = 6.4$) and assume zero stress log time intercepts of 9 (optimistically), 15 and 28 for the three fibers respectively, the "tensile strengths" would be 29%, 57%, and 77% of the 1 second tensile strengths, respectively. Alternatively, for one-second testing time, if the tensile strength were assessed at 100° C, the relative strength would be about 40%, 79%, and 88% respectively of the room temperature values. A note of caution for a fiber such as Kevlar aramid is that at temperatures $\geq 300°$ C tensile strength will fall off more quickly than expected because of the onset of various degradation modes, while for nylon 66 there is a sharp change in slope at ca. 150° C because of a crystal transition.[15] Another caution is that the normal variations in tensile strength can result in lifetime variations of several decades, depending on the slope of the log-time-to-break versus stress relationship.

There are some reports (e.g., Prevorsek and Kwon[10]) which contradict the predictions of the absolute rate theory as decribed above and indicate a more complex behavior, particularly at low stresses where lifetimes are predicted to increase drastically. One equation which has been proposed[10] is

$$\ln t_b = B_1 - B_2 \sigma^2 + B_3 \sigma^{-4} \tag{22.3.6}$$

The inflection in the *stress-lifetime curve* was correlated with fracture surface energy, fiber modulus, and ΔG^* (in the symbolism used here). On the other hand, other studies[51] of polyamide and cellulosic fibers have contradicted the above conclusion and obtained a linear relationship between log-time-to-break and load and have concluded that there is no stress corresponding to a metal elastic limit, below which the fibers will indefinitely sustain loads.

The formulation of the absolute rate theory provides a semiempirical basis for estimating "ultimate tensile strengths," that is when $w_l = 1$, when the zero stress intercept is known. Such an estimate is

$$\sigma_{f,u}(\text{dN/tex}) \sim 11.2 \, a \, d_1 / A_c \varrho_f \tag{22.3.7}$$

where d_1 and A_c are in nm and $(\text{nm})^2$ respectively and ρ_f is fiber density in g/cm^3. This type of treatment has previously been used by this author.[52] While A ranges only from 0.174 to 0.204 $(\text{nm})^2$ for the three fibers discussed above, some judgement must be used in assigning a value to d_1. Contrary to Schaefgen and his coauthors,[52] it now appears that a better estimate for fibers such as nylon and Kevlar aramid would be about 0.13 nm (rather than the 0.85 or 0.64 previously assumed) so that both fibers would have potentially very high strengths of 110–140 dN/tex if all the chains were bearing the load equally. Since the value of a is a key property, it means, also, that ΔG^* and T_0 are correspondingly important in determining ultimate strength potential.

In contrast to the above, the bond breaking approach has been used to predict that "ultimate" tensile strength equals $0.5 \, aD$ where a is the characteristic Morse curve constant and D is the dissociation energy. This is obtained by determining the inflection point in the Morse curve, where the force is a maximum. Such calculations usually lead to *"ultimate" strengths* of ≥ 175 dN/tex.[53, 54]

The tensile strength of a bundle of fibers is generally less than the average strength of the filaments comprising the bundle. The extent of the negative deviation depends on the *coefficient of variation (CV)* of filament strength. In one theoretical treatment[55] for an "infinite" bundle of fibers which have no time dependence of their tensile strength and which obey a Weibull distribution of strengths, average bundle strength decreases to about 85% of the average filament strength when CV is 5%, and to 70% when CV is 17%. When there is a strong mechanical interaction between the fibers (e.g., from friction), the effects on bundle strength have been calculated to be even greater[56] so that a 17% CV has given a calculated median strength conversion of the bundle as low as 30%.

22.4 Compressive Failure

For high modulus, high strength fibers such as Kevlar aramid, where use in resin reinforcement is indicated, we must also be concerned with compressive failure. Thus, with Kevlar 49 reinforced epoxy composites (60 vol% fiber), yielding occurs at about 0.3% compressive strain at a stress which is only about 15% of the tensile failure stress with an ultimate compressive stress of only 18% of the tensile stress reached at 0.6% strain.[15] Compressive strength is also temperature[15] and time dependent, but the extrapolated zero strength temperature of $\leq 400°$ C (versus 640° C for tensile failure) indicates a different mechanism than for tensile failure, possibly buckling of the amorphous regions, which should be related to the glass-transition temperature T_g.[15] This kind of relation to T_g has also been postulated for a wide variety of fibrous materials[57] with compressive strength increasing as the 1.85 power of T_g(K) in the range of about 150 to 3000 K. Compressive failure is manifested in formation of kink bands which are frequently at angles of about 60° from the fiber aixs.[15] Structural studies of kink band formation have indicated that slip of crystal planes containing hydrogen bonded sheets occurs as well as intermicrofibrillar slip.[58]

22.5 Flex Fatigue and Abrasion

One way of assessing compressive fatigue behavior of fibers is to flex single fibers around a wire (e. g., 0.076 mm diameter) under some fixed load (0.53 dN/tex is normal).[15, 17] In such tests, one can obtain drastically different results by increasing the stress or the diameter of the filaments, each of which causes a large decrease in lifetime. For example, increasing filament cross-sectional area of Kevlar aramid fibers by a factor of two reduced single fiber flex life by a factor of about 3, while doubling the stress from the normal level also reduced flex life by about the same factor. Thus, a need to increase resistance to flexing will lead one to reduce filament diameter. However, since this can often result in reduced abrasion resistance, this factor must be taken into account in seeking a desired balance of properties. In general, both flex and abrasion resistance depend strongly on fiber molecular weight. For example, a 50% increase in number average chain length has given an approximately twofold increase in abrasive cycles to failure.[17] Perhaps relevant to this is the more than fourfold increase in fracture surface energy of nylon 6 oriented bars measured perpendicular to the direction of orientation as M_n increased from 15,600 to 27,000.[11]

An increase in testing temperature can have a significant adverse effect on abrasion resistance if, in the temperature range involved, modulus decreases significantly and coefficient of friction increases.[44, 59] Abrasion resistance is, of course, also highly dependent on the normal force and the axial tensile force.[15, 59]

Flex fatigue and abrasion are factors of concern in the durability of tire cords. Studies of the mechanisms of these effects[15, 44] have shown that these mechanical degradative effects only occur when the twisted tire cords go into compression in critical regions of the tire (particularly in bias ply constructions), and they can be controlled by suitable choice of twist level (i. e., helix angle, θ) of the cord. Increasing twist, while beneficial to fatigue, causes reductions in both tensile strength (varies as $\cos^2 \theta$)

and modulus (varies as $\cos^4 \theta$). The modulus is particularly important for such properties as inflation growth (all tire types) and treadwear (belts of radial tires). An increase in filament diameter frequently improves abrasion resistance but at the expense of flex life if twist levels are marginal. However, use of very heavy nylon 66 monofils, which have high compressional stiffness, has reportedly eliminated the compression which normally occurs with twisted cords resulting in essentially no strength loss.

22.6 Fiber Degradation

Very often, fiber durability is strongly affected by degradation (i.e., molecular weight reduction) in use. This can be caused by such chemical stresses as hydrolysis; oxidative degradation initiated by heat, light, or bleach and high energy radiation. It often turns out that strength drops off much more rapidly with molecular weight loss than would be expected from the normal strength versus molecular weight dependence for the as-prepared fibers.[17] For example, with textile nylon 66, a 20% reduction in molecular weight resulting from photodegradation in a weatherometer caused a 50% reduction in breaking strength while one would have expected only about a 10% reduction if initial polymer molecular weight had been similarly reduced. This indicates that this sort of degradation may be causing somewhat localized defect areas. Skin-core effects resulting from Beer's Law or diffusional considerations could also be responsible, but probably not entirely, since similar effects are observed with hydrolytic degradation where diffusional effects are less likely in many cases.

In photodegradation of nylon 66 containing 0.3% TiO_2 delustrant, a linear increase in end-group concentration (deduced from solution viscosity) versus time was observed up to about 50% loss in molecular weight or about 85% strength loss.[17] The strength loss data were correlated by the following relationships:

$$\sigma_b^{-1} = \sigma_{b,0}^{-1} + k_d \Delta g \tag{22.6.1}$$

or

$$\sigma_b^{-1} = \sigma_{b,0}^{-1} + k_d' t \tag{22.6.2}$$

where σ_b is the tensile strength at break cross-sectional area, Δg is the increase in end group concentration resulting from degradation and t is time. A plot of $\sigma_{b,0}/\sigma_b$ versus end-group concentration (in equivalent/10^6 g) is shown in Fig. 22.6.1 for the nylon yarn described above. End-group concentration changes were deduced indirectly from changes in solution viscosity.

Unprotected aliphatic polyamides are particularly vulnerable to oxidative degradation because the methylene group adjacent to the amide nitrogen is especially prone to hydrogen abstraction by free radicals.

For thermal degradation of nylon 66 at 180 to 230° C in air, the reaction rates are much faster than in typical photodegradation, and the result[17] is an oxygen diffusion limited process in which the reciprocal of σ_b is linear with $t^{0.5}$ rather than t (Fig. 22.6.2). From these results, the apparent overall activation energy of thermal oxidative degradation was 113 kJ/mol, corresponding to an 80% increase in rate for a 10° C temperature increase.

Addition of relatively small amounts of copper salts as antioxidants can have a marked effect on the degradation rate, indicating a fairly long kinetic chain length for the degradation process. As shown in Fig. 22.6.3 at very low Cu^{++} levels (e.g., 3 parts per million), antioxidant activity was lost after 1–2 hours at 180° C in air, and then the rate was the same as that of the unprotected fiber. With amounts as low as 15 ppm Cu^{++}, there was no loss in protection even after 24 hours. Commercial

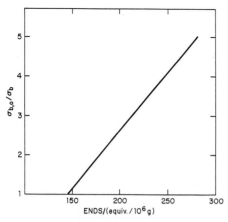

Fig. 22.6.1. Photodegradation of textile nylon 66 containing 0.3% TiO_2 — reciprocal of retained strength versus end group concentration.

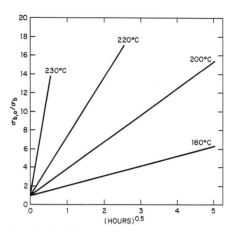

Fig. 22.6.2. Thermal degradation in air of unprotected nylon 66 — reciprocal of retained strength versus (time)$^{0.5}$.

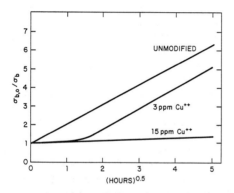

Fig. 22.6.3. Thermal degradation at 180° C in air of nylon 66 — effects of copper antioxidants.

nylon tire yarns containing as much as 60 ppm of Cu^{++} are more than 40 times as stable as unprotected yarns. Polyester yarns with no antioxidant are much more stable than unprotected nylon 66 with respect to thermal oxidative degradation and are about the same as Cu^{++} protected nylon.[13] Aramid fibers such as Nomex or Kevlar are inherently more stable to air and heat than aliphatic polyamides and require no antioxidant to equal or exceed Cu^{++} protected nylon.[13]

In photodegradation, light can be absorbed by the polymer itself to produce free radicals which initiate an oxidative chain reaction or which simply react with oxygen to break a single chain per initiation. Frequently, impurities in the polymer or a delustrant such as TiO_2 absorb the light to produce an initiating species which causes a chain reaction. There has been some speculation[60] that singlet oxygen may be involved in the initiation phase, being produced perhaps by reaction of an excited carbonyl group with molecular oxygen. It can then participate in hydrogen abstraction reactions to produce hydroperoxides. For aliphatic polyamides, as we have mentioned above, the greatest point of vulnerability to hydrogen abstraction and chain reactions is the methylene group adjacent to the amide nitrogen.[61] Good reviews of the UV degradation of nylon 66 and other aliphatic polyamides have been noted.[62, 63] The mechanisms by which TiO_2 promotes photodegradation and how stabilization by manganese ions occurs are discussed in the literature.[62-65] Increased relative humidity can have a significant effect on increasing the photodegradation of nylon 66[66] but has less effect on polyethylene terephthalate, as might be expected from the lower absorption of water by the latter.

An extensive study of aramid photodegradation[67−69] has shown that it proceeds by cleavage of the amide bond to form free radicals which can rearrange, in the absence of oxygen, to form a carbonyl (photo-Fries) product without breaking the chain or react with oxygen to break the chain. Thus, degradation and loss of strength for Nomex and Kevlar aramid fibers are much more severe in air than when oxygen is excluded. In UV exposure of yarns, fabrics, or ropes of Kevlar aramid fibers, the high UV absorption naturally occurring with this composition results in variable exposure between the filaments and a great deal of self-shielding. Thus in one report,[70] a fabric lost about 50% of its strength in the Florida sun in five weeks while a one-half-inch thick rope retained 90% of its initial strength after six months. Such nonuniformity may be one of the reasons for the reported[71] small change in viscosity average molecular weight accompanying rather large strength losses in photochemical aging of fabrics from Kevlar 49 aramid yarns.

In comparing the relative stability to photodegradation of different fiber types, one can get radically different results depending on the wavelength distribution of the impinging radiation. For example,[72] when nylon 66 (containing no TiO_2) was compared with polyester in direct exposure to Florida sunlight, nylon degraded at a slower rate. On the other hand, when the samples were placed under glass, the polyester was superior because of elimination of wavelengths which are highly absorbed by the polyester. A more detailed description of the effects of wavelength on the durability of bright nylon 66 yarns[73] shows a major acceleration in degradation rate as wavelength is reduced into the near UV (i.e., from 415 nm cut-off to 340 nm), and another sharp increase as minimum wavelength is reduced to 270 nm where amide absorption becomes significant. Studies of fluorescence and phosphorescence for aliphatic nylons indicate a high sensitivity to the degree of oxidation and indicate the importance of carbonyl groups conjugated with an ethylene bond.[63]

Polypropylene fibers and films are quite unstable with respect to UV degradation in outdoor exposure to sunlight, deteriorating badly in times as low as 25 hours. Many stabilizers have been found which significantly improve weathering performance, the most effective of which are hindered amines which have increased durability by as much as tenfold. Progress in understanding the performance of these stabilizers has been discussed in detail at the Spring 1984 American Chemical Society Meeting and reported in the literature.[72−75] It is generally thought that nitroxyl radicals are formed on these stabilizers. These can scavenge peroxy radicals which would otherwise participate in an oxidative chain reaction.

When aliphatic polyamides are subjected to high energy radiation, both degradation and cross-linking occur. Degradation is much more severe when oxygen is present. For example, in irradiation with 2 MeV electrons, a high tensile strength nylon tire yarn retained only about 20% of its strength after a 200 Mrad dose in air as compared with 65% retained when oxygen was excluded.[78] By contrast, aramid fibers such as Nomex or Kevlar are much more resistant to high energy radiation and have retained 75 to 90% of their strength after 600 Mrad in air, depending on the dose rate.[17,79] Irradiation of aramid/epoxy composites with up to 3000 Mrad of electron beam radiation had no apparent effect on tensile properties.[80] For nylon 66, the ratio of chain breaks to cross-links occurring on irradiation in the absence of oxygen has been reported to be in the range of 2.3–3.5, decreasing slowly with increasing dose.[78] The degree of cross-linking in nylon, even after 200 Mrad of irradiation — with no oxygen present — is not sufficient to increase modulus significantly because of the much larger effect of hydrogen bonding and crystallinity.

In a review[81] on effects of high energy radiation on polymers, polyester was reported to be significantly more stable in air than aliphatic polyamides (by a factor of more than 3), reflecting again its overall greater stability to oxidative chain reactions versus unprotected nylon.

Hydrolytic degradation can be a use-limiting factor for some fibers, depending on the nature of the environment, polymer type, and, of course, temperature. Aliphatic polyamides are susceptible to acid catalyzed hydrolysis at elevated temperatures but much less to base catalysis. For example,[17] at 121° C, nylon lost little of its strength after 10 hours when the surrounding water had pH values of 5–13 but, in 1% H_2SO_4 only about 10% of its initial strength was retained. One reason for this difference is the low solvation of inorganic bases by nylon and a high affinity for acids which hydrogen bond to the amide groups. On the other hand, polyethyleneterephthalate fibers also have low affinity for inorganic bases but are more sensitive to base catalyzed hydrolysis and degrade by progressive hydrolysis and removal of the fiber surface. Thus, breaking strength loss correlates with weight loss while there is little or no loss of molecular weight of the remaining fiber out to high extents of degradation. This etching process is controlled to a certain extent by diffusional considerations so that random copolymers, which have more open structures, lose weight and strength at a much greater rate than do homopolymers.

Kevlar aramid fibers have reasonably good hydrolytic stability in saturated steam as long as the pH is kept betwen 4 and 8. Estimated time to 50% strength retention was reported to be 100 hours at 160° C and 4400 hours (183 days) at 100° C in this pH range[82] with the rates following the kind of behavior described in Eq. (22.6.2). Activation energy was about 85 kJ mol. Another study of hydrolytic degradation of Kevlar 49 aramid fibers[83] has reported about the same degradation rate at 100° C but postulated a linear decrease in strength with time.

Hydrolysis and aminolysis are concerns with respect to durability of polyester tire cords in use.[13] Both reactions are catalyzed by carboxyl end groups in the polymer while the amines present in the tire rubber contribute to degradation via aminolysis. In practice, these can be controlled through the use of rubber stocks with lower hysteresis and/or reduced amine content and the use of minimal weights of rubber and fiber consistent with other durability needs. Reduction in carboxyl end-group content of the polyester fibers also has a marked beneficial effect on both types of reactions.[44] For heavier tires, where higher temperatures are encountered, polyamide and aramid fibers are probably more suitable since they are much less subject to these degradative reactions and, in many cases, have less hysteresis than polyester.

References

1. E.A. Tippetts and J. Zimmerman, *J. Appl. Polymer Sci.* 1964, **8**, 2465.
2. D.G. Pfeiffer, *Polymer Eng. Sci.* 1980, **20**, 167.
3. P.G. Simpson, J.H. Southern and R.L. Ballman, *Text. Res. J.* 1981, **51**, 97.
4. R.J. Samuels, *J. Polymer Sci. A-2* 1972, **10**, 781.
5. R.J. Samuels, *Structured Polymer Properties*, p. 196–219, Wiley, New York 1974.
6. P. Smith and P.J. Lemstra, *Makromol. Chem.* 1979, **180**, 2983.
7. P. Smith and P.J. Lemstra, *J. Polymer Sci. Phys.* 1981, **19**, 1007.
8. B. Kalb and A.J. Pennings, *J. Mater. Sci.* 1980, **15**, 2584.
9. J. Smook e.a., *Polymer Bull.* 1980, **2**, 293.
10. D.C. Prevorsek and Y.D. Kwon, *J. Macromol. Sci. Phys.* 1976, **B 12**, 447.
11. D.C. Prevorsek, Y.D. Kwon and J.L. Land, *Amer. Chem. Soc. Org. Coat. Plast. Prepr.* 1978, **38**, 404.
12. P. Smith, P.J. Lemstra and J.P.L. Pijpers, *J. Polymer Sci. Phys.* 1982, **20**, 2229.

13. R.E. Wilfong and J. Zimmerman, *J. Appl. Polymer Sci.* 1973, **17**, 2039.
14. E.E. Magat, *Philos. Trans. Royal Soc. London* 1980, **A 294**, 463.
15. R.E. Wilfong and J. Zimmerman, *J. Appl. Polymer Sci. Symp.* 1977, **31**, 1.
16. B.D. Coleman and A.G. Knox, *Text. Res. J.* 1957, **27**, 393.
17. J. Zimmerman, *Text. Manuf.* 1974, **101**, 19.
18. S.N. Zhurkov, *Internat. J. Fracture Mech.* 1965, **1**, 311.
19. S.N. Zhurkov e.a., *J. Polymer Sci. A-2* 1972, **10**, 1509.
20. S.N. Zhurkov and V.E. Korsukov, *J. Polymer Sci. Phys.* 1974, **12**, 385.
21. G.S.P. Verma and A. Peterlin, *J. Macromol. Sci. Phys.* 1970, **B 4**, 589.
22. A. Peterlin, *J. Phys. Chem.* 1971, **75**, 3921.
23. F. Szocs, J. Becht and H. Fischer, *Europ. Polymer J.* 1971, **7**, 173.
24. T.C. Chang and J.P. Sibilia, *J. Polymer Sci. Phys.* 1972, **10**, 2249.
25. B.A. Lloyd, K.L. DeVries and M.L. Williams, *J. Polymer Sci. A-2* 1972, **10**, 1415.
26. V.I. Vettegran and I.I. Novak, *J. Polymer Sci. Phys.* 1973, **11**, 2135.
27. L.N. Shen, *J. Polymer Sci. Lett.* 1978, **15**, 615.
28. H.A. Gaur, *Colloid Polymer Sci.* 1978, **256**, 949.
29. J.P. Park, K.L. DeVries and W.O. Statton, *J. Macromol. Sci. Phys.* 1978, **B 15**, 205.
30. T.M. Stoeckel, J. Blasius and B. Crist, *J. Polymer Sci. Phys.* 1978, **16**, 485.
31. D. Klinkenberg, *Colloid Polymer Sci.* 1979, **257**, 351.
32. B.M. Fanconi, K.L. DeVries and R.H. Smith, *Polymer* 1982, **23**, 1027.
33. G.A. George, G.T. Egglestone and S.Z. Riddell, *J. Appl. Polymer Sci.* 1982, **27**, 3999.
34. H.H. Kausch, *J. Macromol. Sci. Rev.* 1970, **C 4**, 243.
35. H.H. Kausch, *Polymer Fracture,* 2nd ed., Springer, Heidelberg-New York 1986; H.H. Kausch, Chapter 5 in this volume.
36. A. Casale and R.S. Porter, *Polymer Stress Reactions,* Vol. I and II, Academic Press, New York 1979.
37. A. Tobolsky and H. Eyring, *J. Chem. Phys.* 1943, **11**, 125.
38. D.L.M. Cansfield e.a., *Polymer Commun.* 1983, **24**, 130.
39. E. Sacher, *J. Macromol. Sci.* 1978, **B 15**, 171.
40. A. Peterlin, *J. Macromol. Sci.* 1981, **B 19**, 401.
41. G.E.R. Lamb, *J. Polymer Sci. Phys.* 1982, **20**, 297.
42. V. Kalouskova and J. Vsiansky, *Angew. Makromol. Chem.* 1983, **117**, 31.
43. A.R. Bunsell, *J. Mater. Sci.* 1975, **10**, 1300.
44. J. Zimmerman, *Text. Manuf.* 1974, **101**, 49.
45. A.R. Bunsell and J.W.S. Hearle, *J. Mater. Sci.* 1971, **6**, 1303.
46. A.R. Bunsell and J.W.S. Hearle, *J. Appl. Polymer Sci.* 1974, **18**, 267.
47. J.W.S. Hearle, B. Lomas and A.R. Bunsell, *Appl. Polymer Symp.* 1974, **23**, 147.
48. L. Konopasek and J.W.S. Hearle, *J. Appl. Polymer Sci.* 1977, **21**, 2791.
49. M.A. Welding and I.M. Ward, *Polymer* 1978, **19**, 969.
50. R.W. Truss e.a., *J. Polymer Sci. Phys.* 1984, **22**, 191.
51. J.D. Boone, *Nat. SAMPE Symp. Exhib.* 1975, **20**, 193.
52. J.R. Schaefgen e.a., *Ultra High Modulus Polymers,* edited by A. Ciferri and I.M. Ward, p. 196–199, Appl. Sci. Publishers, London 1979.
53. K.E. Perepelkin *Mekh. Polim.* 1966, **2**, 846.
54. T. Ohta, *Polymer Eng. Sci.* 1983, **23**, 697.
55. B.D. Coleman, *J. Mech. Phys. Solids* 1958, **7**, 60.
56. R.E. Pitt and S.L. Phoenix, *Text. Res. J.* 1981, **51**, 408.
57. M.G. Northolt, *J. Mater. Sci. Lett.* 1981, **16**, 2025.
58. T. Takahashi, M. Miura and K. Sakurai, *J. Appl. Polymer Sci.* 1983, **28**, 579.
59. B. Miller, H.L. Friedman and R. Turner, *Text. Res. J.* 1983, **53**, 733.
60. J.F. Rabek and B. Rånby, *Polymer Eng. Sci.* 1975, **15**, 40.
61. W.H. Sharkey and W. Mochel, *J. Amer. Chem. Soc.* 1959, **81**, 3000.
62. B.S. Stowe, R.E. Fornes and R.D. Gilbert, *Polymer-Plast. Technol. Eng.* 1974, **3**, 159.
63. N.S. Allen and J.F. McKellar, *J. Polymer Sci. Rev.* 1978, **13**, 241.
64. H.A. Taylor, W.C. Tincher and W.F. Hamner, *J. Appl. Polymer Sci.* 1970, **14**, 141.

65. N.S. Allen e.a., *J. Polymer Sci. Lett.* 1974, **12**, 723.
66. L.M. Lock and G.C. Frank, *Text. Res. J.* 1973, **43**, 502.
67. D.J. Carlsson, R.D. Parnell and D.M. Wiles, *J. Polymer Sci. Lett.* 1973, **11**, 149.
68. D.J. Carlsson e.a., *J. Polymer Sci. Lett.* 1973, **11**, 683.
69. D.J. Carlsson, L.H. Gan and D.M. Wiles, *J. Polymer Sci. Chem.* 1978, **16**, 2353; 2366.
70. Kevlar 49 Data Manual, E.I. du Pont de Nemours & Co. Inc., Wilmington, DE.
71. J.R. Brown e.a., *Text. Res. J.* 1983, **53**, 214.
72. T. Toda, T. Kuramada and K. Murayama, *Amer. Chem. Soc. Polym. Prepr.* 1984, **25(1)**, 19.
73. H.K. Muller, *Ibid*, 21.
74. D.J. Carlsson and D.M. Wiles, *Ibid*, 24.
75. R.J. Tucker and P.V. Susi, *Ibid*, 34.
76. Du Pont Tech. Info. Bull., X-203 1966.
77. A. Anton, *J. Appl. Polymer Sci.* 1965, **9**, 1631.
78. J. Zimmerman, *Radiation Chemistry of Macromolecules,* Vol. 2, edited by Malcolm Dole, Chapter 7, Academic Press, New York 1973.
79. L.K. McCune, *Text. Res. J.* 1962, **32**, 262.
80. V.F. Mazzio and G. Huber, *Nat. SAMPE Tech. Conf.* 1983, **15**, 234.
81. H.A. Rutherford, *Radiation Effects on Organic Materials,* edited by R.O. Bolt and J.G. Carroll, Academic Press, New York 1963.
82. J. Zimmerman, Symposium on *Chemical & Physical Lifetime Limits of Macromolecular Materials,* Amer. Chem. Soc. Fall Meeting, 1978.
83. R.J. Morgan, C.O. Pruneda and F. Kong, *Amer. Chem. Soc., Polymer Prepr.* 1984, **25(1)**, 189.

23   Failure Processes in Fiber Composites

Michael R. Piggott

Department of Chemical Engineering and Applied Chemistry, University of Toronto
Toronto, Ontario, Canada

23.1   Introduction

The literature in this field is very extensive; in the space of this Chapter it will be necessary to concentrate on a few important ideas rather than review the wealth of interesting theoretical concepts that have burgeoned in the last 40 years.

Progress in the field has been hampered by a scarcity of good experimental data. In the early days, tensile tests were difficult to perform, and the only data available came from flexure tests, which produced unreliable data for both tensile strength and Young's modulus. Today, this situation is much improved, but there are still problems in determining shear and compressive strengths. In fact, it is still thought by some authorities that there may be no such thing as a unique compressive strength for a fiber-reinforced plastic.

To understand the *mechanical properties of fiber composites,* we need to examine the sources of these properties, which are to be found in the *micromechanical interactions* between the fibers and the plastic matrix. Here we are faced with an even greater difficulty; until very recently there has been almost no data available on the important micromechanical parameters needed for analysis, mainly because these parameters are extremely difficult to measure.

In this Chapter we will concentrate on *aligned fiber composites,* since these are the most easily understood. In our analysis we will use data which have recently become available.[1] The treatment will be extended to *random fiber composites,* and laminates, where appropriate.

23.2   Strength and Failure of Aligned Fiber Composites

In this section, we will proceed from a very approximate treatment, adequate for many purposes, to more complete treatments obtained by successive relaxations of basic assumptions.

We start by using the early *theory of slip*[2] which can easily be extended to predict approximate stress-strain curves[3] and to estimate the Young's modulus. Our assumptions at this stage are (1) random distribution of aligned fibers of uniform strength, (2) constant shear stress, τ_i, at the fiber-matrix interface, in the slip region near the fiber ends, and (3) no slip in the region near the fiber center section. Figure 23.2.1 shows, schematically, the interfacial stresses envisaged and the resulting stress distribution along the fiber.

We first note that assumption 3 signifies that the matrix and fiber strains are the same for $-L(1-m) \leq x \leq L(1-m)$. We will denote this strain by ε_1 and equate it with the matrix strain. Thus the maximum fiber stress, $\sigma_{f\,max}$, is given by

$$\sigma_{f\,max} = E_f\,\varepsilon_1 \tag{23.2.1}$$

where E_f is the Young's modulus of the fiber.

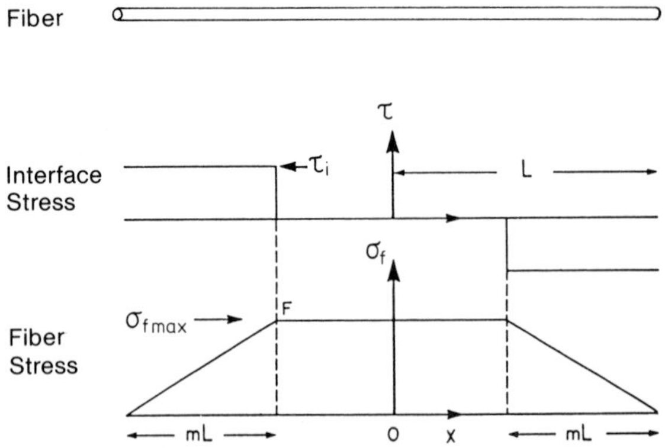

Fiber

Fig. 23.2.1. Fiber-matrix interface stress and corresponding fiber stress.

Near the fiber ends, the stress varies linearly with x, and the force equilibrium gives the slope

$$\frac{d\sigma_f}{dx} = -\frac{2\tau_i}{r} \tag{23.2.2}$$

for $x > L(1-m)$; $2r$ is the fiber diameter. Since the matrix has slipped, there is a small void at the end of the fiber,[4] and $\sigma_f = 0$, at $x = L$. Since $\sigma_f = E_f \varepsilon_1$ at $x = L(1-m)$ we can integrate Eq. (23.2.2), and using these boundary conditions obtain

$$\sigma_f = E_f \varepsilon_1 (1 - x/L) \tag{23.2.3}$$

and

$$m = E_f \varepsilon_1 / 2s\, \tau_i \tag{23.2.4}$$

m is the fraction of fiber surface which has undergone slip. This increases linearly with composite strain; s is the fiber aspect (length/diameter) ratio; $s = L/r$.

We have two possible failure modes in this simple model. (1) When m reaches 1 we have *gross slip*, i.e., the matrix can slide continuously past the fiber without changing the fiber stress. Let the strain at the onset of gross slip be ε_{1p}. This is given by $m = 1$; thus, from Eq. (23.2.4)

$$\varepsilon_{1p} = 2s\, \tau_i / E_f \tag{23.2.5}$$

(2) *Fiber failure.* This occurs when $\varepsilon_1 = \varepsilon_{fu}$ where the u in the subscript indicates the ultimate value. In this case we substitute $\sigma_{f\,max} = \sigma_{fu}$ in Eq. (23.2.1), and write

$$\varepsilon_{1u} = \sigma_{fu} / E_f \tag{23.2.6}$$

Which failure mode is activated in a particular case is determined by the fiber aspect ratio. There will be a critical aspect ratio, s_c, where either mode is equally likely. In this case $\varepsilon_{1p} = \varepsilon_{1u}$, and using Eqs. (23.2.5) and (23.2.6) we find that

$$s_c = \sigma_{fu} / 2\tau_i \tag{23.2.7}$$

For $s < s_c$ we find that $m = 1$ before $\sigma_{f\,max}$ reaches σ_{fu}, so that gross slip is the failure mode. For $s > s_c$ fiber failure dictates composite failure.

Using this simple approach we can draw approximate stress-strain curves. We notice that the average fiber stress $\bar{\sigma}_f$ is given by

$$\bar{\sigma}_f = E_f \varepsilon_1 (1 - m/2) \tag{23.2.8}$$

and we can estimate the average matrix stress, $\bar{\sigma}_m$, from

$$\bar{\sigma}_m = E_m \varepsilon_1 \tag{23.2.9}$$

where E_m is the Young's modulus of the matrix. The loads are distributed between fibers and matrix acording to their *relative volume fractions,* $\bar{V}_f$ and $\bar{V}_m$, i.e.

$$\sigma_1 = \bar{V}_f \bar{\sigma}_f + \bar{V}_m \bar{\sigma}_m \tag{23.2.10}$$

where σ_1 is the stress applied to the composite in the fiber direction. Substituting for $\bar{\sigma}_f$ and $\bar{\sigma}_m$ using Eqs. (23.2.8) and (23.2.9), and for m using Eq. (23.2.4) we find that

$$\sigma_1 = (\bar{V}_f E_f + \bar{V}_m E_m)\varepsilon_1 - \bar{V}_f E_f^2 \varepsilon_1^2/4s\,\tau_i \tag{23.2.11}$$

Notice that we can only use this equation for plastics which have not yielded (i.e. matrix yield strain $\varepsilon_{my} > \varepsilon_{fu}$). Most plastics that are reinforced with fibers have quite high yield strains; for those that do not, it usually makes little difference, unless $\bar{V}_f$ is very low (<0.2). With metal and ceramic matrices this is not the case.

For fibers with $s > s_c$, substituting ε_{fu} into Eq. (23.2.11), and rearranging, using Eq. (23.2.7), we find

$$\sigma_{1u} = \bar{V}_f \sigma_{fu}(1 - s_c/2s) + \bar{V}_m E_m \varepsilon_{fu} \tag{23.2.12}$$

For fibers with $s > s_c$, we assume that the matrix can withstand gross slip until its ductility is exhausted, i.e., for $\varepsilon_1 > \varepsilon_{1p}$ we write

$$\sigma_1 = \bar{V}_f s\,\tau_i + \bar{V}_m E_m \varepsilon_1 \tag{23.2.13}$$

up to some maximum value of ε_1.

We can now draw approximate stress-strain curves. We use the manufacturer's values for strengths and moduli. For s_c we use values obtained in experiments in which single fibers are embedded in resins, and the resin extended until the fiber breaks up into short pieces.[5] $s_c = 70$ is a typical value for carbon fibers in epoxy resins. The curves so obtained are parabolic, as shown in Fig. 23.2.2. Parabolic regions have been observed in stress-strain curves of *short fiber composites.*[6]

For continuous fibers the composite Young's modulus, E_1, obeys the rule of mixtures: $E_1 = \sigma_1/\varepsilon_1$; substituting this, and $1/s = 0$ into Eq. (23.2.11) gives the rule of mixtures expression

$$E_1 = \bar{V}_f E_f + \bar{V}_m E_m \tag{23.2.14}$$

For short fibers, Eq. (23.2.11) gives a curved stress-strain trajectory indicating that no true Young's modulus exists. Both strength and "modulus" are reduced when

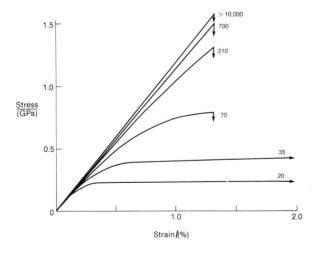

Fig. 23.2.2. Theoretical stress-strain curves (simple slip theory) for aligned carbon fiber reinforced epoxy. Figures on curves indicate fiber aspect ratios. $s_c = 70$, $V_f = 0.5$.

aspect ratio is small, and even when the fibers are continuous, Eq. (23.2.12) indicates that the *rule of mixtures strength* is not achieved

$$\sigma_{1u} = \bar{V}_f \, \sigma_{fu} + \bar{V}_m \, E_m \, \varepsilon_{fu} \tag{23.2.15}$$

However, unless $\bar{V}_f$ is small (<0.2) this is not very different from the rule of mixtures

$$\sigma_{1u} = \bar{V}_f \, \sigma_{fu} + \bar{V}_m \, \sigma_{mu} \tag{23.2.16}$$

A better estimation of the Young's modulus is obtained if we look more closely at the conditions required for slip to occur. This introduces a new failure process, i.e., *fiber-matrix interface failure*.

We will use *Cox's*[7] *treatment,* which indicates that before interface failure occurs, there is a "shear lag" between fibers and matrix, which produces an interface shear stress, τ_e, given by

$$\tau_e = n \, E_f \, \varepsilon_1 \, \sinh(nx/r)/2 \, \cosh(ns) \tag{23.2.17}$$

where n is a dimensionless constant given by

$$n^2 = 2 E_m/E_f (1 + v_m) \, \ln(P_f/\bar{V}_f) \tag{23.2.18}$$

and v_m is the matrix Poisson's ratio. P_f is a fiber packing factor equal to $2\pi/\sqrt{3}$ for hexagonally packed fibers. For $\bar{V}_f$ in the range 0.2 to 0.7, it is adequate to use

$$n \simeq \sqrt{E_m/E_f} \tag{23.2.19}$$

τ_e is clearly a maximum ($= \tau_{e\,max}$) at the fiber end:

$$\tau_{e\,max} = n \, E_f \, \varepsilon_1 \, \tanh(ns)/2 \tag{23.2.20}$$

and varies linearly with ε_1. Interface failure occurs when $\tau_{e\,max} = \tau_{iy}$, the shear yield strength of the interface. If $\varepsilon_1 = \varepsilon_{1s}$ at this point, called the *"slip point"*, then

$$\varepsilon_{1s} = 2\tau_{iy} \coth(ns)/n \, E_f \tag{23.2.21}$$

or

$$\varepsilon_{1s} \simeq 2\tau_{iy}/\sqrt{E_f \, E_m} \tag{23.2.22}$$

for fiber aspect ratios greater than about 30. Up to the slip point, behavior is entirely elastic, and the stress-strain curve is linear,[6] given by

$$\sigma_1 \simeq \bar{V}_f \, E_f (1 - 1/ns)\varepsilon_1 + \bar{V}_m \, E_m \, \varepsilon_1 \tag{23.2.23}$$

for $s > 30$, so that the modulus is

$$E_1 = \bar{V}_f \, E_f (1 - 1/ns) + \bar{V}_m \, E_m \tag{23.2.24}$$

Again for continuous fibers, E_1 is given by the rule of mixtures, Eq. (23.2.14). With short fibers the modulus is lower. Experimental results in qualitative agreement with this have been obtained[8]; the lack of quantitative agreement could well be due to imperfect alignment of the short fibers.

Values of τ_{iy}, obtained from very careful tests, in which 0.1 to 1.0 mm lengths of fiber are embedded in polymers, and then pulled out, are given in Table 23.2.1. Corresponding ε_{1s} values are also given in the table, and it can be seen that small values are obtained for glass fibers ($\varepsilon_{1s} \ll \varepsilon_{fu}$, which is typically 4.2%), and extremely small values are obtained for carbon. It is clear, therefore, that this failure takes place at an early stage in testing a fiber composite and that most of the stress-strain curve is governed by slip, even when the fibers are continuous. Although this important point seems to have been almost entirely neglected in the literature, strain limited design with fiberglass uses a maximum strain of 0.2%, well within the limit set by

Table 23.2.1. Interfacial yield stresses and works of fracture determined from single fiber pull out tests,[1] together with corresponding slip point strains

Fiber	Coating	Resin	Post (°C)	Cure (hr)	G_i (kJ m^{-2})	τ_{iy} (MPa)	ε_{1s} ($s \to \infty$) (%)
Glass	None	Polyester		None	–	10	0.15
Glass	None	Polyester	80	6	0.38	43	0.64
Glass	Silicone	Polyester		None	0.14	13	0.19
Glass	None	Epoxy		None	0.14	8.5	0.127
Glass	None	Epoxy	60	24	–	8.9	0.133
Glass	Silicone	Epoxy		None	0.08	6.8	0.101
Carbon	None	Epoxy	60	24	–	3 ± 1	0.025

ε_{1s}. With carbon the slip point is at a very low strain indeed, and so it must be concluded that this type of failure takes place in normal use.

Beyond the slip point, sliding is governed by friction; thus

$$\tau_i = \mu_f P \tag{23.2.25}$$

where μ_f is the coefficient of friction, and P is the pressure between the fibers and the matrix at the interface. Only recently have reliable values for P and μ_f been obtained. They were also measured by the single fiber pull out measurements mentioned above.[1] P has three components.

$$P \simeq P_0 + v_1 E_m \varepsilon_m - v_2 E_m \sigma_f / E_f \tag{23.2.26}$$

P_0 is the stress arising from the differential shrinkage between fibers and matrix that occurs when the composite is manufactured. The other terms are Poisson's shrinkages for matrix, with $v_1 \simeq v_m/(1 + v_m)$, and fibers, $v_2 \simeq v_f/(1 + v_m)$.[9] With reinforced plastics, for which $E_m \ll E_f$ we can neglect the fiber shrinkage term.

The fraction of fiber which has slipped is

$$m = \frac{E_f \varepsilon_1 - 2 \tau_{iy} \coth(n\bar{s})/n}{2 \mu_f s (P_0 + v_1 E_m \varepsilon_1)} \tag{23.2.27}$$

which, since $\bar{s} = s(1 - m)$, can only be solved by numerical methods. However, if the fibers are fairly long, with $s > 2s_c$ we can put $\coth(n\bar{s}) \simeq 1$ and write explicit stress-strain relationships:

$$\sigma_1 \simeq (\bar{V}_f E_f + \bar{V}_m E_m) \varepsilon_1 - \frac{\bar{V}_f}{s} \left\{ \frac{(E_f \varepsilon_1)^2 - (2 \tau_{iy}/n)^2}{4 \mu_f (P_0 + v_1 E_m \varepsilon_1)} + \frac{2 \tau_{iy}}{n^2} \right\} \tag{23.2.28}$$

For $s < 2s_c$ we must use

$$\sigma_1 = \bar{V}_f \left\{ E_f \varepsilon_1 (1 - m) - \frac{2 \tau_{iy}}{n^2 s} + \mu_f m^2 s (P_0 + v_1 E_m \varepsilon_1) \right\} + \bar{V}_m E_m \varepsilon_1 \tag{23.2.29}$$

after solving Eq. (23.2.27) for m.

We can write an approximate equation for the composite strength for $s > 2s_c$.

$$\sigma_{1u} = \bar{V}_f \sigma_{fu} + \bar{V}_m E_m \varepsilon_{fu} - \frac{\bar{V}_f}{s} \left\{ \frac{\sigma_{fu}^2 - (2 \tau_{iy}/n)^2}{4 \mu_f (P_0 + v_1 E_m \varepsilon_{fu})} + \frac{2 \tau_{iy}}{n^2} \right\} \tag{23.2.30}$$

We note that s_c is now fiber breaking strain dependent

$$s_c = \sigma_{fu} / 2 \mu_f (P_0 + v_1 E_m \varepsilon_{fu}) \tag{23.2.31}$$

Table 23.2.2. Coefficients of friction and shrinkage pressures from single fiber pull out experiments[1]* and corresponding critical aspect ratios

Fiber	Resin	Post (°C)	Cure (hr)	Fiber coating	μ_f	P_0	s_c [a]
Glass	Polyester	—		None	0.28 ± 0.08	23 ± 10	108
Glass	Polyester	80	6	None	0.23 ± 0.04	44 ± 10	93
Glass	Polyester	—		Silicone	0.10 ± 0.02	25 ± 6	292
Glass	Epoxy	—		None	0.54 ± 0.09	37 ± 10	44
Glass	Epoxy	60	24	None	0.76 ± 0.11	42 ± 7	29
Glass	Epoxy	—		Silicone	0.07 ± 0.01	35 ± 10	349
Carbon	Epoxy	60	24	None	0.04	42 [b]	769

[a] Glass; $\varepsilon_{fu} = 4.2\%$, $\sigma_{fu} = 3.0$ GPa; carbon, $\varepsilon_{fu} = 1.32\%$, $\sigma_{fu} = 3.1$ GPa; $v_1 = 0.254$.
[b] Estimated from experiments with glass fibers in epoxy which have the same cure.

and that we can simplify Eq. (23.2.30)

$$\sigma_{1u} = \bar{V}_f \sigma_{fu}(1 - s_c/2s) + \bar{V}_m E_m \varepsilon_{fu} + 2 \bar{V}_f \tau_{iy}(1 - \tau_{iy} s_c/\sigma_{fu})/n^2 \qquad (23.2.32)$$

This more comprehensive treatment thus does not change the earlier results very much. The most important new phenomenon recognized by this approach is the slip point, discussed above, and observed experimentally by Curtis e.a.[6] The strain effect on s_c (Eq. (23.2.31)) can have an important effect on high strain to failure fibers. P_0 for epoxies and polyesters which have been high temperature cured is about 40 MPa. $v_1 E_m \varepsilon_{fu}$ can reach 40 MPa also, for carefully handled glass fibers (with strengths of about 4 GPa, for fiber lengths of 1 mm); thus, taking into account the matrix shrinkage can reduce s_c by 50%. In the case of carbon, the maximum reduction is about 25%. Table 23.2.2 gives typical values for s_c estimated from fiber pull out data.[1]*

Another feature that is predicted by this treatment is a drop in stress at the slip point for very short fiber composites ($s < s_c$). Figure 23.2.3 shows the theoretical stress-

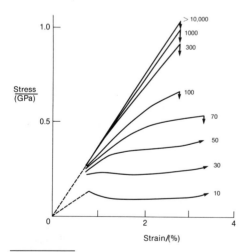

Fig. 23.2.3. Theoretical stress-strain curves for aligned glass fiber reinforced post cured polyester. Interface data given in Tables 23.2.1 and 23.2.2. $V_f = 0.5$.

* Note added in proof. A systematic error has been found in these results. All P_0 values should be divided by 2 and all μ_f values multiplied by 2. The reader is recommended to estimate the effect this has on s_c.

strain curves obtained with glass fiber reinforced post cured polyesters, using the data given in Tables 23.2.1 and 23.2.2 with $\bar{V}_f = 0.5$. The sudden decrease in stress with very low aspect ratios could account for the failures at low strains that are observed in some short, aligned fiber composites. Since the effect is much reduced at low fiber volume fractions, this could explain to some extent, the decrease in failure strain with increasing fiber volume fractions with these types of composites.[6]

Matrix failure also plays an important role in the failure of very short fiber composites, such as extruded glass and carbon reinforced thermoplastics. These commonly have very short fibers, due to the extrusion process causing fiber breakup. Detailed analysis in this case is difficult, because important data, such as the actual fiber strength after extrusion and fiber-matrix interface pressures and friction coefficients, are not available. Much theoretical work on this important problem is in progress.[10, 11]

23.3   Random Fiber Composites

Sheet molding compounds are used to make composites which have short fibers randomly oriented in the plane of the sheet. Three dimensional *random short fibers* can be present in composites made with bulk molding compounds (*fiber-reinforced thermosets*) or with pellets (*fiber-reinforced thermoplastics*).

In both cases, the fibers normal to an applied stress act as stress raisers and can promote early failure of the matrix, which is similar in nature to that observed with cross ply laminates.[12]

The fibers not aligned in the stress direction contribute little to the load bearing capacity, and it is generally accepted that the breaking stress, σ_{cu}, is given by[6]

$$\sigma_{cu} = \bar{V}_f \chi_1 \chi_2 \sigma_{fu} + \bar{V}_m \sigma_{mu} \qquad (23.3.1)$$

where χ_1 is a *fiber orientation factor* which has limiting values of 1.0 for perfectly aligned fibers, 0.375 for fibers directed randomly in two dimensions, and 0.2 for fibers randomly directed in three dimensions. χ_2 is a *fiber length factor*, governed by fiber slip, and is given approximately by

$$\chi_2 \simeq 1 - s_c/2s \qquad (23.3.2)$$

for fibers with $s > s_c$. For fibers with $s < s_c$

$$\chi_2 \simeq s/2s_c \qquad (23.3.3)$$

derived from Eqs. (23.2.12) and (23.2.13). For more accurate estimations of χ_2, use may be made of Eq. (23.2.32).

It should be noted that the minimum volume fraction for reinforcement, $\bar{V}_{f\,min}$, can be so high that it is not practical to get reinforcement at all. First we note that

$$\bar{V}_{f\,min} = \frac{E_f \sigma_{mu} - E_m \sigma_{fu}}{(\chi_1 \chi_2 E_f - E_m) \sigma_{fu}} \qquad (23.3.4)$$

for fibers with $s > s_c$. (The equation developed by Piggott,[13] is here modified by the inclusion of χ_2 as well as χ_1.)

For $\chi_1 = 0.2$ and $\chi_2 = 0.75$, for carbon reinforced polymers, with $\sigma_{fu} = 2.3$ GPa, $E_f = 377$ GPa, $\sigma_{mu} = 70$ MPa and $E_m = 2.5$ GPa, $\bar{V}_{f\,min}$ comes to 0.17. Unless the fibers have very low aspect ratios, this volume fraction is difficult to achieve in three dimensional random composites.[14] This accounts for the weakness of composites made from *bulk molding compounds*.

Sheet molding compounds can be used to make quite strong samples; strengths of 400 MPa and moduli of nearly 40 GPa have been claimed for carbon reinforced polyester resins made in this fashion.[15] They obey Eq. (23.3.1), with $\chi_1 \simeq 0.32$, and $\chi_2 \simeq 0.71$ with the modulus given by the analogous equation

$$E_c = \chi_1 \bar{V}_f E_f + \bar{V}_m E_m \tag{23.3.5}$$

As in the case of the aligned fiber composite, much more work is needed on matrix failure processes. Again, we can expect little progress until good interface data are available, and fiber strengths have been measured on fibers subjected to the processing involved in making the composites.

23.4 Shear Failure and Off-Axis Failure

Two basis types of *shear failure* are possible in *transversely isotropic fiber composites*, 1) parallel to the fibers and 2) perpendicular to the fibers. This can most easily be seen if we examine the three simple shears shown in Fig. 23.4.1. The first two (a and b) are equivalent to the shear τ_{13}, and produce a failure displacement parallel to the fibers. The third, equivalent to τ_{23}, produces a failure displacement perpendicular to the fibers. By symmetry τ_{12} and τ_{13} are equivalent. For orthotropic materials, such as laminates, τ_{12} and τ_{13} are not necessarily equal, and if the plane of the laminate is normal to the 3 axes (this is usual) then τ_{13} is the so-called interlaminar shear stress for unidirectional laminates. It should be appreciated that τ_{23} is also an interlaminar shear stress for laminates. It is incorrect, but unfortunately common, to use the expression *"interlaminar shear strength"* for failure resulting from the stress τ_{13}, when applied to pultrusions. "In plane" shear strength is also often used, and there is an ASTM standard for it.[16]

For pultrusions we can distinguish the two types of failure by the appelations 1) shear strength parallel to the fibers and 2) shear strength normal to the fibers. These could well be abbreviated to "parallel shear strength" and "normal shear strength", and denoted by τ_{13u} and τ_{23u} respectively. *"Laminate shear strength"* should be reserved for τ_{12u}, since in laminate stress analysis, the out-of-plane stresses are usually neglected.[8]

If we have a perfect composite, with straight fibers, neatly lined up in rows in a square array embedded in a matrix with perfect adhesion, it is reasonable to expect that

$$\tau_{12u} = \tau_{13u} = \tau_{23u} = \tau_{mu} \tag{23.4.1}$$

where τ_{mu} is the matrix shear strength. This, however, is not the case with real composites; fibers are not straight, and they cross the shear failure plane. Thus fibers are

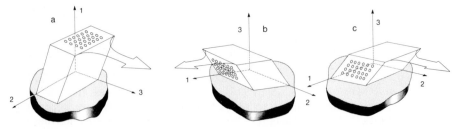

Fig. 23.4.1. Three simple shears in transversely isotropic aligned fiber composites, such as pultrusions.

broken, when the composite fails in shear and consequently

$$\tau_{13u} > \tau_{23u} > \tau_{mu} \tag{23.4.2}$$

We expect τ_{13u} to be greater than τ_{23u} because it is easier to strip fibers from the opposing face when the shear displacement is perpendicular to the fiber direction.

If we have a poorly made composite we expect

$$\tau_{13u} = B\bar{V}_f \tau_{iy} + C\bar{V}_m \tau_{mu} \tag{23.4.3}$$

where τ_{iy} is the interface strength, which for poorly made composites, with voids, imperfect wetting etc. is considerably smaller than τ_{mu}. B and C are constants determined by the geometry of the fiber packing. The shear test has been widely used as a quality control test for composites, since it can reveal how well the fibers and matrix are joined together.

For well made composites the "parallel" shear strength can be somewhat higher than the matrix strength,[17] but is quite commonly close to it.[15,18]

In the case of composite failure normal to the fibers, we expect the strength also to be governed by the matrix and the interface, so that σ_{2u} and σ_{3u} are not likely to be greater than σ_{mu}. For composites with less than perfect adhesion

$$\sigma_{2u} = \sigma_{3u} = B'\bar{V}_f \sigma_{iu} + C'\bar{V}_m \sigma_{mu} \tag{23.4.4}$$

where B' and C' are geometrical constants analogous to B and C, and σ_{1u} is the strength of the interface.

23.5   Failure at Notches

We will concentrate in this section on *aligned fiber composites,* since failure processes are much more readily identified in these cases. In such composites, failure parallel to the fibers can occur with ease. This requires very little work of fracture, since such failure is dominated by the matrix and the interface, which have works of fracture in the region of 100–400 J m^{-2}.

It is much more difficult to get a fracture to propagate so that it is normal to the fibers. To do this, long fibers must be broken, while short fibers must be pulled out of the matrix, as shown in Figs. 23.5.1 and 23.5.2.

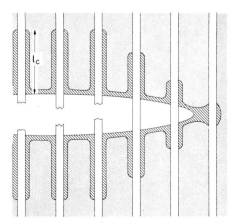
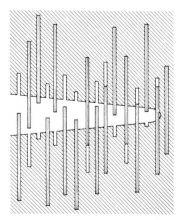

Fig. 23.5.1. A propagating crack in an aligned continuous fiber composite. The fibers have uniform strength.

Fig. 23.5.2. A propagating crack in aligned short fiber composites. $s < s_c$.

Consider, first, the short fibers, Fig. 23.5.2. If all the fibers have the same aspect ratio, and $s < s_c$, no fibers will be broken, and the work of fracture will be G_{fp} where[19]

$$G_{fp} = \bar{V_f} \, d\tau_i \, s^2/6 \tag{23.5.1}$$

Here d is the fiber diameter, and we have assumed a constant interfacial shear stress, τ_i. G_{fp} has a maximum value, $G_{fp\,max}$, when $s = s_c$. Using our original definition of s_c, Eq. (23.2.7)

$$G_{fp\,max} = \bar{V_f} \, d\sigma_{fu} \, s_c/12 \tag{23.5.2}$$

The carbon fiber composite with stress-strain curves shown in Fig. 23.2.2, with $s = s_c = 70$, should have $G_{fp\,max} \simeq 72 \text{ kJ m}^{-2}$.

Before the fibers can slide out, the interface must fail. This introduces a small amount of additional work, G_{if}. For $s < s_c$ (so that all fibers pull out)

$$G_{if} = \bar{V_f} \, s \, G_i \tag{23.5.3}$$

where G_i is the work of fracture of the interface between fibers and matrix. This has also been examined in single fiber pull out experiments. In some cases, the plot of force required to initiate fiber pull out versus fiber embedded length has a plateau region, corresponding to a fiber stress, σ_{fi}. This force or stress plateau can be explained if it is assumed that fracture of the interface starts at the surface of the resin, and propagates along the interface. A simple energy balance then gives[20]

$$G_i = d\sigma_{fi}^2/8E_f \tag{23.5.4}$$

In Table 23.2.1 values of G_i are given for the cases where a force plateau was observed. The highest value, 0.38 kJ m^{-2} was obtained for glass in post cured polyester resin. This result gives $G_{if} = 17$ kJ m^{-2}, using $s = 93$ (the s_c value in Table 23.2.2 for this material). This work is only a small fraction ($<10\%$) of the work of fracture of glass fiber composites.[21]

Interfacial debonding can be important with continuous fibers. Here the process is more complex. We will consider, first, the condition for the interface to fail. This requires a fiber stress σ_{fi}, given by Eq. (23.5.4). If $\sigma_{fi} > \sigma_{fu}$, then the fiber will fail, rather than the interface. Thus for

$$\sigma_{fu} < \sqrt{8E_f \, G_i/d} \tag{23.5.5}$$

interface failure will be suppressed.

Before going any further in this treatment of contiuous fibers, and fibers with $s > s_c$, we need to distinguish between two very different types of fibers 1) the ductile fibers, like Kevlar, which have uniform strength; and 2) the brittle fibers, like glass, carbon, and boron, which have very variable strengths, which are controlled by minute cracks or flaws on the fiber surfaces.

In the case of the Kevlar fibers, failure occurs close to the crack plane, where the fiber stress is highest, as shown in Fig. 23.5.1. The fiber has a significant *work of fracture*, G_f, and this contributes an amount of work, G_{fd}, to the work of fracture of the composite, given by

$$G_{fd} = \bar{V_f} \, G_f \tag{23.5.6}$$

Debonding may be unimportant in this case, since G_f is quite large.

Another source of work in the failure of these composites comes from the stretching of the fibers bridging the crack, as they are stressed to their breaking point. This deforms the matrix at the interface, and gives an amount of work, G_{fb}, given by[22]

$$G_{fb} = \bar{V_f} \, d \, s_c \, \sigma_{fu}^2/3E_f \tag{23.5.7}$$

This work contributes about 30 kJ m^{-2}, out of a total of about 240 kJ m^{-2}, for Kevlar composites.

With *brittle fibers* G_{fb} is only significant if we can make them behave as though they have uniform strength. This can be effected by using bundles of fibers, preimpregnated with resin, so that they behave as reinforcing rods. Failure then occurs close to the *crack plane*,[22] and since d for the rods can be as great as 1 mm, G_{fb} can be very substantial.

When brittle fibers behave independently of each other, we have to take the fiber flaws into account. This affects the value to be used for σ_{fu} in Eq. (23.5.5), to determine whether the interface fails. For example, extremely short lengths of fiber ($\simeq 0.02$ μm) can have strengths which are three times the value obtained with 25 mm lengths.[24] If we assume a value for σ_{fu} of 6 GPa for the few diameters in the crack plane which are unsupported, Fig. 23.5.2, we conclude that for glass, debonding occurs for $G_i <$ 0.6 kJ m^{-2} ($E_f = 72$ GPa, $d = 10$ μm). This is safely below the values given in Table 23.2.1, so we conclude that debonding can occur. For carbon with $\sigma_{fu} = 6$ GPa, debonding occurs for $G_i < 0.15$ kJ m^{-2} ($E_f = 234$ GPa, $d = 8$ μm). Unfortunately, we do not have a value for G_i for carbon, since with the only successful tests with carbon, Table 23.2.1, no force plateau was observed.

When debonding does take place with brittle fibers, fiber failure does not generally occur at the crack plane.[25] The process can be analyzed[26] by making three simple assumptions.

1) The flaws in the fibers give a fiber length (l)-strength relationship that can be represented by

$$\sigma_{fu} = A l^{-w} \tag{23.5.8}$$

where A and w are constants. Experiments with both glass and carbon[28] fibers show that this is true, but that A and w are different for different ranges of l, see Fig. 23.5.3. For glass, for example, with $l > 5$ mm $A = 1.9$ GPa $m^{1/8}$ and $w = 1/8$. (It cannot be assumed that when the fiber is inside the composite these values are maintained; manufacture of the composite usually damages the fibers.

2) We can neglect the work required to start fiber sliding (i.e., the only work considered is frictional).

3) We can neglect all stresses remote from the crack faces.

The frictional sliding process is governed by the matrix pressure, P, and the coefficient of friction, μ_f as in Eqs. (23.2.25) and (23.2.26). The fiber stress in the matrix near the crack face varies with distance from the *crack face*, x, according to the equation

$$\frac{d\sigma_f}{dx} = -2\mu_f (P_0 - v_2 E_m \sigma_f / E_f)/r \tag{23.5.9}$$

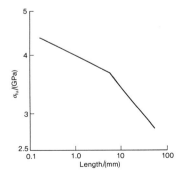

Fig. 23.5.3. The strength of glass fibers versus fiber length.[27]

so long as $P > 0$, i.e.

$$\sigma_f < P_0\, E_f / v_2\, E_m \qquad (23.5.10)$$

since, otherwise, the fibers and matrix would lose contact altogether. (Note that assumption 3 above means that $\varepsilon_1 \simeq 0$.)

Equation (23.5.9) integrates to give

$$\sigma_f = P_0\, E_f / v_2\, E_m - b\, e^{p\,x/r} \qquad (23.5.11)$$

where b is a constant and

$$p = 2\mu_f\, v_2\, E_m / E_f \qquad (23.5.12)$$

Figure 23.5.4 shows schematically the fiber stress for two different values of b, and we have written σ_r^* for $P_0\, E_f / v_2\, E_m$. The fiber length-strength relationship is also shown. We can increase the crack opening, and hence σ_f, until the two curves in Fig. 23.5.4 touch. Then the slopes are the same, so that

$$-w\, A\, x_b^{-w-1} = -2\mu_f (P_0 - v_2\, E_m\, \sigma_f / E_f)/r \qquad (23.5.13)$$

where x_b is the value of x where the curves touch.

Since $w \ll 1$ we can assume $w + 1 \simeq 1$, and hence

$$x_b \simeq w\, A\, r / 2\mu_f (P_0 - v_2\, A\, E_m / E_f) \qquad (23.5.14)$$

The work of fracture was originally calculated[23] assuming that the fibers break at an average distance x_b from the crack face and then pull out under a constant shear stress. The assumption of constant shear stress is not necessary. The shear stress is $\mu_f (P_0 - v_2\, E_m\, \sigma_f / E_f)$, and the fiber stress, from Eq. (23.5.11) with $\sigma_f = 0$ at $x = x_b$ gives

$$\sigma_f = (P_0\, E_f / v_2\, E_m)(1 - e^{p(x - x_b)/r}) \qquad (23.5.15)$$

(the fiber stress falls to zero at the point where the fiber breaks) so that the interface stress, τ_i, is

$$\tau_i \simeq \mu_f\, P_0\, e^{p(x - x_b)/r} \qquad (23.5.16)$$

and for any embedded length l, we replace x_b with l in Eq. (23.5.16).

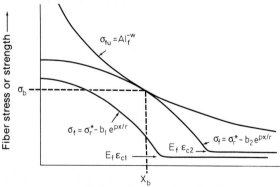

Fig. 23.5.4. Fiber stress as a function of distance from the crack face, for two values of stress at the crack tip, and fiber strength as a function of fiber length. $\sigma_r^* = P\, E_f / v_2\, E_m$.

Doing the appropriate integrations for the total pull out work, G_{bf}, assuming that all fibers pull out over a distance x_b

$$G_{bf} = \frac{4\bar{V}_f}{\pi d^2} \int_0^{x_b} \int_0^l \tau_i \pi d \, dx \, dl \qquad (23.5.17)$$

which gives

$$G_{bf} = \frac{\bar{V}_f d \mu_f P_0}{p^2} \left\{ \frac{px_b}{r} + e^{-px_b/r} - 1 \right\} \qquad (23.5.18)$$

and for $px_b/r \ll 1$ this reduces to the more familiar form

$$G_{bf} \simeq \frac{\bar{V}_f dw^2 A^2 P_0}{4\mu_f (P_0 - v_2 AE_m/E_f)^2} \qquad (23.5.19)$$

G_{bf} increases monotonically with x_b, the *mean fiber pull out length,* which for carbon reinforced polyesters and epoxy resins is only a few diameters.[25] It can be increased by decreasing P_0, and becomes indefinitely large as P_0 approaches $v_2 AE_m/E_f$. Experiments in which P_0 is decreased show that the pull out length and toughness are indeed increased.[29] Figure 23.5.5 shows the effect of P_0 on G_{bf} for carbon fiber reinforced epoxies with $d = 8$ μm, $w = 1/8$, $A = 2$ GPa $m^{1/8}$, $\mu_f = 0.2$, $v_2 = 0.34$, $E_m = 2.5$ GPa, and $E_f = 250$ GPa. Note however, that when we decrease P_0 we increase the adhesion between fibers and matrix. This limits the maximum value of x_b (and hence G_{fb}) that can be achieved, since improving the adhesion increases G_i and τ_{iy}. Nevertheless, a 100 to 200% improvement in G_{fb} appears to be possible.

Composites made with long fibers, $s > s_c$ will not all break at or near the crack plane; those with ends close to the crack plane will pull out instead. In this case we get some contribution from G_{fp}, and some from G_{fb} (or G_{bf}). For example, for fibers of uniform strength, we have[19]

$$G_{fpb} = G_{fp\,max}(s_c/s + 4\varepsilon_{fu}(1 - s_c/s)) \qquad (23.5.20)$$

This equation is plotted in Fig. 23.5.6. Also shown is G_{fp} which is used for $s < s_c$. The plot is rendered dimensionless by dividing by the maximum value of the work of fracture, $G_{max} = G_{fp\,max}$. Analogous results are obtained with flawed fibers.

In the case of fibers crossing cracks obliquely, the fibers assume an s shape,[30] so that brittle fibres like glass, fail at lower stresses. This causes a decrease in work

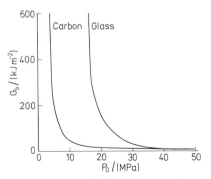

Fig. 23.5.5. Work of fracture as a function of matrix cure shrinkage stress.

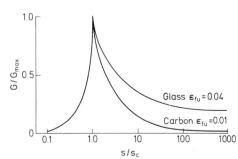

Fig. 23.5.6. Work of fracture versus aspect ratio for fibers of uniform strength.

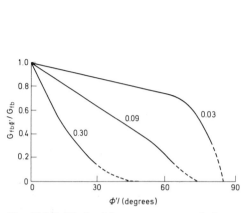

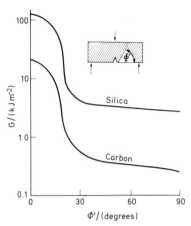

Fig. 23.5.7. Work of fracture versus angle between the fiber and crack plane normal, for brittle fibers such as glass, for ratios of matrix yield strength to fiber strength of 0.3, 0.09, and 0.03.

Fig. 23.5.8. Work of fracture as a function of fiber orientation for reinforced epoxy resins.[31]

of fracture, as the angle between the fiber and the crack plane normal decreases, as shown in Fig. 23.5.7. However, at quite small angles, splitting parallel to the fibers can occur,[30] and, as Fig. 23.5.8[31] shows, this leads to a very rapid decrease in the work of fracture.

This brief review of *failure at notches* suggests that the work of fracture can be improved by a number of methods. These methods do not, unfortunately, make the material more ductile, and some of them introduce a new problem.[32] The work of fracture that we estimate in our equations, requires a certain *crack opening displacement* (*COD*) before the work is completely done. Thus, if a fiber pulls out over a distance l, the crack must open by an amount of at least l for the pull out work to be done. If we compare the fracture work/COD ratios we find that $G_{fp\,max}/\text{COD} = \bar{V_f}\,\sigma_{fu}/6$, while $G_{fb}/\text{COD} = 2\bar{V_f}\,\sigma_{fu}/3$. On this basis, fiber pull out is much less effective than fiber stretching.

23.6 Compression Failure

The development of our understanding of *compression failure* of composites has been hampered by a) the difficulty of measuring the compression strength and b) the failure to recognize that imperfections and weaknesses in fibers, interface, and matrix govern the compression strength, rather than purely elastic effects.

The compression strength is difficult to measure because of the many different modes of failure that are possible. For example, when the fibers are aligned, and the ends not suitably confined, failure can occur at low stress due to splitting of the specimens at the ends. In order to avoid this a waisted specimen may be used, or the ends of the specimen may be confined by special loading pieces. Testing methods have recently been reviewed.[33] It is very important that the aspect ratio of the specimen be kept very small, no more than about 5 to 1, to avoid elastic buckling.[34] (This occurs for much lower aspect ratios than indicated by the Euler buckling analysis, due to the anisotropy of the fiber composite.)

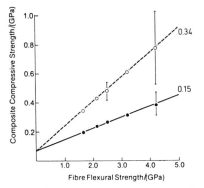

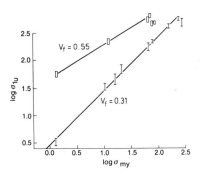

Fig. 23.6.1. Compressive strength of aligned steel reinforced epoxy resins.

Fig. 23.6.2. Logarithmic plot of compressive strength of glass reinforced polyester resins versus resin yield stress (0.5 has been added to $\log \sigma_{my}$ for $V_f = 0.31$, in order to separate the curves).

Fibers that are relatively weak in compression have been shown to be the cause of composite compression failure. Thus with steel, the yield stress can be varied by heat treatment. When this is done, polymer composites, carefully made using very straight steel wires, fail at stresses which are governed by the yield stress of the steel, Fig. 23.6.1.[35] With Kevlar, compression failure of the fibers governs the compression strength of the Kevlar-reinforced polymer. The fiber has an apparent compression strength of only about 0.17 GPa,[29] compared with a tensile strength of 3.5 GPa. Its modulus in compression is also low -21 GPa compared with a Young's modulus of 130 GPa. It is probable that some *carbon fiber composites* also fail by fiber failure; in this case it appears to be a 45° shear failure of the fibers.

The matrix yield strength controls the compression strength of low volume fraction glass fiber pultrusions ($\bar{V}_f = 0.31$) in a relatively simple way, Fig. 23.6.2[35]: up to $\sigma_{my} \simeq$ 60 MPa, the compression strength, σ_{1cu} is given by

$$\sigma_{1cu} = 9\sigma_{my} \qquad (23.6.1)$$

It is clear from Fig. 23.6.2, however, that for $\bar{V}_f = 0.55$, the behavior is more complex (it gives a 0.6 power law relation between composite strength and matrix yield strength).

In the case of both Kevlar and glass fibers, there is a change in failure process, as matrix yield strength is increased.[37] Thus, when $\sigma_{my} \simeq 10$ MPa the Kevlar fibers themselves fail in compression, precipitating composite failure, Fig. 23.6.3. In the case of glass, however, the change in failure process takes place at 60 MPa.

The glass and Kevlar results can be explained if we assume that the fibes are not straight. An analysis of fibers curved so that their axes are sinusoidal[37] based on a modification of an earlier analysis[38] gives

$$\sigma_{1cu} = \frac{2\lambda^2}{\pi^3 a} (\bar{V}_f - \bar{V}_m E_m/E_f) \sigma_{my} \qquad (23.6.2)$$

where λ is the wavelength of the sine wave, divided by fiber diameter, and a is the amplitude, also divided by fiber diameter. The *composite modulus is*

$$E_1 = \bar{V}_f (1/E_f + 1/E_{f1}) + \bar{V}_m E_m \qquad (23.6.3)$$

where

$$E_{f1} = \lambda^4 E_m/\pi^5 a^3 \qquad (23.6.4)$$

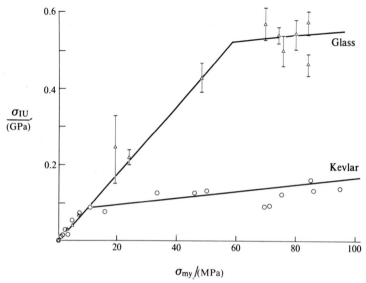

Fig. 23.6.3. Compressive strengths of glass and Kevlar reinforced polyester resins versus resin yield stress.

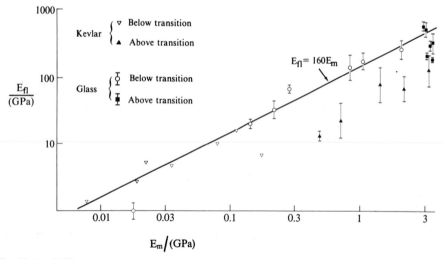

Fig. 23.6.4. Stiffness associated with buckling of sinusoidal fibers (E_{fl}) as a function of matrix modulus.

Equation (23.6.2) fits both glass and Kevlar results close to the origin. Equation (23.6.4) also fits both; Fig. 23.6.4 shows E_{fl} plotted versus E_m. Below the appropriate transition

$$E_{fl} = 160\, E_m \qquad\qquad (23.6.5)$$

Using Eqs. (23.6.2) and (23.6.4) we estimate that $a \simeq 4$ and $\lambda \simeq 43$ fiber "diameters." However, these results are too small to agree with results of experiments in which the fibers were artificially curved.[39] They can be reconciled, however, if the "diameter" used refers to the effective diameter of the fiber roving, rather than the diameter of the individual fibers.

This analysis leads to the conclusion that fiber straightness is an important factor in controlling compression strength. Figure 23.6.4 also gives a clue as to the failure process that supervenes in the use of glass. Above the transition at 60 MPa (Fig. 23.6.3), the results are very approximately given by

$$E_{fl} \simeq 80 \ E_m \tag{23.6.6}$$

This suggests that interface failure has taken place on the inside of the sine curve (the concave side).

This interface failure will be governed by the adhesion between the fibers and matrix. Adhesion is known to be an important factor governing the compressive strength[40]; fiber curvature thus provides the link between adhesion and compressive strength.

In the case of *hybrid composites*, complex interactions are involved when both types of fiber used are brittle.[41] However, when Kevlar, which is ductile, is used, we find that the compression strength is given by

$$\sigma_{1cu} = \bar{V}_{fk} \ \sigma_{fku} + \bar{V}_{fg} \ \sigma_{f \ max} + \bar{V}_m \ E_m \ \varepsilon_{1u} \tag{26.6.7}$$

where the subscript k refers to Kevlar, and g to glass (or carbon). $\sigma_{f \ max}$ is the maximum fiber stress at failure of an all glass (or all carbon) composite. Thus in these hybrids the Kevlar takes a maximum load, equal to its ultimate compressive strength at all fiber loadings.

23.7 Fatigue Failure

Fatigue failure of composites involves three possible failure processes 1) the fibers themselves, 2) the interface, and 3) the matrix. Although a great deal of fatigue data is now available, the role of these elements is not yet entirely clear, at least in part because there are so many different types of fiber reinforced plastics.

In the case of chopped strand mat glass reinforced polyesters, it was known as long ago as 1969[42] that debonding occurred at an early stage of the fatigue process. After about a thousand times as many cycles, some cracking became apparent, but complete separation took about ten thousand times as long again, at a given stress. These observations apply only to this type of composite, and aligned fiber composites, for example, behave somewhat differently. In most cases, their *fatigue properties* are superior to those of metals, and in particular to aluminum.

The fatigue properties of fibers in the absence of the matrix depend greatly on fiber type. Thus, glass fibers only have limited fatigue resistance, and they fatigue much more severely in the presence of moisture than under dry conditions.[43] Kevlar fibers fatigue, as sailors with Kevlar sails well know; the effect appears to be associated with Kevlar's weakness in compression mentioned earlier (Section 23.6). There seems to be little direct information on the fatigue properties of the fibers alone, except in the case of glass.

In the composite, the *fiber modulus* appears to play an important role. Thus in the normalized S–N curves (in accordance with a frequent usage, here S represents the stress amplitude) shown in Fig. 23.7.1, glass fibers ($E_f = 72$ GPa) produce composites with the steepest slope, boron fibers ($E_f = 420$ GPa) produce composites with the least steep slope, while carbon fiber ($E_f = 250$ GPa) composites occupy an intermediate position.[44] These are all fiber reinforced epoxy laminates and indicate that, for this

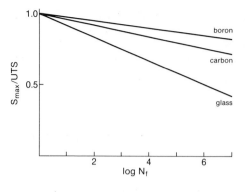

Fig. 23.7.1. Comparison of fatigue behavior of glass, carbon, and boron reinforced epoxy resins.

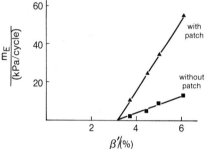

Fig. 23.7.2. Rate of loss in flexural modulus versus slope of S–N curve for aligned carbon epoxy. $R = 0.1$ (tension-tension fatigue).

type of composite, the strain is an important factor. Thus, the lower the failure strain of the composite, the better is its fatigue resistance. This strongly suggests that the fatigue resistance in these composites is matrix dependent. Experiments on short fiber reinforced thermosets lead to the same conclusion.[43]

In some very recent work,[45] in which the matrix shrinkage pressure was varied, the slope of the S–N curve, β', has been found to correlate with P_0 and with the rate of loss of flexural modulus, as shown in Fig. 23.7.2. The flexural modulus decreased continuously during fatigue, at a rate m_e per cycle until just before the specimen failed. The tensile modulus, in contrast, was not affected by fatiguing, until the fatigue life was almost exhausted. The Poisson's ratio of the composite increased approximately linearly with time during fatigue, as did the hysteresis loop energy. When a hole was drilled in the composite, it increased in width transverse to the fibers (and the applied stress) at an approximately constant rate during the fatigue process. The rate of change of all these factors (Poisson's ratio, hysteresis loop energy, and hole size) all correlated with β' in the same way that the flexural modulus did. These results were interpreted in terms of matrix cracking parallel to the fibers and/or interface failure.

Acknowledgement

I wish to thank Mrs. Jenny Clifford for her unstinting help in organizing and typing this manuscript.

References

1. P.S. Chua, *Interface Measurement Using Single Production Fibre Pull Out Tests*, MA Sc. Thesis 1982, University of Toronto.
2. A. Kelly and G.J. Davies, *Metall. Rev.* 1965, **10**, 1.
3. M.R. Piggott, *Load Bearing Fibre Composites*, p. 62, Pergamon, Oxford 1980.

4. M.G. Bader and G.F. Collins, *Proceedings of ICCM4*, 1982, edited by T. Hayashi, K. Kawata and S. Umekawa, p. 1067, Japan Society for Composite Materials.

5. L.J. Drzal, M.J. Rich, J.D. Camping and W.J. Park, *Proceedings of the 35th Annual Technical Conference of the Reinforced Plastics/Composites Institute of the Society of the Plastics Industry* 1980, section 20 C.

6. P.J. Curtis, M.G. Bader and J.E. Bailey, *J. Mater. Sci.* 1978, **13**, 377.

7. H.L. Cox, *Brit. J. Appl. Phys.* 1952, **3**, 72.

8. F. Ramsteiner, *Composites* 1981, **12**, 65.

9. M.R. Piggott, *J. Mater. Sci.* 1978, **13**, 1709.

10. M.G. Bader; T.W. Chou and J.J. Quigley, *Proceedings of the TMS-AIME Meeting, St. Louis, Miss.* 1978, p. 127.

11. H. Ishikawa, T.S. Chou and M. Taya, *J. Mater. Sci.* 1982, **17**, 832.

12. K.W. Garrett and J.E. Bailey, *J. Mater Sci.* 1977, **12**, 157.

13. M.R. Piggott, *Load Bearing Fibre Composites,* p. 100, Pergamon, Oxford 1980.

14. H. Krenchell, *International Symposium on Fibre Reinforced Concrete,* p. 45, American Concrete Institute, Detroit 1973.

15. N. Tsuchiyama, *Proceedings of ICCM4,* edited by T. Hayashi, K. Kawata and S. Umekawa, p. 497, Japan Society for Composite Materials 1982.

16. American Society for Testing and Materials, *Annual Book of Standards* 1983, **D 8.03**, D 3914.

17. D.A. Scola, *Composite Materials,* Vol. 6, *Interfaces in Polymer Matrix Composites,* edited by E.P. Pluedemann, p. 217, Academic Press, New York 1974.

18. M.F. Markham and D. Dawson, *Composites* 1975, **6**, 173.

19. M.R. Piggott, *J. Mater. Sci.* 1974, **9**, 494.

20. J.O. Outwater and M.C. Murphy, *Mod. Plastics* 1970, **7**, 160.

21. P.W.R. Beaumont, P.G. Reiwald and C. Zweben, *ASTM STP 568,* p. 134, American Society for Testing Materials, Philadelphia 1974.

22. M.R. Piggott, *J. Mater. Sci.* 1970, **5**, 669.

23. M. Fila, C. Bredin and M.R. Piggott, *J. Mater. Sci.* 1972, **7**, 983.

24. M.R. Piggott and J. Yokom, *Glass Technol.* 1968, **9**, 1972.

25. B. Harris, P.W. Beaumont and E.M. de Ferran, *J. Mater. Sci.* 1971, **6**, 238.

26. M.R. Piggott, *Proceedings of ICF5,* p. 465, edited by D. Francois, Pergamon, Oxford 1981.

27. A.G. Metcalf and K.G. Schmitz, *ASTM Proceedings* 1974, **64**, 1075.

28. J.W. Hitchon and D.C. Phillips, *The Dependence of the Strength of Carbon Fibres on Length,* Atomic Energy Research Establishment (UK), Report R 9132, 1978.

29. J.T. Lim and M.R. Piggott, *Proceedings of CFC8,* edited by M.R. Piggott, R.T. Woodhams and D. McCammond, p. 56, Canadian Committee for Research on the Strength and Fracture of Solids, 1982.

30. M.R. Piggott, *J. Mech. Phys. Solids* 1974, **22**, 457.

31. C.D. Ellis, PhD Thesis, University of Bath, 1974.

32. M.R. Piggott, *Proceedings of ICCM2,* edited by E. Scala, E. Anderson, I. Toth and B.R. Noton, p. 579, Metallurgical Society of the American Institute of Mining Metallurgical and Petroleum Engineers, New York 1978.

33. J.M. Whitney, I.M. Daniel and R.B. Pipes, *Experimental Mechanics of Fibre Reinforced Materials,* p. 175, Society for Experimental Stress Analysis, Brookfield Center, CT, 1982.

34. M.R. Piggott and B. Harris, *J. Mater. Sci.* 1980, **15**, 2523.

35. M.R. Piggott and P. Wilde, *J. Mater. Sci.* 1980, **15**, 2811.

36. R.K. Clark and W.B. Lisagor, *NASA Technical Note 81796,* National Aeronautics and Space Administration, Washington, D.C., 1980.

37. M.R. Piggott, *J. Mater. Sci.* 1981, **16**, 2837.

38. D.G. Swift, *J. Phys. D.* 1975, **8**, 223.

39. G. Martinez, D.M. Bainbridge, M.R. Piggott and B. Harris, *J. Mater. Sci.* 1981, **16**, 2831.

40. N.L. Hancox, *J. Mater. Sci.* 1975, **10**, 234.

41. M.R. Piggott and B. Harris, *J. Mater. Sci.* 1981, **16**, 687.
42. M.J. Owen, T.R. Smith and R. Dukes, *Plastics Polymers* 1969, **27**, 227.
43. J.F. Mandell, F.J. McGarry, D.D. Huang and C.G. Li, *Research Report R81-6,* School of Engineering, Massachusetts Institute of Technology, Cambridge, MA, 1981.
44. M.R. Piggott, *Proceedings of ICCM4,* 1982, edited by T. Hayashi, K. Kawata and S. Umekawa, p. 193, Japan Society for Composite Materials.
45. P. Lam, Fatigue Properties of Unidirectional Carbon Fibre Composites PhD Thesis, University of Toronto, 1983.

24 Creep and Fracture Initiation in Fiber Reinforced Plastics

Jan-Fredrik Jansson
Henrik Sundström

Department of Polymer Technology, The Royal Institute of Technology
S-10044 Stockholm, Sweden

24.1 Introduction

Although most fibers used as reinforcing constituents in polymer-based composites have reasonably simple deformation and fracture properties, the polymer matrices are anelastic and even nonlinear viscoelastic with fracture properties which are strongly dependent on temperature, deformation rate, loading time, and so forth. It is, therefore, often convenient to characterize the mechanical behavior of the composites as being either *fiber or matrix dominated,* i.e., as being more or less determined by only one of the constituents.

Only in a few cases, however, can the properties be regarded as completely *fiber dominated,* as for instance in the purely theoretical case when under uniaxial tension the fibers, at a high volume content, are oriented strictly in the stress direction.

In all engineering applications, the influence of the matrix on the mechanical behavior has to be taken into account. This is especially important when the direction of the stresses does not coincide with the direction of the reinforcement, for laminates built up of *multidirectional reinforcements,* under compression, bending and shearing stresses, for long term loads, and so forth.

There is no method available at present which is capable of predicting long-term behavior from the properties and composition of the constituents.

The problems have been observed and discussed by many authors. Yet, in spite of its immense practical importance, the connection between *creep and fracture initiation* has been studied very little.

We, therefore, now present a review and brief analysis of the present state of knowledge concerning the relations between creep and creep-fracture phenomena in *thermoset-based* fiber composites with special regard to fracture initiation.

24.2 Creep in Polymer Based Fiber Composites – The Appearance of Nonlinear Viscoelastic Behavior

Creep in polymer-based fiber composites has been studied extensively both from a phenomenological and mechanistic as well as from a more physical and chemical point of view. A very extensive literature exists. Therefore, only a few remarks are presented here as a background to the more complete discussion of the creep fracture initiation phenomena.

The creep of a composite is usually considered in terms of creep in single constituents (fibers and matrix); rupture in fibers, matrix, or interface; and slippage of the fibers in the matrix.

24.2.1 Influence of Type of Fiber and Fiber Arrangement

Kevlar is one exception to the rule that fibers can be regarded as being elastic. It is remarkable in that its creep rate is even higher than that for epoxy resin at low stresses, yet the reverse is true at high stresses.[1] At low initial strains the load carried by the fibers is, therefore, transferred into the matrix which leads to an increasing logarithmic creep rate, whereas at high initial strains the load carried by the matrix is transferred into the fibers giving a decreasing logarithmic creep rate.

Sturgeon e.a.[2,3] have summarized results from creep measurements on different types of unidirectional and cross-ply carbon fiber/epoxy laminates:

— Creep strains are small at (0), (0; 90), and (0; ±45) fiber orientation for stresses up to 80% of the short time tensile strength.

— The creep strain increases on changing the laminate construction from unidirectional, (0), to cross-ply (0; 90) to multi-ply (0; ±45) *fiber reinforcement.*

— Creep in (0) and (0; 90) laminates at 1000 hours is said to be almost independent of stress up to approximately 80% of the short time tensile strength. Multi-ply (0; ±45) materials show a slight increase in creep strain with increase in stress.

— Angle-ply (±45) composites behave like a typical polymeric material. Creep strain is strongly dependent on stress.

— The influence of additional ply orientations on the creep behavior of (±45) laminates is considerable. Adding 90° laminae reduces the creep strain at 1000 hours by about an order of magnitude. The addition of 0° fibers gives a reduction of approximately two orders of magnitude.

Similar results have also been reported by e.g., Soliman.[4]

For undirectional laminates built from ideally elastic fibers, the small amount of creep, noticed also for stresses in the fiber direction, is caused by fiber straightening and by the time-dependent properties of the matrix and the redistribution of stresses within the composite.

As a result of its relaxation, part of the load carried by the matrix is transferred into the fibers, extending them by a further small increment. This phenomenon has been called *"relaxation creep."*[3]

The behavior is dependent on the volume content of the fibers and on the moduli of the two constituents. If the matrix relaxes completely in a *unidirectional composite,* transferring all its load into the fibers, the "relaxation creep" is proportional to $E_m V_m/E_f V_f$ where E is the Young's modulus, V is the volume content, and the indexes m and f refer to the matrix and fibers respectively.

Shear loading of the matrix is one of the coupling mechanisms which transfers loads between individual fibers and thus is the general way of distributing the load through the material. Like the transverse behavior of *unidirectional laminates* the visco-elastic shear behavior is dictated by the resin.[5] In structures which have been carefully designed the shear creep resulting from tensile or compressive loading can be reduced to a minimum by adding extra fibers at ±45°.

24.2.2 Influence of Type of Interface

It is well-known from experience that adhesion between the fibers and the matrix plays an extremely important role for the long-term properties of the composites. Therefore, a large number of different bonding agents have been developed suitable for different types of fibers and matrices.

As an example, different types of *silanes* give acceptable bonding between glass fibers and polyester matrices. New coupling agents based on titanates[6] have shown very good properties for both polyesters and epoxy matrices. To achieve better adhesion to carbon fiber, the most common of several methods is to oxidize the fiber surface although this reduces its strength.[7]

The detailed behavior of the interface is far from understood, however, and is the subject of many present studies.[8] The adhesion is thought to originate from various interactions between the constituents[9]:

— electrostatic interaction between the two electric layers formed during the contact between the surfaces.

— interactions resulting from the mutual difusion of the molecules from the two surfaces through the interface.

— physical interactions of the van der Waals type.

— the creation of true chemical bonds.

However, it is also evident that many other factors are likely to influence the properties of the bonding.

In polymer-based fiber composites, the formation of an interfacial layer with properties different from those of the polymer as well as the changes of the properties of a thin polymer surface layer due to the influence of the interfacial contact, is thought to play a very important role in the adhesion.[8]

24.2.3 Influence of Type of Matrix

As shown by e.g. Jerina e.a.,[10] creep data for polymer-based fiber composites for different times and different temperatures cannot generally be related through a single reduced-time parameter. Furthermore, data for i.a. SMC laminates indicate a decreasing creep rate at high stresses or high temperatures or both.

Nevertheless, the creep behavior of the composite can be described, at least at low stresses, by a master curve for the viscoelastic response of the matrix alone. Thus, the creep compliance of the composite, D_c, can be determined approximately from the relation:

$$D_c = (B_1 + B_2 D_m^{-1})^{-1} \tag{24.2.1}$$

where B_1 and B_2 are constants and D_m is the creep compliance of the matrix.

Experimental compliance data at higher stress levels indicate, however, that nonlinear stress-dependent factors need to be included in the equation if a more complete description of the behavior is required.

On the basis of a nonlinear viscoelastic characterization method introduced by Schapery, a more complete constitutive creep equation has been developed by Brinson e.a.[11]

$$\varepsilon = g_0 D_0 \sigma + B_3 g_1 g_2 (t/a_\sigma)^{B_4} \sigma \tag{24.2.2}$$

where g_0, g_1, and g_2 are stress-dependent material properties, D_0 is the initial linear creep compliance, B_3 and B_4 are materials constants, and a_σ is a stress-dependent time-scale factor.

In laminates built up from woven roving, chopped strand mat, and so forth the creep is always substantial and must be taken into consideration in design work.

Figure 24.2.1a shows creep curves for a chopped strand mat reinforced polyester laminate with 33% (by weight) of glass fiber at room temperature.[12] A considerable amount of creep is observed.

Common creep curves, as plotted in Fig. 24.2.1a, include all information needed for design calculation of a product at the actual testing conditions. The curves contain,

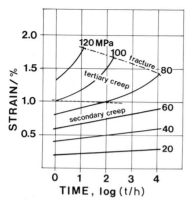
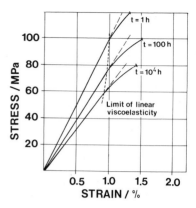

Fig. 24.2.1 a. Uniaxial tensile creep for a glass/polyester, CSM laminate (33% glass by weight) at room temperature.[12]

Fig. 24.2.1 b. Isochronous stress-strain diagram for the laminate in Fig. 24.2.1 a.

however, more information about the material, which is hidden in Fig. 24.2.1 a. A common procedure is, therefore, to present the data with respect to constant loading time, rather than constant stress in isochronous stress-strain diagrams. This is shown in Fig. 24.2.1 b for the same laminate.

The curves are linear for small stresses and strains, where the deformation is also reversible, showing a *linear anelastic behavior*. Within a well-defined stress-strain region the curves level off markedly, however, towards the strain axis, indicating a reduced resistance towards deformation. Simultaneously, the behavior becomes nonreversible and thus the material *nonlinear viscoelastic*.

An additional effect of the transition to nonlinear behavior is demonstrated in Fig. 24.2.2,[12] which shows in principle the *isochronous stress-volume strain* (stress dilatation) behavior for a polymer based fiber composite. At the same level, as that at which the transition to the nonlinear stress-strain behavior occurs, the dilatation accelerates. This means that an additional increase in volume is introduced.

It is generally accepted that the accelerated dilatation and the irreversible deformation in the nonlinear region are caused by dewetting and formation of cracks.

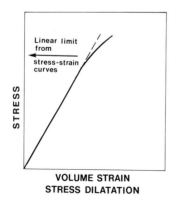

Fig. 24.2.2. Isochronous stress dilatation curve for a polymer based fiber composite.[12]

24.2.4 Viscoelastic Behavior of the Polymer Matrix

From what has been said above, it is clear that the viscoelastic and fracture properties of the matrix play an extremely important role in the creep as well as in the creep rupture processes in composite materials. It is, therefore, necessary to discuss the time-dependent behavior of the matrix in more detail in order to understand the behavior of the composite.

For most plain glassy polymers, including polyesters and epoxides, the degree of *nonlinear anelastic behavior* increases when the strain is increased and becomes pronounced at about the same uniaxial strain level, 1%,[13] although measurements indicate that the materials are slightly nonlinear at much lower strains, Fig. 24.2.3.[14,15]

For glassy, noncrystalline, linear polymers it has been suggested that the behavior originates from stress-activated acceleration and changes in the relaxation processes and possibly also from the introduction of new mechanisms.[16] Thus, creep measurements, infrared dichroism studies,[17] and so forth indicate accelerated main chain motions and conformational changes for increasing stresses, even at times and temperatures for which these mechanisms are usually negligible, see also Chapter 6.3.

At still higher stresses, glassy linear polymers become nonlinear viscoelastic resulting in an increasing degree of nonrecoverable strain.

On the basis of the generally accepted idea of the existence of fluctuations in the chain density distribution also in the amorphous, glassy state of linear polymers,[18] it has been suggested that nonlinear behavior starts in the most heavily strained molecules in regions which contain the smallest number of molecules per unit area (the low density regions) and in areas of stress concentrations formed by inhomogeneities, and so forth.[19] At still higher stresses these molecules begin to orientate and to slip along each other, producing local plastic deformation and a nonlinear viscoelastic behavior.

As a consequence of the slippage and the locally restrained, lateral contraction of the material, "holes" are formed in the structure. It is suggested that these holes,

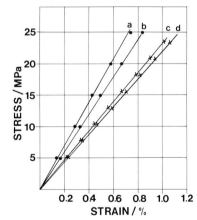

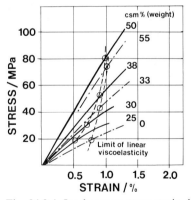

Fig. 24.2.3. Isochronous stress-strain diagram for a) and b) epoxy resin, TGMDA/DDS 1.0/0.6, at 30 and 70° C respectively[14] and c) and d) polycarbonate at 23 and 70° C respectively.[15] Represented by the second order relation $\varepsilon(t) = D_1(t)\,\sigma + D_2(t)\,\sigma^2$, the nonlinear compliance component $D_2(t)$ is about 10 times larger for polycarbonate than for the epoxy.

Fig. 24.2.4. Isochronous stress-strain diagram for glass/polyester CSM laminates in uniaxial tension at room temperature. Solid curves represent materials with undercured matrices.[12]

together with the oriented molecules in the high density regions, initiate the crazes formed in many polymers at about 1–1.5% uniaxial strain[19] and the microfibrous structures, often noticed in *"brittle" fracture surfaces.*[20]

In stiff polyesters and epoxides, the main chain motions and thus the orientation of the molecules are restricted by extensive cross-linking. This also results in a smaller deviation from linearity for stresses close to fracture as shown in Fig. 24.2.3 and in the absence of any greater extent of craze formation. It also explains the lower strain at the initiation and growth of catastrophic cracks.

However, in both polyester and epoxy composites and in plain resins, pronounced globular structures, due to fluctuations in the cross-link density, have been observed and reported by many authors.[21–23]

It is reasonable to expect that the molecules will be more flexible; thus, the nonlinear behavior of the polymer will start in regions of low cross-link density. This also seems to be a plausible explanation for the "fibrous" structures often noticed in the fracture surfaces of plain polyester, epoxy resins, and laminates.[20–23]

Figure 24.2.4 shows isochronous stress-strain diagrams for two laminates composed of the same polyester and CSM glass fiber reinforcement, but with the polyester slightly undercured. As can be seen in the curves for the well-cured laminates both the strain and the stress values at the transition to the nonlinear behavior are enhanced by the reinforcement. For the poorly cured laminate the situation is quite different. At low fiber content the nonlinear transition appears at a strain level which is even lower than that for the well-cured plain resin. However, it is not possible to make a similar comparison with the behavior of the undercured polyester, because it is very difficult to reproduce the same degree and type of undercuring in a solid polyester specimen as in the laminate.

24.3 Creep Rupture

As can be seen in Fig. 24.2.1a the strain at total failure decreases with decreasing long-term stressing and increasing time to failure. In many cases a linear relation is observed between the fracture stress σ_b and the logarithmic time to failure t_b at constant temperature T.[24]

$$T \log t_b / B_3 = B_1 + B_2 \, \sigma_b. \tag{24.3.1}$$

The retention of strength is dependent on the type of matrix, the type of fiber, and reinforcement; the bonding agent; and the processing conditions (i.e. curing procedure). Figure 24.3.1 shows the creep rupture for different glass fiber/polyester laminates.[25]

In Eq. (24.3.1) the fracture is described as a result of an Arrhenius type activation process, as was originally shown by Zhurkov e.a.,[26] Fig. 24.3.2. A large number of studies have been reported confirming the relation in Eq. (24.3.1). It should be noticed, however, that there is usually a considerable scatter in data of this kind. Furthermore, in many cases the behavior is influenced by active environments, slow post curing, and so forth which might give a different temperature dependence as well as nonlinear curves.

The relatively simple temperature dependence demonstrated in the equation has also been used as a basis for the development of accelerated testing and extrapolation methods for fiber composites and other polymeric materials. This subject has recently been discussed by Brinson e.a.[27, 28] who also proposed a flow scheme for accelerated laminate characterization and failure prediction.

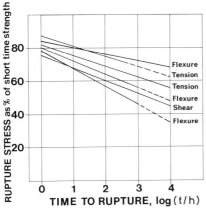

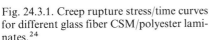

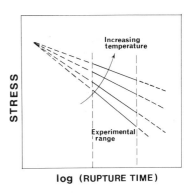

Fig. 24.3.1. Creep rupture stress/time curves for different glass fiber CSM/polyester laminates.[24]

Fig. 24.3.2. Zhurkov plot for creep rupture.

24.3.1 Failure Sequences in Fiber Composites

The initiation and growth of the fracture include fiber and matrix fracture, failure at the interface, and slippage of the fibers in the matrix. The situation is extremely complex, and in spite of the large number of very extensive studies reported in the literature, including experimental and phenomenological analyses, micromechanical treatment (see Chapter 23), studies of the physical and chemical conditions for fracture etc., detailed knowledge of the initiation and growth of fracture during creep is still far from complete.

To give a few examples of the different phenomena that arise, a brief review is presented describing fracture processes reported to have appeared in *chopped strand mat* (CSM)/polyester laminates; *sheet molding compound (SMC)* laminates, stressed in uniaxial tension; and in laminates built up of woven roving and loaded in bending.

Owen[29] reported results from studies of the fracture sequences in chopped strand mat/polyester laminates. Two features are noticeable in cut edges from nonbroken specimens which have been loaded in uniaxial tension, first, "there are groups of closely spaced fibers (strand groups) isolated in fields in resin" and second,

> among fibers that are normal to the cut surface (that is, perpendicular to the line of load) it will be seen that some fibers have separated from the resin matrix. Usually these *"debonds"* extent across the width of the strand group.

If the loading is increased, the debonding intensifies and starts to affect fibers that are at angles other than 90° to the line of load. Depending on the resin formulation, at approximately 70 percent of the ultimate tensile strength (UTS) some of the debonds extend out into the resin rich areas as resin cracks that can reach the specimen surface. Similar damage occurs under static loading, long-term creep loading, and fatigue loading.

Owen has also shown that resin cracking under static loading can be suppressed by increasing the flexibility of the resin. It does not suppress the debonding, however, which occurs at about the same strain and thus at slightly lower stress, due to the reduced modulus of the resin. On the other hand, the failure mechanisms changed for very flexible resins in that damage was initiated at the ends of the aligned strands shortly before transverse fiber debonding was initiated.

Slightly different results have been reported for CSM/Polyester laminates at a fiber content of about 20% by volume in creep and repeated loading by Lyons and Phillips.[30]

They found two distinct regions in their study of fracture surfaces. In the first, the pull out length was so "great that the matrix surface was mostly obscured."

The other part of the surface was clear and

the short pull-out length of the fibres allows features of the resin fracture to be clearly discerned. This region, commonly adjacent to one edge of the fracture surface, is identified as that where fracture originates. Resin cracks have been observed to originate (during tensile tests) at points of fibre debonding, and to propagate rapidly to the surface if this is not too far distant from the point of origin.

Under constant-load conditions, the fracture is thought to grow by an

intermittent process comprising load transfer to an adjacent fibre bundle, followed by environmental stress cracking of the glass under the influence of atmospheric moisture, followed by a further increment in resin cracking.

Sheet molding compound (SMC) composites usually consist of a polyester resin containing a combination of different particulate fillers and randomly oriented chopped glass fiber strands. During uniaxial tension, damage is initiated in the matrix rich regions.[31] At low stress levels the growth of these cracks is arrested by the glass fiber bundles. At higher stresses, however, the fiber bundles cannot resist the matrix crack propagation and the material fails by fiber-matrix debonding. The unhomogeneities in the material were found to play an important role with large local variations in fiber content and so forth. Resin-rich regions of the specimen were identified as being likely failure sites.

These results agree with general data from tensile tests for SMC laminates, as presented by, e.g., Denton,[32] showing two distinct knees in the stress-strain curves which indicate crack initiation at about 0.4% strain.

A still more complex situation appears in the bending of woven roving laminates as has been reported by Diggwa and Norman.[33] A series of fracture modes has been defined at different locations in the bent specimen and after different loading times, Fig. 24.3.3.

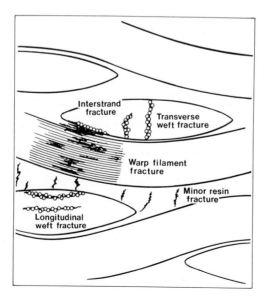

Fig. 24.3.3. Failure mechanisms in woven roving laminates in bending.[33]

— *Warp filament fracture* occurs either in isolated filaments or in small groups of filaments the warp strand.

— *Transverse weft fracture* includes resin cracks between and along the contours of the filaments. The cracks run perpendicular to the warp strands in those zones of the bent sample which are in longidutinal tension.

— *Longitudinal weft fractures* appear in the matrix in highly sheared areas of the sample.

— Continued weft fractures occur as combinations of the longitudinal and transverse fractures.

— *Interstrand fracture* occurs between the warp and weft strands and may follow their contour.

— *Minor resin fractures* appear in the resin-rich areas between the warp and weft strands. They usually appear in the stretched part of the sample and may be strongly affected by the curing conditions and shrinkage during curing.

— *Warp strand shear* and buckling occur in regions of longitudinal compression. They are likely to be the main failure mechanism on the compression surface of the sample.

The total failure of the laminate involves combinations of all modes of fracture mechanisms defined above.

24.3.2 Fracture Initiation in Creep – The Transition to Tertiary Creep

As has been pointed out above, the appearance of nonlinear viscoelastic behavior, which is caused by the initiation and growth of cracks and which results in increasingly pronounced irreversible deformation, coincides with transition to the tertiary creep region. In this way, the limit of the marked nonlinearity constitutes a measure of the long-term load bearing capacity of the material and might, therefore, be defined as the *"critical strain"* for long-time fracture. It has been intensively studied, by for instance, Menges e.a.[34-36] As can be expected the "critical strain" is dependent on the type of reinforcement and matrix, stress direction, loading time, and so forth.

24.3.3 Studies of Fracture Activities by Acoustic Emission

Most of the fracture activities (initiation and growth) can be detected nicely by *acoustic emission.* This technique is developing very rapidly at the moment, and it also forms the basis for the quality control methods developed by CARP (Committee on Acoustic Emission from Reinforced Plastics within the Society of the Plastics Industry in USA).[37-41]

One part of the elastically stored mechanical energy which is released during the formation and growth of cracks is transformed into a shock wave which travels through the laminate and creates sound. The type of fracture will influence the energy of the pulse, its frequency, amplitude, and so forth, and will give a characteristic pattern.

Vallentin and Bunsel[42] have developed a model relating the fracture process in carbon fiber/epoxy composites to the acoustic emission activity during steady loading. Similar acoustic behavior was observed in $(0; 90)_s$, $(\pm 15)_s$ laminates and in $(\pm 88)_s$ and $(\pm 77)_s$ laminates respectively. An overload was shown to be equivalent to an acceleration of time at the lower load as the rate of damage accumulation was increased. With $(\pm 45)_s$ specimens of sufficient width for the fibers to be continuous over the whole gauge length of the sample, the failure was found to be controlled by two distinct mechanisms of *inter- and intralaminar shear* of the matrix.

Very few measurements have been reported, however, relating the acoustic emission intensity to the fracture processes during creep, in spite of the advantages this method might offer for the continuous, "nondestructive" studies of long time fracture kinetics. Some recent data are shown, however, in principle in Fig. 24.3.4 for a chopped strand mat laminate.[12] As can be expected the very low acoustic emission intensity, i.e., the number of acoustic events per logarithmic time unit, in the secondary creep region is followed by a dramatic increase in the tertiary creep zone, constituting the initiation and growth of the ultimate, catastrophic failure.

Also the peak of the acoustic emission activity in the primary creep region, indicating fracture processes already in this very early stage of the creep rupture lifetime, is of interest. This peak has been found, to a different extent, for many composites built up from roving, i.e., in laminates from chopped strand mat, woven roving, (0, 90) cross-ply, and so forth.

The nature of the primary creep fracture mechanisms, the reason for the decay of the acoustic emission intensity in the secondary creep region, and the relation between the primary creep and the final fracture processes in the tertiary creep region have been studied by Sundström and Jansson[43] for chopped strand mat glass fiber/polyester laminates.

Figure 24.3.5 shows the decay of the primary creep acoustic emission intensity for a chopped strand mat/polyester laminate with 25% (by volume) glass fiber at different uniaxial stress levels.[43] The early fracture activities, as measured by acoustic emission, were found only for stress levels above 33 MPa which gave a strain in the primary creep region of more than approximately 0.3%. If, however, this strain was reached, as for instance for 33 MPa, in the secondary creep region no acoustic emission activity was noticed.

As it can be expected, that the fracture processes of matrix dominated composites will be strongly dependent on the strain rate, tensile tests were done at different rates.

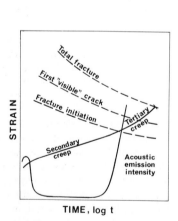

Fig. 24.3.4. Acoustic emission from uniaxial creep rupture in glass fiber/polyester. CSM laminates.[12]

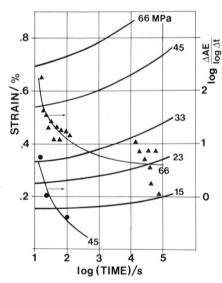

Fig. 24.3.5. Decay of the acoustic emission intensity, the number of acoustic emission events per logarithmic time unit, in the secondary creep region for a glass fiber/polyester, CSM laminate, fiber content 25% by volume.[43]

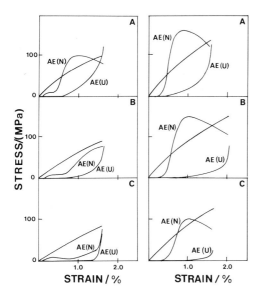

Fig. 24.3.6. Stress-strain behavior and acoustic emission intensity, number of acoustic emission events and acoustic emission energy per strain unit, at room temperature for a glass fiber/polyester, CSM laminate in uniaxial tension at different tensile rates. A 20% per hour, B 2% per hour, and C 0.2% per hour. Glass content: a) 17% and b) 25% by volume.[43]

Figure 24.3.6 shows the stress-strain behavior and the intensity of fracture activities, represented by the number of *acoustic emission events,* AE(N), and the *acoustic emission energy,* AE(U), per unit strain, for two CSM/polyester laminates with different glass content, 17 and 25% by volume respectively.

At high strain rates, a pronounced increase in $A(N)$ appears at about 0.5% strain for the 25% laminate. Visual inspection by microscope showed that this is caused mainly by debonding within fiber bundles, which are approximately perpendicular to the stress direction. Due to the strong binder in the roving, the chopped fibers are not properly dispersed in the composite. This results in large variations in the fiber concentration and in the stress-strain distribution in the laminate.

Only very little debonding was noticed between the fiber bundles and the matrix. This can be explained by the relaxation of the stresses around the bundles caused by the iternal bundle cracks.

At higher strains the bundle debonding was followed by the formation of a small number of cracks in the matrix rich regions between the strands, which also gave very high acoustic emission energy. These cracks were initiated at the split ends of the bundles. The different processes were accompanied by a decrease in the stiffness of the laminate and consequently in *"knees"* in the stress-strain curves at corresponding strain levels.

At reduced strain rate the acoustic emission intensity changed dramatically. Both the number of fracture events indicated by the acoustic emission and the dissipated acoustic emission energy were diminished considerably. A marked increase in the acoustic emission intensity was noticed first at strains very close to the total failure of the specimen. Nevertheless, visual inspection revealed also that in this case, cracks were formed at small strains causing a decrease in the slope of the stress-strain curve.

The behavior was explained by matrix relaxation, which reduces the level of the local stresses and might also give a more even stress distribution in the material. As a result, the strain required to initiate the first crack increases, the rate of the crack growth decreases, and the shape of the large matrix cracks preceding final failure changes as monitored by microscope.

As can be expected due to the lower degree of strain amplification, the bundle cracks appear at higher strains for the 17% laminate, Fig. 24.3.6a. For this laminate, an additional fracture mechanism appears at very low strains as a weak shoulder in the acoustic emission. The nature of this phenomenon is not known at present. It was assumed, however, to originate from the formation of small cracks in areas of high local stresses which are caused by the higher curing shrinkage in this laminate (due to its lower fiber content) as compared to the 25% laminate.

Thus, the following conclusions can be drawn from the creep studies and the tensile tests described above, concerning the observed fracture activities during creep.

— The fracture processes in the *primary creep region* comprise mainly debonding within the strands. This debonding will occur only if the strain at the end of the primary creep region exceeds a critical value. Because of the high initial strain rate, the debonding process will release enough energy for a pronounced acoustic emission activity to appear.

— The decay of the acoustic emission intensity in the *secondary creep region* is due to a decrease in the activity of the fracture processes. This occurs in two ways. First, a "saturation" of the primary creep fracture mechanisms occurs when the interstrand debonding has reached the ends of the fiber bundles. Second, the decrease of strain rate in the secondary creep region causes cracks to propagate more slowly, releasing less energy.

— In laminates based on roving reinforcements the fracture processes appearing in the primary creep region have been found to initiate the final creep rupture mechanisms.

— For the development of the total failure, the stresses concentrated at the ends of the fiber bundles have to exceed the strength of the matrix. This situation can be reached in different ways. First, there is a general decrease of the fracture resistance of the matrix during long term loading, caused by nonlinear viscoelastic processes, formation of micro cracks, and so forth. Second, the stress concentration at the ends of the fiber bundles will increase due to the increase of the overall laminate creep strain.

In general these results agree with the findings reported by Owen[29] and by Lyons and Phillips[30] which have been mentioned above.

24.4   Acknowledgments

The authors wish to thank Dr. D. Katz from Technion — Israel Institute of Technology, Haifa, Israel and Dr. M. Phillips from University of Bath, England for valuable discussions during the preparation of this Chapter.

References

1. A.H. Ericksen, *Composites* 1976, July 189.
2. J.B. Sturgeon, R.I. Butt and L.W. Larke, *Royal Aircraft Establishment, Technical Report* 76168, 1976.
3. J.B. Sturgeon, in *Creep of Engineering Materials,* edited by C.D. Pomeroy, Chapter 10, Mech. Engineering Publ. Ltd., 1977.
4. F.Y. Soliman, *Composite Materials: Testing and Design,* ASTM STP 460, 1968.
5. N.L. Hancox and D.C.C. Minity, *J. Mater. Sci.* 1978, **13**, 797.
6. P.D. Calvert and R.R. Lalanandham, *28th IUPAC Macromolecular Symposium,* 462, Amherst, MA, USA, 1982.

7. O.P. Bahl, R.B. Mathur and T.L. Dhami, *28th IUPAC Macromolecular Symposium*, 461, Amherst, MA, USA, 1982.
8. D. Katz, Dept. of Materials Engineering, Technion — Israel Institute of Technology, Haifa, Private communications
9. J. Schultz and A. Carry, *IUPAC Macromolecules*, edited by Benoit and Remp., New York 1982.
10. K.L. Jerina, R.A. Schapery, R.W. Tung and B.A. Sanders, *Short Fibre Reinforced Composite Materials*, p. 225, ASTM STP 772, 1982.
11. D.A. Dillard, D.H. Morris and H.F. Brinson, *Composite Materials Testing and Design*, 357, ASTM STP 787, 1982.
12. J.-F. Jansson and H. Sundström, *US-Italy Joint Symposium*, Italy, 1981.
13. I.V. Yannas, *J. Makromolecular Sci. Phys.* 1972, **B 6**, 91.
14. Å. Delmar, Master Thesis, The Royal Institute of Technology, Stockholm 1983.
15. M. Robertsson, The Royal Institute of Technology, S-100 44 Stockholm, Sweden, *Private communication.*
16. R.H. Boyd, M. Robertsson and J.-F. Jansson, *J. Polymer Sci. Phys.* 1982, **20**, 73.
17. J.-F. Jansson and I.V. Yannas, *J. Polymer Sci. Phys.* 1977, **15**, 2103.
18. A.V. Sidorovich and Y.U.S. Nadezin, *J. Macromolecular Sci. Phys.* 1979, **B 16(1)**, 35.
19. J.-F. Jansson, To be published.
20. L. Engel, H. Klingele, G.W. Ehrenstein and H. Schaper, *An Atlas of Polymer Damage*, Carl Hanser, Munich 1978.
21. D. Katz, A. Buchman and S. Gonen, *Eighht Intern. Congress on Rheology* 1980, **3**, 249.
22. P.J. Aspbury and W.C. Wake, *Brit. Polymer J.* 1979, **11**, 17.
23. D. Katz, *28th IUPAC Macromolecular Symposium*, p. 691, Amherst, MA, USA, 1982.
24. A.F. Johnson, *Engineering Design Properties of GRP*, British Plastics Federation and National Physics Laboratories, Publ nr 215/2 6–79, 1979.
25. Gibbs and Cox Inc., *Marine Design Manual for Fibreglass Reinforced Plastics*, McGraw Hill, New York 1960.
26. S.N. Zhurkov, *J. Fracture Mech.* 1965, **1(4)**, 311.
27. H.F. Brinson, W.I. Griffith and D.H. Morris, *Exper. Mech.* 1981, Sept. 329.
28. D.A. Dillard, D.H. Morris and H.F. Brinson, *Composite Materials: Testing and Design*, 357, ASTM STP 787, 1982.
29. M.J. Owen, in *Short Fibre Reinforced Composite Materials*, edited by B.-A. Sanders, p. 64, ASTM STP 772, 1982.
30. K.B. Lyons and M.G. Phillips, *Composites* 1981, Oct., 256.
31. R.A. Kline, *Composite Materials: Quality Assurance and Processing.*
32. D.L. Denton, Soc. Automotive Engineers, Detroit, 1979, Paper 790671.
33. A.D.S. Diggwa and R.H. Norman, *Plastics Polymers* 1972, Oct., 263.
34. *IKV Veröffentlichungen*, 1969–1975, 1976, 3.
35. H. Brintrup, *IKV Forschungsdokumentation*, 1977.
36. B. Meffert, *IKV Forschungsdokumentation*, 1977.
37. A. Rotem, *Fibre Sci. Technol.* 1977, **10**, 101.
38. A. Rotem, *Composites* 1978, Jan. 33.
39. J.H. Williams and S.M.Samson, *J. Compos. Mater.* 1978, **12**, 348.
40. D. Laroch and A.R. Bunsel, in *Adv. in Composite Materials* edited by A.R. Bunsel e.a. p. 985, Pergamon Press, New York 1980.
41. T.F. Drouillard and M.A. Hamstad, A Comprehensive Guide to the Literature on Acoustic Emission from Composites, from the *Proceedings from the First International Symposium on Acoustic Emission from Reinforced Plastics*, San Francisco, USA, 1983.
42. D. Vallentin and A.R. Bunsel, *J. Reinf. Plast. Compos.* 1982, **1**, 314.
43. H. Sundström and J.-F. Jansson, to be published.

Subject Index